Accession no.
01019153

Soil Erosion on
Agricultural Land

British Geomorphological Research Group Symposia Series

Geomorphology in Environmental Planning

Edited by
J. M. Hooke

Floods
Hydrological, Sedimentological and Geomorphological Implications

Edited by
Keith Beven and **Paul Carling**

Soil Erosion on Agricultural Land

Edited by
J. Boardman, I. D. L. Foster and **J. A. Dearing**

Soil Erosion on Agricultural Land

Edited by
J. Boardman
Department of Humanities and Countryside Research Unit,
Brighton Polytechnic

I. D. L. Foster

and

J. A. Dearing
Department of Geography, Coventry Polytechnic

John Wiley & Sons
CHICHESTER · NEW YORK · BRISBANE · TORONTO · SINGAPORE

Copyright © 1990 by John Wiley & Sons Ltd
Baffins Lane, Chichester
West Sussex PO19 1UD, England

Reprinted February 1996

All rights reserved.

No part of this book may be reproduced by any means,
or transmitted, or translated into a machine language
without the written permission of the publisher.

Other Wiley Editorial Offices

John Wiley & Sons, Inc., 605 Third Avenue,
New York, NY 10158–0012, USA

Jacaranda Wiley Ltd, G.P.O. Box 859, Brisbane,
Queensland 4001, Australia

John Wiley & Sons (Canada) Ltd, 22 Worcester Road,
Rexdale, Ontario M9W 1L1, Canada

John Wiley & Sons (SEA) Pte Ltd, 37 Jalan Pemimpin #05–04,
Block B, Union Industrial Building, Singapore 2057

Library of Congress Cataloging-in-Publication Data:
Soil erosion on agricultural land / edited by J. Boardman,
  I. D. L. Foster, and J. A. Dearing
      p.   cm.—(British Geomorphological Research Group symposia series)
    Papers presented at a workshop sponsored by the British
  Geomorphological Research Group as its contribution to the Annual
  Conference of the Institute of British Geographers, held at Coventry
  Polytechnic in January 1989.
    Includes bibliographical references.
    ISBN 0 471 92602 7
    1. Soil erosion—Congresses. 2. Soil conservation—Congresses.
  3. Land use, Rural—Congresses. I. Boardman, John, 1942–
  II. Foster, Ian (Ian D. L.) III. Dearing, J. A. (John A.)
  IV. British Geomorphological Research Group. V. Institute of
  British Geographers. Conference (1989 : Coventry Polytechnic)
  VI. Series.
  S622.2S66   1990
  631.4´5—dc20   89-39107
                                                                CIP

British Library Cataloguing in Publication Data:
Soil erosion on agricultureal land.
  1. Great Britain. Agricultural land. Soils. Erosion
  I. Boardman, John, *1942–*   II. Foster, I. D. L. (Ian D. L.)
  III. Dearing, J. A.   IV. Series
  631.4´5´0941

ISBN 0 471 92602 7

Phototypeset by Dobbie Typesetting Service, Plymouth, Devon
Printed and bound in Great Britain by Antony Rowe Ltd, Chippenham, Wiltshire

Contents

Series Preface ... ix

Preface ... xi

List of contributors ... xii

PART 1 EROSION PROCESSES: PAST AND PRESENT

Experimental Plot and Laboratory Studies

1. The Response of Tilled Soils to Wetting by Rainfall and the Dynamic Character of Soil Erodibility ... 3
   A. C. IMESON and F. J. P. M. KWAAD
2. Plot Studies Comparing Water Erosion on Chalky and Non-calcareous Soils ... 15
   G. M. MUTTER and C. P. BURNHAM
3. Design of a Double-split Divisor for Runoff Plots ... 25
   S. SOMBATPANIT, S. JAI-AREE, P. SERMSATANASUSDI, S. HIRUNWATSIRI and C. POONPANICH

Soil Redistribution

4. Use of Caesium-137 to Investigate Patterns and Rates of Soil Erosion on Arable Fields ... 33
   D. E. WALLING and T. A. QUINE
5. Gully Erosion on Agricultural Lands in Romania ... 55
   I. ICHIM, G. MIHAIU, V. SURDEANU, M. RADOANE and N. RADOANE
6. Land-use Controls on Sediment Production in a Lowland Catchment, South-west England ... 69
   A. L. HEATHWAITE, T. P. BURT and S. T. TRUDGILL
7. Soil Erosion on the South Downs: A Review ... 87
   J. BOARDMAN
8. Water Erosion on Arable Land in Southern Sweden ... 107
   K. ALSTRÖM and A. BERGMAN
9. The Calculation of Net Soil Loss Using Caesium-137 ... 119
   R. J. LOUGHRAN, B. L. CAMPBELL and G. L. ELLIOTT

Slope-stream Linkages

10 Linking the Field to the River: Sediment Delivery from Agricultural Land
   D. E. WALLING — 129

11 Magnitude and Frequency of Sediment Transport in Agricultural Catchments: A Paired Lake-catchment Study in Midland England — 153
   I. D. L. FOSTER, R. GREW and J. A. DEARING

12 Recent and Long-term Records of Soil Erosion from Southern Sweden — 173
   J. A. DEARING, K. ALSTRÖM, A. BERGMAN, J. REGNELL and P. SANDGREN

13 Valley Sedimentation at Slapton, South Devon, and Its Implications for the Estimation of Lake Sediment-based Erosion Rates — 193
   P. N. OWENS

14 Lake Sediment-based Studies of Soil Erosion — 201
   F. OLDFIELD and R. L. CLARK

Historical Surveys

15 Soil Erosion: Its Impact on the English and Welsh Landscape since Woodland Clearance — 231
   R. EVANS

16 Some Magnetic and Geochemical Properties of Soils Developed on Triassic Substrates and their Use in the Characterization of Colluvium — 255
   J. P. SMITH, M. A. FULLEN and S. TAVNER

17 Late Bronze Age–Iron Age Valley Sedimentation in East Sussex, Southern England — 273
   C. SMYTH and S. JENNINGS

18 Relationships between Catchment Characteristics, Land Use and Sediment Yield in the Midland Valley of Scotland — 285
   R. W. DUCK and J. MCMANUS

19 Soil Erosion in Cavernous Limestone Catchments — 301
   P. HARDWICK and J. GUNN

PART 2 ASSESSMENT AND PREDICTION

Soil Deterioration

20 Sediment-associated Phosphorus Transport from Two Intensively Farmed Catchment Areas — 313
   B. KRONVANG

21 Soil Erosion and Organic Matter Losses on Fallow Land: A Case Study From South-east Spain — 331
   C. FRANCIS

Contents vii

22 The Influence of Forest-clearance Methods, Tillage and Slope Runoff on Soil Chemical Properties and Banana Plant Yields in the South Pacific 339
S. G. REYNOLDS
23 Phosphorus Transport in Agricultural Runoff: The Role of Soil Erosion 351
A. N. SHARPLEY and S. J. SMITH

Risks and Costs

24 Some Costs and Consequences of Soil Erosion and Flooding around Brighton and Hove, Autumn 1987 369
D. A. ROBINSON and J. D. BLACKMAN
25 An Approach to the Assessment of Erosion Forms and Erosion Risk on Agricultural Land in the Northern Paris Basin, France 383
A. V. AUZET, J. BOIFFIN, F. PAPY, J. MAUCORPS and J. F. OUVRY
26 A Rule-based Expert System Approach to Predicting Waterborne Soil Erosion 401
T. M. HARRIS and J. BOARDMAN

Modelling

27 Model Building for Predicting and Managing Soil Erosion and Transport 415
W. T. DICKINSON, G. L. WALL and R. P. RUDRA
28 Process-based Modelling of Soil Erosion by Water on Agricultural Land 429
G. R. FOSTER
29 Prediction of Ephemeral Gully Erosion on Cropland in the Southeastern United States 447
C. R. THORNE and L. W. ZEVENBERGEN
30 Assessing the Impact of Erosion on Soil Productivity using the EPIC Model 461
J. R. WILLIAMS, A. N. SHARPLEY and D. TAYLOR
31 THEPROM—An Erosion Productivity Model 465
Y. BIOT
32 Probability Distribution of Event Sediment Yields in the Northern Negev, Israel 481
J. B. LARONNE

PART 3 CONSERVATION AND POLICY

Conservation Practices

33 Improved Management of Drylands by Water Harvesting in Third World Countries 495
A. RAPP and A. HÅSTEEN-DAHLIN

Contents

34 Gully Erosion in the Loam Belt of Belgium: Typology and Control Measures 513
J. POESEN and G. GOVERS

35 Experimental Study of Erosion and Crop Production on Bench Terraces on Sloping Land 531
M. DJOROVIĆ

36 Degradation of Dambo Soils and Peasant Agriculture in Zimbabwe 537
N. ROBERTS and R. LAMBERT

37 Erosion Control for the UK: Strategies and Short-term Costs and Benefits 559
C. A. FROST, R. B. SPEIRS and J. MCLEAN

38 Soil Water Management and the Control of Erosion on Agricultural Land 569
A. C. ARMSTRONG, D. B. DAVIES and D. A. CASTLE

39 A Field Study to Assess the Benefits of Land Husbandry in Malawi 575
M. B. AMPHLETT

Politics and Policy

40 Issues on Soil Erosion in Europe: The Need for a Soil Conservation Policy 591
R. P. C. MORGAN and R. J. RICKSON

41 Soil-conservation Policy and Practice for Croplands in Hungary 605
Á. KERTÉSZ, D. LÓCZY and I. OLÁH

42 Identification and Reclamation of Erosion-affected Lands in the Emilia–Romagna Region, Italy 621
F. GUERRIERI and G. VIANELLO

43 The Evolution of US Soil-conservation Policy: From Voluntary Adoption to Coercion 627
T. L. NAPIER

44 Problems of Land Reclamation: Pothwar Loess Plateau, Pakistan 645
M. J. HAIGH

CONCLUSION

45 Soil Erosion Studies; Some Assessments 659
J. BOARDMAN, J. A. DEARING and I. D. L. FOSTER

Place Names Index 673

Subject Index 677

Series Preface

The British Geomorphological Research Group (BGRG) is a national multi-disciplinary Society whose object is 'the advancement of research and education in geomorphology'. Today, the BGRG enjoys an international reputation and has a strong membership from both Britain and overseas. Indeed, the Group has been actively involved in stimulating the development of geomorphology and geomorphological societies in several countries. The BGRG was constituted in 1961 but its beginnings lie in a meeting held in Sheffield under the chairmanship of Professor D. L. Linton in 1958. Throughout its development the Group has sustained important links with both the Institute of British Geographers and the Geological Society of London.

Over the past three decades the BGRG has been highly successful and productive. This is reflected not least by BGRG publications. Following its launch in 1976 the Group's journal, *Earth Surface Processes* (since 1981 *Earth Surface Processes and Landforms*) has become acclaimed internationally as a leader in its field, and to a large extent the Journal has been responsible for advancing the reputation of the BGRG. In addition to an impressive list of other publications on technical and educational issues, BGRG symposia have led to the production of a number of important works including *Nearshore sediment Dynamics and Sedimentation* edited by J. R. Hails and A. P. Carr; *Geomorphology and Climate* edited by E. Derbyshire; *River Channel Changes* edited by K. J. Gregory, and *Timescales in Geomorphology* edited by R. Cullingford, D. Davidson and J. Lewin. This sequence of books culminated in 1987 with publication of the *Proceedings of the First International Geomorphology Conference* edited by Vince Gardiner. This international meeting, arguably the most important in the history of geomorphology, provided the foundation for the development of geomorphology into the next century.

This open-ended BGRG Symposia Series has been founded and is now being fostered to help maintain the research momentum generated during the past three decades, as well as to further the widening of knowledge in component fields of geomorphological endeavour. The series consists of authoritative volumes based on the themes of BGRG meetings, incorporating, where appropriate, invited contributions to complement chapters selected from presentations at these meetings under the guidance and editorship of one or more suitable specialists. Whilst maintaining a strong emphasis on pure geomorphological research, BGRG meetings are diversifying, in a very positive way, to consider links between geomorphology *per se* and other disciplines such as ecology, agriculture, engineering and planning.

The first volume in the series was published in 1988. *Geomorphology in Environmental Planning*, edited by Janet Hooke, reflects the bent towards applied studies. The second volume, edited by Keith Beven and Paul Carling, *Floods—Hydrological, Sedimentological and Geomorphological Implications*, focuses on a traditional research theme. *Soil Erosion on Agricultural Land* reflects the international importance of this topic for researchers during the 1980s. The volume, edited by John Boardman, Ian Foster and John Dearing, forms the third in the series.

The BGRG Symposia Series will contribute to advancing geomorphological research and we look forward to the effective participation of geomorphologists and other scientists concerned with earth surface processes and landforms, their relation to Man, and their interaction with the other components of the Biosphere.

September 1989

Geoffrey Petts
BGRG Publications

Preface

The chapters in this book were presented at a workshop on Soil Erosion on Agricultural Land sponsored by the British Geomorphological Research Group (BGRG) as its contribution to the Annual Conference of the Institute of British Geographers (IBG). The conference was held at Coventry Polytechnic in January 1989. The four-day workshop included 35 oral presentations and 20 posters, plus two field excursions, and was attended by about 120 people from 20 countries.

The aim of this book is to provide a representative selection of aspects of the soil erosion problem presented at the conference. Soil erosion is a broad topic which cannot be comprehensively covered in a single book, let alone one based on a specific conference. However, several contributors were invited to review broad areas in a series of keynote papers.

A secondary aim of the workshop was to bring together erosion researchers from different disciplines rather than confining discussions to geomorphological aspects of the topic. This proved to be difficult to achieve, but to the extent that the volume contains contributions from computing, agronomy, soil science, sedimentology, geology and agricultural economics, the organizers were partially successful.

The contributions to the book are arranged in three sections: 1, Erosion Processes: Past and Present; 2, Assessment and Prediction, and 3, Conservation and Policy. These section titles represent the division of the topic into broad areas corresponding to assessment, prediction and management. Placement of some contributions proved difficult because they span more than one of these areas.

Many people have contributed to the production of this volume. First, the local organizing committee of the IBG conference and Professor D. E. Smith, Head of the Department of Geography, Coventry Polytechnic, facilitated the smooth running of the workshop. The World Association of Soil and Water Conservation sponsored a session on 'Soil Conservation, Policies and Practices' and stimulating field excursions were organized by Drs I. D. L. Foster and J. A. Dearing (Coventry Polytechnic) and Dr M. A. Fullen (Wolverhampton Polytechnic). The BGRG provided funding for several overseas contributors. Finally, we would like to thank the many referees who have reviewed these chapters, often under severe time constraints.

John Boardman
Ian Foster
John Dearing

List of Contributors

K. Alström, Department of Physical Geography, University of Lund, Solvegatan, Lund S-223 62, Sweden.

M. B. Amphlett, Overseas Development Unit, Hydraulics Research, Wallingford, Oxon OX10 8AA.

A. C. Armstrong, ADAS Field Drainage Experimental Unit, Trumpington, Cambridge CB2 2LF.

A. V. Auzet, CEREG-URA 95 CNRS, Université Louis Pasteur, 3 rue de l'Argonne, F-67083 Strasbourg cedex, France.

A. Bergman, Department of Physical Geography, University of Lund, Solvegatan, Lund S-223 62, Sweden.

Y. Biot, School of Development Studies, University of East Anglia, Norwich NR4 7TJ.

J. D. Blackman, Geography Laboratory, University of Sussex, Falmer, Brighton BN1 9QN.

J. Boardman, Department of Humanities and Countryside Research Unit, Brighton Polytechnic, Falmer, Brighton, East Sussex BN1 9PH.

J. Boiffin, INRA, Station d'agronomie de Laon, BP 101, F-02004 Laon, France.

C. P. Burnham, Department of Agriculture, Horticulture and the Environment, Wye College, University of London, Wye, Ashford, Kent TN25 5AH.

T. P. Burt, School of Geography, University of Oxford, Mansfield Road, Oxford.

B. L. Campbell, Australian Nuclear Science and Technology Organization, Lucas Heights, NSW 2234, Australia.

D. A. Castle, ADAS Field Drainage Experimental Unit, Trumpington, Cambridge CB2 2LF.

R. Clark, CSIRO, Division of Water Resources, Australia.

D. B. Davies, ADAS Soil Science, Brooklands Avenue, Cambridge CB2 2DR.

List of Contributors

J. A. Dearing, Centre for Environmental Science Research and Consultancy, Department of Geography, Coventry Polytechnic, Priory Street, Coventry CV1 5FB.

W. T. Dickinson, School of Engineering, University of Guelph, Guelph, Ontario, Canada N1G 2W1.

M. Djorovic, Faculty of Forestry, University of Belgrade, Belgrade, Yugoslavia.

R. W. Duck, Department of Geography and Geology, The University, St Andrews, Fife KY16 9ST.

R. Evans, Department of Geography, University of Cambridge, Cambridge CB2 3EN.

G. L. Elliott, Soil Conservation Service of NSW Research Centre, Gunnedah, NSW 2380, Australia.

G. R. Foster, Department of Agricultural Engineering, University of Minnesota, St Paul, Minnesota, 55108, USA.

I. D. L. Foster, Centre for Environmental Science Research and Consultancy, Department of Geography, Coventry Polytechnic, Priory Street, Coventry CV1 5FB.

C. Francis, Department of Geography, University of Bristol, University Road, Bristol BS8 1SS.

C. A. Frost, East of Scotland College of Agriculture, West Mains Road, Edinburgh EH9 3JG.

M. A. Fullen, School of Applied Sciences, The Polytechnic, Wolverhampton, WV1 1LY.

G. Govers, Laboratory for Experimental Geomorphology, KU Leuven, Redingenstraat 16 bis, B-3000 Leuven, Belgium.

R. Grew, Centre for Environmental Science Research and Consultancy, Department of Geography, Coventry Polytechnic, Priory Street, Coventry CV1 5FB.

F. Guerrieri, CSSAS—Experimental Centre for Analysis and Study of the Soil, University of Bologna, Via le Berti Pichat, 10-40128 Bologna, Italy.

J. Gunn, Limestone Research Group, Department of Environmental and Geographical Studies, Manchester Polytechnic, Chester Street, Manchester M1 5GD.

M. J. Haigh, Geography Unit, Faculty of Environment, Oxford Polytechnic, Headington, Oxford OX3 0BP.

List of Contributors xv

P. Hardwick, Limestone Research Group, Department of Environmental and Geographical Studies, Manchester Polytechnic, Chester Street, Manchester M1 5GD.

T. M. Harris, Department of Geology and Geography, West Virginia University, Morgantown, West Virginia, 26506, USA.

A. Håsteen-Dahlin, Department of Physical Geography, University of Lund, Solvegatan, Lund S-223, Sweden.

A. L. Heathwaite, School of Geography, University of Oxford, Mansfield Road, Oxford. Present address: Department of Geography, University of Sheffield, Sheffield.

S. Hiranwatsiri, Soil and Water Conservation Division, Department of Land Development, Bangkhen, Bangkok 10900, Thailand.

I. Ichim, Research Station 'Stejarul', Piatra Neamt 5600, Romania.

A. C. Imeson, Fysische geographie en Bodemkunde, University of Amsterdam, Dapperstraat 115, 1093 BS Amsterdam, The Netherlands.

S. Jai-aree, Soil and Water Conservation Division, Department of Land Development, Bangkhen, Bangkok 10900, Thailand.

S. Jennings, Department of Geography, The Polytechnic of North London, 383 Holloway Road, London N7 0RN.

A. Kertesz, Geographical Research Institute, Hungarian Academy of Sciences, PO Box 64, H-1388, Hungary.

B. Kronvang, National Agency of Environmental Protection, The Freshwater Laboratory, Lysbrogade 52-DK 8600 Silkeborg, Denmark.

F. J. P. M. Kwaad, Fysische geographie en Bodemkunde, University of Amsterdam, Dapperstraat 115, 1093 Amsterdam, The Netherlands.

R. Lambert, Department of Civil Engineering, Loughborough University, Loughborough, Leicestershire.

J. B. Laronne, Geography Department, Ben Gurion University, Beer Sheva 84 105, PO Box 653, Israel.

D. Loczy, Geographical Research Institute, Hungarian Academy of Sciences, PO Box 64, H-1388, Hungary.

R. J. Loughran, Department of Geography, University of Newcastle, NSW 2308, Australia.

J. Maucorps, INRA, Service d'étude des sols et de la carte pédologique de France, Ardon, F-45160 Olivet, France.

J. McLean, East of Scotland College of Agriculture, West Mains Road, Edinburgh EH9 3JG.

J. McManus, Department of Geography and Geology, The University, St Andrews, Fife KY16 9ST.

G. Mihaiu, Research Station 'Stejarul', Piatra Neamt 5600, Romania

R. P. C. Morgan, Silsoe College, Silsoe, Bedfordshire MK45 4DT.

G. M. Mutter, Department of Agriculture, Horticulture and the Environment, Wye College, University of London, Wye, Ashford, Kent TN25 5AH.

T. L. Napier, Department of Agricultural Economics and Rural Sociology, Ohio State University, 2120 Fyffe Road, Columbus, Ohio, USA.

I. Olah, Komaron County Plant Protection and Soil Conservation Service, Tata, PO Box 50, H-2891, Hungary.

F. Oldfield, Department of Geography, University of Liverpool, Roxby Building, PO Box 147, Liverpool L69 3BX.

J. F. Ouvry, AREAS (Association regionale pour l'étude et l'amélioration des sols), Mairie, F-76740 St-Valery-en-Caux, France.

P. N. Owens, Department of Geography, University of British Columbia, Vancouver, BC, Canada V6T 1W5.

F. Papy, INRA, Unité de Recherche Systèmes Agraires et le Développement, F-78850 Thiverval-Grignon, France.

C. Poonpanich, Soil and Water Conservation Division, Department of Land Development, Bangkhen, Bangkok 10900, Thailand.

T. A. Quine, Department of Geography, University of Exeter, Amory Building, Rennes Drive, Exeter EX4 4RJ.

J. Poesen, Laboratory for Experimental Geomorphology, KU Leuven, Redingenstraat 16 bis, B-3000 Leuven, Belgium.

M. Radoane, Research Station 'Stejarul', Piatra Neamt 5600, Romania.

N. Radoane, Research Station 'Stejarul', Piatra Neamt 5600, Romania.

A. Rapp, Department of Physical Geography, University of Lund, Solvegatan, Lund S-223, Sweden.

J. Regnell, Department of Quaternary Geology, University of Lund, Tornvagen 13, Lund S-223, Sweden.

S. Reynolds, FAO, United Nations, Via delle Terme di Caracalla, 00100 Rome, Italy.

List of Contributors xvii

R. J. Rickson, Silsoe College, Silsoe, Bedfordshire MK45 4DT.

N. Roberts, Department of Geography, Loughborough University, Loughborough, Leicestershire.

D. A. Robinson, Geography Laboratory, University of Sussex, Falmer, Brighton BN1 9QN.

R. P. Rudra, School of Engineering, University of Guelph, Guelph, Ontario, Canada N1G 2W1.

P. Sandgren, Department of Quaternary Geology, University of Lund, Tornvagen 13, Lund S-223, Sweden.

P. Sermsatanasusdi, Soil and Water Conservation Division, Department of Land Development, Bangkhen, Bangkok 10900, Thailand.

A. N. Sharpley, USDA-ARS, Water Quality and Watershed Research Laboratory, PO Box 1430, Durant, Oklahoma 74702-1430, USA.

J. P. Smith, School of Applied Sciences, The Polytechnic, Wolverhampton WV1 1LY.

S. J. Smith, Water Quality and Watershed Research Laboratory, PO Box 1430, Durant, Oklahoma 74702-1430, USA.

C. Smyth, Department of Geography, The Polytechnic of North London, 383 Holloway Road, London N7 0RN.

S. Sombatpanit, Soil and Water Conservation Division, Department of Land Development, Bangkhen, Bangkok 10900, Thailand.

R. B. Speirs, East of Scotland College of Agriculture, West Mains Road, Edinburgh EH9 3JG.

V. Surdeanu, Research Station 'Stejarul', Piatra Neamt 5600, Romania.

S. Tavner, School of Applied Sciences, The Polytechnic, Wolverhampton WV1 1LY.

D. Taylor, USDA-ARS, Grassland, Soil and Water Research Laboratory, Temple, Texas, USA.

C. R. Thorne, Department of Geography, University of Nottingham, University Park, Nottingham NG7 2KD

S. T. Trudgill, Department of Geography, University of Sheffield, Sheffield.

G. Vianello, CSSAS—Experimental Centre for Analysis and Study of the Soil, University of Bologna, Via le Berti Pichat, 10-40128 Bologna, Italy.

G. J. Wall, Agriculture Canada, Department of Land Resource Science, University of Guelph, Guelph, Ontario N1G 2W1, Canada.

D. E. Walling, Department of Geography, University of Exeter, Amory Building, Rennes Drive, Exeter EX4 4RJ.

J. R. Williams, USDA-ARS, Grassland, Soil and Water Research Laboratory, Temple, Texas, USA.

L. W. Zevenbergen, Water Engineering and Technology, 419 Canyon, Suite 225, Fort Collins, Colorado 80521, USA.

PART 1
EROSION PROCESSES: PAST AND PRESENT

Experimental Plot and Laboratory Studies

1 The Response of Tilled Soils to Wetting by Rainfall and the Dynamic Character of Soil Erodibility

A. C. IMESON and F. J. P. M. KWAAD
*Laboratory of Physical Geography and Soil Science,
University of Amsterdam*

INTRODUCTION

If a soil, especially a freshly tilled one, is exposed to rainfall it does not remain unchanged. Interactions between the soil system and rainfall take place, which alter the state of the soil system. This is evident from changes in a number of soil physical properties. Some of these changes are irreversible, and this means that successive rainfall events act upon a soil with different initial hydraulic properties. This leads to different responses of the soil to rainfall, in terms of runoff and erosion, and it may explain the large degree of scatter which often is observed in relationships between rainfall, runoff and erosion.

In this chapter a simple conceptual model of the evolution of a tilled soil in the course of a year is described. The evolution is related to the dynamic character of soil properties which have an influence on infiltration and the generation of overland flow. From the model it will appear that the tilled layer can follow evolutionary pathways which may differ from year to year, depending on the pattern of rainfall. The model is still provisional, as it is based on ongoing field and laboratory work, which was started in 1983. A main fieldwork area is Dutch South-Limbourg, which is part of the West European loess belt.

A central notion in the evolution of a tilled soil is the 'response to wetting'. This term was introduced by Imeson and Verstraten (1986) to include the various reactions of a soil when it is exposed to wetting by rainfall, with or without the effect of raindrop impact. It may be useful to make a distinction between

(a) surface response to wetting and (b) subsurface or internal response. The response to wetting of a soil includes processes such as compaction, shrinking and swelling, mellowing, slaking and dispersion. A typical surface response is crusting or sealing. These processes are well described in the literature (Baver *et al.*, 1972; Callebaut *et al.*, 1985; Edwards and Bremner, 1967; Emerson, 1959, 1983; Utomo and Dexter, 1982).

All processes mentioned have an influence on soil porosity, and thus on soil hydraulic properties, such as saturated and unsaturated hydraulic conductivity, water-retention characteristics, sorptivity, time to ponding at a given rainfall intensity and infiltration capacity.

Other soil properties which are influenced by the response to wetting of a soil are bulk density, surface roughness, resistance to penetration and mechanical resistance of the soil surface to detachment by drop impact and/or overland flow.

An important control of the way(s) in which a soil responds to wetting is the stability in water of the structural elements of the soil. These elements may be either naturally formed (peds) or artificially created by tillage (clods). Aggregate stability, as it is generally termed, depends on the presence and proportion of various soil constituents (e.g. clay, organic matter) and on the mode of formation of the aggregates (a) by biotic action or (b) by physical or mechanical processes, such as freezing, shrinking or tillage. Biotic processes produce a granulation structure, such as is found in the topsoil of forest soils and under grass. The crumbs or granules are relatively stable in water. Physical and mechanical processes result in a fragmentation structure, the elements of which are relatively unstable in water. Aggregate stability changes with soil moisture content (Hofman and de Leenheer, 1975) and exhibits seasonal variations.

THE EVOLUTION OF A TILLED SOIL UNDER RAINFALL

A loess soil, such as occurs in north-west Europe, is taken as an example of the soils to which the model applies. The model also relates to soils with a 'duplex' character, i.e. those with a sharp texture contrast between topsoil (poor in clay) and subsoil (rich in clay). Characteristics of these soils are a low clay content, a high silt content and a low aggregate stability of the Ap horizon (tilled layer). A freshly tilled condition is taken as a starting point for the description of the evolution of the soil.

For convenience, three periods are distinguished in the evolution of the tilled layer, which correspond with three stages in its development. A comparable subdivision was made by Boiffin (1984) for the evolution of the soil surface. The three periods are: (A) a short period with freshly tilled soil; (B) a period in which rainfall-induced processes lead to a stepwise degradation of soil structure and a stepwise decline of various soil physical properties; and (C) a period in

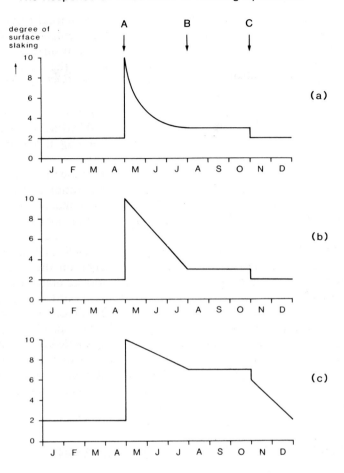

Figure 1.1 Schematic representation of decline of structural status of tilled layer under corn. A, B and C denote times of tillage, 100 per cent crop cover and harvest. (a) Rapid decline by heavy rains in May; (b) gradual decline by continued wet weather in May–July; (c) incomplete decline during dry weather in May–July and continued decline after harvest

which a continuous crust is present at the soil surface and in which no further changes of soil physical properties take place except by biological activity. Schematic illustrations of the evolution of the tilled layer are given in Figures 1.1 and 1.2.

Period A

Period A immediately follows tillage. This may be either autumn or spring tillage. Various tillage actions may be applied, with various tillage objectives

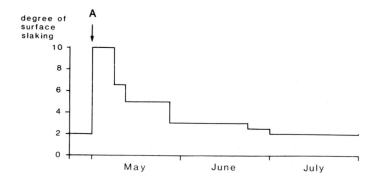

Figure 1.2 Schematic representation of stepwise decline of structural status of tilled layer under corn caused by discrete rainfall events: A denotes time of tillage

and tillage tools, such as ploughing, harrowing, chiselling, disking, rolling and hoeing. Tillage depth may vary. Soil moisture content at the time of tillage influences the results which are attained. It must not be too high, or puddling will occur, nor should it be too low, or tillage will have no effect at all or produce dust. A relatively small range of soil moisture contents remains for the farmer to choose from. However, soil moisture contents at the time of tillage may vary slightly from year to year. This can cause appreciable differences in results of the same tillage operation (e.g. in size of clods). This is a difficulty in obtaining reproducible conditions on runoff plots from year to year.

The effect of primary tillage (ploughing) on the soil is the formation of a fragmentation structure with relatively coarse clods which are loosely packed. Often it is desirable to reduce the size of the clods in the upper part of the tilled layer before seeding (seedbed preparation). This is done by a secondary tillage operation (harrowing), which may be followed by rolling to achieve a certain degree of compaction of the seedbed.

The size of the clods which lie at the soil surface after tillage determines the random roughness of the soil surface. Roughness increases with clod size. Random roughness determines the storage capacity of the soil surface for rainwater. It also influences the velocity of overland flow by exerting a friction resistance on that flow.

The coarser the clods are at the soil surface, the longer it takes before they are broken down under rainfall, other things being equal. The clods of a freshly tilled soil are generally loosely packed with large voids between them. Interped macroporosity is high, bulk density is low, resistance to penetration is low and saturated hydraulic conductivity is high, due to the presence of a system of interconnected macropores. The macropores drain at relatively low suctions and this causes a large decrease in hydraulic conductivity if the soil water suction rises above zero, i.e. if the soil becomes unsaturated (Klute, 1982). Unsaturated flow takes place in the matrix of the clods through the intraped pores, and water under suction will only flow from one clod to another at the points of contact between adjacent clods. In a freshly tilled soil the total contact area between clods will be relatively small, and this explains the low unsaturated hydraulic conductivity of such a soil.

Infiltration capacity (i.e. maximum rate of infiltration of rainwater) will be high in period A. This is due to the presence of macropores which are open to the soil surface (Edwards, 1982). At high rainfall intensities water falling on clods will first be taken up by the clods, but sorptivity decreases and soon rainfall intensity exceeds the infiltration capacity of the clods. Water accumulates on the clod surfaces and flows off them into the macropores between the clods, where it flows down the walls of the pores as free water. This flow of water in macropores is known as bypass flow or short circuiting, as the matrix of the soil is bypassed by the water (Bouma *et al.*, 1977; Beven and Germann, 1982). Macropores also provide important pathways for the escape of air during infiltration of rainwater. Roughness elements protruding above the general soil surface may act as vents for the escaping air (Dixon and Peterson, 1971).

The contact between the tilled layer and the untilled soil underlying it is abrupt. A plough pan or plough sole may be present there. There is a sudden change in many physical soil properties at the plane of contact between the tilled and the untilled soil and this strongly influences the movement of water from the tilled layer into the underlying soil. Percolation is slowed down and this may give rise to the development of a perched water table in the tilled layer and to lateral subsurface flow of water (interflow, throughflow). When and where a perched water table reaches the soil surface, saturated overland flow occurs.

Period A generally does not last very long, depending on the timing of rainfall after tillage. Immediately after tillage the tilled layer begins to settle and to subside under its own weight. Due to this compaction, bulk density increases, porosity decreases and the soil surface sinks. More significant changes take place with the onset of rainfall.

Period B

During period B compaction and subsidence of the tilled layer continues, aided by the impact force of falling raindrops. The most important changes, however, take place at the soil surface. During rainfall the clods and peds, which are exposed at the soil surface, start to break down. Fine material is released, which fills the depressions between the clods and blocks the entrances of the macropores between them. Clods are reduced in size and lose their identity by a process of fusion with neighbouring clods. Remaining parts of clods are immersed in a mass of fine soil material. Random roughness decreases and the soil surface takes on a closed appearance.

The degradation of the soil surface takes place in a stepwise manner, each period of breakdown of the soil structure being triggered by rainfall. Intervals between rainfall events will vary in length, and this leaves the soil in different initial conditions, regarding soil moisture content and aggregate stability, before the start of a rainfall event.

The processes which cause the breakdown of peds and clods at the soil surface are slaking and dispersion. Slaking is the disintegration of structural elements into micro-aggregates and skeleton grains (Edwards and Bremner, 1967). Slaking of initially dry aggregates is ascribed to rapid wetting, causing compression of entrapped air in the aggregates to the point of the aggregate's explosion, when the pressure of the air exceeds the cohesion of the aggregate (Baver et al., 1972). The aggregate may be weakened by stresses, set up by unequal swelling of different parts, when water is rapidly taken up by it. Slow wetting of dry aggregates does not produce the explosion effect. Slaking of initially wet aggregates is caused by raindrop impact. There is evidence (McCalla, 1944) that raindrop impact is not instrumental in the breakdown of dry aggregates. Increasing fall heights of raindrops did not increase the rate of breakdown of initially dry aggreates. Farres (1978), however, works on the assumption that aggregates do not spontaneously break down under addition of water. He observed that they remained intact until raindrop impact accompanied the addition of water. Broken-down aggregates were only found by Farres (1978) at the soil surface and not below the surface crust. Once a crust has developed, it protects aggregates below the surface from the impact of raindrops, thus preventing aggregate breakdown below the surface (Farres, 1978). De Ploey and Mücher (1981) describe unstable loamy soils with a low internal stability of the clods. Under simulated rainfall the clods of the unstable soils (field behaviour) showed the collapse of microstructures and liquefaction of the matrix with a strong reduction of the number of originally present pores (observations in thin sections). It is not explicitly stated by De Ploey and Mücher (1981) whether the collapse of the clods of the unstable soils is due to wetting, to raindrop impact or to the combined effect of wetting and raindrop impact.

Dispersion affects the clay which is present in soil aggregates. Clay domains are broken up and primary clay particles are suspended in the soil solution. If this happens, aggregates fall apart. Dispersion of clay is controlled by chemical conditions in the soil. It occurs when sodium is present as an exchangeable cation above a certain threshold concentration and when the electrolyte concentration of the soil solution is not too high (Agassi et al., 1981), and is often observed under semi-arid climatic conditions. According to Emerson (1983) not only sodium-containing soil aggregates will disperse when immersed in water but also wet soil aggregates with divalent ions only. This occurs after mechanical reworking (remoulding, shearing) of the soils containing divalent ions. This is called mechanical dispersion. In Dutch South-Limbourg there is evidence of dispersion during the winter months. The runoff water in the storage tanks of runoff plots remains turbid during winter whereas sediment in the runoff water settles rapidly during summer.

In the course of period B a layer of fine fragments of aggregates and primary silt and sand-sized particles is formed at the soil surface as a consequence of advanced aggregate breakdown. The particles of this layer are closely packed by the beating action of the falling raindrops and form a crust. On top of this slaking crust thin layers of fine sediment may be deposited by overland flow.

Crust formation may take place in winter (after autumn tillage) or in summer (after spring tillage). It may be caused by high- or low-intensity rainfalls and may be due to breakdown of initially dry or initially wet aggregates. These different conditions of crust formation are reflected in the physical properties of the crust (Römkens *et al.*, 1985).

The alteration in the state of the soil surface during period B greatly affects a number of soil physical properties, such as infiltration capacity, random roughness of the soil surface and mechanical resistance of the soil surface to detachment by drop impact and flowing water. The decrease in porosity and change in pore-size distribution of the tilled layer affect bulk density, penetration resistance, saturated and unsaturated hydraulic conductivity and water-retention function. Some results of measurements of these properties will be presented in the next section.

As mentioned above, the pathway of the evolution of a tilled soil during period B, and the duration of period B, may differ from year to year, according to differences in the pattern of rainfall. In Figure 1.1 three evolutionary pathways are sketched for the same soil, following ploughing in the spring. In Figure 1.1(a) a high-intensity thunderstorm falls on a dry soil in April or May. This causes a strong response from the soil. Extreme slaking occurs and overland flow takes place with deposition of sediment. In case (b) the soil remains wet during the spring and early summer by frequent and prolonged rainy periods. Slaking takes place slowly and gradually, and no sedimentary crust is formed. In case (c) a dry spring and early summer is followed by a rainy August and September. No appreciable degradation of soil structure occurs, at first by the lack of rainfall and later by the protection of the soil surface by the fully developed crop.

Some recovery of roughness, porosity and infiltration capacity may take place during the growing season due to biological activity in the soil (e.g. by earthworms). From studies on untilled soils it is known that macropores formed by earthworms and crop roots contribute to the relatively high infiltration capacity and hydraulic conductivity of these soils (Edwards *et al.*, 1988).

Period C

During period C no further changes of the tilled layer and its physical properties take place. Crusting of the soil surface and compaction and subsidence of the tilled layer are at a maximum. This condition is illustrated by the stubble field, which remains after the harvest of corn in autumn. Period C extends from midsummer all through autumn and winter until the spring of the next year if no tillage is carried out after harvest and/or no winter crop is sown. This can cause appreciable runoff during winter from rain as well as from melting snow. Runoff coefficients of 100 per cent have been measured on runoff plots in South Limbourg during some wet winter months on corn stubble fields. In fact, winter runoff amounts were observed to be much higher than summer runoff if period C lasted throughout the winter.

FLUCTUATIONS IN SOIL PROPERTIES RELATED TO SOIL ERODIBILITY DURING THE YEAR

To illustrate the conceptual model of the evolution of a tilled soil outlined above, some results of measurements of a number of soil physical properties during the year will be presented. Continuous records of changes at one site are not yet available as the work is still continuing.

The degree of surface slaking can be assessed by visual observation and expressed on a scale of 1 to 10. A freshly tilled soil is rated with 10 and a completely crusted surface has a rating of 1. To obtain reproducible results by different observers, field soils are compared with a series of reference photographs (Boekel, 1973). Degree of slaking can be related to rainfall amount and intensity, and can be used to rapidly evaluate the effect of different tillage systems on the response to wetting of a given soil. An example is given in Table 1.1 (Van Mulligen, 1988).

Surface roughness changes can be evaluated with a chain of standard length, which is laid out on the soil surface in such a way that it follows the microrelief

Table 1.1 Evolution of surface slaking of a tilled loess soil. Evolution of surface slaking under different cropping systems of corn on a loess soil in Dutch South-Limbourg (Van Mulligen, 1988). Slaking is rated on a scale of 1–10 by comparison with 10 reference photographs; 10 is freshly tilled soil (Van Boekel, 1973). Cropping systems are: (A) Tillage in October; winter rye from October to April, killed by spraying; corn sown between residue of winter rye in untilled soil; harvest of corn in October. (B) Tillage in October; tilled fallow from October to March; tillage in March; summer barley in March–April, killed by spraying; corn sown in barley seedbed; harvest of corn in October. (C) Corn stubble field from October to April; tillage in April; corn from April to October, (C') Bare fallow all year, tilled in April (reference condition)

Cropping system	Date of last tillage	Degree of slaking			
		28 April 1988	19 May 1988	10 June 1988	24 June 1988
A	30 October 1987	3	4	4	2
B	5 April 1988	9	9	9	4
C	4 May 1988	4	7	7	3
C'	4 May 1988	4	7	7	3

Table 1.2 Changes in bulk density and total pore volume of a tilled loess soil. Bulk density and pore volume of tilled layer of some loess soils in Dutch South-Limbourg (Tiktak, 1983)

	28 March 1983		17 April 1983		5 May 1983		11 July 1983	
	Bulk density (kg m^{-3})	Pore volume (%)	Bulk density (kg m^{-3})	Pore volume (%)	Bulk density (kg m^{-3})	Pore volume (%)	Bulk density (kg m^{-3})	Pore volume (%)
Site 3	1.293	50.4	1.397	46.5	1.401	46.3	1.416	45.7
Site 4	1.160	55.6	1.430	45.2	1.486	43.1	1.371	47.5
Site 5a	1.182	54.7	1.480	43.3	1.507	42.3	1.421	45.6
Site 5b	1.205	53.8	1.460	44.0	1.577	39.6	1.423	45.5
Site 6a	1.243	52.4	1.480	43.3	1.605	38.5	1.481	43.3
Site 6b	1.246	52.3	1.630	37.5	1.537	41.1	1.311	49.8

as closely as possible. A ratio is calculated between the length of the chain and the distance between the two ends of it when it is laid out on the soil surface. A ratio of 1 is found for a completely even soil surface. In Figure 1.3 results are presented of measurements under different cropping systems of corn in South-Limbourg (De Hoog, 1988). Data on bulk density and total pore volume of some loess soils in Dutch South-Limbourg are given in Table 1.2 (Tiktak, 1983). Changes in pore size distribution can be evaluated from soil moisture characteristics by comparing water-retention curves of a given soil on different dates during the growing season (Hill *et al.*, 1985). Curves are given in Figure 1.4 of an autumn-tilled soil. Sampling dates are 28 March and 5 May.

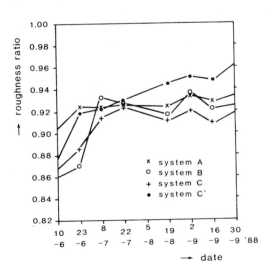

Figure 1.3 Decline of random roughness of soil surface of a loess soil under different cropping systems of corn (De Hoog, 1988). For a description of the cropping systems see Table 1.1

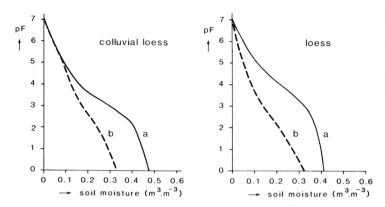

Figure 1.4 Shift of pF curves of tilled layer of autumn-tilled loess soils in Dutch South-Limbourg between 28 March 1983 (curve a) and 5 May 1983 (curve b) (Tiktak, 1983)

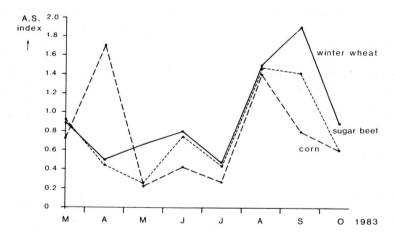

Figure 1.5 Aggregate stability changes under different crops on a loess soil in Dutch South-Limbourg (Van Eijsden, 1986). The aggregate stability index is based on the drop test (Low, 1954)

The curves show a lower moisture content at all suction values on 5 May than on 28 March. This means a reduction in the volume of pores of all size classes between 28 March and 5 May, which must be due to compaction and subsidence of the tilled layer (Tiktak, 1983). Fluctuations in aggregate stability during the growing season under different crops are shown in Figure 1.5 (Van Eijsden, 1986), and the pattern is rather complex. Generally speaking, aggregate stability seems to be at a low in May–June–July and at a high in August–September. This may be related to bacterial activity in the soil.

CONCLUSION

It is clear that soil erodibility, which is controlled by a number of soil physical properties, is far from constant on tilled land during the year. There is a need for a more systematic study of the changes which take place during the year and which, of course, are known in a qualitative way to farmers all over the world. Changes in soil physical properties must be related to the processes which take place in tilled soils and to the specific conditions of soil, weather, tillage system, crop and crop management under which the changes take place.

The dynamic character of soil physical properties, which control runoff and soil loss from agricultural land, must be given full consideration in event-based models of runoff and erosion. Soil erodibility cannot be represented in such models by long-term indices, such as the K-factor of the Universal Soil Loss Equation (Römkens, 1985). At present, the K-factor is used in the ANSWERS model for soil-loss predictions of discrete rainfall events.

Soil-erosion control will also benefit from a greater knowledge of the magnitude of the changes in soil properties in the course of a year and timing of control measures (e.g. tillage operation) will thus be improved.

An improved knowledge of the hydraulic characteristics and of water movement in the tilled layer may lead to an understanding of the threshold conditions for the initiation of rills. Such initiation by shallow subsurface flow through macropores in the tilled layer (micropipe flow) and the possibility of breakdown of clods and peds by wetting of the tilled layer below the soil surface (i.e. outside the reach of raindrop impact) are underresearched areas at present.

REFERENCES

Agassi, M., Shainberg, I. and Morin, J. (1981). Effect of electrolyte concentration and soil sodicity on infiltration rate and crust formation, *Soil Science Society America Journal*, **45**, 848–51.

Baver, L. D., Gardner, W. H. and Gardner, W. R. (1972). *Soil Physics*, 4 edition, John Wiley, New York.

Beven, K. and Germann, P. (1982). Macropores and water flow in soils, *Water Resources Research*, **18**, 1311–25.

Boekel, P. (1973). De betekenis van de ontwatering voor de bodemstructuur op de zavel- en lichte kleigronden, *Rapport Instituut voor Bodemvruchtbaarheid, Haren*, No. 5-1973.

Boiffin, J. (1984). *La dégradation structurale des couches superficielles du sol sous l'action des pluies*, Thésis, Paris.

Bouma, J., Jongerius, A., Boersma, O., Jager, A. and Schoonderbeek, D. (1977). The function of different types of macropores during saturated flow through four swelling soil horizons, *Soil Science Society America Journal*, **41**, 945–50.

Callebaut, F., Gabriels, D. and De Boodt, M. (eds) (1985). Assessment of soil surface sealing and crusting, *Proceedings of the symposium held in Ghent, Belgium, State University of Ghent, Department of Soil Physics, Ghent, Belgium*.

De Hoog B. (1988). *De invloed van vier verschillende teeltsystemen op structuurverval en korstvorming*, Thesis, Lab. Physical Geography and Soil Science, Amsterdam.

De Ploey, J. and Mücher, H. J. (1981). A consistency index and rainwash mechanisms on Belgian loamy soils, *Earth Surface Processes and Landforms*, **6**, 319-30.

Dixon, D. M. and Peterson, A. E. (1971). Water infiltration control: a channel system concept, *Soil Science Society America Proceedings*, **35**, 968-73.

Edwards, A. P. and Bremner, J. M. (1967). Microaggregates in soils, *Journal of Soil Science*, **18**, 64-73.

Edwards, W. M. (1982). Predicting tillage effects on infiltration, *ASA Special Publication*, **44**, 105-15.

Edwards, W. M., Norton, L. D. and Redmond, C. E. (1988). Characterizing macropores that affect infiltration into nontilled soil, *Soil Science Society America Journal*, **52**, 483-7.

Emerson, W. W. (1959). The structure of soil crumbs, *Journal of Soil Science*, **10**, 235-44.

Emerson, W. W. (1983). Inter-particle bonding. In *Soils, an Australian viewpoint*, CSIRO, Division of Soils, Melbourne, pp. 477-98.

Farres, P. (1978). The role of time and aggregate size in the crusting process, *Earth Surface Processes*, **3**, 243-54.

Hill, R. L., Horton, R. and Cruse, R. M. (1985). Tillage effects on soil water retention and pore size distribution of two mollisols, *Soil Science Society America Journal*, **49**, 1264-70.

Hofman, G. and De Leenheer, L. (1975). Influence of soil prewetting on aggregate instability, *Pedologie*, **25**, 190-98.

Imeson, A. C. and Verstraten, J. M. (1986). Erosion and sediment generation in semi-arid and Mediterranean environments: the response of soils to wetting by rainfall, *Journal of Water Resources*, **5**, 388-417.

Klute, A. (1982). Tillage effects on the hydraulic properties of soil: a review, *ASA Special Publication*, **44**, 29-43.

Low, A. J. (1954). The study of soil structure in the field and in the laboratory, *Journal of Soil Science*, **5**, 57-78.

McCalla, T. M. (1944). Water drop method of determining stability of soil structure, *Soil Science*, **58**, 117-21.

Römkens, M. J. M. (1985). The soil erodibility factor: a perspective. In El-Swaify, Moldenhauer and Lo (eds), *Soil Erosion and Conservation*, Soil Conservation Society of America, Ankeny, Iowa, pp. 445-61.

Römkens, M. J. M., Baumhardt, R. L., Parlange, J. Y., Whisler, F. D., Parlange, M. B. and Prasad, S. N. (1985). Effect of rainfall characteristics on seal hydraulic conductance. In Callebaut, Gabriels and De Boodt (eds), *Assessment of Soil Surface Sealing and Crusting*, Ghent, pp. 228-35.

Tiktak, A. (1983). *Onderzoek naar een aantal bodemfysische parameters van enkele lössgronden in het Randsdalerveld (Zuid-Limburg)*, Thesis, Lab. Physical Geography and Soil Science, Amsterdam.

Utomo, W. H. and Dexter, A. R. (1982). Changes in soil aggregates water stability induced by wetting and drying cycles in non-saturated soil, *Journal of Soil Science*, **33**, 623-37.

Van Eijsden, G. G. (1986). *Bodemerosie op landbouwpercelen in een loessgebied*, Thesis, Lab. Physical Geography and Soil Science, Amsterdam.

Van Mulligen, E. (1988). *Schijngrondwater en effekten van maisteeltsystemen (Zuid-Limburg)*, Report Roermond, Thesis Amsterdam.

2 Plot Studies Comparing Water Erosion on Chalky and Non-calcareous Soils

G. M. MUTTER and C. P. BURNHAM
Wye College, University of London

INTRODUCTION

There is a widespread view that the presence of calcium carbonate in soil flocculates colloids and improves aggregation (Tisdall and Oades, 1982; Rimmer and Greenland, 1976), and this should increase resistance to erosion. Confirmation might be found in the fact that seven of the eight soil associations in south-east England on which Jarvis and others (1984) mention a water-erosion problem involve non-calcareous soils. However, water erosion of shallow soils on chalk in England has become widespread (e.g. Foster, 1978; Boardman and Robinson, 1985), and Boardman and Hazelden (1986) imply that erosion is equally common on the calcareous Coombe as on similar but non-calcareous Hamble soils.

Recent unpublished laboratory studies (Mutter, 1989) suggest that calcium carbonate may actually increase the dispersability of soils. This is supported by Al Ani and Dudas (1988), who found that calcareous soils in Iraq were more easily capped and dispersed than similar non-calcareous soils, and by Silleos (1981) in Greece and De Meester and Eppink (1980) in Spain, who recorded more erosion on calcareous soils. The work reported here put this question to the test by monitoring erosion from a set of plots on which four different topsoils had been spread.

EXPERIMENTAL METHODS

Plots

Ten plots, enclosed by vertical boards, were set up on a uniform 12 per cent slope. Each consisted of a rectangular area 4.8 m long and 1.5 m wide, tapering over a further length of 1.0 m to a spout and a large bucket for the collection of runoff and eroded soil. The site is within the perimeter fence of Sidelands Laboratory, 800 m from the Wye College meteorological station and its recording rain gauge, which has a similar aspect and comparable environment.

Soils

Two plots were chosen at random for each of the following five treatments.

Grass: The thick grass sward originally on the site was left undisturbed. On the other plots the turf was removed.

CU: 0.15 m of chalky topsoil, previously uncultivated for many years, was dug from other plots on the site and freshly spread. The soil on the site is about 40–50 cm to chalk, and resembles the Panholes series (Jarvis *et al.*, 1984).

CA: 0.15 m of chalky topsoil was taken from South Sidelands field, a few metres from the site; it is somewhat shallower and more chalky, probably due to erosion, and is an Upton–Panholes intergrade. This was downland before 1914, but has been in regular arable cropping for most of the subsequent period. Structural conditions are typical for the soil type.

NU: 0.15 m of non-calcareous topsoil (Hook series) was taken from Oaks Wood on Wye College Farm. The canopy is open, and the source site carried grass of similar appearance to that associated with CU.

NA: 0.15 m of non-calcareous topsoil (Hook series) was brought from Oaks field, about 100 m from NU. This is an old arable field, and the topsoil is noticeably compact and poorly structured.

After hand spreading (in April 1988), all the soils were given one pass with a garden roller.

Measurement of Soil Properties

Organic carbon was determined by the Walkley–Black method and calcium carbonate by calcimetry. Stones were determined on large samples using a 2 mm sieve. The particle size analyses were made by the pipette method after removing calcium carbonate with dilute hydrochloric acid. Avery and Bascomb's (1974) procedure was followed, except that 2 and 20 μm equivalent settling diameters were used to define the silt fraction. Bulk density and moisture content were obtained from core samples taken in September so that the soil had settled. The sampling holes were filled with care. Moisture retention was determined after equilibration on a ceramic pressure plate. Infiltration was measured in mid-September using a double-ring infiltrometer. Capacity represents total infiltration in 90 min. Infiltration rate was measured when the rate had become constant.

Dispersion ratio was measured by a method modified from Gupta *et al.* (1984). One gram portions of soil aggregates of diameter between 0.5 and 2.0 mm were placed in each of two centrifuge tubes, one with 30 ml of de-ionized water and the other with a little hydrogen peroxide. After 12 h the latter was made up with de-ionized water and 1 ml of 10 per cent Calgon solution to a total volume of 30 ml. Both were then shaken on a horizontal shaker for 24 h, and immediately centrifuged at 300 r.p.m. for 5 min. The absorbance of the still turbid supernatants was measured without delay at 420 μm wavelength with a spectrophotometer and the ratio calculated.

Runoff and Erosion Parameters

The buckets, which were covered to limit evaporation, were checked after every significant period of rain, the runoff measured in millilitres and the eroded soil dried at 105°C and weighed. For easier discussion, the eroded soil was expressed in tonnes lost per hectare (Table 2.3).

Rainfall Parameters

After every erosion event, reference was made to the rainfall record, and the following quantities were calculated: (a) total rainfall during the storm provoking the erosion, (b) maximum intensity during the storm (mm h^{-1}), (c) rainfall kinetic energy, summing the values of the following relationship for the full duration of the storm:

$$KE = 2.93 + 12.46 \log I$$

where I = intensity. This equation was developed by Mutter (1989) on the basis of the equation $KE = \frac{1}{2}MV^2$, and measurements of sand splash from cups on the same experimental site used for the work reported here.

RESULTS

The experiment was monitored for seven months from the beginning of May to the end of November 1988. Runoff occurred from the bare soil plots on sixteen occasions, but was only detectable from the undisturbed grass plots on ten of these. There were no major erosion events during December 1988 or from January to March 1989. A rainstorm on 4 and 5 April 1989 caused considerable runoff and erosion, and this was measured. The rainfall parameters and water runoff during these events are recorded in Table 2.1 and soil loss in Table 2.2. It will be noted that in April and early May and again in October and November, when the soil was relatively moist, the chalky arable soil had most runoff and erosion, while from mid-July to September, when it was relatively dry, the non-calcareous arable soil had the most runoff and erosion. If the bare

Table 2.1 Rainfall parameters compared with mean runoff in millilitres from plots with different soils. Treatment with highest runoff shown by italics

Date (day(s)/month)	Total rain (mm)	Max. intensity (mm h^{-1})	Rain KE (J m^{-2})	Grass	Calc. uncult.	Calc. arable	Non-calc. uncult.	Non-calc. arable
7/5	25.1	17	432.9	150	12 750	*15 250*	1030	4950
25/5	7.8	10	50.8	24	45	57	250	*580*
28–30/5	14.0	4	139.8	0	63	210	210	*460*
1–2/7	18.4	16	290.7	0	825	1400	650	*2530*
3–4/7	8.9	4	67.8	0	38	*270*	75	115
16–17/7	11.4	5	53.5	23	37	53	121	*190*
22–23/7	8.0	16	121.3	0	8	22	120	*125*
27/7	14.5	15	245.6	0	220	350	490	*2350*
31–1/9	19.0	14	80.0	6	73	145	400	*1400*
2–4/9	5.8	4	52.2	0	28	168	34	*228*
4–5/10	16.0	16	188.7	29	65	*173*	58	155
8–9/10	22.9	14	224.6	53	253	*593*	233	580
11–12/10	17.1	15	193.8	130	1150	*2150*	1030	1950
8–11/11	18.0	16	163.2	120	580	*975*	430	775
19–20/11	12.4	5	71.7	20	590	1140	540	*1550*
29/11	15.5	4	135.4	220	1150	*1350*	950	1050
Total	234.8	—	—	775	17 875	24 306	6621	18 988
4–5/4	32.1	5	207.3	800	6750	*13 700*	6300	10 750

Figure 2.2 Rainfall parameters compared with soil eroded in grams from plots with different soil. Treatment with the highest soil loss shown by italics

Date (day(s)/month)	Total rain (mm)	Max. intensity (mm h^{-1})	Rain KE (J m^{-2})	Grass	Calc. uncult.	Calc. arable	Non-calc. uncult.	Non-calc. arable
7/5	25.1	17	432.9	0.16	718.9	*969.1*	12.9	62.7
25/5	7.8	10	50.8	0.21	3.0	3.4	1.2	*5.7*
28–30/5	14.0	4	139.8	0	1.6	*4.0*	1.1	2.2
1–2/7	18.4	16	290.7	0	7.1	*21.4*	1.3	20.2
3–4/7	8.9	4	67.8	0	1.9	*3.4*	1.8	2.3
16–17/7	11.4	5	53.5	0.10	1.4	3.1	2.6	*5.1*
22–23/7	8.0	16	121.3	0	0.4	0.5	0.5	*1.3*
27/7	14.5	15	245.6	0	4.8	12.4	12.0	*104.6*
31–1/9	19.0	14	80.0	0.03	2.5	3.4	4.8	*10.0*
2–4/9	5.8	4	52.2	0	0.7	3.2	0.9	*3.5*
4–5/10	16.0	16	188.7	0.21	3.6	*4.8*	2.2	3.8
8–9/10	22.9	14	224.6	0.06	3.0	*3.9*	2.1	3.1
11–12/10	17.1	15	193.8	0.25	16.7	*35.0*	13.5	32.4
8–11/11	18.0	16	163.2	0.25	7.8	*11.8*	5.3	7.7
19–20/11	12.4	5	71.7	0.08	4.0	*5.2*	0.9	4.0
29/11	15.5	4	135.4	0.43	3.3	*5.7*	3.1	4.8
Total	234.8	—	—	1.78	780.7	1090.7	66.2	273.4
4–5/4	32.1	5	207.3	0.52	15.2	*24.1*[a]	13.1	14.6

[a] Some weed growth inhibiting erosion.

Table 2.3 Soil properties in relation to erosion rate (R = correlation)

Treatment	OM (%)	CaCO$_3$ (%)	Stones (%)	Sand (%)	Silt (%)	Clay (%)	BD (g cc^{-1})	Moisture (%) 0.3 bar	Moisture (%) 15 bar	Infiltration Capacity (mm/h)	Infiltration Rate (mm/h)	Dispersion ratio	Erosion (t ha^{-1})
Grass	8.88	44.6	4.5	37.8	31.1	31.1	0.850	30.16	16.71	38.6	20.4	2.99	0.002
CU	4.58	46.3	10.6	30.5	27.8	41.7	1.100	27.75	14.73	29.5	16.8	9.11	0.964
CA	2.65	60.0	17.3	20.3	29.9	49.8	1.170	28.04	13.18	20.4	15.6	23.24	1.347
NU	4.62	0.0	0.4	41.7	35.0	23.3	1.215	24.75	13.64	35.3	19.2	10.53	0.082
NA	3.13	0.0	1.5	36.8	28.1	35.1	1.265	28.64	12.96	21.0	10.8	17.45	0.338
R	0.63	0.65	0.93	−0.95	−0.52	0.94	0.27	0.03	−0.40	−0.68	−0.32	0.67	—

uncultivated soils are considered on their own, almost the same conditions are observed. Overall, there is decisively more runoff and much more erosion from each chalky soil than from the comparable non-calcareous soil. The effects of arable cultivation in increasing and of a grass cover in greatly decreasing runoff and erosion are also obvious (Tables 2.1 and 2.2).

In Table 2.3 soil properties are related to erosion rate. The number of observations is very small, but the relationship to stoniness and clay content (positive) and to sand content (negative) is significant at 1 per cent. Infiltration capacity, dispersion ratio, organic matter and calcium carbonate content are also strongly correlated, but their significance is not established.

Table 2.4 compares runoff and rainfall parameters with erosion rate for each treatment. As there are 15 degrees of freedom, all treatments show a highly significant correlation between erosion rate and runoff, and all except grass between erosion rate and kinetic energy, much the most successful of the rainfall parameters tested. Of the others, total rain is better correlated with erosion on chalky soils, but maximum intensity is better on non-calcareous soils.

Table 2.4 Correlation (R) between erosion rate and runoff and rainfall parameters

Treatment	Runoff (litres)	Total rain (mm)	Maximum intensity (mm h^{-1})	Rain KE (J m^{-2})
Grass	0.863	0.314	0.032	0.072
CU	0.994	0.538	0.303	0.709
CA	0.990	0.543	0.308	0.716
NU	0.679	0.491	0.484	0.611
NA	0.709	0.364	0.427	0.626

DISCUSSION

The salient conclusion from this work must be that chalky soils are not specially resistant to water erosion: overall, indeed, they eroded more. One factor may be their alkaline reaction, for calcium carbonate is less soluble in water at pH 7–8 (see e.g. Richards, 1954), so that the abundant calcium is not active in flocculation. Another (Silleos, 1981) may be that the calcium carbonate often occurs in fine particles which are not sticky and hence are easily eroded. Note that the calcareous soils in this experiment have a very high calcium carbonate content (>45 per cent), mostly in the form of fine chalk particles. Thus much of what is being eroded is chalk.

The experimental data clearly confirm that the bare chalky soils are more easily dispersed under field conditions than the bare non-calcareous ones. Indeed, the effect seems to be much more marked than the modest differences found in the laboratory (Table 2.3). The mean weight of dry soil eroded per litre of runoff from the chalky soils is 43.7 g off bare, uncultivated soil and 44.9 g off arable soil. From non-calcareous soils the eroded soil is $10.0 \, \text{g} \, \text{l}^{-1}$ where uncultivated, and $14.4 \, \text{g} \, \text{l}^{-1}$ where arable. It will be noted that the effect of calcium carbonate in encouraging dispersion is much greater than that of favourable structure and humus content in resisting it. However, the effect of plant cover is even greater. Not only is runoff reduced to 4 per cent of that from the same soil with turf removed, but the water from the grass plots is clean, with only $2.3 \, \text{g} \, \text{l}^{-1}$ of soil, so that erosion is only 0.2 per cent of that from the same soil bared.

These factors cannot explain why in drier summer conditions the non-calcareous soils were eroded more when a downpour occurred but in wetter parts of the year the chalky soils eroded more. Indeed, so marked is the latter effect that it may be the sole immediate cause of the apparently greater overall erodibility of chalky soils. There is some anecdotal evidence for this special vulnerability of wet chalky soils. The disastrous erosion at Rottingdean in October 1987 was fuelled by further downpours falling on already wet soils, and an earlier exceptional erosion episode on chalk (Kennet of the Dene, 1940) followed snowmelt on already wet soil. Two vegetable farms in East Kent overlying chalk experienced greatly increased erosion following the introduction of irrigation. Downpours then began with the soil already comparatively moist.

What is the critical difference between the water relations of a soil which is half chalk and a non-calcareous fine sandy clay loam? It is not surface permeability, nor overall available water-holding capacity, nor water held at 15 bar tension (Table 2.3). It is more likely size of pores, which are almost entirely around 0.5 μm in chalk (Price *et al.* 1976), but mostly wider than 1 μm in an incompact loamy soil without chalk. Thus, chalk fragments readily become saturated, like blotting paper, but then hold the water not only against gravity but even against evaporation and transpiration unless conditions are really dry (Anderson, 1927; Bunting and Elston, 1964; Davy and Taylor, 1974). Silleos (1981) found that calcareous soils dried more slowly than equivalent non-calcareous ones. In conditions of fluctuating wetness, as are common in autumn, winter and spring, the soil which is half chalk has a considerably lower capacity to take up extra moisture, for the chalk is already saturated. Smith (1980) and Boardman and Robinson (1985) noted that chalk soils may lie wet in winter. Two plots of Robinson and Boardman (1988) reached

Table 2.5 Weight percentage water needed to bring the experimental soils to saturation from various moisture tensions

Treatment	0.33 bar	2.0 bar	5.0 bar	10 bar	15 bar	Oven dry
Grass	20.5	24.9	28.3	29.6	33.9	50.6
CU	19.8	21.5	24.3	29.9	32.8	47.6
CA	18.4	22.3	25.9	31.8	33.3	46.5
NU	26.6	30.9	34.5	35.5	37.7	51.4
NA	17.6	24.9	30.2	31.3	33.3	46.3

40 per cent moisture by weight. An argument along these lines would explain why the total rain in a wet period is an important trigger for erosion on chalky soils, while maximum intensity is more important for non-chalky ones.

This hypothesis was tested by more detailed work on the water-holding capacity of the experimental soils. The parameter most likely to affect soil erosion is the amount of water needed to saturate the soil. As Table 2.3 shows, the chalky soils have about the same overall water-holding capacity as the non-calcareous soils, but Table 2.5 indicates that they hold considerably more water at tensions between 5 and 15 bar. The effect is that the chalky soils at 2 to 5 bar tension, representative of the wind-dried surfaces of otherwise moist soils in the winter and spring, require less water to saturate them than the non-calcareous soils, and so start eroding more quickly. In the summer when the soil surface is more thoroughly dried (say, around 15 bar tension), there is little difference between the chalky and the non-calcareous soils in the amount of water required to saturate them, and any difference in erodibility under summer conditions must be due to other factors.

REFERENCES

Al-Ani, A. N. and Dudas, M. J. (1988). Influence of calcium carbonate on mean weight diameter of soil, *Soil Tillage Research*, **11**, 19–26.

Anderson, V. L. (1927). Studies of the vegetation of the English Chalk. V. The water economy of the chalk flora, *Journal of Ecology*, **15**, 72–129.

Avery, B. W. and Bascomb, C. L. (1974). *Soil Survey Laboratory Methods*, Soil Survey Technical Monograph **6**, Harpenden, Herts.

Boardman, J. and Hazelden, J. (1986). Examples of erosion on brickearth soils in east Kent, *Soil Use and Management*, **2**, 105–8.

Boardman, J. and Robinson, D. A. (1985). Soil erosion, climatic vagary and agricultural change on the Downs around Lewes and Brighton, autumn 1982, *Applied Geography*, **5**, 243–58.

Bunting, A. H. and Elston, J. (1964). Water relations of crops and grass on chalk soil, *Scientific Horticulture*, **18**, 116–20.

Davy, A. J. and Taylor, K. (1974). Water characteristics of contrasting soils in the Chiltern Hills and their significance for *Deschampsia caespitosa* (L.) Beauv, *Journal of Ecology*, **62**, 367–78.

De Meester, T. and Eppink, L. A. A. J. (1980). Difficulties in estimating soil loss on heavy calcareous clay soils near Cordoba, Spain. In De Boodt and Gabriels, D. (eds), *Assessment of Erosion*, John Wiley, Chichester, pp. 407–14.

Foster, S. (1978). An example of gullying on arable land on the Yorkshire Wolds, *Naturalist*, **103**, 157–61.

Gupta, R. K., Bhumbla, D. K. and Abrol, I. P. (1984). Effect of sodicity, pH, organic matter and calcium carbonate on the dispersion behavior of soils, *Soil Science*, **137**, 245–51.

Jarvis, M. G., Allen, R. H., Fordham, S. J., Hazelden, J., Moffat, A. J. and Sturdy, R. G. (1984). *Soils and their Use in South East England*, Bulletin **15**, Soil Survey of England and Wales, Harpenden, Herts.

Kennet of the Dene, Lord (1940). Chalk landscape, *Nature, London*, **145**, 466.

Mutter, G. (1989). *Water Erosion of Calcareous Soils in South East England*, Unpublished PhD thesis, Wye College, University of London.

Price, M., Bird, M. J. and Foster, S. S. D. (1976). Chalk pore-size measurements and their significance, *Water Services* for 1976, 596–600.

Richards, L. A. (1954). *Diagnosis and Improvement of Saline and Alkali Soils*, US Dept Agric. Handbook **60**, Washington, DC.

Rimmer, D. and Greenland, D. J. (1976). Effect of calcium carbonate on the swelling behaviour of a soil clay, *Journal of Soil Science*, **27**, 129–39.

Robinson, D. A. and Boardman, J. (1988). Cultivation practice, sowing season and soil erosion on the South Downs, England: a preliminary study, *Journal of Agricultural Science (Cambridge)*, **110**, 169–77.

Silleos, N. G. (1981). The effect of calcium carbonate on soil erodibility in a survey area in northern Greece *ITC Journal*, **4**, 418–34.

Smith, C. J. (1980). *Ecology of the English Chalk*, Academic Press, London.

Tisdall, J. M. and Oades, J. M. (1982). Organic matter and water-stable aggregates in soils, *Journal of Soil Science*, **33**, 141–63.

3 Design of a Double-split Divisor for Runoff Plots

S. SOMBATPANIT, S. JAI-AREE, P. SERMSATANASUSDI,
S. HIRUNWATSIRI and C. POONPANICH
Department of Land Development, Ministry of Agriculture and Co-operatives, Thailand

INTRODUCTION

In constructing runoff plots to study soil erosion it is normal to have a series of runoff- and sediment-collecting tanks with some kind of divisor between any two of them. The divisor's function is to divide water from one tank into several equal parts, and directing only one (or sometimes two) amounts into the next container. Such an installation eases the operation substantially, for it can be quite costly to construct one tank to accommodate all the runoff water, especially after a heavy storm.

Today, divisors in use in Thailand are those made from galvanized iron sheet and divide water at ratios ranging from 1:2 to 1:20. To use a conventional divisor with higher rates is not practical, as it would be too large to be properly installed in a normal-sized sediment tank (in most cases the tank diameter is about 1 m).

For runoff plots in Thailand it is necessary to install three tanks, with two divisors, usually at the ratio of 1:5, which render the amount of water in the second and third tanks to be one fifth and one twenty-fifth of spill-over from the first, respectively. The types of divisors in use elsewhere have been described by Hudson (1981).

The installation of several collecting tanks, with one divisor between any two of them, is both expensive during plot establishment and time consuming during operation as many sediment samples have to be collected for drying and weighing throughout the rainy season. The authors have concluded that if a divisor of higher dividing ratio, with acceptable accuracy, could be built, its use in the runoff plots would greatly ease the burden of sediment sampling and analysis and make research less onerous to field workers, resulting in lower costs. We believe that a divisor which can divide water twice within its body (in other words a double-split divisor) may be satisfactory.

Soil Erosion on Agricultural Land
Edited by J. Boardman, I. D. L. Foster and J. A. Dearing
©1990 John Wiley & Sons Ltd

26   Soil Erosion on Agricultural Land

MATERIALS AND METHODS

The materials employed in this study include stainless steel sheet No. 28 and electrical soldering equipment. A prototype of the double-split divisor has been made (Figure 3.1), the body of which has a dimension of $10 \times 14 \times 50$ cm ($h \times w \times l$), with a protruding part of $2 \times 10 \times 20$ cm. The number of sediment tanks necessary in most cases is two, the first of which receives runoff water

Figure 3.1 The prototype divisor with 1:49 ratio

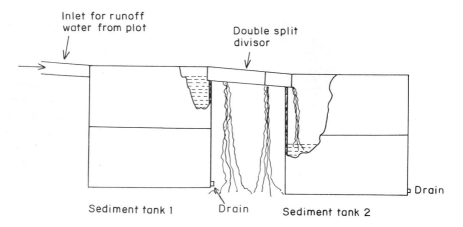

Figure 3.2 Position of double-split divisor in runoff collecting system

Design of a Double-split Divisor for Runoff Plots

through either a tube or an open channel from an erosion plot. The divisor is placed between the first and second tanks, with a gradient of approximately 10 per cent, which is enough to induce sufficient flow and to avoid deposition of suspended particles (Figure 3.2). The front part of the divisor, which is the front splitter, is equally divided into seven channels where water leaving the first tank and entering the central one is allowed to flow into the middle part of the device, while water passing through six others drops away. While moving within the divisor body, the turbulent water is obstructed by a series of barriers (Figures 3.3 and 3.4), which cause the flow to slow down and become uniform across its width. Subsequently, water is allowed to flow over the last barrier which lies at the rear splitter of equal cross-sectional area as the front one. Then only one seventh of the water is allowed to go into the central channel which then flows through the protruding part into the second tank. Holes on the thresholds of barriers (Figure 3.4) enable all water and suspended materials to leave the divisor after each runoff event ceases. Assuming the splitting of water at the front and rear splitters to be exactly 1:7, the ratio of water flowing out of the protruding part is then 1:49.

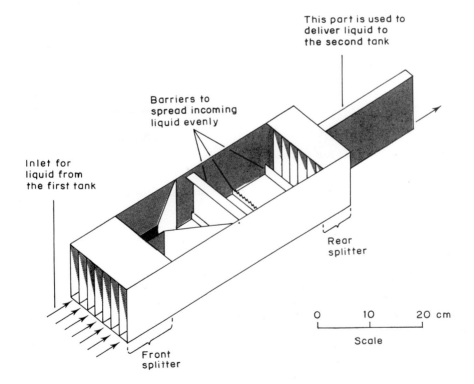

Figure 3.3 Schematic view of the double-split divisor

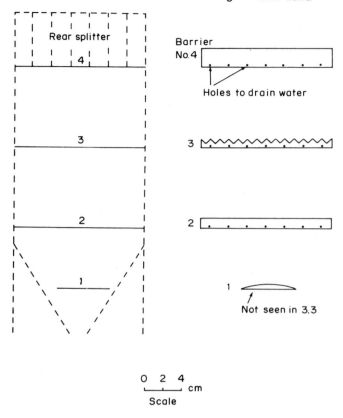

Figure 3.4 Position, size and shape of the barriers

The construction of the first batch of divisors at a tinsmith cost 800 Baht (US $32) each.

DISCUSSION

Figure 3.3 shows a schematic view of the prototype divisor. Due to possible human error in constructing the device and hydraulic properties of liquid flowing through small channels of different lengths, the attainment of exactly a 1:49 ratio for this prototype device is unlikely. A calibration of each divisor in circumstances similar to those in the field may then yield a constant figure which can be subsequently used when it is installed at the experimental site.

The prototype divisors produced in a batch of eight will be subject to thorough testing with a flume of appropriate size by allowing water to pass through at varying velocities. Some modification may have to be made in order to overcome discrepancies, as mentioned earlier.

It is expected that when this double-split divisor functions properly, each runoff plot would require no more than two sediment tanks and one divisor. (To use only one sediment tank by placing the divisor straight from a runoff plot is not feasible, as water and soil material will not spread at the front splitter evenly, resulting in a major error.) If needed, this divisor can be built larger and accommodate more than two splitters, each of which may constitute more than seven channels. Such divisors which can divide water to a very high ratio will find much use in soil- and water-conservation research, for it can be used in large runoff plots, like mini-watersheds, or when field operators wish to sample at low frequency (e.g. once a month).

CONCLUSION

The design of a double-split divisor has originated from the fact that more research on soil erosion and conservation needs to be done quickly to cope with the land degradation that is increasing daily. The present use of a series of sediment tanks, with one divisor ranging from 1:2 to 1:20 ratio between any two of them, is expensive and time consuming. The use of a double-split divisor with a ratio of 1:49 can substantially decrease construction cost and workload during the rainy season. This principle can be used for constructing larger divisors with other dividing ratios and with more splitting successions, which will suit the needs of large plots or avoid sampling too frequently.

ACKNOWLEDGEMENTS

The authors wish to thank Professor N. W. Hudson for his valuable advice and the Director-General of the Department of Land Development, Bangkok, for his permission to present this research at the Workshop.

REFERENCE

Hudson, N. W. (1981). *Soil Conservation*, 2nd edition, Batsford, London.

Soil Redistribution

4 Use of Caesium-137 to Investigate Patterns and Rates of Soil Erosion on Arable Fields

D. E. WALLING and T. A. QUINE
Department of Geography, University of Exeter

SOIL EROSION ON ARABLE LAND IN BRITAIN AND ITS ASSESSMENT

The last decade has seen a growing awareness of the potential hazard of soil erosion on arable land in Britain. However, a dichotomy in the perception of the problem exists between those studying soil erosion and the farmers who experience the impact. It appears that the majority of farmers view soil erosion as being of minimal significance to productivity, although it may be a management hazard, particularly when harvesting crops. As a result, soil erosion is rarely seen to be of sufficient importance to warrant change in soil use or management. In contrast, those studying soil erosion have examined the balance between soil loss and formation, and some predict a long-term problem. If there is a serious imbalance between loss and formation this should be of concern to the farming community. However, the existence of this imbalance must first be demonstrated. The development of a reliable and generally applicable method of estimating rates of soil erosion provides a relevant and important challenge to the geomorphologist.

In order to achieve an accurate assessment of soil erosion, consideration must be given to all the relevant processes. Evans and Cook (1986) have suggested that the four most widely recognized processes involved in water erosion in the UK are rain-splash, inter-rill or sheet erosion (termed 'overland flow erosion' by Morgan, 1979), rill erosion and gully erosion. The most obvious sign is clearly the gully, although in Britain rilling is the most frequent visible indicator of erosion on arable land. The frequency of occurrence and the importance of rain-splash and sheet erosion are less easy to assess. The contrasting patterns and attributes of these processes impose constraints upon the design of an effective

---

Soil Erosion on Agricultural Land
Edited by J. Boardman, I. D. L. Foster and J. A. Dearing
©1990 John Wiley & Sons Ltd

method for monitoring soil erosion. These constraints can be illustrated by consideration of the advantages and limitations of some of the established methods of documenting soil erosion. Several of these will be discussed briefly, before examining the potential for using caesium-137 (^{137}Cs) measurements.

Field erosion plots physically isolate a small area of known physical characteristics in which the land use can be controlled and rates of soil movement, particularly that associated with rain-splash and sheet-wash, may be monitored at the downslope margin. The primary advantage of the technique is therefore the access to the detailed event-based information afforded by intensive long-term investigations. However, this level of investigation imposes limitations. Assessment of the rate of soil erosion under two crop types with two soil variables, for example, would require eight erosion plots (Morgan, 1979). Furthermore, the long-term monitoring is labour intensive. Other problems are associated with the use of bounded areas. First, the measured erosion rate provides a net figure for the entire plot, which may be used to estimate soil losses from a field but provides few data relating to soil redistribution within the field. Second, most erosion plots must be limited in extent to portions of a slope or, at best, slope transects. This may limit the range of potential erosive environments and exaggerate the impact of 'edge effects'. One further problem relates to management. In many cases the soil surface is kept bare and is therefore not comparable with a cultivated soil. Both the size and the management limitation introduce problems when extrapolating from the study site to the agricultural context.

Collecting troughs placed across the study slope may provide data relating to sediment transport by sheet erosion. Their flexibility and lower cost introduce advantages over erosion plots. However, they have a number of limitations, the most important being the definition of sediment-contributing area. Although this is often calculated by using the width of the trough and the upslope length of the field, this will rarely provide an accurate assessment. Second, placement of the troughs inevitably disrupts the slope environment. In addition to influencing surface flow, the troughs may have an impact on the formation of rills and gullies. Finally, the length of time over which the troughs may be operated is frequently limited and only short-term data may be provided.

The problems associated with these techniques and the infrequency of their application in Britain has led to a search for new methods of erosion measurement. In the early 1980s the Ministry of Agriculture, Fisheries and Food (MAFF), in association with the Soil Survey for England and Wales (SSEW), established an aerial photographic survey of soil erosion on 17 flight paths over southern Britain (Evans and Cook, 1986). Rills and gullies, identified using aerial photographs, were measured *in situ* allowing calculation of the volume of eroded soil. This survey has provided valuable information over a wide area for a time period in excess of five years. However, a problem with the technique is the bias towards large-scale erosive features. As only rills and gullies may be identified, no measure of the impact of splash and sheet erosion is gained and the extent of redeposition is difficult to assess.

THE POTENTIAL FOR USING ^{137}Cs MEASUREMENTS

The potential for using ^{137}Cs in studies in soil erosion has now been demonstrated in several areas of the world, including the USA (Ritchie et al., 1974; McHenry and Ritchie, 1977), Canada (De Jong et al., 1983) and Australia (Longmore et al., 1983; Campbell, 1983; Campbell et al., 1986). However, little work has been undertaken in the UK, although early studies in Devon, reported by Walling et al. (1986) and Loughran et al. (1987), appear to reinforce the results from overseas and confirm this potential.

The basis of the technique is essentially simple and is illustrated in Figure 4.1 ^{137}Cs was produced as a by-product of the atmospheric testing of thermonuclear weapons from the 1950s to the 1970s. It was distributed globally in the stratosphere and deposited by rainout and washout. In the UK the resultant regional distribution of ^{137}Cs shows variation which may be related to annual rainfall (Cawse and Horrill, 1986). However, at the field level deposition appears to have been relatively uniform. In most environments ^{137}Cs is strongly bound to clay minerals in the soil, and the available empirical evidence suggests that subsequent redistribution of the ^{137}Cs takes place in association with soil particles (Figure 4.1). Figure 4.2 provides a schematic representation of the possible impact of various agricultural activities upon the distribution of ^{137}Cs within the soil. The sharp decline in ^{137}Cs concentration below the soil surface at the grassland site is indicative of the strong binding of the radionuclide by the fine fraction. The presence of some ^{137}Cs at depth reflects the long period of time over which deposition has occurred. In the 34 years which have elapsed

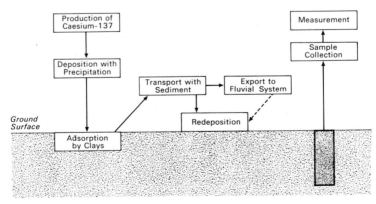

Figure 4.1 The basis of the ^{137}Cs technique for studying erosion and sediment redistribution

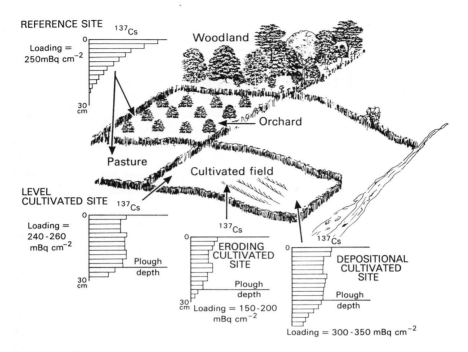

Figure 4.2 Schematic representation of the impact of various agricultural environments upon the loading and profile distribution of ^{137}Cs

since the deposition of ^{137}Cs began some downward movement may be expected as a result of the activity of surface-casting earthworms. In addition, some movement may be due to physical mobilization and translocation of clay particles within the soil. The role of chemical mobilization has not been quantified but is expected to be relatively minor.

The contrasting profiles from within the arable field show the mixing and homogenizing effect of cultivation within the plough zone. Consequently, both the level site and the eroded slope site have similar shaped depth distributions of ^{137}Cs, with the maximum extent reflecting the depth of ploughing. However, a very clear difference is seen in the total ^{137}Cs loading for the two sites. The relatively level site has almost the same ^{137}Cs loading as the grassland location, indicating minimal loss of soil. In contrast, the loading of the eroded site has been reduced by almost 40 per cent. At the depositional site, at the base of the slope, differences are seen in both the shape of the ^{137}Cs distribution and the total loading. The homogenizing impact of cultivation may be seen above the plough depth, but a difference is seen below. Surface accretion has led to the burial, beneath the present plough depth, of a layer of soil which was previously within the plough zone. As this buried soil material was formerly incorporated

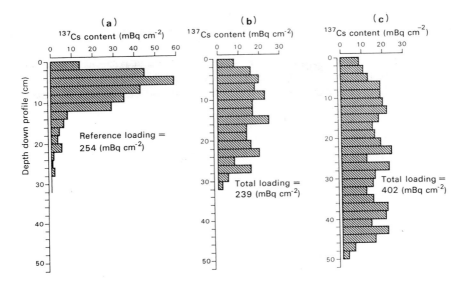

Figure 4.3 Profile distributions of ^{137}Cs from agricultural land in Devon. (A) Undisturbed pasture; (B) cultivated site with low erosion rate; (C) cultivated site with high deposition rate

into the plough zone it includes some ^{137}Cs and a 'stretched' ^{137}Cs profile is produced. Of equal significance is the greatly elevated loading, reflecting the input of ^{137}Cs-bearing soil to the profile in addition to the atmospheric deposition at the site. These expected results have been validated by empirical data in Devon (Walling et al., 1986). Figure 4.3 illustrates profiles from undisturbed grassland (a), a cultivated site subject to mild erosion (b) and a depositional cultivated site (c), identified during ongoing research in Devon.

On this basis it is suggested that a distinction may be made between eroded, uneroded and depositional sites by inspection of the ^{137}C profile, and also by comparison of the total ^{137}Cs loading (per unit surface area) at each point of interest with a reference loading. This reference loading is obtained by sampling an area of level, undisturbed grassland in the vicinity of the study site. It should therefore be possible to identify the pattern of erosion and deposition across a field by examining the pattern of ^{137}Cs loading. If a quantitative relationship can be established between ^{137}Cs gain and loss and the amount of erosion and deposition, it should be possible to estimate the pattern and rate of soil erosion within a field and the proportion of the eroded soil transported beyond it. If it can be demonstrated that ^{137}Cs may be used in this way it will provide a means of investigating rates of soil erosion which offers the following clear advantages:

(1) The erosion rate measured represents the sum of all erosive processes.
(2) The whole field may be studied without disturbance to the slope environment.
(3) The erosion rate measured is an average for the last 25-30 years, and is therefore less influenced by extreme events.
(4) The technique is capable of providing a quantitative measure of both the

Figure 4.4 Location of sampling sites. Rufford Forest Farm = site 7. Brook End Farm = site 8

pattern and the rate of soil erosion, and of the proportion of the eroded soil which is transported beyond the field.
(5) Only a single visit is required to each study site.

In order to confirm this potential a study is being undertaken which aims to apply the ^{137}Cs technique at a number of locations in the UK with contrasting soil types (Figure 4.4 and Table 4.1). Results from two of these sites, at Brook End Farm in Bedfordshire and Rufford Forest Farm in Nottinghamshire, will be discussed below.

THE APPLICATION OF THE ^{137}Cs TECHNIQUE

Sample Collection

As the distinction between eroded and uneroded sites is based upon the deviation of measured values of ^{137}Cs loading from the expected input value it is essential to establish the reference loading for ^{137}Cs in the locality. Samples from both the study field and an undisturbed site must therefore be collected and analysed. Those required for determining ^{137}C profile distributions are collected using a 'scraper-plate' (cf. Campbell *et al.*, 1988), while individual loading determinations are based upon core samples. The 'scraper-plate' yields depth incremental samples of 800 cm^2 at 1 cm or 2 cm intervals. Cores with a cross-section of 38 cm^2 are taken to a depth of 60 cm in order to be certain of retaining the full ^{137}Cs profile. It is possible to check that the full profile has been obtained by separate analysis of the basal 2 cm of the core, and this subsample is therefore stored separately.

In order to establish the reference or input value of ^{137}Cs loading, samples were collected from level, undisturbed grassland in the locality of each of the study sites. Undisturbed grassland may include permanent pasture, parkland, mature gardens and orchards. Woodland sites are not used for collection of reference samples due to the potential for local variability produced by stemflow and canopy interception. At the chosen undisturbed site, both a 'scraper-plate' profile and replicate cores were collected. At Rufford Forest Farm (Figure 4.7) the intensity of agricultural activity enforced the use of a reference sampling site on rough grassland 1.7 km from the study field. The distance was only 0.5 km at Brook End Farm, where reference samples were collected from an area of ridge and furrow immediately to the south of the farm buildings (Figure 4.8).

Within each study field samples were collected on a 20 m grid over an area of 2–4 ha, including as much of each field as was feasible. This provided a sampling density of 117 samples from 3.8 ha in Nottinghamshire and 84 samples from 2.4 ha in Bedfordshire. In addition to collecting full cores at each sample site, surface samples were collected to a depth of 5 cm and the height of each point, above a site datum, was recorded. The locations of the sampled areas

Table 4.1 Sampling sites and soil associations

Site number	County	Site name	Grid reference	Soil association	Soil type
1	Devon	Yendacott, Efford	SS 898 007	541e Crediton	Typical brown earth
2	Somerset	Mountfields, Shepton Beauchamp	ST 401 168	541m South Petherton	Typical brown earth
3	Dorset	Higher Farm, Langton Herring	SY 617 824	411a/b Evesham 1/2	Typical calcareous pelosol
4	Gwent	Fishpool Farm, Dingestowe	SO 446 100	571b Bromyard	Typical argillic brown earth
5	Herefordshire	Wooton Farm, Bishops Frome	SO 643 487	571b Bromyard	Typical argillic brown earth
6	Shropshire	Dalicott Farm, Claverley	SO 773 944	551a Bridgnorth	Typical brown sand
7	Nottinghamshire	Rufford Forest Farm, Rainworth	SK 614 584	551b Cuckney 1	Typical brown sand
8	Bedfordshire	Keysoe Park Farm, Keysoe	TL 063 625	411d Hanslope	Typical calcareous pelosol
8	Bedfordshire	Brook End Farm, Keysoe	TL 075 632	411d Hanslope	Typical calcareous pelosol
9	Norfolk	Manor House Farm, South Creake	TF 856 365	343g Newmarket 2	Brown rendzina
10	Norfolk	Hole Farm, Baconsthorpe	TG 115 358	551g Newport 4	Typical brown sand
11	Kent	West Street Farm, Ham	TR 328 541	511f Coombe 1	Typical brown calcareous earth
12	Sussex	Housedean Farm, Falmer	TQ 367 090	343h Andover 1	Brown rendzina

at the two sites are illustrated in Figures 4.7 and 4.8. As is shown in Figure 4.7, the sampled area at Rufford Forest Farm extended from hedge to hedge on the east–west axis, and from the hedge at the south of the field to the crest of the ridge in the north. At Brook End Farm, the length of the slope (400 m) constrained the proportion of the field which could be sampled. A strip 60 m wide including the full length of the slope was therefore studied.

The use of a detailed sampling strategy in conjunction with recording of elevation data for each sample point provides the necessary basis for comparison of ^{137}Cs loadings with topography. This will be discussed in the following section.

Data Collection and Analysis

All samples collected were prepared and analysed using the procedures summarized in Figure 4.5. Laboratory analysis of each sample yields two measures of ^{137}Cs content, the total activity of the sample (AS mBq) and the activity of the sample per unit mass (AM mBq g^{-1}). For core samples the total

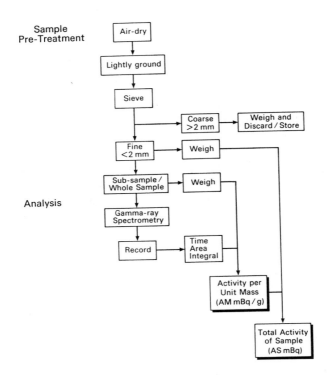

Figure 4.5 The procedure employed in sample preparation and analysis

activity is the most useful measure, as this can be simply converted into a measure of loading per unit area (AA mBq cm^{-2}) by dividing by the internal cross-sectional area of the tube, i.e.

$$AA = \frac{AS}{CA}$$

where AA = ^{137}Cs loading (mBq cm^{-2}),
AS = ^{137}Cs activity of core sample (mBq), and
CA = cross-sectional area of core tube (cm^2)

For profile sequences the same calculation may be performed for each incremental sample, using the area of the excavated hole:

$$AA_i = \frac{AS_i}{PA}$$

where AA_i = ^{137}Cs loading of the ith increment (mBq cm^{-2}),
AS_i = ^{137}Cs activity of the ith increment (mBq), and
PA = area of excavated hole (cm^2)

The total loading of the profile site is equal to the sum of the individual increment loadings:

$$AA_t = \sum_{i=1}^{n} AA_i$$

where AA_t = total ^{137}Cs loading of the profile sequence (mBq cm^{-2}),
AA_i = ^{137}Cs loading of the ith increment (mBq cm^{-2}), and
n = number of increments.

However, for profile sequences the activity per unit mass (AM mBq g^{-1}) may provide a more useful measure of ^{137}Cs. Variation in activity per unit mass through the profile sequence provides valuable evidence for the depth distribution of ^{137}Cs, the significance of which has been illustrated in Figure 4.2. In particular, such profile distributions demonstrate the effective binding of ^{137}Cs close to the surface of soils under undisturbed grassland. This is seen very clearly in the profiles from undisturbed sites on both the sandy soil at Rufford Forest Farm and the clay soil at Brook End Farm (Figure 4.6). As was indicated in the preceding section, replicate cores were also taken at these undisturbed sites and these provide additional evidence for the local reference loading of ^{137}Cs (CS[L]). In the second section of the chapter, it was suggested that deviation in sample loading from the reference level may reflect erosion and deposition. In order to quantify this deviation and examine its pattern ^{137}Cs residuals (CS[R]) are calculated:

$$CS[R]_i = CS[P]_i - CS[L]$$

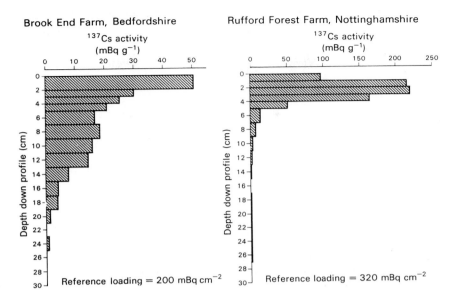

Figure 4.6 Profile distributions of ^{137}Cs from the reference sites at both sampling locations

where $CS[R]_i$ = calculated ^{137}Cs residual for point i,
$CS[P]_i$ = measured ^{137}Cs loading for point i,
$CS[L]$ = local ^{137}Cs reference loading for site.

Negative ^{137}Cs residuals indicate that ^{137}Cs loss has taken place and therefore, by implication, soil has been eroded. Conversely, positive residuals are indicative of ^{137}Cs gain and therefore deposition.

In order to examine the pattern of ^{137}Cs loss and gain the GINOSURF package was used to plot isolines of ^{137}Cs residuals based upon the individual data points. These patterns are illustrated in Figures 4.7 and 4.8. On the sandy soil at Rufford Forest Farm there is a clear correspondence between the ^{137}Cs residuals and the topography. Figure 4.7 indicates that almost all the sloping ground within the field has a negative residual in excess of -50 mBq cm^{-2}. Furthermore, the high rates of ^{137}Cs loss, and, by implication, soil loss, from the slopes is reflected in the large area with negative ^{137}Cs residuals in excess of -100 mBq cm^{-2}. Further confirmation of the relationship between topography and ^{137}Cs distribution is seen in the zones with the highest negative residuals, which lie on the steepest ground on both the south- and north-facing slopes. The positive residuals also occur where expected, at the base of the slopes on the lowest lying, relatively level ground. This is consistent with redistribution of ^{137}Cs associated with deposited sediment. However, zones of positive ^{137}Cs residuals are also seen on the eastern boundary of the field above the base of the depression and, in the north-east corner, at the crest of the ridge. These

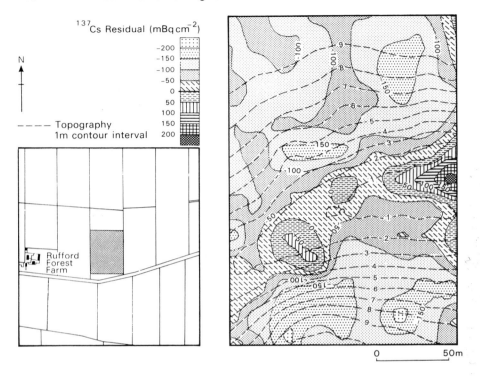

Figure 4.7 ^{137}Cs residuals at Rufford Forest Farm, Nottinghamshire

localized extensions of the positive zone may be the result of wind erosion, known to be a hazard in the region, leading to deposition along the hedge line.

More detailed interpretation of features of the pattern must be tentative. However, it may be significant that the slight depression, near the centre of the south-facing slope, coincides with an area of slightly lower negative residuals. This may reflect localized redeposition of eroded material from further upslope, but at a lower rate than downslope loss. However, the opposite pattern of residuals is seen towards the base of the adjacent depression, approximately 40 m to the west.

As might be expected, the redistribution of ^{137}Cs on the clay soil of Brook End Farm is much less extensive (Figure 4.8). The dominant feature of the pattern of ^{137}Cs residuals is the large proportion of the sampled area with values between $+50$ and -50 mBq cm^{-2}. Relatively little soil movement will have occurred in this zone. There is a zone of higher ^{137}Cs gain at the base of the field where deposition would be expected, and the areas with low positive residuals along the north-east edge of the field may be related to the depression in that area. The zone of high positive residuals at the top of the slope and

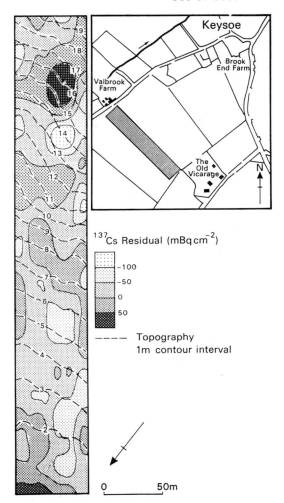

Figure 4.8 ^{137}Cs residuals at Brook End Farm, Bedfordshire

adjacent to the highest negative residual may be the result of localized earth movement, such as might be associated with the infilling of a hollow. Localized soil truncation and surface soil deposition could produce the pattern observed. The remaining zones with negative residuals exhibit no clear relationship with the topography, although the higher negative residuals are associated with the steeper slopes. As noted above, the dominant feature of the pattern is the low level of variability across field.

The ^{137}Cs residual data may also be used to provide an estimate of the redeposition of eroded material within the sampled area and a qualitative

Table 4.2 ^{137}Cs residual budgets

Site	Negative residuals Sum	n	Positive residuals Sum	n	Percentage redeposition	Mean residual	Mean percentage ^{137}Cs loss
Brook End Farm	1 860	53	755	31	40.6	−13	6.5
Rufford Forest	11 104	105	709	12	6.4	−89	27.8

comparison of erosion rates between the two sites. By calculating the sums of the negative and positive residuals for each site a ^{137}Cs budget may be calculated (Table 4.2). As the number of samples and the reference loadings vary between the sites, comparison is best undertaken using the data for percentage redeposition and mean percentage ^{137}Cs loss. The plots of ^{137}Cs residuals, discussed above, have already highlighted the difference in extent of ^{137}Cs loss between the two sites. The summary data for redeposition demonstrate a further contrast. Approximately 40 per cent of the ^{137}Cs eroded on the clay soil has been redeposited within the field, while on the sandy soil over 93 per cent has been removed from the field. The effects on net ^{137}Cs loss and, by implication, net soil loss are summarized in the values of mean percentage loss of ^{137}Cs for each site. The mean percentage loss for the sandy soil is over four times that for the clay soil, and the sandy soil has lost, on average, over 25 per cent of the ^{137}Cs input.

These two examples illustrate the potential for using ^{137}Cs measurements to determine patterns of ^{137}Cs redistribution and, by implication, soil erosion and deposition. The contrast between the two sites also provides a qualitative comparison of erosion rates. However, it is desirable to progress from discussion of ^{137}Cs redistribution to interpretation of such data in terms of quantitative rates of soil erosion and deposition. This interpretation demands the establishment of relationships between the magnitude of the ^{137}Cs residuals and rates of soil erosion and deposition.

THE RELATIONSHIP BETWEEN ^{137}Cs LOADING AND RATES OF EROSION AND DEPOSITION

A number of methods for relating ^{137}Cs losses to rates of soil erosion have been proposed by workers outside the UK. These methods fall into two categories, those based on theoretical models (e.g. Kachanoski and de Jong, 1984) and those based on erosion plot data (e.g. Kachanoski, 1987; Cambell *et al.*, 1986). No calibration relationship has been proposed for the UK, so a procedure based on a theoretical model or accounting system, similar to that employed by

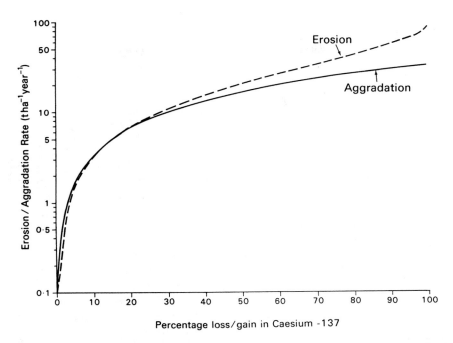

Figure 4.9 Estimation of annual erosion and aggradation rates from percentage loss and gain of ^{137}Cs. Calibration curves for Brook End Farm, Bedfordshire

Kachanoski and de Jong (1984), has been developed (Quine, 1989, in press). The model predicts the resultant ^{137}Cs loading for sample sites subject to different erosion and deposition rates, using data for annual deposition of ^{137}Cs, ploughing depth and depth of ^{137}Cs penetration. The data which the model provides may be used to produce calibration relationships for the estimation of erosion and deposition rates from the percentage loss or gain of ^{137}Cs. Examples of estimated calibration curves, for both erosion and deposition rates, calculated for Brook End Farm using this model are illustrated in Figure 4.9.

The calibration relationships have been used to estimate erosion and deposition rates from the ^{137}Cs loadings at both Brook End Farm and Rufford Forest Farm. These erosion and deposition rates have also been plotted using the GINOSURF package, and the plots are illustrated in Figures 4.10 and 4.11. In this case the isolines correspond to points with the same estimated erosion or deposition rate. Naturally, these plots show a pattern very similar to those based upon the ^{137}Cs residuals. In this case, however, the two sites may be compared directly since the isoline intervals are identical. As noted above, the field on the clay soil at Brook End Farm is characterized by low variability and

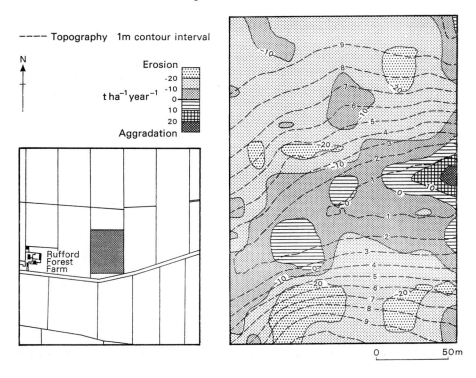

Figure 4.10 Estimated erosion and aggradation rates for Rufford Forest Farm, Nottinghamshire

low erosion and deposition rates. The erosion and deposition rates for the greater part of the field do not exceed 10 t ha^{-1}yr^{-1}. In contrast, almost the entire area of the sloping ground at Rufford Forest Farm is characterized by erosion rates in excess of 10 t ha^{-1}yr^{-1}, and on the steeper north-facing slope there are large areas with erosion rates in excess of 20 t ha^{-1}yr^{-1}.

Further quantitative comparison of the fields may be made by using the erosion and deposition data to calculate a sediment budget for each sampled area. These budgets are defined in Table 4.3. These summary data again illustrate the contrast between the two sites. In addition to the marked difference in both net and gross erosion rates there is a clear difference between the sites in the proportion of sediment redeposited. The significance of these contrasts and the validity of the suggested erosion rates will be discussed in the following section.

THE VALIDITY OF THE RESULTS

As has been demonstrated, ^{137}Cs provides a valuable tool for determining the pattern of soil movement within cultivated fields. This may assist interpretation

Use of Caesium-137   49

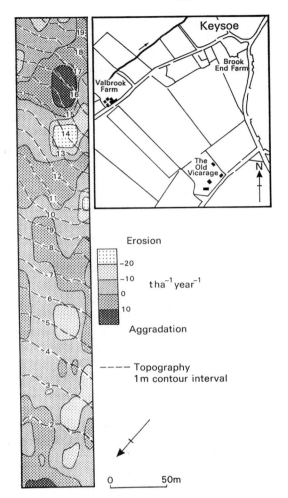

Figure 4.11 Estimated erosion and aggradation rates for Brook End Farm, Bedfordshire

Table 4.3 Annual sediment budgets for the fields

Site	Gross erosion rate (t ha^{-1})		Gross aggradation rate (t ha^{-1})		Net erosion rate (t ha^{-1})	Percentage redeposition
	Eroding sites	Whole field	Aggradation sites	Whole field		
Brook End Farm	6.2	3.9	4.1	1.5	2.4	38.7
Rufford Forest Farm	12.6	11.3	6.8	0.7	10.6	6.1

of the interaction between topography and erosive processes. However, one principal stated aim has been the assessment of rates of soil erosion. Use of a specifically developed model has permitted the estimation of erosion and deposition rates from the ^{137}Cs loadings. The validity of these results must now be assessed.

As was indicated in the opening discussion, evidence for soil-erosion rates in Britain is sparse. However, the sites were chosen to lie on flight paths of the MAFF/SSEW survey, so comparison with data from that survey is possible. Evans and Cook (1986) provide a summary of results up to 1984, in which they discuss erosion in Nottinghamshire but do not give an estimated range of erosion rates. However, both a mean and range of erosion rates for the clay soils of Cambridge and Bedfordshire were suggested. Comparable data were also given for sands and light loams in Staffordshire and Worcestershire, and Shropshire. These soils are texturally similar to those at Rufford Forest Farm, which may be expected to have erosion rates of a similar order of magnitude. The rates quoted by Evans and Cook (1986) are for a number of fields, but refer to only a single year (1982). This proviso must be taken into account when comparison is made with rates, based upon ^{137}Cs data, which represent an average for a period of approximately 30 years. Table 4.4 summarizes the data given by Evans and Cook (1986), and the erosion rates estimated using ^{137}Cs.

The erosion rate estimated for Rufford Forest Farm, using ^{137}Cs, compares well with the rates estimated by Evans and Cook (1986) for soils of similar textural properties. In this case the caesium-based rate is close to the mean rates proposed by Evans and Cook (1986). As has been indicated, the technique of erosion assessment employed by Evans and Cook (1986) is based entirely upon visible traces of erosion such as rilling. In contrast, the ^{137}Cs technique provides an estimate of the net effect of all erosive processes. Consequently, similarity in erosion-rate estimates for the two techniques may be expected for soils where rilling is the dominant erosive process. However, the caesium-based rates may be overestimates since the rates quoted by Evans and Cook (1986) are for eroded fields, and not all fields erode in any one year. In the Nottinghamshire study Evans and Cook (1986) give the percentage areas of the flight path subject

Table 4.4 Comparison with published erosion rates

Site	Erosion rate (m^3 ha^{-1} yr^{-1})		
	Aerial survey—1982 (Evans and Cook, 1986)		1954-87 ^{137}C estimate
	Range	Mean	
Cambs/Beds	0.5–2.5	1.1	
Brook End Farm			2.4
Shropshire	2.0–36.8	11.3	
Staffs/Worcs	0.4–42.5	12.7	
Rufford Forest Farm			10.6

to erosion as 1.1, 18.0 and 4.3 for 1982, 1983 and 1984, respectively. This could suggest that a 30-year average for an individual field, which is close to a single-year mean for only those fields which erode, may be an overestimate. However, Evans and Cook (1986) also note that 21 per cent of eroded fields were subject to erosion twice in the three-year period. As this field was chosen for its erosion hazard, it might be expected that the field would erode with a greater frequency than the average.

Comparison of the rates for Brook End Farm indicates that although the caesium-based rate is low in absolute terms, it is at the top end of the range quoted by Evans and Cook (1986) and twice their mean. While this may be the result of overestimation by the calibration relationship, two other factors should be considered. First, as in the case of the Nottinghamshire site, this field was not randomly selected, but was chosen with the expectation that it would be subject to erosion. A higher than average erosion rate may therefore be expected. Second, the differences between the two methods, discussed above, must be taken into account. The caesium-based erosion rate represents the net effect of all erosive processes operating over the last 30 years. In contrast, the MAFF/SSEW method provides an estimated rate based upon rills visible at the time of photography. The divergence between the rates estimated by the two techniques may reflect the importance of erosive processes other than rilling on this fine-grained soil. It is not inconceivable that the higher rate of erosion suggested by the ^{137}Cs technique may reflect the impact of rain-splash and sheet erosion, which are not detectable by aerial photography.

Particle size selectivity may also influence the estimation of erosion rates from ^{137}Cs data. As noted in the introductory discussion, ^{137}Cs is adsorbed by clay minerals in most environments. Movement of ^{137}Cs is therefore in association with the fine fraction, and the ^{137}Cs loadings reflect the redistribution of this fraction. As detachment and erosion of soil material is expected to occur primarily in aggregate form (cf. Foster et al., 1985; Walling, 1988), loss of the fine fraction should reflect movement of the entire soil. However, subsequent redeposition may be more strongly influenced by particle size selectivity. This is of particular relevance on the sandy soil of Rufford Forest Farm. The ^{137}Cs budget for this site (Table 4.2) indicates that only 6 per cent of the eroded ^{137}Cs has been redeposited, and this is reflected in the estimate of only 6 per cent redeposition of sediment in the sediment budget (Table 4.3). If preferential deposition of the coarser fractions occurs, 6 per cent may be a valid estimate of the redeposition of the fine fraction but an underestimate of total redeposition. If this is the case, the gross erosion rate for Rufford Forest Farm is reasonable but the net erosion rate will be an overestimate, due to unidentified aggradation of coarse material. To correct this aspect of the calibration procedure it will be necessary to include a particle-size selectivity factor into the deposition model.

The relatively high erosion estimate for the clay soil in Bedfordshire may also reflect the influence of particle size. In this very fine-grained soil, the abundance of binding sites for the ^{137}Cs close to the soil surface may have led to reduced initial penetration of the ^{137}Cs. Reduction of this factor in the model would produce lower estimates of soil erosion rate for given losses of ^{137}Cs.

Both these examples nevertheless indicate that the order of magnitude of the caesium-based erosion rate estimates are consistent with the limited available data. Divergence between estimates based upon the MAFF/SSEW survey data and the ^{137}Cs data may reflect the need for refinement of the calibration procedure. However, in considering this divergence it is important to take into account the role of site selection and the difference between the techniques in relation to the erosive processes measured.

CONCLUSION

In the context of the continuing debate over the severity of the soil erosion problem there is a need for empirical data concerning rates of erosion. Many of the existing techniques for soil-erosion assessment possess significant limitations. In contrast, the use of ^{137}Cs measurements has many important advantages. These have been demonstrated in the application of the technique to a study of soil erosion on two contrasting soil types under arable cultivation in the UK. Measurement of the spatial distribution of ^{137}Cs has permitted interpretation of the pattern of soil erosion and deposition within the fields studied. These detailed patterns may be related to topography in order to aid the modelling of soil erosion. The ^{137}C data have also been used to estimate the rates of soil erosion and deposition within the fields and the net loss from them. The rates of net loss compare well with data obtained by MAFF/SSEW through a combination of aerial survey and field measurement (Evans and Cook, 1986). Current research indicates that in areas which were recently subject to high levels of ^{137}C deposition, as a result of the Chernobyl accident, the technique may not be applicable. Despite this problem, there are many areas in Britain and outside it, where soil-erosion monitoring is of fundamental importance, in which the technique may provide a valuable tool in soil-erosion assessment.

ACKNOWLEDGEMENTS

This research was undertaken as part of the NERC/NCC 'Agriculture and the Environment' initiative. The authors gratefully acknowledge this support and the collaboration of staff of both ITE and SSEW. Discussions with Drs R. Evans, B. L. Campbell and R. J. Loughran have proved very profitable.

Finally, we are indebted to the landowners who have given us access to their property, particularly Mr Ward (Brook End Farm) and Mr King (Rufford Forest Farm).

REFERENCES

Campbell, B. L. (1983). Applications of environmental caesium-137 for the determination of sedimentation rates in reservoirs and lakes and related catchment studies in developing countries. In *Radioisotopes in Sediment Studies. International Atomic Energy Technical Document*, **298**, 7–30.

Campbell, B. L., Loughran, R. J., Elliott, G. L. and Shelly, D. (1986). Mapping drainage basin sediment sources using caesium-137. In Hadley, R. F. (ed.), *Drainage Basin Sediment Delivery*, IAHS Publication, **159**, 437–46.

Campbell, B. L., Loughran, R. J. and Elliott, G. L. (1988). A method for determining sediment budgets using caesium-137. In Bordas, M. P. and Walling, D. E. (eds), *Sediment Budgets*, IAHS Publication, **174**, 171–9.

Cawse, P. A. and Horrill, A. D. (1986). *A Survey of Caesium-137 and Plutonium in British Soils in 1977* AERE Harwell report R-10155, HMSO, London.

de Jong, E., Begg, C. B. M. and Kachanoski, R. G. (1983). Estimates of soil erosion and deposition for some Saskatchewan soils, *Can. J. Soil Sci.*, **63**, 607–17.

Evans, R. and Cook, S. (1986). Soil erosion in Britain. SEESOIL, **3**, 28–59.

Foster, G. R., Young, R. A. and Niebling, W. H. (1985). Sediment composition for nonpoint source pollution analyses, *Trans. Am. Soc. Agric. Eng.*, **28**, 133–9.

Kachanoski, R. G. and de Jong, E. (1984). Predicting the temporal relationships between soil cesium-137 and erosion rate, *J. Environ. Qual.*, **13**, 301–4.

Kachanoski, R. G. (1987). Comparison of measured soil 137-cesium losses and erosion rates, *Can. J. Soil. Sci.*, **67**, 199–203.

Longmore, M. E., O'Leary, B. M., Rose, C. W. and Chandica, A. L. (1983). Mapping soil erosion and accumulation with the fallout isotope caesium-137, *Aust. J. Soil Res.*, **21**, 373–85.

Loughran, R. J., Campbell, B. L. and Walling, D. E. (1987). Soil erosion and sedimentation indicated by caesium-137: Jackmoor Brook catchment, Devon, England, *Catena*, **14**, 201–12.

McHenry, J. R. and Ritchie, J. C. (1977). Estimating field erosion losses from fallout caesium-137 measurements. In *Erosion and Sediment Transport in Inland Waters*, IAHS Publication, **122**, 26–33.

Morgan, R. P. C. (1979). *Soil Erosion*, Topics in Applied Geography, Longman, Harlow.

Quine, T. A., (1989, in press). Use of a simple model to estimate rates of soil erosion from caesium-137 data, *Hydrological Problems of Arid and Semi Arid Regions*, proceedings of the Third Iraqi Hydrological Conference, Baghdad, 13–16 March 1989.

Ritchie, J. C., Spraberry, J. A. and McHenry, J. R. (1974). Estimating soil erosion from the redistribution of caesium-137, *Soil Sci. Soc. Am. Proc.*, **38**, 137–9.

Walling, D. E. (1988). Erosion and sediment yield research—some recent perspectives, *J. Hydrol.*, **100**, 113–41.

Walling, D. E., Bradley, S. B. and Wilkinson, C. J. (1986). A caesium-137 budget approach to the investigation of sediment delivery from a small agricultural drainage basin in Devon, U.K. In Hadley, R. F. (ed.), *Drainage Basin Sediment Delivery*, IAHS Publication, **159**, 423–35.

5 Gully Erosion on Agricultural Lands in Romania

I. ICHIM, G. MIHAIU, V. SURDEANU, M. RADOANE and N. RADOANE
Piatra Neamt, Romania

INTRODUCTION

The Moldavian Tableland is situated to the east of the Oriental Carpathian and Subcarpathian and occupies around 10 per cent of Romania's territory (Figure 5.1). The tableland has an average altitude of around 250 m and its highest point just exceeds 600 m. This study focuses on the Moldavian Hills which lie in the north-east of the region.

This is an area of Neogene–Quaternary deposits. These strata, a series of slight-to-non-cohesive marls, clays and sands, dip to the south-east at a gentle 5–8 m per 1000 m. The high silt/clay content (60–70 per cent) favours the development of large gullies.

The climate is temperate continental. The mean annual temperature ranges from 7 to 8°C in the north-east to 9 to 10°C in the south-west. Annual precipitation is 450–550 mm rising to about 600 mm in the central and north-western parts of the tableland.

The hydrothermal coefficient is regarded as a climatic parameter which correlates closely with erosion (Zachar, 1982). It is calculated from the formula:

$$\text{HTK} = \frac{\Sigma R}{\Sigma t} \cdot 10$$

where ΣR is the aggregate precipitation and Σt is the summation of air temperatures above 10°C.

Most gullies occur in regions with HTK values between 1.25 and 2.50. In the Moldavian Tablelands, HTK values range from 1.5 in the east up to 2.5 in the north-west. This indicates a propensity for gully erosion.

Traditionally, the Moldavian Tableland was a heavily forested region. At the beginning of the nineteenth century the forest cover was 50 per cent. This

---

Soil Erosion on Agricultural Land
Edited by J. Boardman, I. D. L. Foster and J. A. Dearing
©1990 John Wiley & Sons Ltd

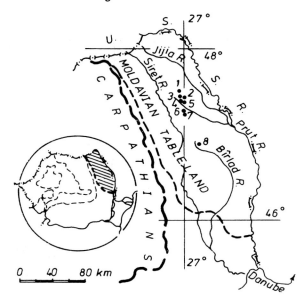

Figure 5.1 Location of the study area. 1, Deleni Gully; 2, Gurguiata Gully; 3, Coada Gistii Gully; 4, Ceplenita Gully; 5, Secaresti I Gully; 6, Secaresti II Gully; 7, Giurgeni Gully

declined to 25 per cent at the beginning of this century and to 13 per cent at present. Today, agriculture is the dominant land use. Some 58 per cent of the Tableland is tilled, 15 per cent is grassland and pasture and just 4 per cent is given to vineyards and orchards.

Deforestation has contributed to the area's soil erosion problems. Some 71 per cent of the Tableland suffers from erosion and major erosion control works are in progress. The problems of soil erosion have been the subject of much research, especially in the last 20–30 years (Simonescu, 1903; Tufescu, 1937; Martiniuc, 1954; Bacauanu, 1968; Harjoaba, 1968; Sficlea, 1972; Popa, 1977; Motoc et al., 1979; Ionita and Ouatu, 1985). The erosion-control research station at Perieni has been studying the region's problems for 30 years.

DATA COLLECTION

This study concerns the particular problems of gully erosion in the central parts of the Jijia–Bahlui basin, an agricultural region of around 3718 km² in the Moldavian Hills. It includes a general examination of the 2600 gullies surveyed in the tract and recorded on topographical map sheets of the 1:25 000 Series (1982/4 edition) and of the 1:5000 Series (1971/9 edition). This general survey collected data on the geological structure, slope form, the length, depth and width of the gully channel and the difference in altitude between the gully's head and foot.

Gully Erosion on Agricultural Lands in Romania

Table 5.1 Data on the study gullies

Gully name	Length (m)	Active area (m)	Relief (m)	Volume removed (m³)
Secaresti I	424	8 807	50.9	22 453
Secaresti II	414	5 809	52.0	15 108
Ceplenita	410	20 076	48.4	55 368
Gurguiata	145	1 493	16.4	7 088
Coada Gistii	173	6 721	15.5	20 764
Deleni	262	3 902	31.9	4 371
Giurgeni	428	6 269	36.4	26 228

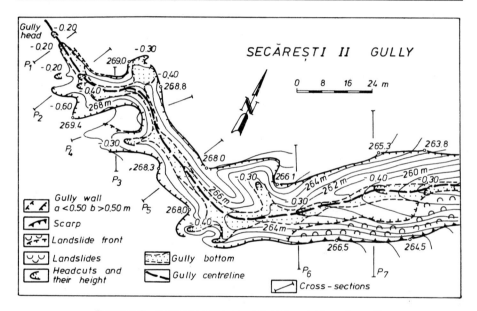

Figure 5.2 Evaluating the geometrical variables of a gully

This survey was supported by some more detailed examinations of seven gullies. These were mapped at scales of 1:200 and 1:400 and the maps recorded details of geomorphological and micromorphological processes (Table 5.1). Gully cross-sectional geometries were measured (Figure 5.2) at intervals down the long profile. The spacing of cross-profile measurement down the thalweg was always less than upstream gully width. A total of 133 cross-sections were surveyed and 31 variables recorded at each cross-section. Preliminary screening of this data set, however, demonstrated that only 19 of these variables might prove to be useful (Table 5.2). These variables may be subdivided into two clusters which include those that describe the gully long-profile and those which characterize the cross-section. Both clusters are considered in relation to the distance from the gully head.

Table 5.2 Variables used in the study

Variable name	Symbol	Units	Range
Distance from gully head	LG	m	2.20–428
Gully relief	RG	m	0.01–54.8
Active area	A	m^2	0.10–412
Side slope	SS	m/100	1.40–13.9
Channel slope	SC	m/100	1.60–25.0
Maximum depth	DX	m	0.10–10.4
Headcut height	HH	m	0.10–2.2
Mean depth	$DM = CS/B$	m	0.10–4.5
Gully width	B	m	1.00–91.6
Cross-section perimeter	P	m	0.90–128
Hydraulic radius	HR	m	2.80–20.8
Cross-section area	CSA	m^2	0.10–412
Shape factor (Heede, 1974)	$SF = DX/DM$	—	1.00–2.9
Width–depth ratio	F	—	0.82–28.0
Cumulated volume of material			
removed	W	m^3	0.90–16 329
by sidewall processes	WS	m^3	0.10–14 450
by downcutting processes	WC	m^3	0.10–3 150
by left sidewall processes	WL	m^3	0.05–7452
by right sidewall processes	WR	m^3	0.01–6998
Sidewall/linar incision	G	—	0.01–27.0

RESULTS AND DISCUSSION

According to the classification of Kalincenko and Ilinski (1976), the Moldavian Tableland is a region of slight to moderate gully erosion. The hillslopes have an average gully frequency of 2–4 gullies per km^2 ranging upwards to a maximum of 16–18 gullies per km^2 while the gully density averages 0.1–1 km km^{-2} ranging upwards up a maximum of 4 km km^{-2} (Figure 5.3).

Discontinuous gullies on hillsides represent 90 per cent of the total population. Their average width ranges between 1 and 50 m. Depths average 7 m and rarely exceed 25 m. They are most numerous on the slopes of consequent valleys (Figure 5.4).

The lower parts of the hillslopes are especially prone to gully erosion. The average distance from the water divide to the gully head is around 327 m and the average length of a gullied slope is 550–600 m (range: 50–1800 m), while most gullies are less than 250 m in length. The relief energy of these gullies, the difference in altitude between the headcut and foot, in 80 per cent of the cases, ranged between 30 and 100 m.

Statistical analysis of the data sets, which include measurements of gully geometry, drainage basin and hillslope characteristics, are presented in Table 5.3. Histograms show that all the variables are approximately log-normally distributed, so logarithms were taken of all variables. Computation of a matrix

Gully Erosion on Agricultural Lands in Romania

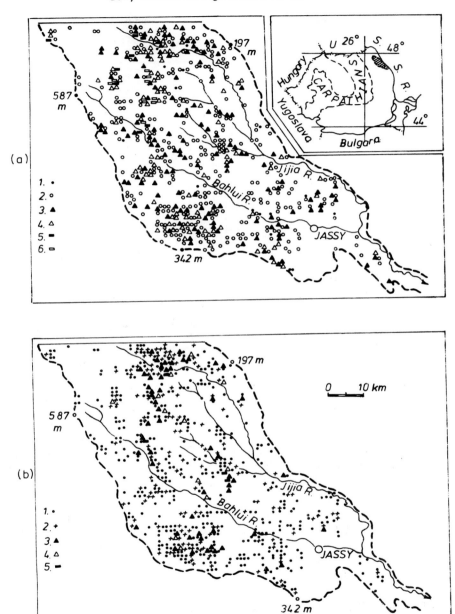

Figure 5.3 Area distribution of gullies from the Moldavian Tableland. (a) Density of gullies (km km^{-2}): 1, Below 0.1; 2, 0.101–0.5; 3, 0.501–1.0; 4, 1.101–2.0; 5, 2.01–3.0; 6, over 3.0. (b) Number of gullies per km^2: 1, 1–3; 2, 4–6; 3, 7–10; 4, 11–15; 5, over 15

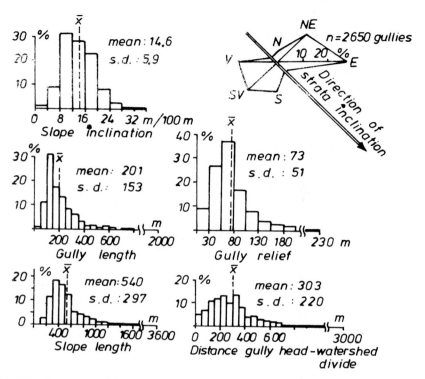

Figure 5.4 Frequency of the geometrical variables related to gullies: slope inclination; gully number in relation to slope orientation and direction of strata inclination; gully length; gully relief (maximum elevation difference in the watershed); slope length; distance from gully head to watershed divide; s.d. standard deviation

of correlation coefficients provides some initial idea of the structure of the data, particularly the high degree of redundancy (Table 5.3). Matrices were constructed for each gully as well as for the whole data set and have the following attributes:

(1) About 80 per cent of the correlation coefficients are positive, a fact that proves the categorical domination of direct correlation among the variables collected.
(2) Over 70 per cent of the correlation coefficients are very strong and are highly significant $(r > 0.7)$.

Elementary linkage analysis was employed to determine the hierarchical structure of the variables. This technique uses correlation coefficients descriptively to build a quantitative measure of similarity between variables. The McQuitty (1957) algorithm involves using the largest correlation coefficient in the matrix to link a first 'reflexive pair' of variables and then using the correlations of the

Table 5.3 Matrix of correlation coefficients

	L_g	R_g	A	S_s	S_c	S_l	S_r	d_{max}	$\bar{d}_m$	w	P	H_r	A_r	F_f	F	W	W_s	W_c	W_l	W_r	V	V_s	V_c	V_l	V_r	R_v
L_g	1.0																									
R_g	0.851	1.0																								
A	0.863	0.905	1.0																							
S_s	0.415	0.486	0.467	1.0																						
S_c	−0.172	0.219	0.135	0.452	1.0																					
S_l	−0.056	−0.199	−0.308	−0.022	−0.135	1.0																				
S_r	−0.033	−0.196	−0.315	−0.109	−0.246	0.775	1.0																			
d_{max}	0.543	0.592	0.651	0.070	0.203	−0.235	−0.264	1.0																		
$\bar{d}_m$	0.525	0.678	0.731	0.285	0.343	−0.404	−0.409	0.629	1.0																	
w	0.432	0.444	0.493	0.186	0.090	−0.292	−0.331	0.696	0.444	1.0																
P	0.602	0.728	0.813	0.189	0.261	−0.467	−0.470	0.78	0.848	0.591	1.0															
H_r	0.060	0.024	0.079	−0.086	−0.223	−0.190	−0.202	−0.235	−0.039	−0.094	0.197	1.0														
A_r	0.400	0.575	0.623	0.219	0.415	−0.472	−0.429	0.729	0.692	0.546	0.721	−0.352	1.0													
F_f	0.158	0.126	0.164	−0.188	−0.046	−0.055	−0.041	0.617	−0.076	0.434	0.248	−0.172	0.301	1.0												
F	−0.16	0.027	0.077	0.027	0.643	−0.354	−0.334	−0.202	−0.092	0.220	0.137	0.643	−0.009	−0.069	1.0											
W	0.640	0.744	0.832	0.242	0.236	−0.376	−0.418	0.769	0.815	0.642	0.917	0.028	0.751	0.235	0.122	1.0										
W_s	0.523	0.680	0.760	0.246	0.350	−0.482	−0.515	0.766	0.882	0.594	0.897	−0.091	0.805	0.177	−0.007	0.922	1.0									
W_c	0.697	0.767	0.819	0.277	0.164	−0.193	−0.265	0.630	0.658	0.547	0.798	0.111	0.636	0.169	0.230	0.883	0.718	1.0								
W_l	0.518	0.692	0.754	0.236	0.351	−0.488	−0.481	0.748	0.861	0.593	0.896	−0.068	0.828	0.181	0.065	0.921	0.983	0.739	1.0							
W_r	0.513	0.668	0.748	0.241	0.358	−0.470	−0.528	0.767	0.862	0.604	0.893	−0.079	0.817	0.190	0.032	0.926	0.992	0.741	0.973	1.0						
V	0.604	0.736	0.818	0.240	0.309	−0.459	−0.470	0.837	0.870	0.688	0.966	0.012	0.810	0.278	0.092	0.949	0.947	0.806	0.944	0.943	1.0					
V_s	0.564	0.688	0.716	0.233	0.295	−0.447	−0.433	0.737	0.847	0.525	0.872	0.060	0.767	0.179	−0.047	0.839	0.937	0.656	0.929	0.928	0.900	1.0				
V_c	0.657	0.785	0.766	0.227	0.199	−0.296	−0.287	0.721	0.759	0.483	0.847	0.014	0.672	0.227	0.013	0.786	0.767	0.776	0.772	0.761	0.843	0.832	1.0			
V_l	0.566	0.695	0.719	−0.022	0.351	−0.453	−0.383	0.713	0.824	0.516	0.857	−0.065	0.776	0.183	−0.002	0.831	0.908	0.666	0.929	0.891	0.889	0.974	0.829	1.0		
V_r	0.562	0.683	0.711	−0.109	0.358	−0.436	−0.428	0.824	0.824	0.516	0.853	−0.076	0.761	0.193	−0.044	0.833	0.917	0.667	0.902	0.918	0.884	0.983	0.823	0.969	1.0	
R_v	0.403	0.550	0.571	0.185	0.353	−0.376	−0.442	0.636	0.778	0.431	0.758	−0.087	0.751	0.113	−0.100	0.724	0.877	0.485	0.866	0.869	0.794	0.941	0.678	0.903	0.915	1.0

61

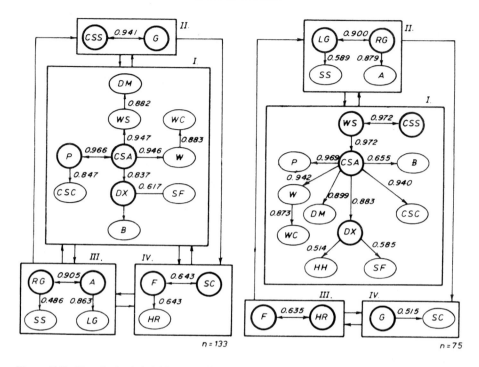

Figure 5.5 Topological classification of the morphometric variables listed in Table 5.2. I–IV classes of morphometric variables for $n = 133$ gully cross-sections and $n = 75$ gully cross-sections where headcuts are developed

reflexive pair variables to construct a ranked list of related variables. Analysis then proceeds to the highest significant correlation which does not involve the first pair of variables. This correlation provides the reflexive pair for a second group and so on (cf. Norcliffe, 1977, p. 193).

In this case the variables group into four classes. In each class, the variables are ranked according to their value for the description of gully evolution (Figure 5.5). In the first class of variables the following are very important: cross-sectional area (*CSA*), cross-section perimeter (*P*) and the maximum depth of the gully (*DX*). The following classes, ordered in relation to the decrease in correlation coefficients, included a smaller number of variables. However, the following variables proved important: gully relief (*RG*), distance from the gully head (*LG*) and the active area (*A*). In sum, the analysis determined the following variables to be of utmost importance for the evaluation of gully morphometry: *CSA*, *P*, *RG*, *LG*, *A*, *DX*. As such, these variables are selected for further study.

Gully expansion has often been modelled as proportional growth. The study of proportional growth as it affects two variables in a system is termed allometry. Allometric growth is characterized by the simple power function:

$$y = aX^b$$

Allometric analysis may be based on the relative changes of variables during the course of the evolution of a system or they may be based on measurements of the same variables in systems at different stages in their growth. The 'static allometry' approach is adopted by this study to demonstrate the adjustment of some key variables with respect to distance from the gully head. Power function equations were determined for 17 variables. Table 5.4 portrays the equations calculated for the Ceplenita Gully. For all 17 equations, the correlations between the two variables were highly significant ($p > 0.001$). The coefficient of determination varied from 50 per cent to 90 per cent with the smaller values for the channel slope variable. In the equations, the coefficients a and b are called the allometric parameters because they indicate the nature and direction of change. The following general observations are possible:

(1) All the equations have a positive value for the b exponent, except channel slope. This indicates a general increase in morphometric dimensions down the gully channel.

Table 5.4 Allometric equations of geometrical variables versus distance from gully head (Ceplenita Gully, $n = 30$ cross-sections)

Independent variable	a	b	r	$r^2 \cdot 100$
Cross-section area (CSA)	$3.08 \cdot 10^{-6}$	3.116	0.947	89.7
Gully relief (RG)	0.043	1.127	0.969	93.9
Active area (A)	0.076	1.636	0.772	59.6
Maximum depth (DX)	0.00069	1.608	0.929	86.3
Gully width (B)	0.00169	1.810	0.916	83.9
Width–depth ratio (F)	0.257	0.605	0.841	70.7
Perimeter (P)	0.00021	2.251	0.914	83.5
Side slope (SS)	21.43	−0.261	0.563	31.7
Channel slope (SC)	20.56	−0.204	0.407	16.6
Hydraulic radius (HR)	0.223	0.697	0.866	74.9
Mean depth (DM)	0.0018	1.308	0.945	89.4
Shape factor (SF)	0.384	0.301	0.679	46.1
Cumulated volume material removed (W)	0.000037	3.151	0.894	79.9
by sidewall processes (WS)	$3.5 \cdot 10^{-7}$	3.924	0.855	73.1
by downcutting processes (WC)	0.00084	2.309	0.860	74.0
Sidewall/linear incision (G)	0.00033	1.683	0.719	51.7

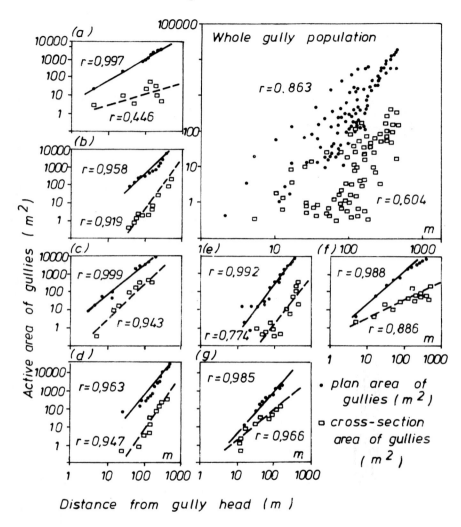

Figure 5.6 Gully active area and gully cross-section area to the distance from gully head. (a) Deleni Gully; (b) Giurgeni Gully; (c) Coada Gistii Gully; (d) Ceplenita Gully; (e) Secaresti I Gully; (f) Secaresti II Gully; (g) Gurguiata Gully

(2) A high value of b indicates more rapid expansion of the morphometric variable downchannel. It is possible to distinguish two groups of gullies. Two, the Ceplenita and Giurgeni gullies, show relatively rapid allometric expansion. The other five show much slower expansion. However, in all cases the cumulated volume of material removed shows the highest value of the b exponent ($b > 2.0$).

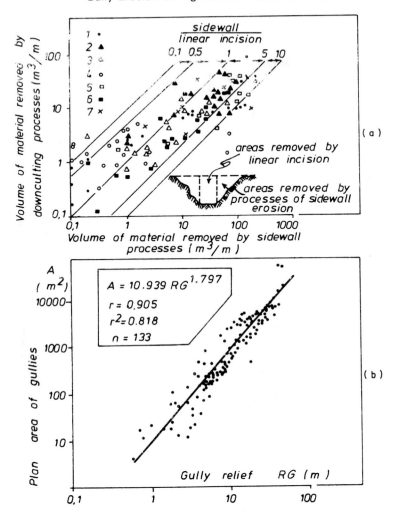

Figure 5.7 (a) Volumes of material removed from cross-sections of studied gullies by linear incision and sidewall erosion processes (1–7 are gullies listed in Table 5.1); (b) plan area of gullies related to gully relief (maximum elevation difference in the watershed)

Some examples of the graphs of these relationships are portrayed as Figures 5.6 and 5.7. These show the increase in the cross-sectional area (*CSA*) and active area (*A*) with increasing distance from the gully head (*LG*). The variable *LG* explains around 90 per cent of the variation in each of these two variables in each of the seven gullies examined. However, an attempt to correlate across the whole gully population produces a reduction in this correlation.

This reflects the fact that not all the gullies have the same rate of allometric expansion.

Vaness's method (Blong *et al.*, 1982) was employed to differentiate between the contributions of sidewall processes and vertical incision to total gully excavation. The ratio between sidewall and downcutting processes varies between 0.5 and 1.0:1 for most gully cross-sections. On average, the gully sidewalls contribute about 62 per cent of the total sediment production. However, the bulk of the sidewall contribution tends to come from one gully sidewall, the left- or the right-hand sidewall alternately. The value of the sidewall/downcutting sediment yield ratio correlates directly with slope aspect and the relief energy of the gully channel.

Heede (1974) recommended the use of the ratio between gully maximum depth and gully average depth to determine whether or not a channel is in equilibrium. In 70 per cent of the cross-sections examined, this ratio had a value lower than 2:1. This is indicative of great hydrological efficiency and rapid gully evolution.

Gully long profiles tend to be convex overall, with a slight concavity at the downstream end. This form is more susceptible to linear and logarithmic modelling rather than the semilogarithmic models specific to larger rivers. In Graf's scheme of gully evolution, these gullies belong to the stage of youth and vigorous expansion (Graf, 1977).

These analyses of gully geometry reveal several key correlations between environmental factors which have implications for field management practice. It may be considered, however, that the reclamation capability (cf. Haigh, 1984) for most of the gullied fields in the Moldavian Hills is good or moderate.

CONCLUSION

Gully erosion problems in the Moldavian Tableland may result from the conversion of forest to agricultural land during the last two centuries. The gullies of the Jijia–Bahlui basins are young and undergoing rapid expansion but they have relatively small dimensions. Ninety per cent are discontinuous gullies which form on the lower parts of the hillside. A survey of 2600 gullies in the Moldavian Hills, average frequency 2–4 gullies per km^2, was supported by detailed morphometric analysis of seven gully systems. This discovered strong positive correlations between gully length and drainage basin area, the length of the gully and its depth, and the size of the gully excavation and the size of the gully basin. The results reported are consistent with an allometric pattern of gully expansion. Gully size also correlated with the relief energy of the gully channel from headcut to foot and the distance downslope from the gully headcut. Sidewalls contribute around 62 per cent of the total sediment yield. The reclamation potential is considered to be good to moderate.

ACKNOWLEDGEMENTS

The authors gratefully acknowledge Martin Haigh, of Oxford Polytechnic, for helpful advice and encouragement to write this chapter. We were assisted in the field by Vasile Frunzeti and Dan Pipirigeanu, to whom many thanks.

REFERENCES

Bacauanu, V. (1968). *Plains of Moldavia: Geomorphological Study* (in Romanian), Editura Academiei, Bucharest.
Blong, R. J., Graham, P. and Veness, J. A. (1982). The role of sidewall processes in gully development: some NWS example, *Earth Surface Processes and Landforms*, **7**, 381–5.
Graf, W. J. (1977). The rate law in fluvial geomorphology, *American Journal of Science*, **227**, 178–91.
Haigh, M. (1984). Ravine erosion and reclamation in India. *Geoforum*, **15**, 543–61.
Harjoaba, I. (1968). *Relief of Tutova Hills* (in Romanian), Editura Academiei, Bucharest.
Heede, B. (1974). Stage of development of gullies in Western United States of America, *Zeit. für Geomorph.*, NF **18**, 3,260–71.
Ionita, I. and Ouatu, O. (1985). Contributions on the soil erosion study from the Tutova Hills (in Romanian), *Agronomical Researches in Moldavia*, **XIII**, 58–68.
Kalinicenko, N. P. and Ilinski, V. V. (1976). *Gully Improvement and Control by Means of Forestry Measures* (in Russian), Izd. Lesnaya promishlennost, Moscow.
Martiniuc, C. (1954). Deluvial slopes. Contributions on the field degradation study (in Romanian), *Problems in Geography*, **III**, 217–22.
McQuitty, L. L. (1957). Elementary linkage analysis for isolating orthogonal and oblique types and typal relevancies *Educational and Psychological Measurement*, **17**, 207–29.
Motoc, M., Taloescu, I. and Negut, N. (1979). Evaluation of the gully growth rate (in Romanian), *Inform. Bull. of ASAS*, **8**, 77–85.
Norcliffe, G. B. (1977). *Inferential Statistics for Geographers*, Hutchinson, London, pp. 192–5.
Popa, N. (1977). *Research on the Soil Erosion and Control Measures from the Moldavian Tableland* (in Romanian), ASAS, Bucarest.
Simionescu, I. (1903). *Geology of Moldavia between Siret and Prut Rivers* (in Romanian), Romanian Academy Publ., IX, Bucharest.
Sficlea, V. (1972). *Covurlui Tableland: Geomorphological Study* (in Romanian), Unpublished thesis, University of Jassy.
Tufescu, V. (1937). Great Hill of Hirlau. Observations on the relief evolution (in Romanian), *Bull. of Romanian Geogr. Soc.*, **LVII**, Bucharest.
Zachar, D. (1982). *Soil Erosion: Development in Soil Erosion*, Amsterdam.

6 Land-use Controls on Sediment Production in a Lowland Catchment, South-west England

A. L. HEATHWAITE
School of Geography, University of Oxford

T. P. BURT
School of Geography, University of Oxford

and

S. T. TRUDGILL
Department of Geography, University of Sheffield

INTRODUCTION

The stream sediment and associated nutrient load is of major importance in water quality control. Although the stream sediment load is derived from non-point inputs from the surrounding catchment, sediment sources may be highly localized (Hodges and Arden-Clarke, 1986; Morgan, 1980; Reed, 1983). Agricultural land use is one of the major controls on the source and magnitude of sediment transfer to the stream. In UK agricultural land use there is evidence for the large-scale conversion of grassland to arable, with increased livestock (mainly dairy) numbers in the remaining grassland areas. Such agricultural intensification may accelerate soil erosion on a large scale through, first, lowered soil organic matter levels in arable areas (Reed, 1979; Morgan, 1980, 1985). Organic matter is critical in soil particle cohesion and in the maintenance of soil structural stability. Once organic matter levels fall below 2 per cent the soil becomes structurally unstable (Greenland *et al.*, 1975) and subject to slaking (Morgan, 1985) or cap formation (Fullen and Reed, 1986). Second, increased livestock numbers may accelerate erosion through compaction of the soil surface and the removal of the vegetation cover in heavily grazed areas.

In the quantification of soil erosion in the UK much of the research has focused on the conversion of grassland to arable as the source of increased sediment transport (Boardman, 1984; Reed, 1979). Of the few authors who have

---

Soil Erosion on Agricultural Land
Edited by J. Boardman, I. D. L. Foster and J. A. Dearing
© 1990 John Wiley & Sons Ltd

examined the role of grazing in sediment transfer, the emphasis has been on upland areas (Evans, 1977). Lowland agricultural catchments in the UK and, in particular, south-west Devon traditionally have grazed permanent grassland on steeper slopes or adjacent to streams. It is therefore important to examine the implications of increased livestock numbers on a reduced grassland area for sediment and associated nutrient transfer to the stream system.

This chapter will establish how different land uses in the UK act as sediment sources. The presence or absence of surface vegetation, grazing animals and the timing of ploughing operations may be particularly important in determining the magnitude of sediment production. The effect of land use on infiltration capacity and bulk density will be linked to measurements of surface runoff and suspended sediment production from hillslope plots using rainfall-simulation experiments.

The objective is to relate sediment production in surface runoff under controlled experimental conditions to stream sediment loads for monitored storm events in small (headwater) drainage basins. The magnitude of storm sediment production at different catchment scales will be examined and the potential links between hillslope sediment production and the stream sediment load explored in the context of catchment land use. Particular emphasis is placed on storm runoff, as a substantial proportion of non-point pollutants will be transported in overland flow associated with storm events (O'Loughlin and Cullen, 1982).

THE STUDY SITE

Slapton Ley (Ordnance Survey Grid Reference SX 825 439), a 0.8 km^2 lake, is the largest natural freshwater body in south-west England, and is the sink for sediment and solute inputs from the surrounding 46 km^2 arable and grassland catchment. Past research has shown the importance of subsurface flow (throughflow or interflow) in delivering high concentrations of nitrate to streams (Burt *et al.*, 1983, 1988; Troake and Walling, 1973). There is little information for the role of surface delivery in stream sediment loads and associated nutrient transport. This chapter will examine the production of suspended sediment in surface runoff in more detail.

Catchment topography consists of flat-topped ridges dissected by narrow deep valleys. The soils are freely drained acid brown earths overlying impermeable Devonian slates and shales (Trudgill, 1983). The area has a mean annual rainfall of 1035 mm and a mean annual temperature of 10.5°C (Ratsey, 1975).

The 46 km^2 catchment may be subdivided into four subcatchments of fairly distinct land use (Figure 6.1). The 23.6 km^2 Gara catchment is primarily permanent and temporary grassland, whereas the 10.8 km^2 Start catchment has a high proportion of arable and temporary grassland. The Slapton Wood (0.9 km^2) and Stokeley Barton (1.5 km^2) catchments are mainly arable with

Land-use Controls on Sediment Production 71

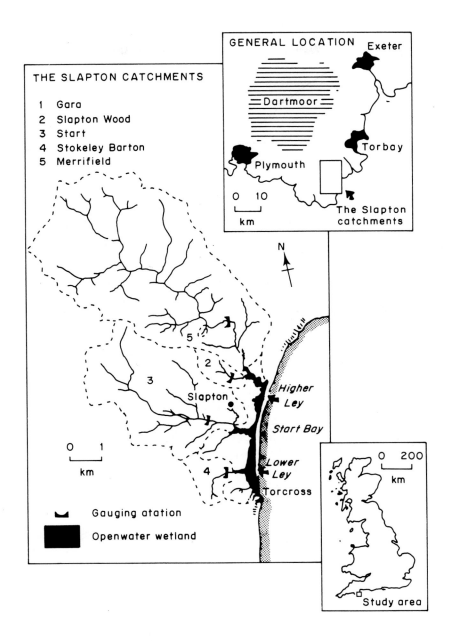

Figure 6.1 Map of the Slapton Catchments

Table 6.1 A comparison of land use in the Slapton Wood and Stokeley Barton catchments for 1974 and 1986

Land use	Slapton Wood 1974	Slapton Wood 1986	Stokeley Barton 1974	Stokeley Barton 1986
	Percentage of total catchment area			
Arable [a]	24	46	52	66
Pasture [b]	65	43	41	32
Wood	11	11	7	2
Catchment area (ha)	93		153	

[a] Includes both cereal and root crops.
[b] Includes both temporary and permanent grass.

woodland and arable with a high proportion of market gardening, respectively. As an indicator of land-use trends in the 46 km² catchment, a comparison of land use in the Slapton Wood and Stokeley Barton catchments for the period 1974–86 was made (Table 6.1). This highlights the shift from pasture to arable land use giving a 22 per cent and 14 per cent increase in arable land for the Slapton Wood and Stokeley Barton catchments, respectively. This is associated with a corresponding decrease in the percentage of grassland in each catchment.

The 0.18 km² Merrifield catchment (Ordnance Survey Grid Reference SX 817 475) was used for the detailed land use and sediment production monitoring discussed below. Its land use, permanent grass, temporary grass and arable land is representative of the Slapton catchment as a whole. The catchment is part of the 27 km² Gara catchment (Figure 6.1) and has been monitored since August 1987. It consists of a small plateau area merging into a 10–15° slope with Merrifield stream at the slope base. This stream, which is gauged, discharges directly to the River Gara. This catchment has the advantage of being small enough to provide a relatively homogeneous environment and yet large enough to encompass the results of the site-specific hillslope experiments into patterns of stream sediment delivery.

METHODS

Field Measurement

(1) *Soil physical parameters*—for each land use: permanent grassland, arable (kale) and recently reseeded grass (previously barley), the bulk density of the upper 5 cm soil, soil moisture and infiltration capacity measured using a constant head infiltrometer (Burt, 1978), were monitored at approximately monthly intervals starting in August 1987. Replicate measurements were made at successive points downslope to give average values for each land use.

(2) *Steam gauging*—all stream inputs from the Slapton catchments to Slapton Ley (see Figure 6.1) were gauged continuously using Ott water-level recorders. Water sampling, using Rock and Taylor water samplers, was on a four-hourly basis, increased to a 15-min interval during storm events. Suspended sediment was measured after filtration under suction through a Whatman GF/C filter paper.

Hillslope Plot Experiments

To quantify surface runoff and suspended sediment production from different land uses a series of rainfall-simulation experiments were conducted at Merrifield in August 1988. Five land-use categories, considered to be representative of land use in the Slapton catchment as a whole, were studied. Two of the hillslope plot experiments were run in the same, permanent pasture field; one on lightly grazed land where the vegetation cover remained intact and one on a heavily grazed area which was trampled to the extent that the vegetation cover had been completely removed. The remaining land uses were cereal, temporary grassland and prepared ground with no vegetation cover (defined as land ploughed and rolled within the previous two months to form a seedbed). Recently ploughed land was also studied but as no runoff was recorded from these plots at the rainfall intensity used, due to their high infiltration capacity (averaging 260 mm h^{-1}), they were not included in the results presented below. Four of the land uses were monitored at Merrifield. For the fifth (cereal) a harvested barley field in an adjacent catchment was used.

For each land use the experiments were run for a 4-h period at a rainfall intensity of 12.5 mm h^{-1}. For the heavily grazed plot, an additional intensity of 3.25 mm h^{-1} was used. In south-west England, 3.25 mm h^{-1} reflects average rainfall intensities. The higher rainfall intensity of 12.5 mm h^{-1} was used here to maximize the amount of runoff from the plots for suspended sediment measurement while still remaining representative of heavy rainfall within the catchment.

Surface runoff was collected by means of a trough from a 0.5 m^2 bound plot. The runoff total was measured at 20-min intervals and retained for suspended sediment determination.

RESULTS AND DISCUSSION

The Effect of Land Use on Infiltration and Bulk Density

Figure 6.2 shows the variation in infiltration capacity from August 1987 to March 1988 for the three land uses, permanent pasture, arable and temporary grass at Merrifield and Figure 6.3 shows the corresponding variation in soil bulk density.

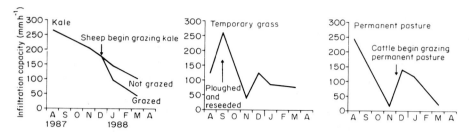

Figure 6.2 The variation in infiltration capacity through time with land use at Merrifield

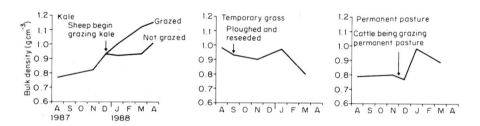

Figure 6.3 The variation in bulk density through time with land use at Merrifield

Prior to the ploughing of the barley field in September 1987 the infiltration capacity of the different land uses was of the order kale > permanent grassland > barley. The infiltration capacity for all land uses was high and ranged from 123 mm h^{-1} in the barley to 264 mm h^{-1} in the kale. The kale had been recently ploughed and sown prior to measurement, whereas harvesting is likely to have compacted the barley field. The infiltration capacity of the barley field, not surprisingly, increased following ploughing (Figure 6.2). Cattle grazing the permanent pasture in November 1987 significantly reduced the infiltration capacity from 244 mm h^{-1} to 16 mm h^{-1}. Although this decrease may in part be due to a high October rainfall, there was no corresponding decrease in the kale, which was not grazed. Bulk density (Figure 6.3) similarly varied with land use and closely follows the pattern shown by the infiltration capacity. A bulk density decrease is closely associated with an increase in the infiltration capacity.

The distinction between the land uses remained of the same order throughout the study period until January, 1988, when sheep began grazing the kale. Grazing in January 1988 reduced the infiltration capacity of the kale from 145 mm h^{-1} in ungrazed sections to 98 mm h^{-1} in grazed ones. A corresponding increase in bulk density from 0.92 g cm^{-3} in ungrazed sections to 1.00 g cm^{-3} in grazed sections was recorded (Figure 6.3). By the beginning of March 1988, when the kale had been grazed for one month, the infiltration capacity was reduced further to 44 mm h^{-1}. The bulk density had, following one month's grazing, increased to 1.12 g cm^{-3} relative to 0.93 g cm^{-3} in ungrazed sections. In the 6-day grazed section the infiltration capacity was only 20 mm h^{-1} (Figure 6.2). Sheep were currently grazing this section at the time of measurement, whereas in the 1-month section the kale had been completely eaten and hence had no sheep in it! This may account for the lower infiltration capacity in the currently grazed section, which may be due to poaching.

In the UK, low rainfall intensities mean that the infiltration capacity of most soils is not usually exceeded (Kirkby, 1978). There is evidence to suggest, however, that infiltration-excess overland flow is becoming more common. Istok and Boersma (1986) concluded that antecedent soil moisture was the most important factor generating surface runoff from low-intensity storms. However, this trend may also be due to changing land-use practices. Continuous arable cultivation, which lowers soil organic matter levels, and a reduced infiltration capacity, through increased compaction by machinery or intensive grazing, means that such low-intensity storms will become more important in surface runoff and sediment transport.

In the Merrifield catchment it was found that although bulk density and infiltration capacity differ between land uses these differences were small in comparison with the compaction caused by grazing animals in the kale and permanent pasture. Subsurface flow would normally dominate flow processes in the Merrifield drainage basin due to the presence of permeable soils over impermeable bedrock. The compaction of the soil surface through overgrazing of the kale in winter and early spring reduced the infiltration capacity to 20–44 mm h^{-1} from an estimated 102 mm h^{-1} in ungrazed areas. Widespread overland flow, estimated at up to 1 mm h^{-1}, was recorded in the grazed kale even for relatively low rainfall intensities of less than 3 mm h^{-1}. Grazing is obviously an important contributor to surface runoff and associated sediment transport to streams during storms.

Land-use Controls on Surface Runoff and Sediment Production

Table 6.2 shows the effect of different land uses on infiltration capacity, bulk density and total surface runoff from the hillslope plots. At a simulated rainfall intensity of 12.5 mm h^{-1} over a 4-h period total surface runoff for the different land uses followed the order:

Heavily grazed permanent pasture > lightly grazed permanent pasture > ground with no crop cover > cereal > temporary grassland

For the heavily grazed plot, surface runoff was 53 per cent of the total rainfall input, whereas surface runoff was only 23 per cent of the total rainfall input for the lightly grazed permanent pasture (Table 6.2). A higher infiltration

Table 6.2 Rainfall simulation of the effect of land use on surface runoff from hillslope plots

Land use	Rainfall intensity (mm h^{-1})	Total runoff (mm)	Percentage runoff as rainfall	Infilt. capacity (mm h^{-1})	Bulk density (g cm^{-3})
Temporary grass	12.50	2.3	5	12.33	0.96
Cereal	12.50	3.7	7	11.04	1.08
Bare ground	12.50	10.6	21	4.00	0.93
Lightly grazed permanent pasture	12.50	11.6	23	5.85	1.12
Heavily grazed permanent pasture	12.50	26.5	53	0.10	1.18
Heavily grazed permanent pasture	3.25	2.6	5	0.10	1.18

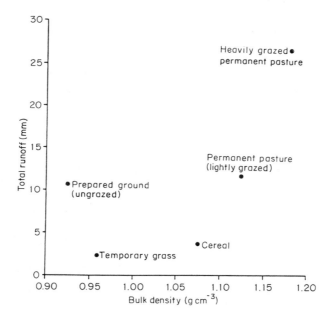

Figure 6.4 The relationship between bulk density and total surface runoff for different land uses measured during rainfall simulation at an intensity of 12.5 mm h^{-1} for 4 h

capacity (5.85 mm h^{-1}) was recorded for the lightly grazed area relative to the 0.1 mm h^{-1} in the heavily grazed area. The bulk density of the heavily grazed land at 1.179 g cm^{-3}, suggests greater compaction than in lightly grazed areas, where the bulk density averaged 1.122 g cm^{-3}.

For ungrazed land (temporary grass and cereal) the total runoff was considerably lower at 5 per cent and 7 per cent of the total rainfall input, respectively. Despite recent ploughing and rolling, surface runoff from the bare ground was still high at 21 per cent (Table 6.2).

Figure 6.4 illustrates the relationship between bulk density and total surface runoff for different land uses. A general trend of increasing surface runoff with increasing bulk density is clear. Grazed land in particular exhibits both high bulk density and high surface runoff. The surface runoff from the prepared ground is high given its bulk density. This may be due to the absence of vegetation cover.

Figure 6.5 shows the runoff and sediment production at 20-min intervals from each land-use plot. This illustrates the close association between surface runoff and suspended sediment production for each land use studied. For all land uses except temporary grass, surface runoff generally increases with time. Sediment production is more variable in magnitude but consistently increases if surface runoff increases. It is possible that a number of threshold points are reached during the 4-h experiment when a large amount of sediment is transferred from the hillslope plot to the collection trough. For example, the permanent pasture, whether lightly or heavily grazed, shows a similar pattern. Here two periods of high sediment production between 40 and 80 min and again at 120 to 180 min are evident. Cereal similarly shows two peaks in sediment production, one at 80 to 100 min and the second at 200 to 220 min. The bare ground plot appears to stabilize at a relatively constant sediment production rate of approximately 700 mg sediment every 20 min following a peak of over 1000 mg at 100 to 120 min (Figure 6.5). Only temporary grass shows decreasing sediment production with time. This is associated with decreasing surface runoff. The formation of a surface cap in this field earlier in the year may be continuing to influence surface runoff.

Table 6.3 shows the total sediment production in surface runoff from the monitored land-use plots together with the average sediment load (mg mm^{-1}) over the 4-h measurement period. Figure 6.6 indicates the relationship between total surface runoff and total sediment production for the land uses studied and shows the general trend of increased sediment production with increasing surface runoff. The total suspended sediment production in surface runoff (Table 6.3) was extremely high from heavily grazed permanent pasture at a rainfall intensity of 12.5 mm h^{-1}. Here over 22 g suspended sediment was delivered in 26.5 mm surface runoff from the 0.5 m^2 plot! However, for the same plot at the lower rainfall intensity of 3.25 mm h^{-1}, suspended sediment was not mobilized. Here less than 1 g of sediment was produced, which forms only 4 per cent of the total sediment production at the higher rainfall intensity.

The intact vegetation cover of the lightly grazed permanent grassland significantly reduced the production of suspended sediment (Table 6.3 and Figure 6.6). Only 0.37 g suspended sediment was recorded in 11.6 mm runoff. In contrast, the absence of vegetation cover in the recently ploughed hillslope

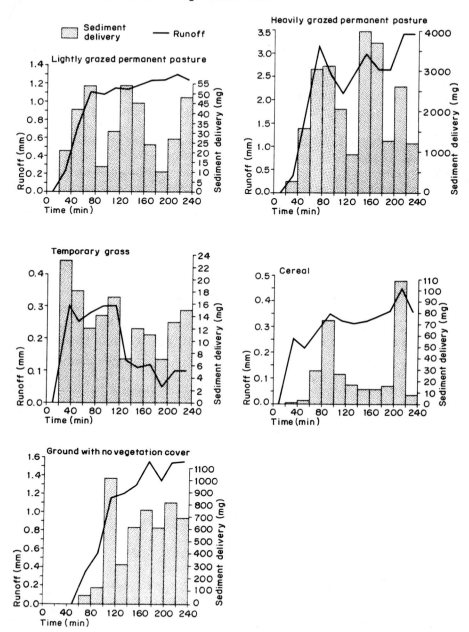

Figure 6.5 The variation in surface runoff and suspended sediment load over a 4-h period at a rainfall intensity of 12.5 mm h^{-1}

Table 6.3 Rainfall simulation of the effect of land use on suspended sediment delivery from hillslope plots

Land use	Rainfall intensity (mm h^{-1})	Total runoff (mm)	Total sediment yield[a] (g)	Sediment delivery rate (mg mm^{-1})
Temporary grass	12.50	2.3	0.15	65
Cereal	12.50	3.7	0.31	84
Bare ground	12.50	10.6	5.10	480
Lightly grazed permanent pasture	12.50	11.6	0.37	32
Heavily grazed permanent pasture	12.50	26.5	22.28	840
Heavily grazed permanent pasture	3.25	2.6	0.89	340

[a] Total sediment yield (g) in runoff over 4-h period.

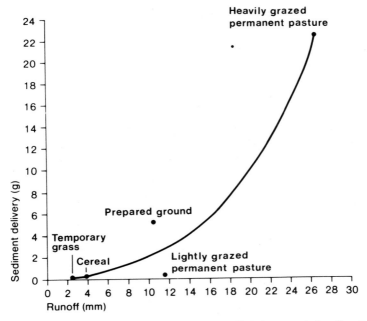

Figure 6.6 The relationship between total surface runoff and suspended sediment delivery for different land uses

plot (bare ground, Table 6.3) resulted in a suspended sediment production of over 5 g in a similar volume of surface runoff (10.6 mm).

For both temporary grass and cereal, the high infiltration rates (11–12 mm h^{-1}) and low surface runoff (5–7 per cent of the total rainfall input) meant that surface production of suspended sediment was minimal

(Tables 6.2 and 6.3). The total production was 0.15 g and 0.31 g suspended sediment for the temporary grass and cereal, respectively. However, the average production rate of 65–84 mg mm^{-1} exceeded that of the permanent pasture (Table 6.3). In the cereal in particular, this is likely to be due to extensive tramlining and a relatively sparse vegetation cover following harvesting within the monitored field.

The results suggest that heavily grazed land may be an important suspended sediment source for the stream system. Surface production will vary with rainfall magnitude and duration, antecedent moisture conditions, the nature of the vegetation cover and the presence of grazing animals. As a general indicator, heavily grazed land will result in high suspended sediment production during storms. Its importance for stream sediment loads will depend on its proximity to the stream or the existence of partial source areas such as animal tracks within the field, acting to channel sediment inputs to the stream.

Predicting the Delivery of Suspended Sediment in Surface Runoff

Figure 6.7 illustrates the predicted surface runoff (m^3 ha^{-1}) and suspended sediment load (g ha^{-1}) from the monitored land uses on the basis of the rainfall-simulation results. Two trends are clear. First, surface runoff in the absence of vegetation cover are up to ten times greater than land with intact crop cover (temporary grass and cereal). Second, the magnitude of runoff increases if grazing animals are present. Sediment production is up to eighty times greater from heavily grazed land in comparison with ungrazed land. However, where grazing is light and crop cover intact, sediment production is of the same order of magnitude as temporary grass and cereal.

It has been shown that different land uses in the Merrifield catchment will contribute surface runoff and suspended sediment of different magnitudes. In extrapolating the results of the simulation experiments to give predicted loads per hectare (Figure 6.7) it was necessary to assume that the measured rate of sediment production will be uniform for the entire land-use area. This is unlikely,

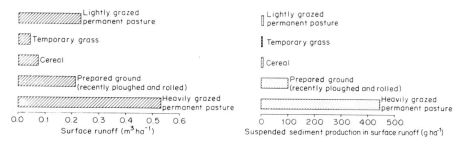

Figure 6.7 The effect of land use on predicted surface runoff (m^3 ha^{-1}) and sediment production (g ha^{-1})

particularly as a very small (0.5 m²) hillslope plot was monitored and localized variations in sediment production within the catchment will occur. Dunne (1983) and Dunne and Black (1970) indicate the importance of localized (variable source) areas adjacent to the stream as contributors of storm runoff. Additionally, partial source areas (Betson, 1964) such as tracks formed by grazing animals are important, particularly when considering the extremely high sediment production from areas that have been heavily grazed. Both form only a small proportion of the total field area at any one time but may be key areas in terms of sediment contribution to the stream system.

Land-use Distribution in the Slapton Catchments

Table 6.4 shows the overall land use for the Slapton catchments in 1986. Arable land, of which cereals form a high percentage of the total land area, tends to be located in the plateau areas of the catchment. Thus its importance as a stream sediment source may be limited except through partial source area contribution. The dominance of grassland, particularly permanent pasture, on slopes and adjacent to streams is confirmed by Table 6.5, which shows the land use in riparian zones of the catchment. For the Gara and Start catchments, permanent pasture forms 65 per cent and 56 per cent, respectively, of riparian zone land use. This is particularly important if areas immediately adjacent to the stream are subject to heavy grazing or trampling if, for example, they serve as access points to drinking water for stock. These riparian zones may be important in regulating sediment and associated nutrient fluxes between the catchment and the stream.

Storm Runoff and Sediment Delivery

To be of predictive value, the results for the hillslope plots which illustrate the effect of different land uses on sediment production in surface runoff need to

Table 6.4 Land use in the Slapton catchments (1986)

Catchment area (ha)		Grassland		Arable woodland		Woodland
		Permanent	Temporary	Cereals	Roots	
			(Percentage of total area)			
Gara	2362	39	34	15	5	4
Slapton Wood	39	21	21	33	9	12
Start	1079	25	37	28	6	3
Stokeley Barton	153					

Permanent grass: permanent pasture and long-term leys (7–10 years).
Temporary grass: grass sown within 4 years of survey date and short-term leys (less than 7 years).
Cereals: wheat, barley, oats, rye and maize.
Roots: potatoes, sugar beet, turnips, swede, fodder beet and mangolds.

Table 6.5 Land use in riparian zones of the Slapton catchment 1986

Catchment	Total stream length (m)	Permanent grass	Temporary grass	Wood	Arable Cereals	Arable Roots	Marsh
			(Percentage of total stream length)				
Gara	20 620	65	19	12	2	2	—
Slapton Wood	1 400	—	29	64	—	7	—
Start	9 860	56	16	16	3	—	9
Stokeley Barton	1 030	8	36	42	—	—	—
Total	32 910	60	19	16	1	1	3

be related to sediment loads for monitored storm events. The stream-suspended sediment load is largely quickflow derived and made up of surface runoff from the surrounding catchment. This rapid surface runoff may have several sources, including variable source areas adjacent to the stream channel (Sklash and Farvolden, 1979), partial source areas such as roads and trackways, and subsurface flow, including flow through soil macropores (Coles and Trudgill, 1985). The catchment area contributing quickflow during storm events is thought to be small given the small percentage of rainfall which generates quickflow (Troake and Walling, 1973). A varying component of the stream suspended sediment load may come from bank erosion or riparian zones.

Two storm events will be discussed here. A storm of 34.3 mm on 20 March 1988 and one of 41.8 mm over the period 27–29 January 1988. For the March storm, measured storm runoff as quickflow was 4.6 mm, 0.3 mm and 2.5 mm for the Gara, Start and Merrifield catchments, respectively. This forms a high percentage of the total storm runoff for the Gara (13 per cent) and Merrifield (7 per cent) catchments but is only 1 per cent of the total storm runoff for the Start catchment. This may be due to the effect of a small marsh zone upstream from the Start gauging station which absorbs much of this rapid surface runoff.

Figure 6.8 illustrates the pattern of stream discharge and suspended sediment delivery for the March storm. All catchments indicate a concentration effect for sediment (Webb and Walling, 1985) with the sediment load peaking at peak storm discharge. The total sediment delivery for this storm was; Gara 82 kg, Start 9 kg and Merrifield 0.1 kg, equivalent to an average catchment yield of 27 g ha^{-1}, 2 g ha^{-1} and 3 g ha^{-1}, respectively. This is comparable with the estimated suspended sediment yield of 3–7 g ha^{-1} for permanent pasture, temporary grass and cereal from the hillslope experiments, although much lower than the 102–446 g ha^{-1} predicted for heavily grazed land or land with no vegetation cover (Figure 6.7).

The sediment delivery for the January storm for the Gara and Start catchments only is shown in Figure 6.9. This 41.8 mm storm, which occurred over a 3-day period, produced an estimated surface runoff (as quickflow) of 2.35 mm and

Land-use Controls on Sediment Production

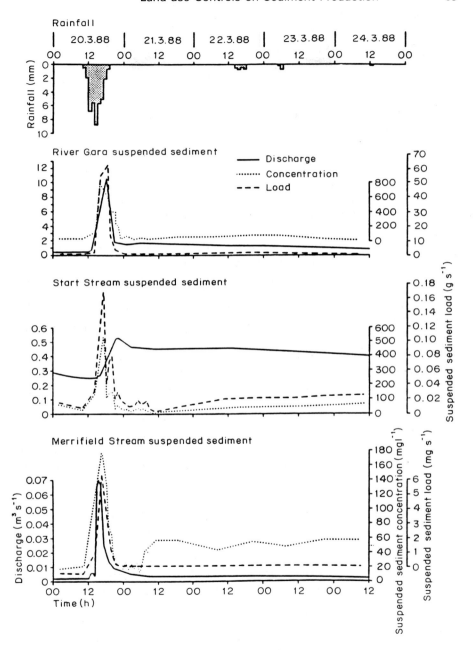

Figure 6.8 Suspended sediment concentrations and loads in the River Gara, Start Stream and Merrifield Stream for the 34.3 mm 20 March 1988 storm

84 Soil Erosion on Agricultural Land

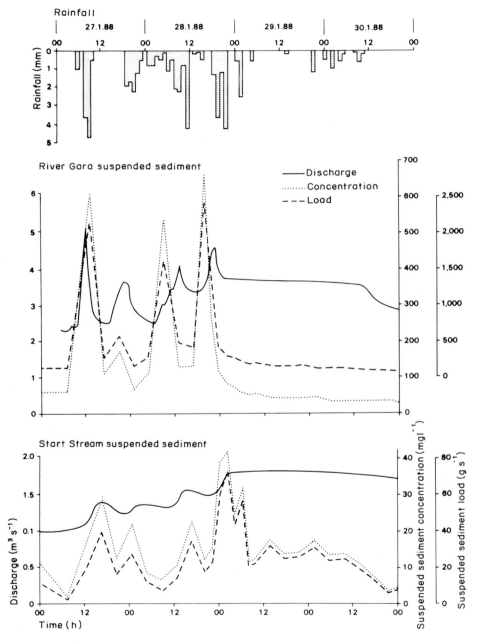

Figure 6.9 Suspended sediment concentrations and loads in the River Gara and Start Stream for the 41.8 mm 27–9 January 1988 storm

1.18 mm for the Gara and Start catchments, respectively, this is, 6 per cent and 3 per cent of the total storm runoff. Although quickflow delivery is lower for the Gara in this January storm, the total sediment delivery, estimated at 54 g ha^{-1}, is double that of the March storm. For the Start catchment, the estimated yield of 5 g ha^{-1} is only slightly higher. The magnitude and duration of this storm may account for the differences in stream suspended sediment load recorded. Again, the inputs are of the same order of magnitude to those recorded in the hillslope experiments.

Prediction of storm sediment loads and therefore catchment erosion for the larger catchments of the Start and the Gara is difficult, as sediment source areas, although localized, may vary for each storm. It is likely that partial area contributions from, for example, riparian zones that have been ploughed or are intensively grazed may produce the high total sediment delivery recorded for the Gara in quickflow. The results of the hillslope experiments identify the importance of grazing and vegetation cover in determining the magnitude of sediment delivery in surface runoff. As both catchments contain such a high proportion of permanent pasture adjacent to the stream (Table 6.5), sediment inputs from this source are likely to be of major importance in contributing to the stream sediment load during storm events.

SUMMARY AND CONCLUSIONS

Infiltration capacity and bulk density vary with season (primarily rainfall input) and land use. These distinctions are, however, greatest where grazing animals trample and compact the soil surface. This can result in up to an 80 per cent reduction in the infiltration capacity of the grazed area. This is in addition to the removal of the 'protective' crop cover which will increase the production in surface runoff of sediment and associated nutrients. The results of the simulation experiments indicate the importance of an intact crop cover in reducing surface runoff and suspended sediment delivery from hillslope plots.

The runoff volume from heavily grazed permanent grassland is at least double that from lightly grazed areas, and nearly twelve times greater than that of ungrazed (temporary grassland) areas. This is comparable with the results of McColl (1979), who found that the runoff volume was seven times greater from grazed pasture when compared with ungrazed pasture.

As a general indicator, heavily grazed land will result in high suspended sediment production during storms. This input may be critical in terms of stream water quality for grazed land located adjacent to the stream. Careful management of these land-use zones is required and, in particular, a limit should be placed on the numbers of grazing animals to avoid poaching around watering points.

REFERENCES

Betson, R. P. (1964). What is watershed runoff? *Journal of Geophysical Research*, **69**, 1541-52.

Boardman, J. (1984). Erosion on the South Downs, *Soil and Water*, **12**, 19-21.

Burt, T. P. (1978). Three simple and low cost instruments for the measurement of soil moisture properties, Huddersfield Polytechnic, Department of Geography Occasional Paper No. 8.

Burt, T. P., Butcher, D. P., Coles, N. and Thomas, A. D. (1983). Hydrological processes in the Slapton Wood catchment, *Field Studies*, **5**, 731-52.

Burt, T. P., Arkell, B. P., Trudgill, S. T. and Walling, D. E. (1988). Stream nitrate levels in a small catchment in southwest England over a period of 15 years (1970-1985), *Hydrological Processes*, **2**, 267-84.

Cullen, P. and O'Loughlin, E. M. (1982). Non-point sources of pollution. In O'Loughlin, E. M. and Cullen, P. (eds), *Prediction in Water Quality*, Australian Academy of Science, Canberra.

Dunne, T. (1983). The relation of field studies and modelling in the prediction of storm runoff, *Journal of Hydrology*, **65**, 25-48.

Dunne, T. and Black, R. D. (1970). Partial area contributions to storm runoff in a small New England watershed, *Water Resources Research*, **6**, 1296-1311.

Evans, R. (1977). Overgrazing and soil erosion on hill pastures with particular reference to the Peak District, *Journal of the British Grassland Society*, **32**, 65-76.

Fullen, M. A. and Reed, A. Harrison- (1986). Rainfall, runoff and erosion on bare arable soils in East Shropshire, England, *Earth Surface Processes and Landforms*, **11**, 413-25.

Greenland, D. J., Rimmer, D. and Payne, D. (1975). Determination of the structural stability class of English and Welsh soils, using a water-coherence test, *Journal of Soil Science*, **26**, 294-303.

Hodges, R. D. and Arden-Clarke, C. (1986). *Soil Erosion in Britain*. The Soil Association, Bristol.

Istok, J. D. and Boersma, L. (1986). Effect of antecedent rainfall on runoff during low-intensity rainfall. *Journal of Hydrology*, **88**, 329-42.

Kirkby, M. J. (1978). Implications for sediment transport. In Kirkby, M. J. (ed.), *Hillslope Hydrology*, John Wiley, Chichester, 325-63.

McColl, R. H. S. (1979). Factors affecting downslope movement of nutrients in hill pasture, *Progress in Water Technology*, **11**(6), 271-85.

Morgan, R. P. C. (1980). Soil erosion and conservation in Britain, *Progress in Physical Geography*, **4**, 24-47.

Morgan, R. P. C. (1985). Assessment of soil erosion risk in England and Wales, *Soil Use and Management*, **1**(4), 127-31.

Ratsey, S. (1975). The climate of Slapton Ley, *Field Studies*, **4**, 191-206.

Reed, A. H. (1979). Accelerated erosion of arable soils in the United Kingdom by rainfall and runoff, *Outlook on Agriculture*, **10**, 41-8.

Reed, A. H. (1983). The erosion risk of compaction. *Soil and Water*, **11**, 29-33.

Sklash, M. G. and Farvolden, R. N. (1979). The role of groundwater in storm runoff, *Journal of Hydrology*, **43**, 45-65.

Troake, R. P. and Walling, D. E. (1973). The hydrology of Slapton Wood stream: a preliminary report, *Field Studies*, **3**, 719-40.

Trudgill, S. T. (1983). The soils of Slapton Wood, *Field Studies*, **5**, 835-840.

Walling, D. E. and Peart, M. R. (1980). Some quality considerations in the study of human influence on sediment yields, *Proceedings of the Helsinki Symposium*, June 1980, IAHS-AISH Publ. No. 130, 293-302.

7 Soil Erosion on the South Downs: A Review

JOHN BOARDMAN
Countryside Research Unit, Brighton Polytechnic

INTRODUCTION

The Erosion Problem: National

In the late 1970s and early 1980s there was a sharp rise in the number of recorded cases of erosion on agricultural land in Britain. This is shown by the data from the Soil Survey of England and Wales/Ministry of Agriculture Fisheries and Food (SSEW/MAFF) air photographic survey of selected areas (Evans and Cook, 1986). Further evidence is provided by studies of discrete, high-magnitude events (e.g. Evans and Nortcliff, 1978; Reed, 1979; Boardman, 1983).

The considerable body of evidence that has now been accumulated may, however, be a function of increased research activity. Although this has certainly occurred there are good reasons to believe that the recent increase in erosion is real. Surveys of current erosion in specific areas have considered the evidence for the recent increase and have concluded that the number of incidents and the rates of erosion have increased perhaps by an order of magnitude (Speirs and Frost, 1985; Boardman and Robinson, 1985; Evans and Cook, 1986). There is also evidence that aerial photographic surveys of erosion lead to underreporting of many minor incidents (Reed, 1983; Evans and Cook, 1986). Finally, large areas of the country have not been monitored in any detailed or systematic manner, and these include areas that in terms of topography, land use and soils would be expected to be subject to erosion (e.g. the Cotswolds; Boardman, 1988a).

Thus, it is suggested that although the full extent and scale of the erosion problem is known only in outline, it is clear that agricultural land in certain areas is eroding regularly and that some of the rates are such as to give cause for concern. The National Soil Map (Soil Survey of England and Wales, 1983) and legend identify areas at risk of both wind and water erosion, and these total about 44 per cent of arable land in the country (ENDS, 1984). There is general agreement that erosion is a regional problem confined at present to certain identifiable areas and that preventative and ameliorative action could be targeted on these areas (Morgan, 1985; Bullock, 1987; Evans and Skinner, 1987; Boardman, 1988a).

The reasons for the increase in erosion differ, to some extent, in different areas. The general explanation is that a series of technological changes occurred in agriculture which are usually referred to as an intensification of the industry. These were accompanied by economic changes largely resulting from European Economic Community (EEC) agricultural policies.

Erosion occurs under a variety of crops. Row crops such as sugar beet and potatoes give high rates of erosion as a result of spring and summer rainfall events (for example, in the West Midlands and Nottinghamshire). Rates of erosion under sugar beet are higher than under other crops but the areas are relatively small (Evans and Cook, 1986). Similarly, high rates have been recorded under salad and vegetable crops (Boardman and Hazelden, 1986), under strawberries (Boardman, 1983) and under maize (Boardman, unpublished). However, of greater significance, because of the large areas involved, has been the sharp rise in erosion associated with the increase in the growing of autumn-planted cereals ('winter cereals'). Evans and Cook (1986) note a threefold increase between 1969 and 1983 in the area under winter cereals. In many parts of the country most erosion occurs during the autumn and winter on wheat or barley fields (Colbourne and Staines, 1985; Boardman and Robinson, 1985; Evans and Cook, 1986; Speirs and Frost, 1985). The role of other factors seems to vary from area to area (for example, the increase in size of fields, the finer tilths, compaction, a move onto steeper slopes, soil erodibility and the declining organic matter content of soils). These factors have been fully debated in the literature.

The aim of this chapter is to review the erosion problem as it affects the South Downs, thus largely covering investigations carried out since 1982. Many aspects of the work have already been published but no overview has been attempted.

The South Downs

The South Downs is a dissected upland area in south-east England rising to just over 200 m. It is composed of chalk, a soft Cretaceous limestone, with patches of sandy and clayey deposits of Tertiary age confined to small areas on the interfluves. Dry valley networks of periglacial origin dissect the chalk landscape giving rise to valley-side relief of up to 150 m. The principal soils are classified as Andover 1 Association (Jarvis *et al.*, 1984). This association includes thin stony rendzinas which are of primary concern in the context of erosion. These soils contain high proportions of silt (60–80 per cent) of loessial origin and stones either of chalk or flint. They are silty clay loams and silt loams, rarely thicker than 25 cm and frequently with A horizons of only 15 cm over bedrock. In valley bottoms superficial deposits of greater than 1 m depth occur; these contain horizons which are extremely stony and others composed of organic-rich silts. Paleosols of Flandrian age are found in these deposits.

Archaeological studies on the Downs show that the area has a history of cultivation of, in places, 3000–4000 years, and that this activity has led to erosion of once thicker loess soils which in the early Flandrian supported deciduous forest. Valley-bottom deposits represent stored soil and parent material eroded from the slopes (Bell, 1983; Allen, 1984; Ellis, 1986), although much has been lost to the flood plains of major rivers such as the Ouse, and to the sea. Apart from that in the Neolithic and Bronze Ages, little is known about erosion until the present day. It is assumed that it increased during periods of population pressure when cereal cultivation spread onto the Downs. Eighteenth- and nineteenth-century erosion is the subject of a current study (Rattenbury, in preparation).

In the first 40 years of the twentieth century farming on the Downs was mainly concerned with the grazing of sheep, but a ploughing-up campaign during the Second World War brought much of the area under arable cultivation. After the war large areas remained under cereals, primarily barley planted in spring. The area under cultivation was also extended as more powerful agricultural machinery was able to work steeper slopes. This was encouraged by the provision of grants to bring under cultivation 'new' land and to clear scrub. In the late 1970s, in common with other areas of the country, a change to autumn-planted or 'winter' cereals took place. In the 1980s these higher yielding cereals have come to occupy about 55 per cent of the farmed area in the eastern South Downs, a trend encouraged by high prices guaranteed under the EEC's agricultural arrangements (Figure 7.1). Slopes of up to 25° are now under cereal cultivation.

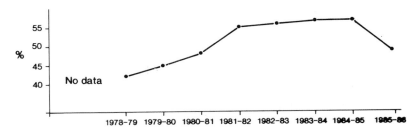

Figure 7.1 Autumn-planted cereals as a percentage of total cultivated area in seven downland parishes that comprise the monitored area. Reproduced by permission of MAFF from Agricultural Census returns

Agricultural practices typical of winter cereal cultivation on the Downs are intimately associated with erosion. Many cultivated fields are under continuous cereals but some farmers use a rotation system with grass, peas, beans or oilseed rape as break crops. A minority of farms have some sheep or cattle. In July and August, following the cereal harvest, excess straw is burnt. This is immediately followed by incorporation of the ash into the soil either by mouldboard or chisel ploughing, or by discing. A fine tilth is often obtained by use of a power harrow. Farmers aim to drill winter cereals as early as possible and some fields are drilled in early September. However, with large areas to be drilled the operation can, on many farms, take up to six weeks. During this, the wettest time of the year, interruptions may occur; late harvests may also delay the onset of drilling. Thus, in a typical year, drilling is not completed until early November and may have to be done on ground that is unsuitably wet, thus producing compacted vehicle wheeltracks ('wheelings'). Figure 7.2 shows the disruptive effect of a very wet October 1987; a similar pattern occurred in 1982. Drilling is followed by rolling of the seedbed which produces a compact, smooth soil surface. Applications of herbicides and insecticides in the winter and nitrogen in the spring give rise to further compaction of wheelings.

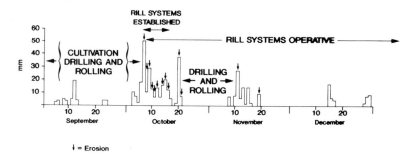

Figure 7.2 Rainfall (Southover), farming operations and erosion, September to December 1987. Erosion continued until the end of January

Mean annual rainfall on the eastern South Downs is between 750 and 1000 mm with an autumn peak (Potts and Browne, 1983). There is uncertainty as to the respective roles of intense and prolonged rainfall in generating erosion. Browne and Robinson (1984) have shown that the autumn of 1982, when severe erosion occurred, was characterized by a large number of short-duration high-intensity (>5 mm h^{-1}) rain events. However, changes in the pattern of rainfall in recent years are difficult to prove because of the short record available from autographic recorders. It is also clear that erosion and flooding incidences on the Downs often occur as a result of low daily rainfall amounts, particularly if networks of rills already exist. Table 7.1 applies to higher, wetter parts of the Downs and may be used as an approximate guide to the return periods of rainfall events of given duration.

Prior to the early 1980s there are very few records of erosion on the South Downs. With the exception of Mile Oak (Robinson and Blackman, 1990), such

Table 7.1 Relationship between rainfall (mm), return period and duration for higher parts of South Downs (Southern Water Authority: data from NERC, 1975)

Duration	Return period (years)					
	1	5	10	20	50	100
5 min	4.6	7.5	9.0	10.4	12.1	14.0
10 min	6.6	10.8	13.1	15.3	17.9	20.7
15 min	7.9	12.9	15.7	18.4	21.6	25.2
30 min	10.5	16.7	20.5	24.1	28.5	33.4
1 h	13.5	21.0	25.8	30.4	36.3	42.6
2 h	17.1	25.8	31.4	36.8	44.3	51.7
6 h	24.5	35.4	42.3	49.2	59.0	68.2
12 h	30.7	43.4	50.9	58.9	70.3	80.9
24 h	38.1	52.6	60.9	69.6	82.5	94.3
48 h	47.2	63.6	72.7	82.5	96.7	109.2

records that do exist refer to runoff of clean water along valley bottoms during very wet periods such as the autumn of 1976. It is significant that in that year erosion was minimal; this is assumed to be because of the low area of winter cereals at that time. In the autumn and winter of 1979 serious erosion occurred on loamy soils 5 km to the north of the Downs at Albourne; erosion was not confined to one site but was relatively widespread on bare erodible soils in that area (Boardman, 1983). On the Downs, there was flooding of houses at Saltdean due to runoff from wheelings in November: this was a repeat of a similar incident in 1976 (Bell, 1982), but no other cases are known. In the autumn of 1980, rainfall of 112 mm (20 September) and 90.2 mm (10 October) led to flooding in the Worthing area (Potts, 1982; Boardman, 1988b) and serious erosion on arable fields at Buddington Bottom, Findon (R. Evans, personal communication). These, however, appear to be isolated incidents rather than part of a pattern of widespread erosion and flooding during wet autumns.

In the majority of cases erosion is due to the establishment of rill systems on valley-side slopes; the rills are of the order of 10×5 cm in cross-section, they are widely spaced and are confined to small areas, often on slope convexities. Where sufficient water is concentrated in the valley floor, wide shallow flow may give rise to incision alternating with areas of deposition. The coarse fraction of the eroded soil is deposited in fans on valley-side footslopes or on the valley floor. The location of areas of deposition may be controlled by ponding of water due to field boundaries.

At sites of major erosion (e.g. $>10\,m^3\,ha^{-1}\,yr^{-1}$) rills tend to be more extensive, or larger dimensions and densely spaced, the distance between compacted vehicle wheeltracks being a major control. The formation of gullies is confined to steep valley sides and valley-side depressions. The largest gully was at Balsdean in 1987, where 1400 m³ of soil and flinty gravel was lost from an 800 m long gully with a depth of 1–2 m in a few days (Boardman, 1989).

RESEARCH METHODOLOGY

Systematic research into erosion began on the South Downs in the autumn of 1982 in response to serious erosion and flooding of property. A monitoring scheme was established on an area of the eastern Downs that was convenient to inspect several times each year. The area comprises about 36 km² of farmland and an arable area of about 30 km² (Figure 7.3). Sites of erosion are identified particularly following rainfall events, and measurements of rill, gully and fan dimensions are made. Erosional and depositional landforms are recorded on 1:10 000 scale maps. In some cases measurements have to be repeated as erosion continues. At sites of small-scale rill erosion detailed measurement of rill lengths, widths and depths can be made at intervals along the rill (e.g. every 10 m). On severely eroded slopes, with rills typically about 1.5 m apart, traverses along the contour are made with the cross-section of every rill being recorded. Traverses may be 50 or 100 m apart. A repeated traverse on a severely eroded slope with a two-week, dry interval between gave a second

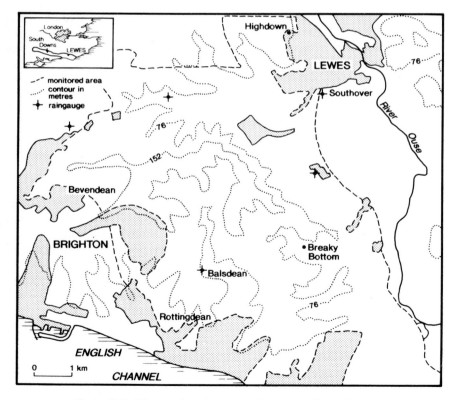

Figure 7.3 The monitored area on the eastern South Downs

result which differed by 0.4 per cent from the first in the estimate of erosion rate.

Crop and farming practices of eroded sites are recorded, and morphometric data are collected from 1:25 000 scale maps. Rainfall data are available from a number of gauges operated by Southern Water Authority and some farms.

Specific sites have been studied in detail and some of the results have been published: Bevendean, Highdown and Breaky Bottom in 1982 (Boardman and Robinson, 1985); Breaky Bottom, Balsdean and Rottingdean in 1987 (Boardman, 1989) (Figure 7.3). Several sites where erosion has been accompanied by flooding of houses and roads have also been the subject of studies for local councils, loss adjusters and solicitors (e.g. Boardman et al., 1983; Boardman, 1988b; Boardman, 1988c; Robinson et al., 1987).

As part of the annual aerial photographic survey of erosion by the Soil Survey of England and Wales/Ministry of Agriculture, Fisheries and Food, a flight line was established in 1983 across the eastern South Downs passing through the monitored area (Evans, 1983). Sites of erosion are identified on the photographs and field checking carried out. The data that have been assembled will allow a comparison to be made of the value of aerial photography as a reconnaisance technique as against the far more costly field-mapping procedures. This project has been carried out in association with Dr R. Evans.

A series of plots was established in 1985-6 to examine erosion rates, various farming practices, crop growth rates and soil conditions throughout autumn and winter (Boardman, 1986a; Robinson and Boardman, 1988).

RESULTS

Expression of Soil Loss

Soil loss may be expressed simply as a volume or weight lost from, or moved on, a field (Speirs and Frost, 1985). It is, however, more useful to express the loss as a rate in which the volume or weight of material lost from a rill or gully system is referred to an area and a unit of time, usually a year. The unit of area may be the field or the catchment. An agricultural field may contain several catchments or may itself be part of a larger one. Field boundaries can be permeable or impermeable, and without checks during high-magnitude rainfall events this can be difficult to establish. Some studies have quoted soil loss with reference to neither field nor catchment but the area on which rilling occurred (e.g. losses of soil by rilling reach $195 \, t \, ha^{-1}$ on parts of a field in north Norfolk; Evans and Nortcliff, 1978), although averaged over the whole field the loss is $11.8 \, t \, ha^{-1}$. Use of the catchment as the unit of area is more meaningful in terms of geomorphological explanation; however, in an agricultural context it is often preferable to use the field as the basic areal unit. Appreciation of spatial variation of rates within a field is of value in the discussion of conservation measures (cf. Evans and Nortcliff, 1978).

The relevant unit of time is the growing season. On the South Downs almost all erosion occurs in the first three months of that period and estimates for that period therefore approximate to the annual rate. At less than six sites out of 300, erosion has occurred on harrowed land and the operation has been repeated in order to remove the rills. Drilling has then been followed by renewed erosion. In such cases, the annual rate is the sum of the two separate incidents.

Soil loss is expressed as $m^3\,ha^{-1}\,yr^{-1}$. Use of volumetric measures of soil loss avoids the problem of varying bulk density. 'Soil loss', or 'erosion rate', refers to removal of soil from rill and gully systems. The soil may or may not be removed—some may be stored within the field. Rills are defined as being small enough to be of no obstacle to tillage operations, whereas gullies are deep enough to interfere with normal tillage operations, but not to be obliterated by them (Soil Science Society of America, 1987).

Rates and Distribution

The total amount of erosion and median rates in the monitored area varies from year to year (Figure 7.4). Rates for each year have a similar distribution, that is, they are strongly positively skewed with a tail of high values. The range of measured values, expressed on a catchment basis, covers four orders of magnitude from 0.01 to over $200\,m^3\,ha^{-1}\,yr^{-1}$.

The spatial distribution of eroded fields is shown in Figure 7.5. Forty per cent of fields which have eroded have done so in more than one year (Table 7.2). The distribution of fields is related to both physical and management factors. There is no evidence that rainfall variations across the area are a major control on erosion rates. Even when, as in 1987, there was a correspondence between the areas to maximum erosion and the area of highest daily rainfall (Boardman, 1989), it is likely that other factors were dominant. There is also little evidence that soil variation across the area is an important control. This statement requires qualification in that soils recently converted to arable from grassland and having high organic matter contents (e.g. >15 per cent), as well as occasional very stony soils (e.g. >60 per cent stones), do not readily erode. Sites of significant erosion have typically been located on the sides of major dry valleys with considerable relief (e.g. >100 m). However, the high relief and long slopes characteristic of fields on these slopes are the result of conscious management decisions on the part of farmers through the removal of walls, hedges, grass banks and lynchets. The growing of cereals on slopes steeper than 15° is a response to economic pressures: farmers feel the need to cultivate all land that modern machinery can traverse. Of equal importance are the large areas of bare ground in autumn associated with winter cereal production, the frequency of wheelings and the fashion for rolling drilled surfaces. Thus consideration of a combination of physical and management factors is necessary in order to explain the distribution of erosion within the area.

Soil Erosion on the South Downs 95

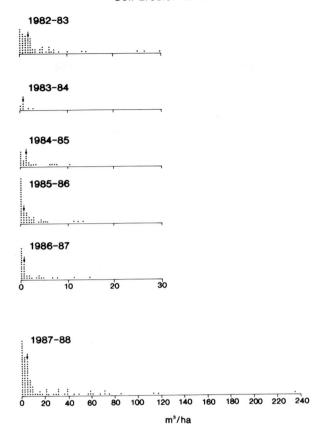

Figure 7.4 Rates of erosion, on a catchment basis, within the monitored area, 1982–7. Median values are indicated by arrows

Table 7.2 Eroded fields in the monitored area: 1982–8

Number of years eroded	Number of fields
1	68
2	24
3	12
4	7
5	3
6	0

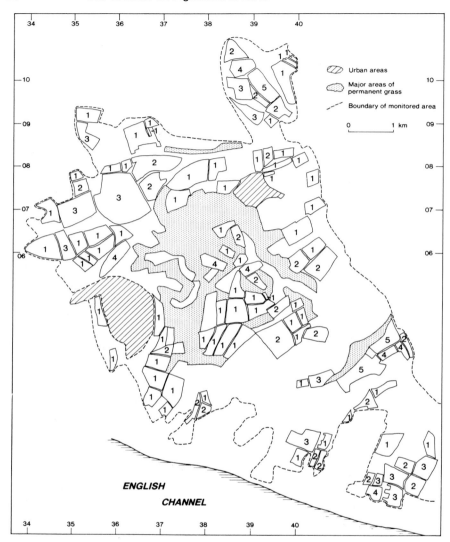

Figure 7.5 Spatial distribution of eroded fields in the monitored area, 1982–7. Figures represent the number of years that each field has eroded out of a possible six

Causes

Major erosion events are initiated on bare or almost bare soils in the months of September, October and November. It is likely that vegetation cover in excess of about 30 per cent inhibits erosion (Elwell and Stocking, 1976; Robinson and Boardman, 1988); the dates at which this is attained vary with climate and drilling

Table 7.3 Erosion in the monitored area, 1982–8

	Number of sites	Total soil eroded (m^3)	Downslope drilling (%)	Autumn-planted cereal and grass ley (%)	Median rate ($m^3\ ha^{-1}$)
1982–3	68	1 816	82	87	1.7
1983–4	7	27	100	86	0.6
1984–5	25	182	84	72	1.1
1985–6	49	541	73	84	0.7
1986–7	34	211	82	85	0.7
1987–8	97	13 529	84	73[a]	5.0

[a] Ploughed land in preparation for drilling not included.

date from about early November to April. Some fields are drilled consistently early and tend to avoid erosion; others are frequently at risk due to late drilling. However, early drilling is not always an effective control measure: for example, serious erosion occurred on a grass ley at Bevendean around 20 September 1982 (Boardman and Robinson, 1985).

The condition of the field surface is of importance. Residue from previous crops is non-existent or minimal. The desire to obtain a fine tilth has already been noted. Surface roughness is further reduced by the rolling of drilled fields, one purpose of which is to press protruding stones into the soil. Prominent wheelings are a feature of many sites of erosion. These provide zones of compaction and low infiltration, and act as efficient channels with low roughness values. Drilling is, in almost all cases, down the line of maximum slope angle or slightly oblique to that direction; wheelings follow the drill lines (Table 7.3).

The silty soils of the Downs are prone to capping or crusting. This may take place as a result of a series of rainfalls or one high-magnitude event. Optimal conditions for erosion are a series of low-magnitude events early in the autumn which cause crusting; these are succeeded by higher-magnitude events in which erosion occurs. The low aggregate stability of the silty soils on the Downs contributes to crusting, but these properties are not simply related to low organic content (Boardman and Robinson, 1985). Temporal and spatial variation in aggregate stability is currently being investigated (Blackman, 1989).

Soils on the Downs reach and remain at, or about, field capacity during the autumn. Soils on experimental plots in 1985–6 reached values of 30–40 per cent moisture content in late November and declined slightly in the following five months (Robinson and Boardman, 1988). At the same time, infiltration declined to a low value, where it remained throughout the winter. Under these conditions rapid response to rainfall events is seen in terms of runoff; however, at the same time the soils developed crusts. It is therefore difficult to separate the effects of high soil moisture content and crusting in generating runoff. Very rapid response during and after major runoff events is undoubtedly due to crusting, rill formation and the presence of wheelings. Rolled inter-rill areas generate runoff but display minimal signs of erosion. Soil is locally redistributed as a result of splash and surface wash but small amounts of soil are moved into rills. Very low rates of erosion were recorded on plots where rills did not form (Robinson and Boardman, 1988). Sheet erosion is extremely rare on these soils, probably because they are sufficiently cohesive to resist entrainment except in areas of concentrated flow, where velocities are high. This is in contrast to the form of erosion which is seen on small areas of sandy soils on the Downs, where sheet erosion does occur.

The chalk surface below the soil layer is frequently a compact, poorly fissured medium composed of cemented soliflucted chalk fragments in a fine chalky matrix. Where thin soils are found, the surface is smooth and the junction between soil and bedrock is abrupt, due to the cutting action of the plough (R. Evans, personal communication). Drainage of soil water into the chalk may, therefore, be inhibited.

For the purpose of characterizing the rainfall of each year a Rainfall Index has been devised. The daily rainfall record for the growing season is inspected and rain events are allotted a value (Table 7.4). These are summed to obtain an index for the year. The approach is based on the observation that rills were

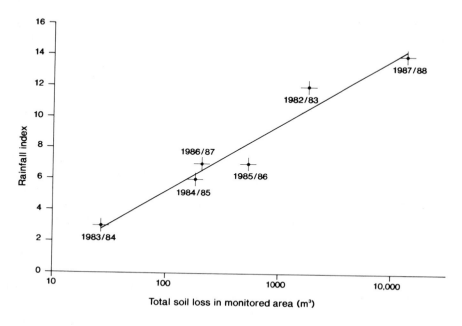

Figure 7.6 The relationship between total erosion in the monitored area and the rainfall index (see text for explanation)

Table 7.4 Values used in Rainfall Index

Rain event	Index value
30 mm in 2 days	1
30 mm in 1 day	2
60 mm in 2 days	3
60 mm in 1 day	4

initiated on soils at field capacity on a 7°, 200 m long slope at rainfall amounts of about 30 mm in 2 days (Boardman, 1986a; Robinson and Boardman, 1988). Erosion recommenced whenever this amount or slightly lower 2-day totals were recorded.

There is a strong correlation between total erosion and rainfall (Figure 7.6). This is because the area of winter cereals planted, the fields and farming methods are approximately the same each year. There is therefore little year-to-year variation in topographic and management factors. Thus the only factor showing major variations is rainfall. However, the latest MAFF data (for 1985–6) show a fall in the percentage of land under winter cereals. If this trend was maintained, then the severe erosion of 1987–8 is surprising, since there would have been less vulnerable land on which erosion could occur (Figure 7.1).

The main problem with this approach is that in any one year total erosion on some fields may result from fewer rainfall events than on others, due principally to different drilling dates. An index value for the whole growing season, as used in Figure 7.6, is therefore an imprecise measure of erosion-creating rainfall.

The relationship of rainfall and erosion is more complex than Figure 7.6 suggests. Two effects are worth noting:

(1) Major erosion events of 1987 and, less so, 1982 were associated with a high daily rainfall amount falling on ground conditions that facilitated runoff. That is, there was little or no vegetation cover, fine seedbeds had been produced and wheelings and rolled fields were commonplace. However, subsequent erosion resulted from much lower amounts because of the establishment of efficient and extensive rill systems. For example, at Breaky Bottom in 1987 rill systems on a 15° slope had drainage densities of 640 km km^{-2}, these having been established on 7 October. Subsequent flow in rill systems sufficient to produce down valley flooding of property resulted from daily totals as low as 7 mm. The threshold at which runoff and erosion commenced had been lowered from around 30 mm in 2 days to 5 mm in a day once rill systems were established and became operative (Figure 7.2 and cf. Ritter, 1988). This is a further reason why the Rainfall Index provides only a crude characterization of relevant rainfall.

(2) It might be expected that increasingly wet and warm autumns would lead to faster crop growth which would, in turn, give earlier erosion-resistant

surfaces. However, wetter autumns such as 1982 and 1987 produce two effects that counterbalance the above. First, early heavy rainfall initiates rill systems. Second, farming operations are delayed and significant amounts of drilling take place late in the season and often on wet ground, thus producing deep, compact wheelings. Autumns wetter than 1987 might well lead to less erosion if much land were to remain undrilled. In 1987, land that was ploughed but not harrowed (and therefore rough) showed losses of up to $30\,m^3\,ha^{-1}$, that is, about an order of magnitude lower than maximum losses on drilled surfaces.

The site, management and rainfall characteristics responsible for major erosion events have been discussed previously (Boardman, 1988a). If major erosion is defined as $>3\,m^3\,ha^{-1}\,yr^{-1}$, a typical site would have a slope length >200 m, maximum slope angle $>10°$ and site relief >30 m. It would be drilled in a downslope direction with a winter cereal and such sites would be at risk in years with rainfall indices >5. With reference to 1987, the concept of major erosion has to be redefined, since there were 28 sites with losses $>10\,m^3\,ha^{-1}$. Of the 28 sites, eight have slightly lower topographic indices than was suggested, but the major inadequacy of the proposal was that it failed to recognize ploughed land as being at risk. In 1987 there were six sites of major erosion on ploughed land. This suggests that with high rainfall indices (the value for 1987 was 14), a geomorphic threshold is crossed and land with much higher surface roughness values becomes at risk.

The recording of erosion at sites on the eastern South Downs including some beyond the monitored area for each year since 1982, has provided both an extensive database incorporating over 330 sites as well as more detailed information on selected sites. The former is utilized in the Expert System approach advocated by Harris and Boardman (1990) as being particularly appropriate to the problem of identifying high-risk sites under given conditions of land use, management and rainfall. The more detailed site information is being used in an attempt to adapt soil-erosion models developed elsewhere to the British situation (Favis-Mortlock, 1989).

THE IMPACT OF EROSION

The impact of erosion on the farming community in the short term has been minimal. The area of individual fields damaged by either erosion or deposition has rarely been in excess of 5 per cent. In the worst case, that of a 10 ha field at Rottingdean in 1987, about 10 per cent of the total volume of the soil was eroded, with the coarse fraction being deposited on valley-side footslopes and most of the fines being carried out of the field. Costs to farmers, even in cases such as these, are low (Robinson and Blackman, 1990).

The longer-term implications of erosion for farming on the Downs are more difficult to evaluate. On a loess-based soil overlying a relatively pure limestone, under an arable cropping regimen, rates of soil formation can be regarded as effectively zero. Thus even modest rates of erosion represent depletion of the soil resource. At the same time, under intensive farming systems with generous applications of fertilizers, soils of only 20 cm depth are capable of giving yields of 6 t ha^{-1} of winter wheat (Robinson and Boardman, 1988), which under current subsidy arrangements is a profitable crop. It is not yet clear if such a regimen is likely to continue even though the cost of surpluses of cereals represents an embarrassment to the EEC. Although schemes which are designed to encourage farmers to move out of arable and into grass are in place, take-up of them has been slow and targeting of high-risk erosion sites is non-existent, this despite the development of a methodology to identify such sites (Boardman, 1986b, 1988; Harris and Boardman, 1990). There is also the problem that American experience suggests that farmers opting into such schemes are very likely to revert to arable cropping on erodible land when that again becomes profitable (Napier, 1990). The future is therefore uncertain, and it is unclear as to whether erosion is likely to continue to deplete these soils to the point at which many slopes become unfarmable. A prudent society would presumably take the view that it is unwise to continue what is an unnecessary experiment.

The on-farm costs of erosion cannot therefore be used as an effective argument for conservation (cf. Evans and Skinner, 1987). The off-farm costs are, however, very considerable (Evans, 1989), and probably exceed on-farm costs by an order of magnitude. Evidence from other parts of Britain emphasizes this point (Evans, 1981; Boardman, 1986c; Boardman and Spivey, 1987) and accords with American findings (e.g. Crosson, 1984). On the South Downs, many communities have suffered from flooding by soil-laden runoff from fields of cereals, particularly in the winters of 1982 (Boardman *et al.*, 1983; Stammers and Boardman, 1984) and 1987 (Boardman, 1988c; Robinson *et al.*, 1987). Some communities have been flooded on several occasions as well as in these two years (Boardman, 1988b). Many cases undoubtedly go unrecorded being regarded as 'acts of God' rather than anything to do with farming practices. Attitudes are, however, changing as media attention has been directed to some of the more spectacular cases (Boardman, 1987). The off-farm costs of one such event, that at Rottingdean in 1987, are reported elsewhere (Robinson and Blackman, 1990). Costs are borne by local authorities, insurance companies and householders. This is significant, since neither central government, who could be held responsible for farming policy, nor the farmers, who translate the policy into practice, are having to consider the true costs of erosion.

Changes in farming practice on those parts of the Downs where property is at risk of flooding may occur as a consequence of current legal action. The repeated flooding of Breaky Bottom farm in October 1987 by runoff from winter cereal fields of a neighbouring farm is the subject of legal dispute. A successful resolution to the case will mean that farmers may be held negligent if they farm in a manner likely to cause erosion and flooding, and therefore responsible for damage. Thus a knowledge of the risk of flooding and erosion and the use of methods that are known to increase that risk would be relevant factors.

In almost every case of erosion on the Downs the eroded soil remains within the dry valley network that characterizes the area. Because of the lack of surface drainage, little soil reaches the main rivers, the Ouse, Adur and Cuckmere that pass through the Downs. Soil erosion is a process of redistribution, giving rise to thicker soils in the valley bottoms and thinner ones on the slopes. In terms of crop production the thicker soils are of little consequence, since they occupy small areas, and the off-farm impacts occur because the redistributive process brings water and soil into contact with valley bottom settlements.

CONCLUSION

In the short period of time (1982–9) erosion and associated flooding on the Downs have become major environmental issues which have engaged the attention of the public, statutory authorities, the media and researchers. To the geomorphologist the primary concern has been to monitor a representative area so that rates, frequencies and causes could be established. Although rates have generally been low, in two years out of six, major events involving losses $\leq 100 \text{ m}^3 \text{ ha}^{-1} \text{ yr}^{-1}$ have been recorded. Erosion in 1987 constitutes the most serious recorded on British farmland both in terms of soil loss on several fields and off-farm impact (Boardman, 1989). However, even these catastrophic rates of landscape change are of little concern to farmers since costs to them are low. There is no evidence that changes in agricultural policy or practices have been provoked by the impact of erosion on the farmer. What has happened, as a result of erosion and flooding of properties in 1982 and 1987, is that some farmers in locations adjacent to housing have been forced to consider their position. This has arisen because of the threat of legal action, but so far the costs of emergency and preventative engineering works have been borne by local councils (Stammers and Boardman, 1984).

ACKNOWLEDGEMENTS

The author thanks Southern Water Authority for access to rainfall records and permission to publish data in Table 7.1 and Figure 7.2. The Ministry of Agriculture, Fisheries and Food kindly allowed the use of data contained in Figure 7.1. I also thank Drs R. Evans, D. A. Robinson and Mr J. D. Blackman for useful comments on the text, as well as Drs T. M. Harris, R. Smith and Mr D. Favis-Mortlock for continued inputs into statistical, pedological and modelling aspects of this work. Mr G. Reeve kindly drew the figures.

REFERENCES

Allen, M. J. (1984). Ashcombe Bottom excavation: First Interim Report, Lewes Archaeological Group.
Bell, M. G. (1982). The effects of land-use and climate on valley sedimentation. In Harding, A. F. (ed.), *Climatic Change in Later Prehistory*, Edinburgh University Press, Edinburgh, pp. 127-42.
Bell, M. G. (1983). Valley sediments as evidence of prehistoric land-use on the South Downs, *Proceedings of the Prehistoric Society*, **49**, 118-50.
Blackman, J. D. (1989). Variation and change in aggregate stability and erodibility of downland soils, Poster paper, British Geomorphological Research Group Workshop, 'Soil Erosion on Agricultural Land', Coventry, January.
Boardman, J. (1983). Soil erosion at Albourne, West Sussex, England, *Applied Geography*, **3**, 317-29.
Boardman, J. (1986a). *Erosion Plots at Houndean, East Sussex, Preliminary Results, Winter 1985-86*, Report to ICI Plant Protection Division, Fernhurst.
Boardman, J. (1986b). *Soil Erosion in a Proposed Environmentally Sensitive Area*, Report for East Sussex County Council.
Boardman, J. (1986c). The context of soil erosion, *SEESOIL*, **3**, 2-13.
Boardman, J. (1987). A land farmed into the ground, *The Guardian*, 18 December.
Boardman, J. (1988a). Public policy and soil erosion in Britain. In Hooke, J. M. (ed.), *Geomorphology in Environmental Planning*, John Wiley, Chichester, pp. 33-50.
Boardman, J. (1988b). *Flooding and Erosion at Shepherds Mead, Worthing*, Unpublished report for Worthing Borough Council.
Boardman, J. (1988c). *The Causes of Flooding and Erosion at Breaky Bottom Farm, October 1987*, Unpublished report to Brocklehursts Loss Adjusters, Guildford.
Boardman, J. (1989). Severe erosion on agricultural land in East Sussex, UK, October 1987, *Soil Technology*, **1**, 333-48.
Boardman, J. and Hazelden, J. (1986). Examples of erosion on brickearth soils in east Kent, *Soil Use and Management*, **2**(3), 105-8.
Boardman, J. and Robinson, D. A. (1985). Soil erosion, climatic vagary and agricultural change on the Downs around Lewes and Brighton, autumn 1982, *Applied Geography*, **5**, 243-58.
Boardman, J. and Spivey, D. (1987). Flooding and erosion in west Derbyshire, April 1983, *East Midlands Geographer*, **10**(2), 36-44.
Boardman, J., Stammers, R. L., and Chestney, D. (1983). *Flooding Problems at Highdown, Lewes: Technical Report*, Report to Lewes District Council.
Browne, T. J. and Robinson, D. A. (1984). Exceptional rainfall around Lewes and the South Downs, autumn 1982, *Weather*, **39**(5), 132-6.
Bullock, P. (1987). Soil erosion in the UK—an appraisal', *Journal of the Royal Agricultural Society of England*, **148**, 144-57.
Colborne, G. J. N. and Staines, S. J. (1985). Soil erosion in south Somerset, *Journal of Agricultural Science, Cambridge*, **104**, 107-12.
Crosson, P. (1984). New perspectives on soil conservation policy, *Journal Soil and Water Conservation*, **39**(4), 222-5.
Ellis, C. (1986). The postglacial molluscan succession of the South Downs dry valleys. In Sieveking, G. de G. and Hart, M. B. (eds), *The Scientific Study of Flint and Chert*, Cambridge University Press, Cambridge, pp. 175-84.
Elwell, H. A. and Stocking, M. A. (1976). Vegetal cover to estimate soil erosion hazard in Rhodesia, *Geoderma*, **15**, 61-70.
ENDS (1984). Disappearing soil, Report 115, Environmental Data Services Ltd, London.

Evans, R. (1981). Potential soil and crop losses by erosion, Unpublished paper SAWMA conference, 3 December.

Evans, R. (1983). Soil erosion, in *Annual Report*, The Soil Survey England and Wales, Harpenden, pp. 14–15.

Evans, R. (1989). Soil erosion—the nature of the problem, *Scottish Geographical Magazine*, in press.

Evans, R. and Cook S. (1986). Soil erosion in Britain, *SEESOIL*, **3**, 28–58.

Evans, R. and Nortcliff, S. (1978). Soil erosion in north Norfolk, *Journal Agricultural Science, Cambridge*, **90**, 185–92.

Evans, R. and Skinner, D. (1987). A survey of water erosion, *Soil and Water*, **15**, 28–31.

Favis-Mortlock, D. (1989). Using EPIC to model erosion on the British South Downs, Poster paper, British Geomorphological Research Group Workshop, 'Soil Erosion on Agricultural Land', Coventry, January.

Harris, T. M. and Boardman, J. (1990). A rule-based Expert System approach to predicting waterborne soil erosion. In Boardman, J., Foster, I. D. L. and Dearing, J. A. (eds), *Soil Erosion on Agricultural Land*, John Wiley, Chichester, pp. 401–412.

Jarvis, M. G., Allen, R. H., Fordham, S. J., Hazelden, J., Moffat, A. J. and Sturdy, R. G. (1984). Soils and their use in South East England, *Soil Survey of England and Wales Bulletin*, **15**.

Morgan, R. P. C. (1985). Assessment of soil erosion risk in England and Wales, *Soil Use and Management*, **1**(4), 127–31.

Napier, T. L. (1990). The evolution of U.S. soil conservation policy: from voluntary adoption to coercion. In Boardman, J., Foster, I. D. L. and Dearing, J. A. (eds), *Soil Erosion on Agricultural Land*, John Wiley, Chichester, pp. 627–644.

NERC (1975). *Flood Studies Report*, Natural Environment Research Council, UK.

Potts, A. S. (1982). A preliminary study of some recent heavy rainfalls in the Worthing area of Sussex, *Weather*, **37**, 220–27.

Potts, A. S. and Browne, T. E. (1983). The climate of Sussex. In Geographical Editorial Committee (ed.), *Sussex: Environment, Landscape and Society*, Alan Sutton, Gloucester, pp. 88–108.

Rattenbury, G. (in preparation). *Soil Erosion on Agricultural Land: Historical and Geomorphological Perspectives*, PhD thesis, Silsoe College, Cranfield Institute of Technology.

Reed, A. H. (1979). Accelerated erosion on arable soils in the United Kingdom by rainfall and run-off, *Outlook on Agriculture*, **10**(1), 41–8.

Reed, A. H. (1983). The erosion risk of compaction, *Soil and Water*, **11**(3), 29–33.

Ritter, D. F. (1988). Landscape analysis and the search for geomorphic unity, *Geological Society America Bulletin*, **100**, 160–71.

Robinson, D. A. and Blackman, J. D. (1990). Some costs and consequences of soil erosion and flooding around Brighton and Hove, autumn 1987. In Boardman, J., Foster, I. D. L. and Dearing, J. A. (eds), *Soil Erosion on Agricultural Land*, John Wiley, Chichester, pp. 369–382.

Robinson, D. A. and Boardman, J. (1988). Cultivation practice, sowing season and soil erosion on the South Downs, England: a preliminary study, *Journal Agricultural Science, Cambridge*, **110**, 169–77.

Robinson, D. A., Williams, R. B. G., Funnell, D. C., Blackman, J. D., Potts, A. S. Browne, T. J. and Boardman, J. (1987). *Flooding and Soil Erosion at Rottingdean, October 1987: Final Report to Brighton Borough Council*, Unpublished report, University of Sussex.

Soil Science Society of America. (1987). *Glossary of Soil Science Terms*, Soil Science Society of America, Madison, Wisconsin.

Soil Survey of England and Wales (1983). *The Soil Map of England and Wales*, Soil Survey of England and Wales, Harpenden.

Speirs, R. B. and Frost, C. A. (1985). The increasing incidence of accelerated erosion on arable land in the east of Scotland, *Research and Development in Agriculture*, **2**(3), 161–7.

Stammers, R. L. and Boardman, J. (1984). Soil erosion and flooding on downland areas, *The Surveyor*, **164**, 8–11.

8 Water Erosion on Arable Land in Southern Sweden

KERSTIN ALSTRÖM and ANN BERGMAN
Department of Physical Geography, University of Lund

INTRODUCTION

There is an increasing interest in soil erosion by water on arable land in Sweden. It is arguable whether or not this is due to an accelerated erosion rate. Many changes in the Swedish agricultural landscapes during the twentieth century (e.g. the enlargement of fields, removal of field boundaries, changes in rotation of crops, land use, drainage pattern and changes in tillage practices) have increased the susceptibility to erosion (Alström and Bergman, 1988a; Anderson, 1981).

Another explanation of the increasing interest in water erosion in Sweden is an awareness of the environmental problems and off-farm effects caused by surface runoff and erosion. Many lakes, rivers and coastal waters suffer from serious problems of eutrophication. Non-point pollution from arable land is considered to be responsible for a large part of the total transport of nutrients. Most of the losses of phosphorus from arable land is transported on occasions with intense surface runoff, both as dissolved phosphate and bound to the soil particles (Alström and Bergman 1988b). This emphasizes the importance of surface runoff and erosion as contributors to non-point pollution from arable land.

The most severe erosion in the south of Sweden occurs during snowmelt and when frozen soil thaws. Results from this and several other surveys confirm that the erosion rate can be high in a climate without high-intensity rainfalls during snowmelt and when a frozen soil thaws (Burwell *et al.*, 1975; McCool *et al.*, 1977; Van Vliet and Wall, 1981; Zuzel *et al.*, 1982; Spomer and Hjelmfelt, 1983; Pikul *et al.*, 1986; Kirkby and Mehuys, 1987; Alström and Bergman, 1988b); especially when the winter has been repeatedly interrupted by warm spells. The macropores of the soil may then be ice filled to a large degree due to the freezing of infiltrated meltwater, which may cause low infiltration and considerable surface runoff (Lundin, 1989). The topsoil is usually saturated

---

Soil Erosion on Agricultural Land
Edited by J. Boardman, I. D. L. Foster and J. A. Dearing
©1990 John Wiley & Sons Ltd

during these conditions and the soil particles often highly erodible because of prior freeze-drying; which leads to decreasing aggregation and particles are easily eroded by running water from the thawing soil, melting snow and/or rain.

INVENTORY OF RILL AND GULLY EROSION

The aim of this survey is to identify the extent of water erosion, as rill and gully erosion, on arable land and to assess the relative importance of various factors that influence soil loss in the south of Sweden. The result will be applied in a mapping method for locating areas prone to soil erosion and surface runoff.

The magnitude and location of rill and gully erosion have been investigated in three agricultural districts over a period of two years. The districts are about 30 km² each and situated in the south of Sweden (Figure 8.1). The topography and soils differ between the three districts.

The district of 'Harlösa' is situated on a valley side, with a relative relief of 110 m. The slopes within this district are relatively steep and long. The dominating soil is sandy till with a clay content of less than 5 per cent. The district of 'Igelösa' is located on a less steep valley side, with a relative relief of 60 m. The soils within the district are mainly clay tills, with a clay

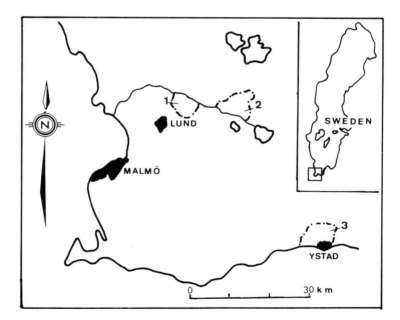

Figure 8.1 Location of the districts, where rill and gully erosion have been assessed. 1, Igelösa; 2, Harlösa; 3, Ystad

Table 8.1 Soil loss due to rill and gully erosion in three agricultural districts

Area	Period	Maximum soil loss in one field (kg ha^{-1})	Mean soil loss within the whole district (kg ha^{-1})^a	Fields affected by erosion (%)
Harlösa	1986/7	50 000	1 200	16
	1987/8	16 000	200	6
Igelösa	1986/7	6 000	500	18
	1987/8	1 000	10	4
Ystad	1986/7	22 000	200	7
	1987/8	200	10	3

Bulk density of the soil is estimated at 1.5 g cm^3.
^{a}The quota between the total amount of assessed rill and gully erosion and all the arable land within the district.

content more than 15 per cent. The topography of the 'Ystad' district is undulating and hummocky, with a relative relief of 60 m. The soils are mainly clay till, with a varying clay content.

Rills and gullies, developed during the period October to April have been measured before they were removed by cultivation and the sediment loss assessed by quantifying the volume of soil excavated by the rills and gullies.

The soil loss during the two years is presented in Table 8.1. There is a clear difference in soil loss and number of eroded fields between the two years due to the different weather conditions. The autumn of 1986 was rainy and was followed by an unusually cold winter, with soils frozen to a depth of more than 1 m and containing much frozen water. The cold winter was interrupted by four major events with snowmelt. The highest level of erosion occurred during those events, when only the upper part of the soil was thawed. The subsequent winter of 1987/8 was mild with no snow and no frozen soils and a much lower erosion rate. The Harlösa area had the largest soil loss during the 2 years, which included both the maximum soil loss in one field and the mean soil loss within the whole investigated area. This difference in soil loss between the three districts is mainly explained by topographical differences, where Harlösa has the combination of steep, long and convex slopes.

Statistical analyses were performed in an attempt to assess the relative importance of various factors influencing rill and gully erosion. The three investigated districts were divided into 136 smaller slope units, ranging in size from 2 to 318 ha, separated by watersheds distinguished from a 1:10 000 map with a contour interval of 5 m. All the units ended either at a head of a stream or along a stream, as shown in Figure 8.2. The analyses were based on data from the first winter season 1986/7, when 46 of the 136 slope units and 135 of the total 935 fields within the districts were affected by rill or gully erosion.

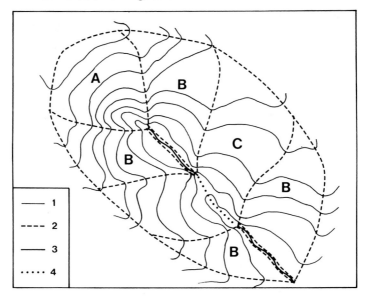

Figure 8.2 Map illustrating slope units. 1, Contours; 2, dividing line between slope units; 3, stream; 4, submerged stream. A, Slope unit defined by watershed and ending at the head of a stream; B, Slope units defined by watershed and ending along a stream; C, Slope unit, where an open stream has been put into covered drains. The slope unit is defined by the watershed and ends at the beginning of the open stream

In an attempt to define the optimal unit of study for explaining water erosion, correlation analyses were performed between sediment loss and different characteristics for both eroded slope units and for eroded fields.

The results in Tables 8.2(a) and 8.2(b) show that the correlation between the sediment loss and slope, slope length, size and arable land is low, both for individual fields and for slope units, but the correlations increase when the variables are combined in algorithms. The algorithms, which had the highest

Table 8.2(a) The correlation between sediment loss (m^3) within a *field* and characteristics of the *field*

Slope	0.06
Length	0.23
Area	0.07
$S^2 \times Z$	0.29

Table 8.2(b) The correlation between sediment loss (m^3) within a *slope unit* and characteristics of the *slope unit*

Slope	0.15
Length	0.27
Area	0.38
Arable land	0.10
$S^2 \times A \times L$	0.87

S = Slope (%). Z = Length (m). A = Area (ha).
L = Arable land (percentage of the total area of the slope unit).

correlation with sediment loss, both in fields and slope units, are shown in Tables 8.2(a) and 8.2(b). The correlation between sediment loss within a slope unit and the tested algorithms based on the characteristics of the slope unit is higher than between equivalent algorithms based on the field characteristics and sediment loss within the field.

The spatial variation of rill and gully erosion is best defined within a slope unit by an equation based on the algorithm presented in Table 8.2(b). Regression analyses, based on 136 slope units, explain 76.8 per cent of the variation in sediment loss. This is significant at the 99 per cent level. The equation is:

$$R = 10.7 \times 10^{-4} (S^2 \times A \times L) - 20.2$$

where R = rill and gully erosion (m³)
 S = slope (%)
 A = area (ha)
 L = arable land (per cent).

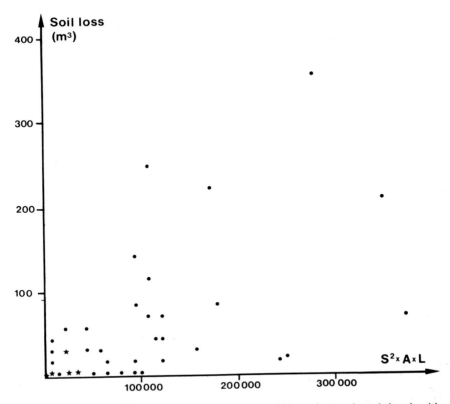

Figure 8.3 The relationship between sediment loss within a slope unit and the algorithm $S^2 \times A \times L$. An extreme value ($x = 688,860$, $y = 951$) is excluded in the figure. *Several points

The data set, on which the analyses are based, includes an extreme value. If this value is excluded the level of explanation (r^2) decreases to 51.6 per cent (significant at the 99 per cent level). The data set, excluding the extreme value, is shown in Figure 8.3.

For regression models based on this algorithm, the level of explanation increases if the 136 slope units are classified into three groups according to the down-slope form of the area; i.e. slopes with straight, concave or convex forms. Slope units with convex slopes were found to be more susceptible to rill and gully erosion than areas with straight or concave slopes. This is shown in Figure 8.4(a), where the regression line has the greatest slope, i.e. slope units with convex form are the most erodible. This confirms earlier work. Slopes below convexities

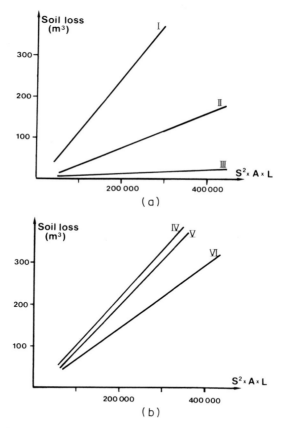

Figure 8.4 The relationship between sediment loss within a slope unit and the algorithm $S \times A \times L$. A, The slope units are grouped according to their form: I convex slopes, II concave slopes, III straight slopes; B, The slope units are grouped according to the dominating soil within the slope unit: IV dominated by soils with a clay content > 15%. V dominated by soils with a clay content between 5 and 15%, VI dominated by soils with a clay content < 5%

have been found to be more erodible by Evans (1980), and Colborne and Staines (1985) and Tregubov (1980).

The level of explanation is also increased if the units are grouped according to the clay content of the dominating soil within the slope unit. Three soil groups were used; soils with a clay content of more than 15 per cent, a clay content between 5 and 15 per cent and soils with less than 5 per cent clay. Units dominated by soils with a clay content of more than 15 per cent are more prone to erosion (Figure 8.4(b)). One explanation of the higher erodibility of clay soils could be the higher water-retention capability, which results in a higher ice content and a lower infiltration capacity, when the soil is frozen. Even the freeze-drying effect increases the erodibility of the clay particles, but we have not yet verified this by experimental measurements. A more detailed report of the methods and results from this survey of water erosion is presented by Alström and Bergman (1988b).

The transport of sediment and nutrients was measured in four rills during the snowmelt of 1987 by quantifying the discharge and analysing water samples. A large quantity of soil and nutrients were transported in the rills investigated, (Table 8.3). Most of the material was further transported in ditches and pipes draining to open streams. The rills were ephemeral and occurred only during a few days and the extreme discharge existed only for a few hours each day, yet the rills caused severe damage to the fields and to the pipes draining them.

Changes in the drainage pattern might have caused erosion. Many ditches and streams have been converted into covered drains, which might increase the risk of water erosion. Due to the topography, much surface water is concentrated in the remains of the former waterway, particularly when the inlets for surface

Table 8.3 Discharge and transport of sediment and nutrients in some rills during the snowmelt of 1987

	Measuring date	Discharge ($l\,s^{-1}$)	Soil loss ($kg\,h^{-1}$)	Losses of nutrients ($g\,h^{-1}$)			
				PO_4-P	Part.-P	Tot.-P	Tot.-N
Rill A	22.3.1987	6.0	32	2.6	16.6	19.2	79.9
	23.3.1987	1.2	6	0.5	3.5	4.0	16.6
	25.3.1987	0.2	1	0.1	0.7	0.8	6.8
	26.3.1987	100.0	576	49.0	642.2	691.2	7 200.0
Rill B	23.3.1987	0.2	<1	0.1	0.9	1.0	5.7
	26.3.1987	18.0	194	54.4	872.2	926.6	3 985.2
Rill C	23.3.1987	120.0	9 180	62.2	4 301.0	4 363.2	85 536.0
	25.3.1987	10.0	180	4.7	50.0	54.7	388.4
Rill D	6.2.1987	170.0	520	185.4	2 002.5	2 187.9	5 569.2

Discharge, litres per second ($l\,s^{-1}$).
Transport of nutrients; phosphate-phosphorous (PO_4-P), particle-bound phosphorus (Part.-P), total phosphorus (Tot.-P), and total nitrogen, (Tot.-N), in grams per hour ($g\,h^{-1}$).
Transport of sediment, kilograms per hour ($kg\,h^{-1}$).

114 Soil Erosion on Agricultural Land

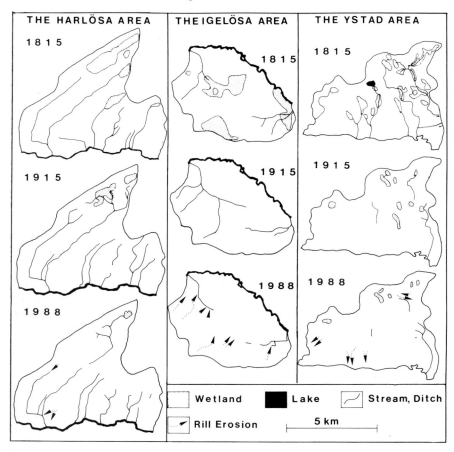

Figure 8.5 The drainage pattern 1815, 1915 and 1988 in the three agricultural districts, Harlösa, Igelösa and Ystad. The rill and gully erosion that occurred during the winters 1986/7 and 1987/8 and were directly caused by drainage operations are shown in the maps presenting the drainage pattern in 1988

water are not correctly positioned in the valley bottom or when they are separated by excessive distances so that the surface water reaches the eroding point. The draining of wetlands might also cause erosion, as wetlands serve as storage areas for surface water, reducing runoff peaks. A study of the changes of the waterways and wetlands within the three areas was made with maps from 1815 and 1915, and by field studies in 1988 (Figure 8.5). The rill and gully erosion that occurred during the winters of 1986/7 and 1987/8, which can be interpreted as effects of changes in the drainage pattern, are also shown in Figure 8.5. An estimate of the amount of sediment loss due to these changes is presented in Table 8.4.

Table 8.4 Soil loss by rill and gully erosion caused by changes in the drainage pattern

District	Period	Total amount of rill and gully erosion caused by drainage operations (m^3)	Percentage of the total amount of rill and gully erosion within the districts
Harlösa	1986/7	94.6	4
	1987/8	54.3	13
Igelösa	1986/7	318.0	50
	1987/8	6.8	70
Ystad	1986/7	113.6	49
	1987/8	1.3	77

DISCUSSION

Our results indicate the importance of the slope unit regarding the location of water erosion. This is explained by the fact that most of the erosion occurred during snowmelt when the soils still had frozen layers. Under these circumstances the infiltration capacity was very low (zero or almost zero) and the surface runoff was very intense. Most of the surface water moved from one field to another and was not restrained by field boundaries. The amount of surface water is more closely related to the slope unit than to the area of a single field. This suggests that an assessment of the areas susceptible to erosion during conditions with frozen soils should be based on the characteristics of the slope unit and not on field characteristics.

It has been found that some changes in the drainage pattern, especially putting streams into covered drains, increases the risk of rill and gully erosion. Nevertheless, it is arguable if this results in a higher sediment and nutrient transport. By converting open streams into covered drains the sediment and nutrient loss by stream bank erosion is eliminated and there is a reduced possibility for field-eroded material to be further transported. It should be emphasized, however, that if there is a possibility for the eroded material from the valley floor to be further transported (by surface water inlets or adjacent streams), there could be considerable losses of nutrients and sediment.

The acceptable soil loss in Sweden is established at $100-200$ kg ha^{-1} yr^{-1} (Swedish Environmental Protection Board; SNV, 1986). Our results show a great spatial and temporal variation of sediment loss. In the district of Harlösa, the mean sediment loss exceeded the tolerance level during both the years, which implies that a soil-conservation programme should be implemented for the whole district and not only in the most affected fields. In the other two districts, the mean sediment loss exceeded the tolerance level only during the first year, when the winter was unusually cold. As long as it is not established how often these erosive snowmelt events occur, it is questionable if there is a need for soil-conservation programmes in those areas. However, even in those districts there are fields with unacceptable losses which require soil-conservation measures.

In many agricultural districts in southern Sweden the nutrient losses by surface water, particularly phosphorous, have a vital importance for the deteriorating water quality of many lakes (Swedish Environmental Protection Board, SNV, 1985). This implies that large areas are in need of soil-conservation programmes to reduce the diffuse pollution by surface water from arable land.

The limited technical and financial resources make it important to focus the management programmes on the critical geographical areas. The results from this erosion survey will be applied to develop a method to assess the areas prone to water erosion. Such a method should be of great interest in environmental planning as a tool for implementation of measures where they have the greatest effect on preventing erosion and non-point pollution.

REFERENCES

Alström, K. and Bergman, A. (1988a). Sediment and nutrient losses by water erosion from arable land in south of Sweden. A problem with non-point pollution? *Vatten*, **44**, Lund, 193-204.

Alström, K. and Bergman, A. (1988b). Vattenerosion och när-saltförluster via ytavrinning i åkermark i Skåne. [Water erosion and nutrient losses by surface runoff on arable land in Skåne] (In Swedish with abstract in English), Lund Universtitets Naturgeografiska institution, *Rapporter och Notiser*, No. 71, Lund.

Andersson, R. (1981). Växtnäringsförluster från åkermark, kunskapsöversikt. In Falkenmark, M. and Johansson, I. (eds), *Hydrologi, Markanvändning, Vattenkvalitet*, SNV PM 1455, Stockholm, pp. 85-101.

Burwell, R. E., Timmons, D. R. and Holt, R. F. (1975). Nutrient transport in surface runoff as influenced by soil cover and seasonal periods, *Soil Science Society of America, Proceedings*, **39**, 523-8.

Colborne, G. J. N. and Staines, S. J. (1985). Soil erosion in south Somerset. *Journal of agricultural Science, Cambridge*, **104**, 107-14.

Evans, R. (1980). Characteristics of water-eroded fields in lowland England. In De Boode, M. and Gabriels, D. (eds), *Assessment of Erosion*, Wiley, Chichester. pp. 77-87.

Kirkby, P. C. and Mehuys, G. R. (1987). The seasonal variation of soil erosion by water in southwestern Quebec, *Canadian Journal of Soil Science*, **67**, 55-63.

Lundin, L-C. (1989). Water and heat flows in frozen soils. Basic theory and operational modeling. Acta Univ. Ups, *Comprehensive Summaries of Uppsala Dissertations from the Faculty of Science 186*, Uppsala.

McCool, D. K., Molnau, M., Papendick, R. I. and Brooks, F. L. (1977). Erosion research in the dryland grain region of the Pacific Northwest: Recent developments and needs. In Foster, G. R. (ed.), *Soil Erosion: Prediction and Control*, The Proceedings of a National Conference on Soil Erosion. Purdue University, West Lafayette, Indiana, Soil Conservation Society of America, Special Publication No. 21, 50-59.

Pikul, J. L., Zuzel, J. F. and Greenwalt, R. N. (1986). Formation of soil frost as influenced by tillage and residue management, *Journal of Soil and Water Conservation*, **41**(3), 196-9.

Spomer, R. G. and Hjelmfelt, A. T. (1983). Snowmelt runoff and erosion on Iowa loess soils, *Transactions of American Society of Agricultural Engineers*, **26**(4), 1109-11, 1116.
Swedish Environmental Protection Board, SNV (1985). *Växtnäringsläckage till yt- och grundvatten från jord- och skogsbruk—orsaker och åtgärder* [Leakage of nutrient from agriculture and forestry to surface water and ground water—causes and measures] (In Swedish), *SNV PM* 1972, Stockholm.
Swedish Environmental Protection Board, SNV (1986). *Jordbruket och miljön. Handlingsprogram* [Agriculture and environment. Plan of action] (In Swedish), Stockholm.
Van Vliet, L. J. P. and Wall, G. J. (1981). Soil erosion losses from winter runoff in southern Ontario, *Canadian Journal of Soil Science*, **61**, 451-4.
Zuzel, J. F., Allmaras, R. R. and Greenwalt, R. (1982). Runoff and soil erosion on frozen soils in North-eastern Oregon, *Journal of Soil and Water Conservation*, **37**(6), 351-4.

MAPS EXAMINED

Skånska rekognoseringskartan, scale 1:30 000, made in 1815 for military purpose, reprinted in 1985 by Lantmäteriverket, Malmö.
Map sheets: VÖ208-209, IVÖ208, IÖ204, IIÖ204, IIÖ205, IIIÖ204, IIIÖ205.
Economic maps, scale 1:20 000, made in 1915, by Rikets Allmänna Kartverk, Lantmäteriverket, Gävle.
Map sheets: Löberöd (M-län), Refvinge (M-län), Harde-berga (M-län), Lund (M-län), Örtofta (M-län), Ystad (M-län).
Economic maps, scale 1:10 000, made in 1967, by Lantmäteriverket, Gävle.
Map sheets: 013 93, 013 94, 013 95, 022 67, 022 68, 022 77, 022 78, 023 04, 023 05, 023 51, 023 60, 023 61.
Topographical maps, scale 1:50 000, made in 1983 by Lantmäteriverket, Gävle
Map sheets: 022 4, 023 3, 013 3.
Soil maps, *serie Ad*, scale 1:20 000, made in 1951 by Sveriges Geologiska Undersökning, SGU
Map sheets: 1, 3, 4, 5.
Serie Ae, scale 1:50 000, made in 1985 by Sveriges Geologiska undersökning, SGU.
Map sheets: 65, 66 and maps unpublished.

9 The Calculation of Net Soil Loss Using Caesium-137

R. J. LOUGHRAN
Department of Geography, University of Newcastle, Australia

B. L. CAMPBELL
Australian Nuclear Science and Technology Organization

and

G. L. ELLIOTT
Soil Conservation Service of NSW Research Centre

INTRODUCTION

The environmental isotope caesium-137 (^{137}Cs) is being widely used to determine the erosion status of soils (Loughran, 1989). Caesium-137, half-life 30 years, is a fallout product of atmospheric thermonuclear weapons testing, and the Chernobyl accident in 1986 provided an additional source of ^{137}Cs in parts of Europe. On reaching the earth's surface, ^{137}Cs is rapidly and firmly adsorbed onto surface soils (Davis, 1963; Tamura, 1964) and becomes an effective tracer of soil movement (McHenry and Ritchie, 1977).

The isotope was first detected in the environment in the mid-1950s, and maximum fallout occurred in the mid-1960s. Data for Brisbane and Sydney, Australia (Campbell *et al.*, 1982; Longmore *et al.*, 1983) were used to construct a graph of annual ^{137}Cs fallout for the Hunter Valley, New South Wales (200 km north of Sydney and 550 km south-south-west of Brisbane). The amount of ^{137}Cs remaining in the soil at stable sites in the Hunter valley corresponds with the total ^{137}Cs fallout shown in Figure 9.1, corrected for decay: 100 mBq cm^{-2}.

Sites undergoing no erosion or deposition usually accumulate ^{137}Cs within the top 5 cm of the profile, losing it only by radioactive decay. These may be considered atmospheric 'input' or 'reference' sites for the locality. Cultivated soils will have ^{137}Cs redistributed within the plough layer, although in

---

Soil Erosion on Agricultural Land
Edited by J. Boardman, I. D. L. Foster and J. A. Dearing
©1990 John Wiley & Sons Ltd

agricultural rotations, ^{137}Cs from fallout may remain on the surface for some time before incorporation.

It has been found that eroded soils lose ^{137}Cs in quantitites related to the severity of erosion (Ritchie and McHenry, 1975; Campbell *et al.*, 1986a). At sedimentation sites ^{137}Cs will be present at depths greater than at reference or slope sites, and the quantity will invariably be greater than input (Loughran *et al.*, 1988a).

Campbell *et al.* (1986b) regressed percentage ^{137}Cs loss against soil loss from plots in eastern NSW. The equation was used to calculate net soil loss (kg ha^{-1} yr^{-1}) from two vineyards at Pokolbin, Hunter valley, NSW, based on a soil-sampling grid 10 m × 10 m. An alternative method for calculating net soil loss from percentage ^{137}Cs loss has been developed by de Jong and colleagues in Canada (de Jong *et al.*, 1982). In this instance, it was assumed that ^{137}Cs was uniformly distributed by ploughing in cultivated soils, and that the amount

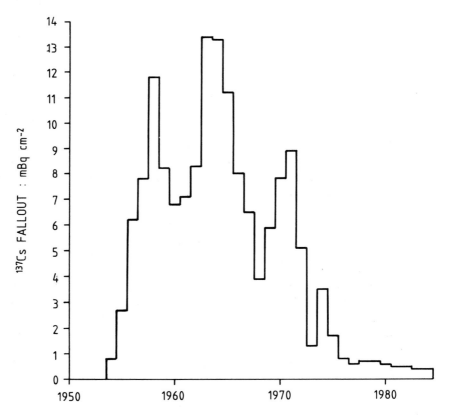

Figure 9.1 Caesium-137 fallout 1954–84 (uncorrected for decay), Hunter Valley, New South Wales, Australia

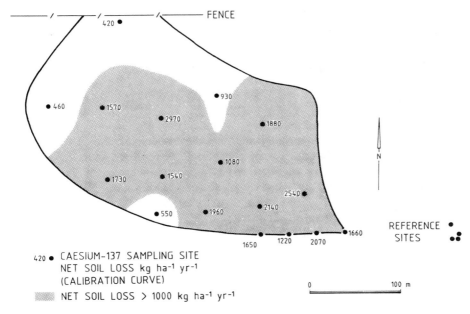

Figure 9.2 Net soil loss from the Merriwa catchment using the calibration curve method (kg ha^{-1} yr^{-1})

of ^{137}Cs lost was directly proportional to the depth of the cultivated layer removed by erosion. This chapter discusses these two contrasting approaches for net soil-loss estimation within a small catchment in the Hunter valley, New South Wales.

METHODS

Soil samples for ^{137}Cs analysis were collected at 2 cm depth increments within a sampling frame (1000 cm^2) or by coring. After laboratory drying and sieving (<2 mm), samples were counted on a hyperpure germanium detector system. The method is described in detail in Campbell *et al.* (1988). The total ^{137}Cs at each site was expressed in millibecquerels per unit area (mBq cm^{-2}). Total ^{137}Cs at the sites within the catchment was compared with an uncultivated reference site 100 m to the east and on the same soil type (Figure 9.2) (Loughran *et al.*, 1988b). The percentage ^{137}Cs loss or gain, compared with the reference value, was calculated for each site.

The study catchment is 6.6 ha in area and is situated 3 km east of Merriwa, New South Wales (32°09′S; 150°21′E). Black earth soils, developed on Tertiary basalts, occur throughout the area. The basin has a mean slope of 6° (relief range 330–55 m). Mean annual rainfall is 590 mm. Since 1966, the entire catchment has been managed as a wheat–pasture rotation (Loughran et al., 1988b).

RESULTS

Twenty-three sites were sampled (including two replicates). A ^{137}Cs reference value of 100 ± 10 mBq cm^{-2} was measured. Seventeen sites across the catchment (Figure 9.2) showed ^{137}Cs variations between 28 ± 6 and 80 ± 10 mBq cm^{-2}, with the most severe losses of ^{137}Cs along the slope crest and through the centre of the basin. A detailed description of the results is contained in Loughran et al. (1988b).

Two methods were used to translate the percentage ^{137}Cs losses (compared with the reference value) into net soil loss:

(1) A calibration curve relating percentage ^{137}Cs loss to net soil loss, derived from measurements of soil-loss from plots on black earth soils; and
(2) A proportional method, which assumes that net soil loss of the cultivated layer is directly proportional to percentage ^{137}Cs loss (Martz and de Jong, 1987).

Calibration Curve

Soil loss data from 28 runoff plots in eastern NSW were correlated and regressed against percentage ^{137}Cs loss. The equation was:

$$Y = 4.35 \ X^{1.526} \qquad (r = +0.85) \qquad (9.1)$$

where Y is net soil loss (kg ha^{-1} yr^{-1}) and X is the loss of ^{137}Cs, expressed as a percentage of the reference input level (Loughran et al., 1988b). The calculated soil losses, with 95 per cent confidence range, are given in Loughran et al. (1988b), and are presented on a map in Figure 9.2. Net soil losses ranged from 420 to 2540 kg ha^{-1} yr^{-1}, and it was estimated that over half the catchment experienced losses greater than 1000 kg ha^{-1} yr^{-1} (Figure 9.2).

Proportional Method

This method assumes that if soil labelled with ^{137}Cs is lost because of sheet and rill erosion, unlabelled subsoil will be incorporated into the cultivated layer. It is also assumed that complete mixing of ^{137}Cs is achieved by cultivation. 'As a result, the cultivation layer is maintained at a constant thickness while its

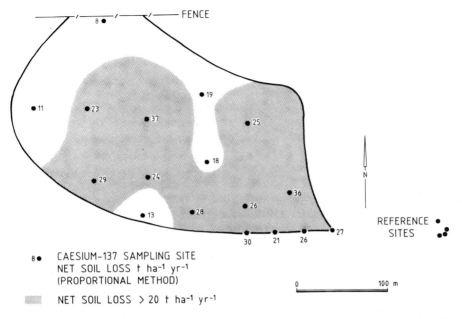

Figure 9.3 Net soil loss from the Merriwa catchment using the proportional method (t ha^{-1} yr^{-1})

^{137}Cs content declines' (Martz and de Jong, 1987, p. 442). Net soil loss for the 30-year lifetime of ^{137}Cs (1955–85, the time of sampling) was calculated from:

$$E = DpC \qquad (9.2)$$

where E is net soil loss over 30 years (t ha^{-1} yr^{-1}), D is the depth of cultivation (14 cm), p is the soil bulk density at each site (g cm^{-3}) and C is the percentage ^{137}Cs loss. Figure 9.3 shows the net soil loss for the catchment (t ha^{-1} yr^{-1}). Estimated losses ranged from 8 to 37 t ha^{-1} yr^{-1}, with over half the basin experiencing losses in excess of 20 t ha^{-1} yr^{-1}.

DISCUSSION

Soil loss estimates by the proportional method were approximately ten to twenty times greater than by the calibration curve. Factors contributing to an underestimation by the calibration curve include:

(1) The plots used in the calibration, surface area 100 m^2, probably did not experience deep rilling; and

(2) The calibration curve is a 'best-fit' regression line, and there were only three points with measured soil-losses exceeding 5 t ha^{-1} yr^{-1} (Loughran et al., 1988).

Overestimation of net soil losses by the proportional method may result from uneven mixing of ^{137}Cs in the cultivated layer. Seasonal fallout and erosion may have occurred before cultivation-mixing, thereby increasing the likelihood of disproportionately higher ^{137}Cs losses. Observations of ^{137}Cs with depth at two sites showed that more ^{137}Cs was present in the upper part of the profile (Loughran et al., 1988).

Implicit within the calibration-curve method was the assumption that plot-soil loss was equivalent to field soil-loss. With the proportional method, no account was taken of temporal variations in fallout (Figure 9.1) nor of soil erosion. For example, Figure 9.4 shows a predicted amount of ^{137}Cs remaining in the soil at the end of the year (1954–84), with annual fallout accounted for (Figure 9.1) and with one centimetre of sheet erosion per year. (It was assumed that cultivation depth was 14 cm and that ^{137}Cs was always evenly mixed in the profile.) Prior to 1974, the model shows that progressively more ^{137}Cs is present in the soil at the end of each year, with the exception of 1968–9, despite erosion of one centimetre depth. Due to the decrease in ^{137}Cs fallout, the model predicts progressively lower annual amounts of ^{137}Cs in the period 1974–84. Further difficulties for interpretation may arise from particle size effects and sediment sorting during erosion, transport and deposition, because ^{137}Cs can be preferentially adsorbed onto the finer fractions.

CONCLUSION

There are no independent measurements of soil loss at the Merriwa sites against which the methods may be assessed. Probably, the amount of soil lost lies between the two estimates. Choice of method for calculating net soil loss from ^{137}Cs measurements must be judged from a knowledge of site history. For example, if land use changed from grazing to cultivation after the period of maximum ^{137}Cs fallout (early 1970s), it may be assumed that the bulk of soil and ^{137}Cs loss occurred in the latter period, and the proportional method could then be more suitable. If plot soil-loss data were short term, or for a restricted number of soil types or managements, or for different climatic zones, they may prove to be unsuitable for calibrating ^{137}Cs losses.

Wise (1980) suggested that ^{137}Cs measurements provided a method for transferring plot data to a wider environment. With the plot-calibration method, assumptions need not be made about complete mixing of ^{137}Cs in the cultivated layer, thereby making it applicable to uncultivated soils, provided the plots themselves can be regarded as representative of field sites.

The Calculation of Net Soil Loss Using Caesium-137

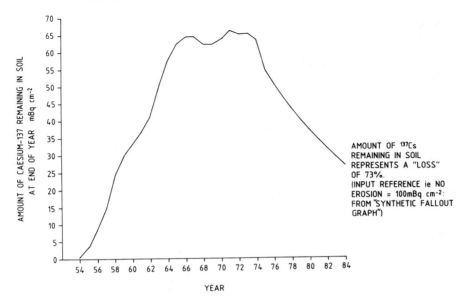

Figure 9.4 A simple soil loss—^{137}Cs loss model for Merriwa, NSW, Australia

Because the ^{137}Cs technique can be used to measure rates of deposition, sediment budgets can provide a means for checking net soil-loss estimates. Even if ^{137}Cs levels are unable to be calibrated to provide estimates of soil erosion, they can be used to determine relative levels of soil erosion status.

ACKNOWLEDGEMENTS

Funds for the research were provided by the Australian National Soil Conservation Program. Darren Shelly gave field and laboratory assistance and Lawrie Henderson prepared the figures.

REFERENCES

Campbell, B. L., Elliott, G. L. and Loughran, R. J. (1986a). Measurement of soil erosion from fallout Cs-137, *Search*, **17**, 148–9.

Campbell, B. L., Loughran, R. J. and Elliott, G. L. (1982). Caesium-137 as an indicator of geomorphic processes in a drainage basin system, *Australian Geographical Studies*, **20**, 49–64.

Campbell, G. L., Loughran, R. J., Elliott, G. L. and Shelly, D. J. (1986b). Mapping drainage basin sediment sources using caesium-137, *International Association of Hydrological Sciences Publ.*, **159**, 437–46.

Campbell, B. L., Loughran, R. J. and Elliott, G. L. (1988). A method for determining sediment budgets using caesium-137, *International Association of Hydrological Sciences Publ.*, **174**, 171–9.

Davis, J. J. (1963). Caesium and its relationship to potassium in ecology, in Schultz, G. and Klements, A. W. Jr (eds), *Radioecology*, Reinhold, New York, 539–56.

de Jong, E., Villar, H. and Bettany, J. R. (1982). Preliminary investigations on the use of Cs-137 to estimate erosion in Saskatchewan, *Canadian J. of Soil Research*, **82**, 673–83.

Longmore, M. E., O'Leary, B. M., Rose, C. W. and Chandica, A. L. (1983). Mapping soil erosion and accumulation with the fallout isotope caesium-137, *Australian Journal of Soil Research*, **21**, 373–85.

Loughran, R. J. (1989). The measurement of soil erosion, *Progress in Physical Geography*, **13**, 216–33.

Loughran, R. J., Campbell, B. L. and Elliott, G. L. (1988a). Determination of erosion and accretion rates using caesium-137, in Warner, R. F. (ed.), *Fluvial Geomorphology of Australia*, Academic Press Australia, pp. 87–103.

Loughran, R. J., Elliott, G. L., Campbell, B. L. and Shelly, D. J. (1988b). Estimation of soil erosion from caesium-137 measurements in a small, cultivated catchment in Australia, *Applied Radiation and Isotopes, Int. J. Radiat. Appl. Instrum. Part A*, **39**, 1153–7.

Martz, L. W. and de Jong, E. (1987). Using caesium-137 to assess the variability of net soil erosion and its association with topography in a Canadian Prairie landscape, *Catena*, **14**, 439–51.

McHenry, J. R. and Ritchie, J. C. (1977). Estimating field erosion losses from fallout caesium-137 measurements, *International Association of Hydrological Sciences Publ.*, **122**, 26–33.

Ritchie, J. C. and McHenry, J. R. (1975). Fallout Cs-137: a tool in conservation research, *J. of Soil and Water Conservation*, **30**, 283–6.

Tamura, T. (1964). Selective sorption of caesium with mineral soil, *Nuclear Safety*, **5**, 262–8.

Wise, S. M. (1980). Caesium-137 and lead-210: a review of techniques and some applications in geomorphology. In Cullingford, R. A., Davidson, D. A. and Lewin, J. (eds), *Timescales in Geomorphology*, John Wiley, Chichester, pp. 109–27.

Slope-stream Linkages

10 Linking the Field to the River: Sediment Delivery from Agricultural Land

D. E. WALLING
Department of Geography, University of Exeter

THE CONTEXT

Information on sediment yields from drainage basins has been widely used as a basis for assessing rates of erosion or land degradation in the upstream basin (cf. Fournier, 1960; Jansson, 1982). Such information clearly has important potential advantages in providing estimates of average erosion rates representative of sizeable areas and therefore in avoiding the need for spatial sampling. Furthermore, sediment yield data have frequently been collected as part of routine hydrological monitoring programmes and may be readily available for such assessments. Care is, however, required in any attempt to interpret sediment yield data in terms of upstream erosion rates. For example, Walling (1987, 1988) has emphasized the problems involved in taking account of the processes of sediment delivery interposed between on-site erosion and downstream sediment yields. It is well known that only a proportion, and perhaps a rather small proportion, of the soil eroded within a drainage basin will find its way to the basin outlet, but there are numerous uncertainties involved in estimating the magnitude of this proportion as represented by the sediment delivery ratio (cf. Walling, 1983). In addition, temporal discontinuities in sediment conveyance may introduce further complexity into the relationship between upstream erosion and downstream sediment yield, since the amount of sediment transported out of a basin may reflect the recent history of erosion and sediment delivery, rather than contemporary erosion within its watershed.

Recent concern for the increasing incidence of soil erosion in areas of arable farming in the UK (cf. Morgan, 1985; Evans and Cook, 1986) has highlighted the lack of quantitative evidence regarding the rates involved. There are very few sites for which representative erosion plot data are available, and other

---

Soil Erosion on Agricultural Land
Edited by J. Boardman, I. D. L. Foster and J. A. Dearing
©1990 John Wiley & Sons Ltd

methods of assessing rates of soil loss possess many limitations and operational constraints. Notwithstanding the many potential problems involved, it would therefore seem worth assessing the evidence afforded by existing information on the suspended sediment loads of British rivers. This chapter reviews this evidence and the associated uncertainties, and attempts to highlight the need for further work to improve our current understanding of the linkage between on-site erosion and downstream sediment yield. An improved understanding of this linkage is essential for more meaningful interpretation of sediment yield data, but is also important in the wider context of non-point pollution from agricultural sources and predicting the likely impact of various scenarios of land-use change on downstream suspended sediment loads.

THE BRITISH SCENE

Although the importance of monitoring suspended sediment loads in rivers has been clearly recognized in many countries (e.g. Bogen, 1986), there has been no attempt to establish a national monitoring programme in Britain. As a result, the only suspended sediment yield data available to date are those afforded by a number of measurement programmes undertaken in association with specific research projects (cf. Walling and Webb, 1987). These data possess a number of important limitations, including short periods of record, the lack of a common period of measurement and the diversity of measurement techniques and monitoring strategies employed. Nevertheless, they provide a basis for a preliminary assesment of the magnitude and pattern of suspended sediment transport by British rivers.

Figure 10.1 provides a map of Britain on which most of the available values of annual suspended sediment yield derived from river-monitoring programmes and some of the data available from reservoir surveys have been superimposed. This information indicates that the suspended sediment yields of British rivers are low by world standards (cf. Walling and Webb, 1986) and lie typically in the range $50-100 \text{ t km}^{-2} \text{ yr}^{-1}$. The highest values are generally located in upland regions, whereas over most of lowland Britain, the area where arable cultivation represents a significant proportion of the land use, yields rarely exceed $50 \text{ t km}^{-2} \text{ yr}^{-1}$ or $0.5 \text{ t ha}^{-1} \text{ yr}^{-1}$.

The low levels of suspended sediment yield associated with rivers draining the agricultural land of lowland Britain could be seen as evidence of low rates of soil erosion, although it must be recognized that the values of sediment yield involved represent areal averages for drainage basins within which areas experiencing appreciable erosion may be juxtaposed with areas experiencing low or negligible rates of soil loss. Furthermore, it is important to take account of the various problems involved in linking on-site rates of erosion to downstream sediment yields outlined previously. On-site rates of erosion could be

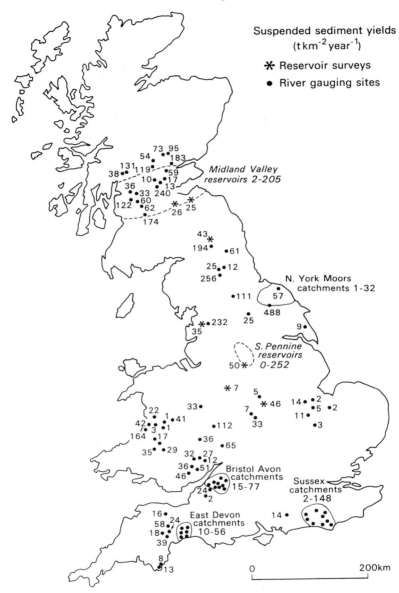

Figure 10.1 Countrywide variation of suspended sediment yields in Britain. (Based on Walling and Webb (1987), and additional data presented by Brown (1987), Butcher, Labadz and Pallister (personal communication), Duck and McManus (personal communication), Foster (personal communication) and Mitchell and Gerrard (1987))

considerably in excess of the typical value of $0.5 \, t \, ha^{-1} \, yr^{-1}$ suggested by the sediment yield data if only a small proportion of the eroded sediment reaches the river channels or if substantial transmission losses are associated with the movement of the sediment through the channel system. To date, however, relatively little is known about the sediment delivery dynamics of agricultural drainage basins in Britain, and the wide range of sediment delivery ratios reported from other environments (cf. Walling, 1983; Ongley, 1987) makes it difficult to provide even a tentative estimate of the likely magnitude of the ratios involved. Some speculation is, however, possible.

Existing evidence concerning rates of soil loss from agricultural land in Britain reviewed by workers such as Boardman (1986) and Evans and Cook (1986) suggest that rates of soil loss from cultivated areas in Britain may be an order of magnitude or more greater than the value of $0.5 \, t \, ha^{-1} \, yr^{-1}$ noted above (cf. Table 10.1). Direct comparison between the two lines of evidence is, however, questionable, since the values of soil loss listed in Table 10.1 generally represent studies of small areas where erosion is known to be occurring, whereas sediment yield data reflect the rates of erosion operating throughout the basin, much of which may be characterized by land use less conducive to erosion. There is, nevertheless, some evidence to suggest that on-site rates of soil loss may be substantially greater than indicated by downstream sediment yields. This suggestion is further supported by consideration of the typical particle size composition of suspended sediment transported by rivers in lowland Britain. The availability of such data is limited, and the use of different analytical

Table 10.1 Some erosion rates reported for arable fields in the UK

Location	Soil type	Annual soil loss ($t \, ha^{-1} \, yr^{-1}$)	Author
Bedfordshire	Chalky	0.6–21.0	Morgan (1985)
Bedfordshire	Sandy Loam	0.6–24.0	Morgan (1985)
Bedfordshire	Clayey	0.3–0.5	Morgan (1985)
Cambridgeshire/ Bedfordshire	Clay	0.7–3.3	Evans and Cook (1986)
Dorset	Clay	0.9–11.8	Evans and Cook (1986)
Hereford	Med. silt	10.–28.6	Evans and Cook (1986)
Gwent	Med. silt	2.2–13.8	Evans and Cook (1986)
Shropshire	Sand/loam	2.6–47.8	Evans and Cook (1986)
Staffordshire/ Worcestershire	Sand/loam	0.5–55.3	Evans and Cook (1986)
Somerset	Light silt	4.9–6.4	Evans and Cook (1986)
Somerset	Light silt	0.1–5.5	Colborne and Staines (1986)
Dorset	Clayey	0–19.5	Colborne and Staines (1986)
Shropshire	Sand/loam	8.0–13.0	Reed (1986)

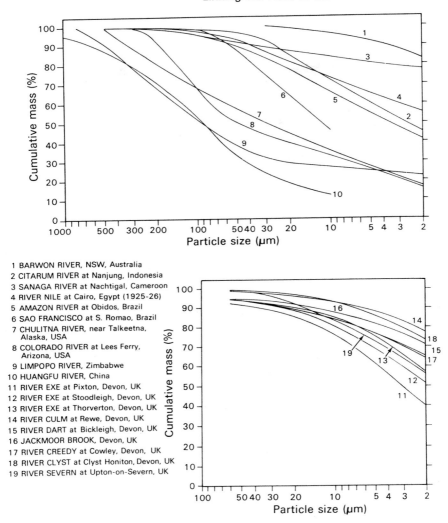

Figure 10.2 A comparison of typical particle size distributions for suspended sediment transported by several British rivers with equivalent distributions for a range of world rivers. (Based partly on Walling and Moorehead, 1989)

procedures in those studies which have been undertaken inevitably introduces uncertainty into any attempt at countrywide generalization. However, Walling and Webb (1981) have noted the fine-grained nature of the suspended sediment transported by British rivers, and this feature is further emphasized by Figure 10.2, which compares typical particle size distributions for several of these rivers studied by the author with a number of distributions more representative of the range exhibited by world rivers. The British rivers represented in Figure 10.2 contrast with most of the other rivers, in that clay- and silt-sized material accounts for nearly all the suspended sediment load. This contrast is undoubtedly in part a reflection of the differing particle size composition of the source materials (cf. Walling and Moorehead, 1989), but it may also provide evidence as to the nature of the sediment delivery systems involved. Analysis of the flow velocities occurring in their channels during flood events indicates that the absence of coarser particles in the loads transported by these British rivers is not the result of a lack of transport capacity but rather of the limited supply of such material to the flow. This could, in turn, point to inefficient delivery systems, wherein most of the coarser material eroded from the slopes of the basin is redeposited before reaching the channels, and therefore indicate low delivery ratios.

Examination of the magnitude of the suspended sediment yields of British rivers in relation to available information on rates of soil loss and the particle size composition of the suspended sediment loads provides some evidence of inefficient systems of sediment delivery to the streams and therefore of low sediment delivery ratios within agricultural drainage basins in Britain. However, such conclusions must remain essentially speculative. Further research is required to elucidate the linkages between the field and the river and the sediment delivery systems involved. It is suggested that such research could usefully address a number of questions and uncertainties, including the provision of more specific data on sediment sources, the tracing of sediment movement and storage within the delivery system, the selectivity of the delivery process and the role of aggregates. These aspects will be considered in turn, both in terms of potential approaches to field investigations and the evidence obtained by the author from work undertaken within the basin of the River Exe in Devon.

SEDIMENT SOURCES

Information concerning the source of the suspended sediment transported by a river must be seen as an essential prerequisite in any attempt to elucidate the sediment delivery dynamics involved. In Britain there is still uncertainty as to whether erosion of cultivated fields and other slope sources or channel erosion provides the dominant source. The former must be shown to be a significant source if the linkage between soil erosion from agricultural land and downstream sediment yield is to be further explored. Elucidation of the dominant source of the suspended sediment transported by a river is, however, a difficult task. Approaches traditionally employed for this purpose have included direct monitoring of potential sources using erosion pins, runoff troughs and related techniques (cf. Imeson, 1974) and more indirect assessments, including comparison of the measured load with an estimate of the magnitude of the contribution associated with sheet and rill erosion derived using a soil loss prediction equation such as the USLE coupled with an appropriate sediment delivery ratio. The former involves many operational difficulties, as well as important sampling problems, and is impractical in anything but small basins, while the latter effectively prejudges the result, since any estimate of the sediment delivery ratio will involve numerous assumptions and uncertainties.

An alternative approach is the use of 'fingerprinting' techniques. In essence, this method involves first, the selection of a physical or chemical property which clearly differentiates potential source materials and, second, comparison of measurements of the same property obtained for suspended sediment with the equivalent values for potential sources (cf. Wall and Wilding, 1976). In the case of a simple distinction between surficial and channel sources, the property could be one that differentiated topsoil (A-horizon) from the underlying parent materials, and Oldfield *et al.* (1979) have, for example, suggested that several mineral magnetic parameters offer considerable scope for fingerprinting purposes.

This approach has been successfully employed by the author to determine the relative importance of the channel and surficial source contributions to the suspended sediment yield of the small (9.3 km^2) agricultural drainage basin of the Jackmoor Brook in Devon. A detailed study of a variety of potential fingerprinting properties was undertaken (cf. Peart and Walling, 1986) and, in order to avoid problems associated with the enrichment of suspended sediment relative to its source material, emphasis was placed on the use of property ratios which were essentially independent of such enrichment effects. The caesium-137 (^{137}Cs) content of sediment and potential source materials was also found to be a valuable fingerprinting tool, since this reflected the relative receipt of atmospheric fallout by the different sources. In this case, however, it was necessary to introduce a correction for the enrichment of suspended sediment in fines relative to the source material, since ^{137}Cs is known to be preferentially associated with the fine fractions. Table 10.2 provides a summary of the results obtained from applying these fingerprinting properties in a simple mixing model

Table 10.2 The relative contribution of bank material and surface soil sources to the total suspended sediment load of the Jackmoor Brook estimated using a simple mixing model applied to a number of 'fingerprint' properties

Fingerprint property	Relative contribution (%)	
	Surface soil	Channels
Carbon/nitrogen ratio	99.6	0.4
SIRM/magnetic susceptibility ratio	93	7
Dithionite iron/magnetic susceptibility ratio	98.5	1.5
Manganese/dithionite iron ratio	89	11
^{137}Cs activity	94	6

to determine the relative contributions of bank and surface erosion to the suspended sediment yield from the basin. These indicate that in this basin, a substantial proportion of which is occupied by arable cultivation, surficial erosion provides the dominant sediment source. The ^{137}Cs measurements also provide a means of estimating the relative importance of permanent pasture and arable areas within these surficial sources, since the mixing associated with arable cultivation distributes the radionuclide throughout the plough depth and reduces the ^{137}Cs content of topsoil from these areas relative to that from areas of permanent pasture where the ^{137}Cs remains concentrated in the upper 10 cm of the soil. These measurements suggested that of the order of 75 per cent of the sediment derived from surficial sources originated from arable fields.

Because of the importance of soil loss from cultivated areas to the overall suspended sediment yield from the Jackmoor Brook basin, the work on fingerprinting suspended sediment sources outlined above was extended to consider the relative importance of sediment-associated nutrients (N and P) to the overall nutrient output from the basin. This involved monitoring the transport of N and P both in solution and in association with sediment. The results summarized in Table 10.3 indicate that in this basin and in the adjacent basins of the River Dart (46 km²) and the River Exe (601 km²) sediment-associated transport contributes about 10 per cent and 50 per cent of the total outut of N and P, and therefore represents an important component of the overall nutrient budget.

Table 10.3 The contribution of sediment-associated transport to the total nutrient loads transported by rivers in the Exe Basin

River	Gauging site	Contribution to total nutrient load 1980–82 (%)	
		N	P
Jackmoor Brook	Pynes Cottage	5.9	54.2
Exe	Thorverton	8.6	50.5
Dart	Bickleigh	16.4	70.6

Based on Walling and Thornton (1984)

TRACING SEDIMENT MOVEMENT AND STORAGE WITHIN THE DELIVERY SYSTEM

If it can be demonstrated that areas of arable cultivation make a substantial contribution to the downstream sediment yield from a drainage basin, it is important to consider the precise relationship between the on-site rates of soil loss and the downstream sediment yield and therefore the magnitude of the sediment delivery ratio. Substantial amounts of eroded sediment could be redeposited before reaching the channel and significant depositional losses could also be associated with its subsequent conveyance through the channel system (cf. Trimble, 1976; Miller and Shoemaker, 1986). Again, however, major problems exist in attempting to quantify this aspect of the delivery system. Measurements should ideally aim to demonstrate both the proportion of the eroded sediment reaching the basin outlet and the location and magnitude of the major sinks involved. Recent work has suggested that ^{137}Cs could provide a useful tracer for use in such investigations (cf. Brown *et al.* 1981a,b; Cooper *et al.*, 1987; Campbell *et al.*, 1986; McHenry and Ritchie, 1977; Ritchie *et al.*, 1982; Walling and Bradley, 1988; Walling *et al.*, 1986).

Caesium-137 is present in the environment primarily as a product of the atmospheric testing of nuclear weapons during the late 1950s and early 1960s, and fallout of this radionuclide was first detected in 1954. Rates of fallout reached a maximum in 1964 and declined rapidly after the nuclear test ban treaty (cf. Ritchie *et al.*, 1975; Pennington *et al.*, 1976), although in many areas of Europe there were further significant inputs in 1986 as a result of the Chernobyl disaster (e.g. Dorr and Munnich, 1987). Existing evidence indicates that fallout reaching the surfaces of most soils is rapidly and strongly adsorbed by the upper horizons of the soil and that further downward translocation by physico-chemical processes is limited (cf. Tamura, 1964, Frissel and Pennders, 1983). Subsequent movement of ^{137}Cs is therefore generally associated with the erosion, transport and deposition of sediment particles (e.g. Rogowski and Tamura, 1970; Campbell *et al.*, 1982). Caesium-137 has a half-life of 30.1 years, and approximately 60 per cent of the total input of this radioisotope since fallout began in 1954 could still remain in the environment. Considerable potential therefore exists for studying the mobilization of sediment by erosion and its movement through the delivery system over the past 30 years by measuring the spatial distribution of ^{137}Cs within a drainage basin. If measured values of ^{137}Cs loading (mBq cm^{-2}) for individual points within a field or drainage basin are compared with an estimate of the overall input, depletion would indicate erosion, whereas areas of deposition would be marked by elevated levels of ^{137}Cs.

Furthermore, comparison of the relative magnitude of the total loss of ^{137}Cs from areas experiencing erosion with the gain associated with those areas experiencing deposition provides a basis for estimating the sediment delivery ratio (cf. Walling and Bradley, 1988). Estimates of the baseline or reference input are commonly obtained by measuring the total activity or loading at undisturbed sites located on an interfluve, where soils are unlikely to have experienced either erosion or deposition. Substantial inputs of Chernobyl fallout will introduce significant modifications to the distribution patterns generated during the past 30 years and therefore complicate their intepretation. However, these inputs were of low magnitude throughout much of lowland Britain, where investigations of soil erosion and sediment delivery from agricultural land are likely to be undertaken.

Further discussion of the potential for using ^{137}Cs measurements to elucidate sediment delivery from cultivated areas can usefully introduce some examples from the work of the author and his colleagues within and adjacent to the drainage basin of the Jackmoor Brook referred to previously, and for which a large proportion of the suspended sediment load has been shown to be derived from cultivated fields (cf. Walling and Bradley, 1988). This catchment, which lies about 10 km north of the city of Exeter, is in an area of intensive mixed farming. It is underlain by sandstones, breccias and conglomerates of Permian age and the characteristic soils are fertile brown earths. Slopes range between about 2° and 12° and the individual fields, which are frequently bounded by hedges and banks, are typically of the order of 12 ha in size. Mean annual precipitation over the area is estimated at 800 mm and more than 50 per cent of the annual precipitation commonly falls between November and March, when many of the fields, which are tilled and sown with cereals in the autumn, have limited vegetation cover. The suspended sediment yields of streams in the area are typically of the order of 50 t km^{-2} and are dominated by clay- and silt-sized fractions.

Within this area the baseline or reference value of ^{137}Cs input has been estimated at 250 mBq cm^{-2} and soil cores have been collected from a large number of locations, in order to investigate the spatial pattern of erosion and deposition. The results of reconnaissance surveys of two fields in the study area are shown in Figures 10.3 and 10.4. Figure 10.3 refers to a field with an average slope of about 12°, which represents one of the steepest fields under arable cultivation in the area and in which rill erosion had been observed on several occasions. Locations in the field that have experienced significant erosion over the past 30 years are indicatd by ^{137}Cs loadings which are more than 20 per cent less than the reference value, while areas of significant deposition are marked by loadings which are more than 20 per cent in excess of the reference value. These margins or confidence limits have been used in order to take account of the precision of the ^{137}Cs measurements and the natural sampling variability evidenced by the soil cores. A tentative map of erosional and depositional areas has been interpolated from the measured values for the sampling points (Figure 10.3) and this suggests that erosion has been active over most of the field. A sizeable deposition zone is also located at the foot of the slope above the hedge boundary, indicating that a significant proportion of the soil eroded from the field was redeposited at the margin of the field and failed to reach the nearby stream.

Further analysis of the levels of ^{137}Cs plotted at the individual sampling sites can provide an estimate of the proportion of the overall input that has been mobilized by erosion, the proportion that is currently stored in the depositional zone and the proportion that has been transported beyond the field. In order to undertake these calculations it has been assumed that the sampled points provide a representative sample of the conditions within the field and that the

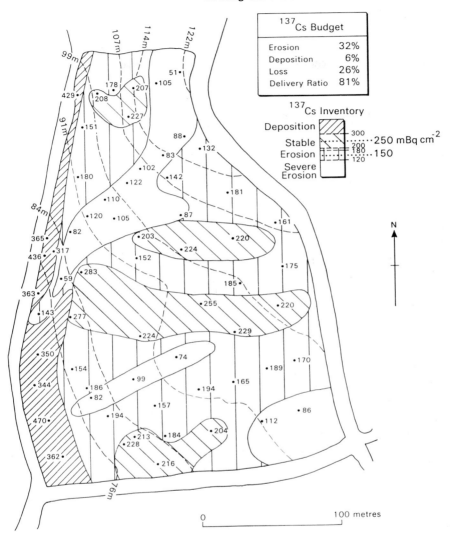

Figure 10.3 The spatial distribution of ^{137}Cs loadings within a cultivated field representative of the steep portions of the study area. The mapping classes have been selected to take account of the overall level of precision of the ^{137}Cs measurements. (Based on Walling and Bradley, 1988)

analytical and sampling variability referred to above is essentially random. The results are presented as a '^{137}Cs budget' in Figure 10.3. They confirm that most of the ^{137}Cs mobilized by erosion has been transported beyond the field and the delivery ratio, which expresses this loss as a proportion of the total erosion, is as high as 81 per cent.

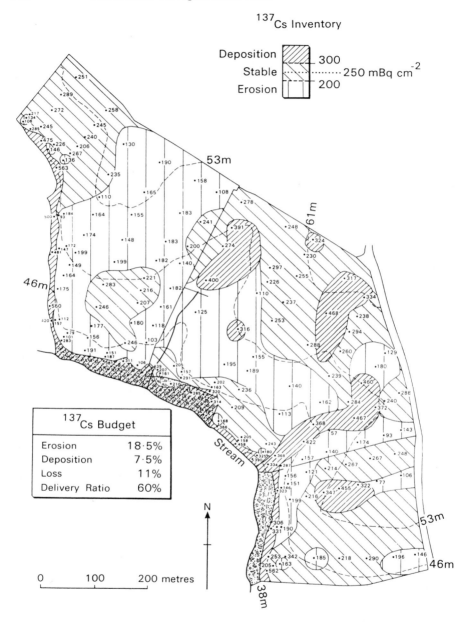

Figure 10.4 The spatial distribution of ^{137}Cs loadings within a group of cultivated fields representative of the portions of the study area with more subdued topography. The mapping classes have been selected to take account of the overall level of precision of the ^{137}Cs measurements. (Based on Walling and Bradley, 1988)

Figure 10.4 provides another example of a survey of ^{137}Cs levels in the soils of a cultivated area. In this case several contiguous fields are involved and they have been selected to be representative of an area with much lower slope angles than those encountered in Figure 10.3. Here the average slope angle is approximately 3°. The same procedure has been used to interpolate the pattern of erosion and deposition and this has been superimposed onto Figure 10.4. Several contrasts with the previous example are immediately apparent. First, there are substantial areas that have experienced negligible erosion. Second, the proportion of the ^{137}Cs lost from those areas experiencing erosion is less than that in Figure 10.3, indicating that the erosion rates on these gentler slopes are less severe. Third, in this area there is evidence of deposition *within* the field, as well as along the downslope field boundary. This indicates that some of the eroded soil has been redeposited before reaching the field margin and that a less efficient delivery system is involved. Most of this within-field deposition is associated with the two small valley-like features which dissect the area. The ^{137}Cs budget for these fields, which has been estimated in the same way as for Figure 10.3, embodies many of the contrasts noted above. The overall proportion of the ^{137}Cs input that has been mobilized by erosion is about half of that mobilized on the steeper field represented by Figure 10.3. In addition, more of the eroded ^{137}Cs has been deposited within the field or along the field boundary and, in consequence, the overall delivery ratio is significantly lower. This lower delivery ratio may reflect both the lower slope angles and the lack of any evidence of significant rill erosion within this group of fields.

The information presented in Figures 10.3 and 10.4 refers specifically to ^{137}Cs movement and therefore provides only essentially qualitative information on sediment delivery from the fields involved. Since ^{137}Cs is known to move in association with sediment, the patterns of erosion and deposition produced will closely reflect those of soil loss and deposition, both in terms of spatial distribution and relative intensity. Any attempt to use the data to estimate absolute rates of erosion and deposition and the sediment delivery ratio must, however, take account of the preferential association of ^{137}Cs with the finer fractions of the soil. If high rates of soil loss or deposition involve the mobilization or deposition of a greater proportion of coarse material, the ^{137}Cs information may underestimate their relative magnitude. The delivery ratio calculated for ^{137}Cs may similarly overestimate the sediment delivery ratio if depositional losses are preferentially associated with the coarser fractions. However, in view of the fine nature of the soils in this study area, the fact that much of the clay fraction is likely to move in association with larger aggregates containing coarser particles, and the limited evidence of contrasts in the particle size composition of topsoil from areas experiencing erosion and deposition, it is thought that the ^{137}Cs data provide a meaningful approximation of the pattern of erosion, deposition and sediment delivery from these areas.

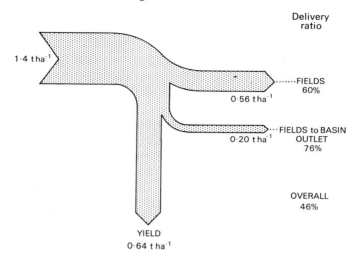

Figure 10.5 A tentative sediment budget for the Jackmoor Brook drainage basin based on evidence provided by ^{137}Cs measurements undertaken at several locations within the basin

Furthermore, possible limitations of the results must be seen in the context of the very great problems involved in obtaining any quantitative information on sediment delivery processes.

More work is required to develop relationships between the magnitude of the loss or gain of ^{137}Cs at a sampling site and the rate of soil loss or deposition at that site, in order to evaluate the sediment budgets for the fields shown in Figures 10.3 and 10.4 and for the overall drainage basin of the Jackmoor Brook. However, on the basis of the ^{137}Cs measurements undertaken to date on a number of fields in this drainage basin, and existing knowledge of the suspended sediment yield from the basin, a tentative sediment budget has been constructed for the basin, and this is illustrated in Figure 10.5. This indicates an average delivery ratio for the basin of about 46 per cent and therefore that in this locality a substantial proportion of the sediment eroded from the fields is reaching the streams.

THE SELECTIVITY OF THE DELIVERY PROCESS

If it can be demonstrated that cultivated fields represent an important sediment source within an area and it is possible to derive a tentative estimate of the sediment delivery ratio and overall sediment budget, it is also important to consider the selectivity of the erosion and delivery system and to investigate the relationship between the physical and chemical properties of the sediment

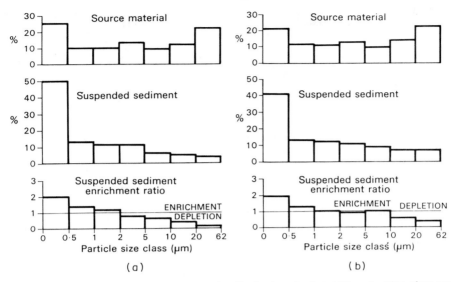

Figure 10.6 Comparison of typical particle size distributions for the <63 μm fraction of source material and suspended sediment for the (b) River Dart and (a) Jackmoor Brook drainage basins. Enrichment ratios have been calculated for individual size fractions as the ratio of the proportion of that size fraction in the suspended sediment to that in the source material

and those of the source materials. Suspended sediment commonly shows evidence for enrichment in fines and the organic fraction relative to the source material, and Figure 10.6 and Table 10.4 illustrate relationships between the particle size and chemical composition of suspended sediment and source materials for the Jackmoor Brook and River Dart drainage basins in Devon, UK. The inherent spatial variability of source material properties inevitably introduces problems into any attempt to characterize such material by a single distribution or concentration, and the data presented in Figure 10.6 and Table 10.4 must therefore be viewed as highly generalized. Figure 10.6 compares the mean particle size characteristics of suspended sediment (chemically dispersed mineral fraction) transported by the two streams with those of typical source materials within their upstream catchment areas. Enrichment ratios have been calculated for individual particle size classes as the ratio of the proportion of a particular particle size fraction in suspended sediment to that in source material. In the case of the Jackmoor Brook, suspended sediment can be seen to be enriched in clay (<2 μm) and depleted in silt (2–62 μm), whereas in the River Dart the degree of enrichment and depletion is less marked and is effectively limited to the <1 μm and >10 μm fractions, respectively.

Table 10.4 compares the organic carbon, nitrogen, total phosphorus and free (dithionite extractable) iron contents of suspended sediment with those of surface soils under arable cultivation or permanent pasture. Enrichment ratios have been

Table 10.4 Enrichment of suspended sediment in organic carbon, nitrogen, total-phosphorus and dithionite-extractable iron relative to typical surface soils

Drainage basin	Land use	Enrichment factor			
		C	N	P	Fe
Jackmoor Brook	Arable	1.85	2.00	1.20	1.23
	Pasture	1.24	1.49	1.17	1.00
River Dart	Arable	4.47	4.00	1.83	1.20
	Pasture	1.25	1.28	1.00	1.15

Based on Walling and Kane (1984)

calculated as the ratio of the concentration of a specific constituent in suspended sediment to that in the soil. In the case of the Jackmoor Brook, areas of arable cultivation have already been shown to represent the dominant sediment source and the enrichment ratios related to this source most closely reflect the processes actually operating in the basin. These range between 1.2 and 2.0, with the wholly organic constituents (C and N) showing the highest degree of enrichment. In view of the very much smaller proportion of land under arable culivation in the drainage basin of the River Dart, pasture areas must represent a more important sediment source in this steeper basin. Actual enrichment ratios could therefore be expected to lie between the values given for arable and pasture soils, and these would therefore be closely comparable with those cited above for the Jackmoor Brook. In considering the relationship between the properties of suspended sediment and those of its source material, it would therefore seem realistic to suggest that enrichment ratios for the organic fraction and for the major nutrients could lie between about 1.2 and 2.0.

The above discussion of enrichment ratios refers to average values associated with a comparison of the average composition of suspended sediment with that of typical source material. Significant temporal variability in enrichment ratios could be expected in response to inter- and intra-storm variability in the dynamics of sediment mobilization and delivery. Figure 10.7, for example, presents plots of the relationship between the particle size composition of suspended sediment (percentage clay and sand) and water discharge for six sites in the Exe basin, including the Jackmoor Brook and the River Dart. Some of the variability in particle size composition associated with different levels of flow will undoubtedly reflect temporal variations in the relative importance of different sediment sources within the individual basins, but the nature and efficiency of the sediment delivery processes could also be expected to vary through time and to influence the relationships. The complexity of the controls involved is, however, emphasized by the considerable diversity of behaviour shown by these six basins. These include cases where the percentage clay decreases (River Exe), increases (Rivers Creedy and Culm and Jackmoor Brook) and remains essentially constant (River Clyst) as water discharge increases and where it initially decreases but

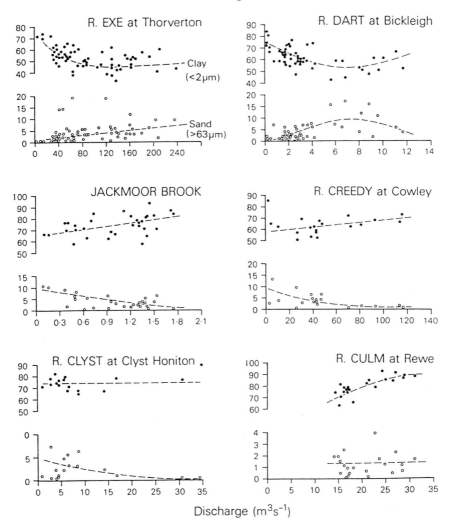

Figure 10.7 Relationships between the particle size composition (% clay and sand) of suspended sediment and flow magnitude for six sites within the Exe Basin. (Based on Walling and Moorehead, 1987)

subsequently increases over the range of flows (River Dart). Equally, the proportion of sand shows a similar diversity of response, with only one of the rivers (River Exe) showing the positive relationship between percentage sand and discharge that might be expected from a simple consideration of transport capacity.

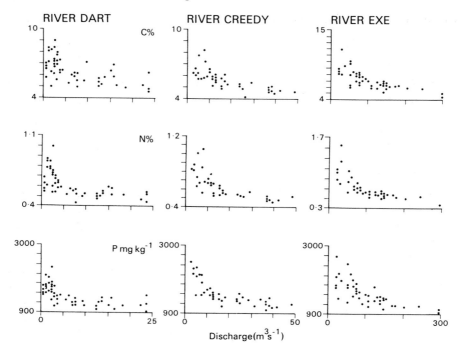

Figure 10.8 Relationships between the organic carbon, nitrogen and total-phosphorus content of suspended sediment and flow magnitude for three sites within the Exe basin. (Based on Walling and Kane, 1982)

Figure 10.8 also illustrates the relationships between the organic matter (C) and nutrient (N and P) content of suspended sediment and water discharge derived for the Rivers Dart, Creedy and Exe. These show more consistency, in that the C, N and P content of suspended sediment shows a similar trend of decreasing concentration with increasing flow in all three rivers. Again this will reflect temporal variation in the relative importance of different sources, as well as in the selectivity of the erosion and delivery processes. However, any attempt at more detailed explanation or modelling of the trends shown in Figures 10.7 and 10.8 must account for the contrast between the similar behaviour of the three rivers evident in Figure 10.8 and the substantial differences in their behaviour shown in Figure 10.7, which in turn suggests that grain size composition and organic matter content are influenced by different controls.

THE EFFECTIVE SIZE OF TRANSPORTED SEDIMENT

The grain size composition of the sediment moving through the sediment delivery system of a drainage basin will clearly exert a fundamental influence on the transport and depositional processes involved, and such information must represent an important requirement in any attempt to understand and model the delivery process. Particle size data are frequently available for soils and sediment, but in most cases these will have been obtained using standard laboratory procedures and therefore relate to the chemically dispersed mineral fraction. Such data may be termed *ultimate* particle size data, since they refer to the discrete particles comprising the sediment. There is, however, increasing evidence to indicate that most of the sediment moving through a drainage basin will be transported as aggregates rather than as discrete particles. In such circumstances it is clearly important to consider what Ongley *et al.* (1981) have termed the *effective* particle size distribution of the sediment, because this will govern the actual behaviour of the transported sediment. Thus in a situation where deposition of primary clay is unlikely to occur, because of its minimal fall velocity, significant quantities of clay may nevertheless be deposited if this is incorporated within larger aggregates.

Foster *et al.* (1985) report results from a variety of soil types in the United States which indicate that only about 25 per cent of the primary clay within a soil will be represented as primary clay in eroded sediment, the remainder being incorporated into larger aggregates. These authors have developed a series of empirical relationships for predicting the likely composition of eroded sediment at the point of detachment, in terms of both primary particles and aggregates, from information on the texture of the matrix soil. Figure 10.9 illustrates a hypothetical example of the application of these relationships to a matrix soil composed of 25 per cent clay, 60 per cent silt and 15 per cent sand. This indicates that the eroded soil will contain a substantial proportion of large aggregates ($>63\ \mu$m) and that these will contain sizeable amounts of clay- and silt-sized particles. Although it is known that much of the material eroded from cultivated fields in the UK will be in the form of aggregates, there is little quantitative information available on the likely size and composition of these aggregates for different soil types or on the degree to which the composition of the eroded soil could vary temporally in response to soil condition and cultivation practices.

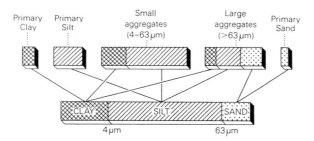

Figure 10.9 A schematic example of the likely aggregate content of eroded soil based on relationships developed by Foster *et al.* (1985) for US soils

The collection of such information must represent an important prerequisite for the development of effective modelling procedures and the improved understanding of the enrichment mechanisms associated with sediment delivery. Maximum enrichment will be associated with poorly aggregated soils, since little of the clay is likely to be lost by deposition during transport. Conversely, sediment characterized by a high degree of aggregation is likely to show little enrichment, since a large proportion of the clay will be associated with aggregates and will therefore be liable to deposition.

Aggregation may also be evident within the suspended sediment transported by a stream, and this will again exert an important influence on its behaviour in depositional environments. An order of magnitude increase in the particle size of transported sediment could, for example, result in an increase in fall velocities by two orders of magnitude. Considerable uncertainty surrounds the selection of a method for measuring the effective particle size characteristics of suspended sediment, since this must reflect *in situ* conditions within the river channel. However, work in the Exe Basin (Walling and Moorehead, 1989) suggests that on-site measurements undertaken immediately after sample collection using a bottom withdrawal sedimentation tube technique (cf. Owen, 1976) provides one means of overcoming many of the problems involved and of obtaining a meaningful representation of the effective grain size distribution. This distribution will, however, relate to the equivalent spherical diameter of the particles, rather than to their actual size. Figure 10.10 compares the mean

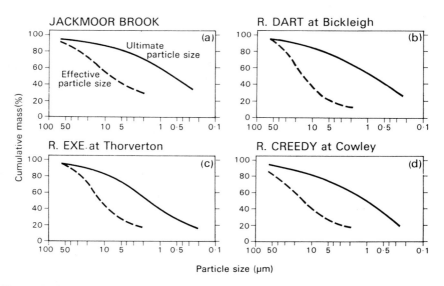

Figure 10.10 A comparison of characteristic ultimate and effective particle size distributions for suspended sediment from four sites within the Exe basin. (Based on Walling and Moorehead, 1989)

effective particle size distributions obtained at several sites on the River Exe and its tributaries, including the River Dart and the Jackmoor Brook, with equivalent mean curves for their ultimate particle size distributions. In all cases there is almost an order of magnitude difference between the median particle size of the two distributions. Maximum contrasts between the two distributions appear to be associated with sites where the suspended sediment contains a high proportion of clay, indicating that such sediment is more prone to aggregation.

Available evidence suggests that much of the aggregation reflected in Figure 10.10 is the result of binding by organic material. Many of the discrete sediment particles exhibit organic coatings which assist such aggregation. In-stream processes are undoubtedly partly responsible for this aggregation, but many of the aggregates may also represent the persistence of the soil-derived aggregates referred to above. More work is required to investigate the persistence, breakdown and formation of aggregates as sediment moves through the delivery system from its point of detachment, since such mechanisms are again central to an understanding of the transport and enrichment processes involved.

PERSPECTIVE

In a recent review of issues and problems in the field of sediment research in the United States, Meade and Parker (1986) highlighted the many uncertainties surrounding the modelling and prediction of sediment delivery when they indicated that:

> Although onsite erosion of specific types of soils under specific conditions of cultivation or other land uses can be predicted by using such tools as the Universal Soil Loss Equation, predicting how much of the eroded soil will be delivered eventually to the channel of a neighbouring stream still remains difficult.

It is clear that similar, if not greater, problems exist in Britain, since methods for predicting on-site erosion are themselves less well developed, and there is a paucity of detailed sediment yield data. As in the United States, this area must represent an important area for future research, particularly in view of the increasing awareness of the role of sediment-associated transport in non-point pollution and in the movement of contaminants through terrestrial and aquatic ecosystems. The development of predictive capabilities is clearly important as an ultimate goal, but as yet there still remains a need for field investigations of the delivery system involved, in order to provide a basis for an improved understanding of the processes that will need to be modelled and to provide data which could be used to parameterize and test such models. Furthermore, more information on on-site erosion rates and downstream sediment yields for specific situations is required, in order to provide the necessary input and output data for model development and verification.

ACKNOWLEDGEMENTS

The author gratefully acknowledges the support of a Natural Environmental Research Council Research Grant for work on sediment delivery processes and the contribution of several postgraduate students and research fellows to the Exe Basin research programme.

REFERENCES

Bogen, J. *ErosJonsprosesser og Sedimenttransport i Norske Vassdrag. Utredning av Forvaltningsansvar, Faglig Status og Forskningsbehov*. Norsk Hydrologisk Komité Rapport nr. 20, Oslo.

Brown, A. G. (1987). Long-term sediment storage in the Severn and Wye catchments. In Gregory, K. J., Lewin, J. and Thornes, J. B. (eds), *Palaeohydrology in Practice*, John Wiley, Chichester, pp. 307-32.

Brown, R. B., Cutshall, N. H. and Kling, G. F. (1981a). Agricultural erosion indicated by ^{137}Cs redistribution: I. Levels and distribution of ^{137}Cs activity in soils, *Soil Sci. Soc. Am. J.*, **45**, 1184-90.

Brown, R. B., Kling, G. F. and Cutshall, N. H. (1981b). Agricultural erosion indicated by ^{137}Cs redistribution: II. Estimates of erosion rates, *Soil Sci. Soc. Am. J.*, **45**, 1191-7.

Boardman J. (1986). The context of soil erosion, *SEESOIL*, **3**, 2-13.

Campbell, B. L., Loughran, R. J. and Elliott, G. L. (1982). Caesium-137 as an indicator of geomorphic processes in a drainage basin system, *J. Aust. Geogr. Studies*, **20**, 49-64.

Campbell, B. L., Loughran, R. J., Elliott, G. L. and Shelly, D. J. (1986). Mapping drainage basin sediment sources using caesium-137. In Hadley, R. F. (ed.), *Drainage Basin Sediment Delivery*, Int. Assoc. Hydrol. Sci. Publ. No. 159, 437-46.

Colborne, G. J. N. and Staines, S. J. (1980). Soil erosion in Somerset and Dorset, *SEESOIL*, **3**, 62-71.

Cooper, J. R., Gilliam, J. W., Daniels, R. B. and Robarge, W. P. (1987). Riparian areas as filters for agricultural sediment, *Soil Sci. Soc. Am. J.*, **51**, 416-20.

Dorr, H. and Munnich, K. O. (1987). Spatial distribution of soil-^{137}Cs and ^{134}Cs in West Germany after Chernobyl, *Naturwissenschaften*, **74**, 249-51.

Evans, R. and Cook, S. (1986). Soil erosion in Britain, *SEESOIL*, **3**, 28-59.

Foster, G. R., Young, R. A. and Neibling, W. H. (1985). Sediment composition for nonpoint source pollution analyses, *Trans. Am. Soc. Ag. Engrs.*, **28**, 133-9, 146.

Fournier, F. (1960). *Climat et Erosion*, Presses Universitaires de France, Paris.

Frissel, M. J. and Pennders, R. (1983). Models for the accumulation and migration of ^{90}Sc, ^{137}Cs, ^{239,240}Pu, and ^{241}Am in the upper layers of soils, *Spec. Pub. Brit. Ecol. Soc.*, No. 3, 63-72.

Imeson, A. C. (1974). The origin of sediment in a moorland catchment with special reference to the role of vegetation. In Gregory K. J. and Walling, D. E. (eds), *Fluvial Processes in Instrumented Watersheds*, Inst. Brit. Geogr. Spec. Pub. No. 6, 69-72.

Jansson, M. (1962). *Land Erosion by Water in Different Climates*, UNGI Rapport No. 57, Department of Physical Geography, University of Uppsala, Sweden.

McHenry, J. R. and Ritchie, J. C. (1977). Estimating field erosion losses from fallout caesium-137 measurements. In *Erosion and Solid Matter Transport in Inland Waters*, *Int. Assoc. Hydrol. Sci. Pub.* No. 122, 26–33.

Meade, R. H. and Parker, R. S. (1986). Issues in sediment research in rivers of the U.S., *Proceedings of the Fourth Federal Interagency Sedimentation Conference*, pp. 4.162–4.168.

Miller, A. J. and Shoemaker, L. L. (1986). Channel storage of fine-grained sediment in the Potomac River. In Hadley, R. F. (ed.), *Drainage Basin Sediment Delivery, Int, Assoc. Hydrol. Sci. Pub.* No. 159, 287–303.

Mitchell, D. J. and Gerrard, A. J. (1987). Morphological responses and sediment patterns. In Gregory, K. J., Lewin, J. and Thornes, J. B. (eds), *Palaeohydrology in Practice*, John Wiley, Chichester, pp. 177–99.

Morgan, R. P. C. (1985). Soil erosion measurement and soil conservation research in cultivated areas of the UK, *The Geographical Journal*, **151**, 11–20.

Oldfield, F., Rummery, T. A., Thompson, R. and Walling, D. E. (1979). Identification of suspended sediment sources by means of magnetic measurements, *Water Resour. Res.*, **15**, 211–18.

Ongley, E. D. (1987). Scale effects in fluvial sediment-associated chemical data. *Hydrological Processes*, **1**, 171–179.

Ongley, E. D., Bynoe, M. C. and Percival, J. B. (1981). Physical and geochemical characteristics of suspended solids, Wilton Creek, Ontario, *Canadian J. Earth Sci.*, **18**, 1365–79.

Owen, M. W. (1976). *Determination of the Settling Velocities of Cohesive Muds*, Report No. IT 161, Hydraulics Research Station, Wallingford, UK.

Peart, M. R. and Walling, D. E. (1986). Fingerprinting sediment source: The example of a drainage basin in Devon, UK. In Hadley, R. F. (ed.), *Drainage Basin Sediment Delivery, Int. Assoc. Hydrol. Sci. Pub.* No. 159, 41–55.

Pennington, W., Cambray, R. S., Eakins, J. D. and Harkness, D. D. (1976). Radionuclide dating of the recent sediments of Blelham Tarn, *Freshwater Biology*, **6**, 317–31.

Reed, A. H. (1986). Erosion risk on arable soils in parts of the West Midlands, *SEESOIL*, **3**, 84–94.

Rogowski, A. S. and Tamura, T. (1970). Erosional behavior of caesium-137, *Health Physics*, **18**, 467–77.

Ritchie, J. C., Hawks, P. H. and McHenry, J. R. (1975). Deposition rates in valleys determined using fallout caesium-137, *Geol. Soc. Am. Bull.*, **86**, 1128–30.

Ritchie, J. C., McHenry, J. R. and Bubenzer, G. D. (1982). Redistribution of fallout ^{137}Cs in Brunner Creek watershed in Wisconsin, *Wisconsin Acad. Sciences Arts and Letters*, **70**, 161–6.

Tamura, T. (1964). Selective sorption reaction of caesium with mineral soils, *Nuclear Safety*, **5**, 262–8.

Trimble, S. W. (1986). Sedimentation in Coon Creek Valley, Wisconsin, *Proceedings of the Third Federal Interagency Sedimentation Conference*, US Water Resources Council, Washington, DC, pp. 5.100–5.112.

Wall, G. J. and Wilding, L. P. (1976). Mineralogy and related parameters of fluvial suspended sediments in Northwestern Ohio, *J. Environ. Qual.*, **5**, 168–73.

Walling, D. E. (1983). The sediment delivery problem, *J. Hydrol.*, **69**, 209–37.

Walling, D. E. (1987). Land degradation and sediment yields in rivers. A background to monitoring strategies, *Monitoring Systems for Environmental Control*, 64–105, SADCC Soil and Water Conservation and Land Utilization Programme, Maseru, Lesotho.

Walling, D. E. (1988). Measuring sediment yield from river basins. In Lal, R. (ed.), *Soil Erosion Research Methods*, Soil and Water Conservation Society, Ankeny, Iowa, pp. 39-73.

Walling, D. E. and Bradley, S. B. (1988). The use of caesium-137 measurements to investigate sediment delivery from cultivated areas in Devon, UK. In Bordas, M. P. and Walling, D. E. (eds), *Sediment Budgets, Int. Assoc. Hydrol. Sci. Pub.* No. 174, 325-35.

Walling, D. E. and Kane, P. (1982). Temporal variation of suspended sediment properties. In Walling, D. E. (ed.), *Recent Developments in the Explanation and Prediction of Erosion and Sediment Yield, Int. Assoc. Hydrol. Sci. Pub.* No. 137, 409-19.

Walling, D. E. and Kane, P. (1984). Suspended sediment properties and their geomorphological significance. In Burt, T. P. and Walling, D. E. (eds), *Catchment Experiments in Fluvial Geomorphology*, Geobooks, Norwich, pp. 311-34.

Walling, D. E. and Moorehead, P. W. (1987). Spatial and temporal variation of the particle size characteristics of fluvial suspended sediment, *Geografiska Annaler*, **69A**, 47-59.

Walling, D. E. and Moorhead, P. W. (1989). The particle size characteristics of fluvial suspended sediment: an overview, *Hydrobiologia*, 176/177, 125-149.

Walling, D. E. and Thornton, R. (1984). The role of suspended sediment in catchment nutrient budgets, *J. Sci. Food Agric.*, **35**, 856-7.

Walling, D. E. and Webb, B. W. (1981). Water quality. In Lewin, J. (ed.), *British Rivers*, George Allen and Unwin, London, pp. 126-69.

Walling, D. E and Webb, B. W. (1987). Material transport by the world's rivers: evolving prespectives. In Rodda, J. C. and Matalas, N. C. (eds), *Water for the Future: Hydrology in Perspective, Int. Assoc. Hydrol. Sci. Pub.* No. 164, 313-29.

Walling, D. E. and Webb, B. W. (1987). Suspended load in gravel-bed rivers: UK experience. In Thorne, C. R., Bathurst, J. C. and Hey, R. D. (eds), *Sediment Transport in Gravel-bed Rivers*, John Wiley, Chichester, pp. 693-732.

11 Magnitude and Frequency of Sediment Transport in Agricultural Catchments: A Paired Lake-catchment Study in Midland England

IAN FOSTER, ROB GREW and JOHN DEARING
Department of Geography, Coventry Polytechnic

INTRODUCTION

Sediment yields respond to both climatic and human impacts, yet the magnitude and type of response may vary considerably. Such response has been defined by geomorphologists in terms of 'Geomorphic effectiveness', although the concept of effectiveness has been applied to both sediment transport rates (Wolman and Miller, 1960) and, more recently, to the ability of an event or a number of events to induce landform response (Wolman and Gerson, 1978). The variability in response is, in part, a function of the scale of the basin but is also related to the relaxation time or sensitivity of the system to disturbance (Trudgill, 1976; Brunsden and Thornes, 1979).

Despite an increasing understanding of the dynamics of soil erosion on hillslopes (cf. Kirkby and Morgan, 1980) links between the hillslope and river channel remain less well understood. This problem was highlighted by Meade and Trimble (1974), who demonstrated continuing high sediment yields, in regions where catchment rehabilitation measures had been introduced, due to the reworking of sediment deposited in the valley bottom. Only recently has it been recognized that the dynamics of sediment delivery, storage and reworking need to be considered in order to place sediment yield data into a sounder theoretical context (Walling, 1988). As Dunne (1984, p. 6) points out: More emphasis needs to be placed upon sound conceptual models of how sediment is being transferred through the basin. . . .

Soil Erosion on Agricultural Land
Edited by J. Boardman, I. D. L. Foster and J. A. Dearing
©1990 John Wiley & Sons Ltd

In evaluating the 'off-farm' impacts of soil erosion, such as channel and reservoir sedimentation, an understanding of sediment delivery and transfer mechanisms is of vital significance. Furthermore, an understanding of sediment budgets will enable better understanding of eutrophication problems due to the close association between sediment and nutrient transport (Kronvang, 1990).

In addition to the problem of identifying sediment source linkages, it is important to place contemporary sediment dynamics in a historical context. This may be achieved by application of the lake-catchment framework, described by Oldfield (1977), which forms a logical extension to the unitary nature of the drainage basin (Chorley, 1966) and the watershed ecosystem model expounded by Bormann and Likens (1979).

Figure 11.1 shows the linkages, in terms of sediment flux and storage, between hillslopes, channels and the lake or reservoir. Any forcing variable, such as agricultural intensification, will be reflected in part by the response of the lake through a greater influx of sediment or through a change in water quality. However, the response of the lake sediments assumes that variations in sediment yield are not absorbed, for example, by a change in the sediment delivery ratio through an increase in valley sedimentation. Reconstruction of sediment yields through lake sediment studies will, therefore, only provide information on the outputs from a catchment system over time. However, analysis of the magnetic properties of the sediments and potential contributing sources may provide an opportunity to reconstruct some of the fluxes indicated in Figure 11.1 over different time periods.

Research conducted in two North Warwickshire experimental catchments aims to quantify particulate fluxes in contemporary and historical periods in order to reconstruct some of the sediment fluxes identified in Figure 11.1 and provide information on the magnitude and frequency of sediment transport over the last 200 years in cultivated and forested basins.

THE CATCHMENTS

Locational and site details for the Merevale Lake and Seeswood Pool experimental catchments are contained in Table 11.1 and Figure 11.2. The catchments were chosen for study on the basis of their similar morphology, geology and regional settings; both catchments also feeding reservoirs of similar size and depth. The catchments contrast significantly in terms of contemporary and historical land use which permits the use of the Merevale site, comprising semi-natural woodland, as a reference site in order to calculate background conditions for comparative purposes. Historical records reveal that this catchment has been forested since AD 1740, with the headwater region currently dominated by an oak woodland (*Quercus petraea*) planted c.1830–40. The lower half of this

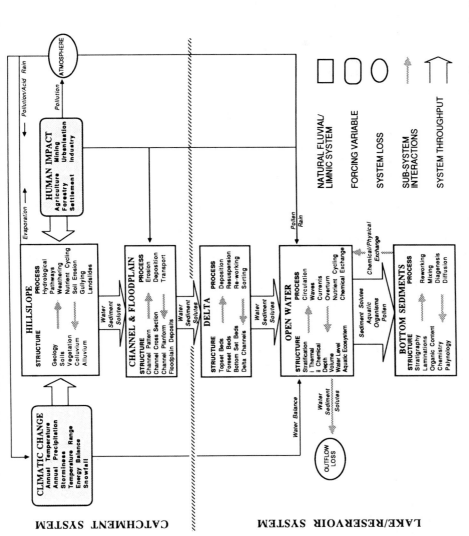

Figure 11.1 A conceptual model of sediment and solute transfer to lake and reservoir basins (after Foster et al., 1988)

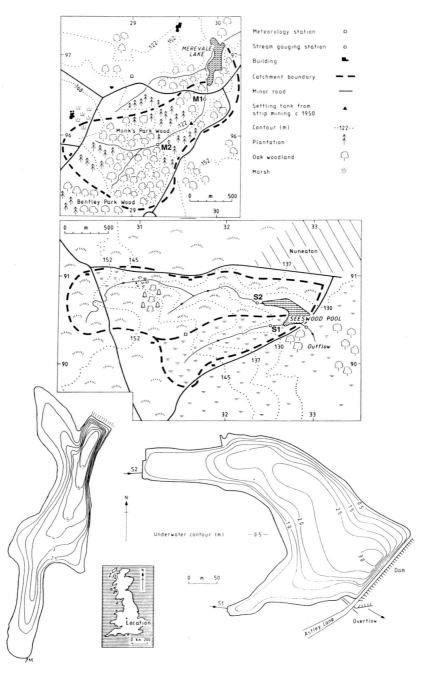

Figure 11.2 Site locations for studies of soil erosion and sediment transport in north Warwickshire

Table 11.1 Site characteristics of the Seeswood and Merevale catchments

Properties	Merevale	Seeswood
Grid ref.	SP300970	SP327905
Lake area (ha)	6.5	6.7
Total catchment area (ha)	201.0	238.0
Area above gauging stations		
M1	195.00	
M2	95.16	
S1		65.40
S2		161.00
Lake:catchment ratio	1:30	1:33
Max. altitude (m)	175	160
Min. altitude (m)	118	125
Relative relief (m)	57	35
Max. lake depth (m)	8	3.5
Mean lake depth (m)	4	1.5
Lake volume (m^3)	154 219	103 241
Date of impoundment	1838	1765
Land use (percentage area)		
Deciduous woodland	78.4	4.0
Coniferous plantation	7.5	0.0
Permanent pasture	14.1	55.0
Grass ley	0.0	11.0
Arable	0.0	30.0
Soil groups (percentage area)		
Stagnogleys	50.0	70.0
Argillic brown earths	15.0	20.0
Reclaimed	20.0	0.0
Other	15.0	10.0

watershed is planted with a range of commercial timbers including European Larch (*Larix europea*) and Scots pine (*Pinus sylvestris*). Outputs of water and sediment were monitored at M1 and M2 (Figure 11.2) to compare coniferous and deciduous woodland behaviour.

Seeswood Pool was constructed in the mid-eighteenth century as a feeder reservoir for the Coventry Canal, and is located in an area known to have been cultivated since Domesday times (twelfth century). At present, the southern tributary stream (S1, Figure 11.2) drains an area planted in a four-year rotation of winter wheat and oilseed rape, although more recently beans have been introduced into the rotation. The northern tributary (S2, Figure 11.2) drains an area mainly of improved pasture. The two lakes are of similar area, although Merevale Lake contains a deep narrow trench which, at a maximum, is some 4.5 m deeper than Seeswood Pool. The bathymetries are fairly simple and show

158 Soil Erosion on Agricultural Land

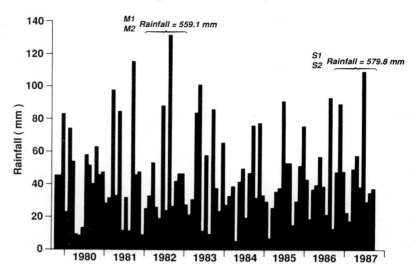

Figure 11.3 Distribution of monthly rainfall, 1979–87

Table 11.2 Rainfall–runoff relationships at Merevale and Seeswood

(a) Merevale 1981–2		
Rainfall (mm)	Runoff (M1) (mm)	Runoff (M2) (mm)
559.1	202.92 (36.3%)	201.61 (36.1%)
(b) Seeswood 1986–7		
Rainfall (mm)	Runoff (S1) (mm)	Runoff (S2) (mm)
579.8	264.00 (45.5%)	244.70 (42.4%)

both basins to reflect the surrounding ground contours (Figure 11.2). Because of its greater depth, however, Merevale Lake has a 50 per cent greater volume (Table 11.1).

The catchments are some 5 km apart and are at similar altitudes. They lie within a climatically homogenous region characterized by mean monthly January air temperatures of 3.1°C and mean monthly August air temperatures of 15.9°C. Rainfall records collected since 1979 at the Merevale site show monthly values ranging from less than 10 mm to over 130 mm with an annual average of 529.8 mm (Figure 11.3). Water balances for the four catchments are given in Table 11.2. At Merevale, the runoff ratio is 36 per cent for both sites. At Seeswood, under cultivated conditions, the ratio is around 45 per cent, whereas runoff from the grassland subcatchment is 42 per cent. The difference between

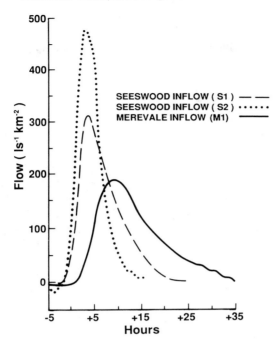

Figure 11.4 Nash Unit Hydrographs for a 10 mm storm of 1 h duration

the two catchments reflects the higher rate of evaporation expected under a forest canopy.

Analysis of storm period response by derivation of unit hydrographs from six to eight single peaked events at stations M1, S1 and S2 reveal important differences in behaviour (Figure 11.4). Time to rise for a 10 mm storm of 1 h duration is 13.1 h at Merevale, reducing to 6.1 h at S1 and 3.9 h at S2. The grassland-dominated S2 subcatchment has a much higher peak flow than the heavily underdrained and cultivated catchment of S1. Storm response at the Merevale catchment is dominated by delayed flow due to high permeabilities and infiltration rates of the soils (Foster *et al.*, 1983).

LONG-TERM SEDIMENT YIELDS

Long-term sediment yields have been reconstructed for both lakes. Multiple undisturbed sediment cores were retrieved from the lake bed using a modified 'Mackereth' corer. Over 70 cores were taken at the intersections of a 25 m^2 grid superimposed on Merevale Lake and over 30 cores at the intersection of a

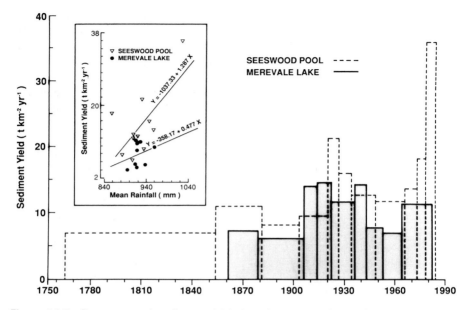

Figure 11.5 Reconstructed sediment yields based on an analysis of reservoir bottom sediments in Merevale Lake and Seeswood Pool. The inset shows the relationship between mean rainfall and sediment yield for the catchments (the sediment yields are derived from reservoir sediment analysis)

50 m² grid on Seeswood Pool. From both lakes, two cores were analyzed for ^{210}Pb (lead-210) and ^{137}Cs (Caesium-137) content. The ^{210}Pb chronology was calculated using the CRS model (Appleby and Oldfield, 1978) in both cases. The other cores were correlated with the dated master cores using a range of mineral magnetic properties which also show good correlation with the master radiometric chronologies. On the basis of the radiometric chronologies and magnetostratigraphic correlations, the lake sediment based estimates of sediment yield were divided into eleven and nine time periods for Seeswood Pool and Merevale Lake, respectively. These loadings are adjusted for atmospheric dust inputs in the local region for the most recent period and for autocthonous organic carbon and biogenic silica content which in part reflect non-catchment-derived denudational components (cf. Foster *et al.*, 1985; 1986).

Adjusted yields in Merevale Lake vary from less than 4 to over 9 t km^{-2} yr^{-1}, whereas in Seeswood Pool they increase from a pre-twentieth-century level of c.8 to over 36 t km^{-2} yr^{-1} in the most recent period. The patterns of sediment yield in Merevale Lake (Figure 11.5) show trends broadly similar to the pattern in Seeswood Pool except for the most recent time period and a significant reduction during the 1950s in Merevale Lake. This reduction is associated with a brief period of strip mining, when settling tanks were installed in the catchment by British Coal in order to minimize sediment transport from the workings. These are now full and the sediment yields have recovered to pre-extraction levels. The most recent time period at Seeswood Pool is associated with very high yields which occurred at a time of extreme annual rainfall in comparison with the past. Attempts to model sediment yield–rainfall relationships for the historical period were based upon correlation of sediment yields with average annual rainfall based on the long-term rainfall series produced by Wigley *et al.* (1984) (Figure 11.5 inset). These relationships are only weakly significant (5 per cent level at Seeswood and 10 per cent level at Merevale), but would indicate that, in the long-term record, sediment movement is limited more by energy availability than sediment supply to the transporting rivers.

The history of land management has also been established for the Seeswood Pool catchment (Owen, 1985), which indicates that since 1765, field boundary lengths have reduced by some 23 per cent, tile drainage has been installed throughout the catchments, particularly in the S1 sub-basin, and that farming practice has shifted from extensive to intensive production. Intensive arable production in the most recent period is found in the southern basin and increased cattle populations in the north.

SEDIMENT YIELD AND TRANSPORT CHARACTERISTICS OF RIVERS

Sediment yields from the inflowing rivers have been calculated on the basis of two-hourly turbidity meter records and continuous-flow records derived from stage measurements and stage discharge relationships established in the field. The sediment yields are given in Table 11.3. Sediment yields from the oak woodland and subcatchment of Merevale are low, whereas the total output of sediment from the basin approaches $10 \, t \, km^{-2} \, yr^{-1}$ and is remarkably similar

Table 11.3 Sediment yields from the catchments and from recent estimates of reservoir sedimentation

Sediment yields ($t \, km^{-2} \, yr^{-1}$)			
Seeswood Pool		Merevale	
S1	S2	M1	M2
8.7	68.9	10.3	6.4
Sediment Inflow to lakes ($t \, km^{-2} \, yr^{-1}$)			
Seeswood Pool		Merevale Lake	
(a)	(b)	(a)	(b)
77.6	36.2	10.3	10.3

(a) Estimated from river inflow.
(b) Estimated from lake sedimentation.

162   Soil Erosion on Agricultural Land

to the yields calculated from reservoir sedimentation in the most recent time period. In contrast, the total inflow of sediment from both subcatchments at Seeswood Pool is almost double the estimates from the most recent period of reservoir sedimentation. This difference appears to relate to the recent development of a deltaic deposit at the inflow of the S2 catchment to the reservoir. A rise in water level since 1982 has induced deltaic sedimentation which, from recent survey estimates, currently holds a sediment mass equivalent to around 10 years' outflow from the subcatchment. Of greater significance

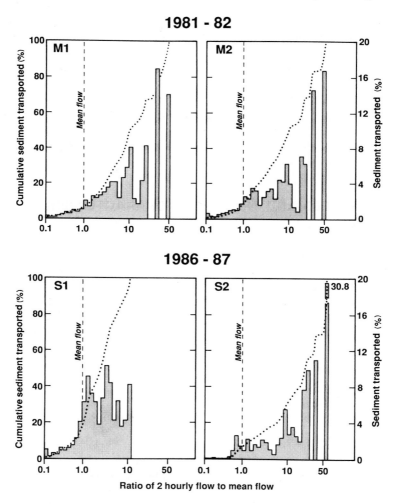

Figure 11.6 Magnitude and frequency of suspended sediment transport at Merevale and Seeswood. Data are plotted as frequency in flow class (histogram) and cumulative frequency (dotted line)

is the fact that the yield from the intensively cultivated subcatchment of S1 is similar in magnitude to the yield of the forested Merevale site.

Analysis of the magnitude–frequency characteristics of sediment transport at all four gauging stations (Figure 11.6) highlights some important differences. At Merevale and at S2, the amount of sediment transported increases as the ratio of two-hourly flow to mean annual flow increases. Most sediment in these systems is transported at flows above a ratio of 20:1. The transport of sediment at S1 appears to be limited at high flows where the ratio of two-hourly to mean annual flow rarely exceeds 10:1 and where, in ratios above 8, sediment transport does not exceed 8 per cent of total transport. Analysis of the amount of sediment transported in storms of over 5 mm reveals a similar pattern of response. At Merevale, around 64 per cent of sediment is transported in these storms. At S2, this figure increases to 86 per cent, whereas at S1 only 58 per cent of sediment is transported in runoff from storms exceeding 5 mm. As will be shown later,

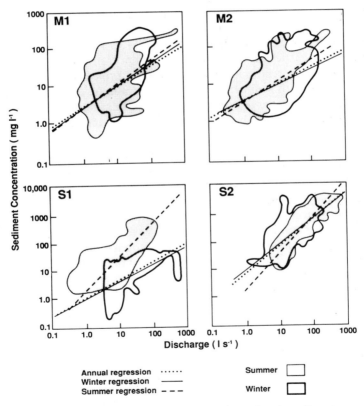

Figure 11.7 Scattergrams of the relationship between two-hourly sediment concentration and discharge for the four gauging stations. Envelope curves show the scatter of the 4368 data points divided into summer and winter seasons

Table 11.4 Regression and correlation analysis for two-hourly streamflow and sediment concentration data analysed by season

Site	Total regression			Summer regression			Winter regression		
	(a)	(b)	r^2	(a)	(b)	r^2	(a)	(b)	r^2
S1	0.82	0.24	11.3	0.25	1.04	36.0	0.64	0.31	25.5
S2	0.91	0.58	23.0	0.28	1.13	37.6	1.08	0.50	25.1
M1	0.48	0.35	15.0	0.48	0.36	14.3	0.43	0.38	10.3
M2	0.46	0.38	25.8	0.46	0.51	40.9	0.52	0.29	8.0

(a) and (b) from $S_c = 10^a \cdot Q^b$
S_c = Sediment concentration
r^2 = Coefficient of determination.
All correlations significant at 99.9%.
$n = 4368$ for total data set.

the extremely high yields from S2 reflect the way in which sediment is delivered to the channel system and becomes incorporated into the estimate of sediment yield.

Correlation and regression analysis (Figure 11.7, Table 11.4) reveals a similar disparity between the characteristics of sediment transport at S1 and the other basins. The winter relationship at S1 shows a typical exhaustion effect (cf. Walling, 1974), with sediment concentrations almost an order of magnitude lower for a given discharge than in the summer relationship. In contrast, there is no significant difference between the winter and summer regression equations for other rivers.

The river monitored at S1 is different from those monitored at other stations for three major reasons. First, flow is intermittent, and the stream rarely flows between the end of April and the end of September. The summer database for regression analysis, therefore, is limited. Second, during periods of no flow, a semi-continuous vegetation cover (mainly grasses) develops in the channel bed. It is thought that this in-channel vegetation cover reduces sediment availability in the subsequent autumn period and the amount of sediment transported at this time of the year. Third, most arable fields in the S1 catchment have a broad (1 m wide) grass strip separating the cultivated slopes from the channel. It is suggested that this strip acts as an effective buffer zone to the transport of hillslope-derived sediment to the river.

SEDIMENT SOURCES

Recent research has been concerned with identifying contemporary sediment dynamics by monitoring sources and sinks within the present drainage basin network at Seeswood Pool. A number of sources contributing directly to the channel have been identified:

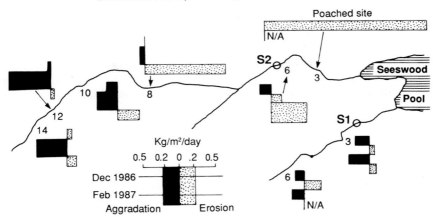

Figure 11.8 Sediment budgeting for selected sites in the Seeswood catchment based on sediment trap and channel resurvey

(1) Adjacent cultivated hillslopes;
(2) Direct runoff from tractor wheelings and trampled areas;
(3) Erosion of stream banks;
(4) Reworking of sediment stored in river channels.

The hillslope subsystem undoubtedly possesses its own temporary sinks but these have yet to be investigated in detail.

The monitoring programme was set up for the year 1986–7 in order to identify and classify potential sediment sources, select representative sections for detailed survey and periodically remeasure erosional and depositional processes in order to calculate a sediment source budget. Methods of analysis included remeasurement of erosion pins located in a grid network over potentially eroding sites and the remeasurement of fixed cross-sections in the channel network. At some sites, sediment traps were installed to collect material either moving as slopewash or delivered directly to the channel by foraging animals.

Results show a complex pattern of erosion and deposition at all sites (Figure 11.8). Aggradation was recorded on many sites from December 1986 to February 1987, when material was released to the channel system by freeze–thaw processes operating upon unvegetated channel banks. This material was incorporated into flow in two different ways. With a frozen river surface, and with vertical banks, sediment was liberated into the channel immediately upon thawing. On complex banks, temporary sediment stores awaited the next rise in river stage to a level where material could be eroded and transported. In-channel vegetation, especially grasses, were found to reduce sediment availability and were thought to be responsible for sediment concentration depletion during successive storm events observed in the autumn period at S1.

Estimates were made of the contribution of these different sources to the total sediment budget of the contemporary river systems at sites S1 and S2. Bank collapse processes contributed only 5.5 t of sediment to S1 but some 195 t of sediment to the S2 sub-basin. In addition to stream bank erosion, sites of eroding lake margins were measured by both erosion pin and resurvey methods. The results indicate an additional and direct contribution to the lake of some 13 t of sediment by this process. Comparison with the most recent lake sediment-based estimate of sediment yield suggests that the sediment delivery ratio for the Seeswood catchment ranges between 37 and 39 per cent.

In order to extend sediment source tracing to past environments, attempts have been made to compare the mineral magnetic properties in lake sediments with potential sediment sources (Dearing *et al.*, 1986; Dearing and Foster, 1987; Foster *et al.*, 1988). Both specific susceptibility and frequency-dependent susceptibility increase towards the surface of soil profiles, indicating the presence of secondary pedogenic ferrimagnetic minerals. These minerals are concentrated in the clay and fine silt fraction. Two attempts have been made to produce a magnetic mixing model for the Seeswood Pool catchment (Figure 11.9 (a) and (b)). The first is based upon frequency-dependent magnetic susceptibility, which has high levels in the topsoil and in the upper 30 cm of lake sediment cores, indicating an increase in topsoil movement since the mid-twentieth century. Plotting influx against sediment yield provides a basis for comparing the magnetic properties of lake sediments with those of soils and stream margin-derived material. The model of Figure 11.9(a) suggests that sediment in the lake has not normally been dominated by either topsoil or stream channel material but rather a combination of the two sources. It is also not possible at this site to distinguish between topsoils derived from overland flow processes and that topsoil which enters rivers from collapse of stream bank material which shows characteristics of magnetic susceptibility enhancement towards the surface.

A more refined model uses the *S*-ratio, a qualitative guide to mineral magnetic assemblage, and SIRM, a quantitative measure of the concentration of ferrimagnetic minerals (Dearing *et al.*, 1986). The two irregular envelopes show the distribution of topsoil and subsoil plots superimposed upon values for lake sediments of different ages. The two dotted lines represent the values derived from the mixing of different proportions of topsoil and subsoil in the laboratory. Comparison of overlaps suggests that topsoil has consistently made a greater than 50 per cent contribution to sediments since 1854. Only the earlier period, 1765–1853, shows a smaller topsoil contribution.

Although the topsoil contribution appears high, especially during the 1960s when it represented between 60 and 80 per cent of total sediment in the lake, this second mixing model is again unable to distinguish between slope-derived topsoil and channel margin topsoil contributions.

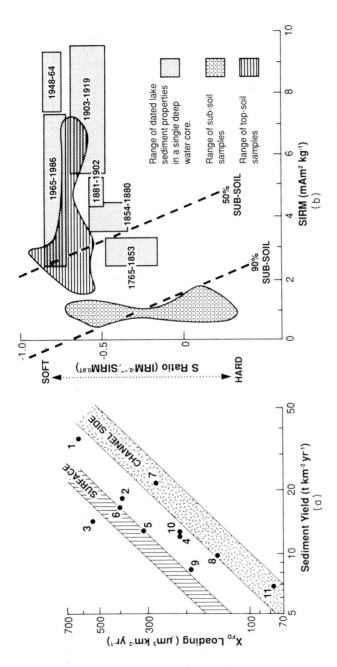

Figure 11.9 Magnetic mixing models for sediment sources and lake sediments. (a) Based on frequency-dependent magnetic susceptibility. Sediment yield influx (X axis) is plotted against the influx of minerals carrying a frequency-dependent signal (Y axis). Shaded areas represent the expected relationship if all material were derived from either channel margins or topsoil. Number of plotted points of the lake sediments represent time zones dated in the lake sediments from oldest (time zone II) to youngest (time zone I). (b) Based on SIRM and the S-ratio. The plot contains field samples of soils and lake sediments and includes parallel lines depicting 50 per cent and 90 per cent subsoil. These are derived from analysis of laboratory-prepared soil mixtures to identify the expected range of SIRM and S-ratio data

A combination of magnetic mixing models and contemporary source area tracing has been used to construct the sediment source area model of Figure 11.10. This illustrates significant changes in source area through time affecting sediment yields entering the lake. In the early period, yields were around 15 t yr^{-1}, with a higher contribution from S2 than S1. The recent period also shows a dominance of sediment derived from the grassland catchment, with most sediment derived from cattle poaching adjacent to river channels. Little change in yield is found in the southern tributary despite the fact that this area is currently under intensive cultivation.

These findings are important in relation to the interpretation of sediment yields and to an understanding of the magnitude and frequency of sediment transport. Increased sediment yields appear to relate less to changes in the hillslope subsystems under cultivation since little opportunity arises for these sediments to reach the river channel. There is undoubtedly a separate set of processes operating in these hillslopes, and the link between sediment source and sediment yield is analagous to the partial area concept in hillslope hydrology. We suggest, therefore, that sediment transport to the channels will only occur in this catchment when there is a direct link between the sediment and hydrological partial contributing areas. Of particular importance at S1 is the presence of a narrow riparian buffer zone along the channel margin, comprising a narrow (up to 1 m wide) undisturbed grassland strip.

DISCUSSION AND CONCLUSIONS

The magnitude and frequency of sediment transport has been examined by comparing the behaviour of forested and cultivated basins in Midland England and the long-term sediment yields derived from lake sediment-based reconstruction. It is clear from the analysis presented above that the dynamics of sediment transport and the product of that transport, the sediment yield, reflects the availability of energy to move the sediment and the existence of opportunities for linking sediment and hydrological partial contributing areas. Indeed, it could be argued that the opportunities for linking the hillslope to channel system are far more important controls on sediment transport than the amount of energy available. This is clear from an analysis of the dynamics of the S1 sub-basin, where the narrow grassland buffer zone prevents the transfer of available sediment from the cultivated fields and is partly responsible for the discernable exhaustion of sediment supply in the magnitude–frequency and regression analysis of Figures 11.6 and 11.7. In contrast, opportunities for transferring sediment to the channel at S2 are provided by intensive poaching by cattle and their physical delivery of sediment to the channel at all flow stages. At Merevale, undercutting, bank collapse and subaerial weathering processes appear to maintain an adequate supply of sediment for fluvial transport in both

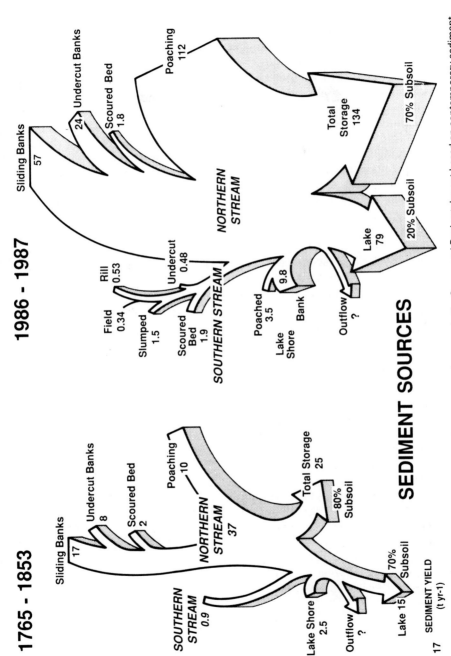

Figure 11.10 Historical and contemporary sediment yields and sources in the Seeswood Pool catchment based on contemporary sediment source monitoring and magnetic mixing models of Figures 11.8 and 11.9

summer and winter periods and both basins provide little evidence of sediment exhaustion. Indeed, it has been calculated that the entire channel system at Merevale contains sufficient sediment to maintain sediment yields at $10 \text{ t km}^{-2} \text{ yr}^{-1}$ for around 30 years.

There is clear evidence from these studies that traditional magnitude–frequency concepts should be re-examined, since sediment yields are not necessarily a direct function of increasing rainfall or runoff. Responses measured in contemporary fluvial systems are clearly complex, with sediment yields in one of the four basins being controlled more by sediment supply than runoff.

The much higher sediment yields in the Seeswood Pool catchment run contrary to expected behaviour in many agricultural ecosystems, yet it must be stressed that the results presented here deal with sediment transport in rivers rather than soil movement on hillslopes. Nevertheless, it is important to stress the significance of these findings in understanding and predicting the movement of soil and associated pollutants from agricultural systems, since this study shows that there is no simple relationship between land use and sediment yield. Furthermore, this analysis underlines the effectiveness of rather simple control measures (the maintenance of a narrow grass strip) and the importance of keeping cattle away from highly sensitive areas such as channel margins.

REFERENCES

Appleby, P. G. and Oldfield, F. (1978). The calculation of lead 210 dates assuming a constant rate of supply of un-supported lead 210 to the sediment, *Catena*, **5**, 1–8.

Bormann, F. H. and Likens, G. E. (1979). The watershed ecosystem concept and studies of nutrient cycles. In Van Dyne (ed.), *The Ecosystem Concept in Natural Resource Management*, Springer-Verlag, New York.

Brunsden, D. and Thornes, J. B. (1979). Landscape sensitivity and change, *Institute of British Geographers, Transactions; New Series*, **4**, 463–84.

Chorley, R. J. (1969). The drainage basin as the fundamental geomorphic unit. In Chorley, R. J. (ed.), *Water Earth and Man*, Methuen, London.

Dearing, J. A. and Foster, I. D. L. (1987). Limnic sediments used to reconstruct sediment yields and sources in the English Midlands since 1765. In Gardiner, V. (ed.), *International Geomorphology*, 1986. Part 1, John Wiley, Chichester, pp. 853–68.

Dearing, J. A., Morton, R. I., Price, T. W. and Foster, I. D. L. (1986). Tracing movements of topsoil by magnetic measurements: two case studies, *Physics of Earth and Planetary Interiors*, **42**, 93–104.

Dunne, T. (1984). The prediction of erosion in forests. In O'Loughlin, C. L. and Pearce, A. J. (eds), *Symposium on the Effects of Forest Land Use on Erosion and Slope Stability*, Environmental Policy Institute, University of Honolulu, Hawaii.

Foster, I. D. L., Carter, A. D. and Grieve, I. C. (1983). Biogeochemical controls on river water chemistry in an afforested catchment, *Hydrological Sciences Publication*, **141**, 241–53.

Foster, I. D. L., Dearing, J. A., Simpson, A. D., Carter, A. D. and Appleby, P. G. (1985). Lake catchment based studies of erosion and denudation in the Merevale Catchment, Warwickshire, UK, *Earth Surface Processes and Landforms*, **10**, 45–68.

Foster, I. D. L., Dearing, J. A. and P. G. Appleby (1986b). Historical trends in catchment sediment yields: a case study in reconstruction from lake sediment records in Warwickshire, UK *Hydrological Sciences Journal*, **31**, 427–43.

Foster, I. D. L., Dearing, J. A. and Grew, R. (1988). Lake catchments: an evaluation of their contribution to studies of sediment yield and delivery processes, *Hydrological Sciences Publication*, **174**, 413–24.

Kirkby, M. J. and Morgan, R. P. C. (1980). *Soil Erosion*, John Wiley, Chichester.

Kronvang, B. (1990). Sediment associated phosphorus transport from two intensively farmed catchment areas. In Boardman, J., Foster, I. D. L., and Dearing, J. A. (eds), *Soil Erosion on Agricultural Land*, John Wiley, Chichester.

Meade, R. H. and Trimble, S.W. (1974). Changes in sediment loads in rivers of the Atlantic drainage of the United States since 1900, *International Association of Hydrological Sciences Publication*, **113**, 99–104.

Oldfield, F. (1977). Lakes and their drainage basins as units of sediment-based ecological study, *Progress in Physical Geography*, **3**, 460–504.

Owen, C. L. (1985). *Lake Sediment Based Erosion Studies at Seeswood Pool: an Appraisal of Historical Data Sources*, Unpublished undergraduate dissertation, Coventry Polytechnic, Coventry, UK.

Trudgill, S. T. (1976). Rock weathering and climate: quantitative and experimental aspects. In Derbyshire, E. (ed.), *Geomorphology and Climate*, John Wiley, Chichester.

Walling, D. E. (1974). Suspended sediment and solute yields from a small catchment prior to urbanisation, *Institute of British Geographers Special Publication*, **6**, 169–92.

Walling, D. E. (1988). Erosion and sediment yield research—some recent perspectives, *Journal of Hydrology*, **100**, 113–41.

Wigley, T. M. L., Lough, J. M. and Jones, P. D. (1984). Spatial patterns of precipitation in England and Wales and a revised, homogenous England and Wales precipitation series, *Journal of Climatology*, **4**, 1–25.

Wolman, M. G. and Miller, J. C. (1960). Magnitude and frequency of forces in geomorphic processes, *Journal of Geology*, **68**, 54–74.

Wolman, M. G. and Gerson, R. (1978). Relative scales of time and effectiveness of climate in watershed geomorphology, *Earth Surface Processes*, **3**, 189–208.

12 Recent and Long-term Records of Soil Erosion From Southern Sweden

J. A. DEARING
Department of Geography, Coventry Polytechnic

K. ALSTRÖM and A. BERGMAN
Department of Physical Geography, University of Lund

and

J. REGNELL and P. SANDGREN
Department of Quaternary Geology, University of Lund

INTRODUCTION

Most field studies of soil erosion on agricultural land are based on records of slope processes or stream sediments spanning relatively short timescales of days to years. In combination, the results from these studies provide much detail about factors and processes of erosion at different spatial scales and over different periods of time, and may also provide inputs into predictive models. However, in terms of analysing sediment movements within an entire catchment, rather than just on slopes, there are two major problem areas which lie in the path of attempts to produce and to test empirical models which are conceptually strong.

First, there is the problem of identifying the response to changes in climate or land use in systems which may have a strong interdependence of successive erosional events. In this sense, short records may be subjected to analyses of magnitude and frequency, but the lack of direct information over long timescales of decades to millennia inevitably introduces a large degree of uncertainty and subjective assessment: there is no assurance that the available record can identify the important threshold conditions or long-term reactions and recoveries (cf. Kelsey, 1982). Second, there is the problem of linking the results of slope process monitoring with those from basin-wide and stream-based sediment yield estimation. Assessment of sediment delivery to the channels throughout a catchment is not easily undertaken by study of a few selected slope segments and, as yet, there is no widely applicable technique for tracing the specific slope or soil origins of trapped suspended sediment in a stream.

The study described here attacks problems of long-term responses and subsystem linkages within a lake-catchment framework in, what is today, an intensively cultivated area of southern Sweden. The framework encompasses calculations of sediment yield from analyses of lake sediments over the timescale 10 000–0 BP, data for contemporary slope processes, historical land-use data and magnetic appraisal of changes in the dominant sediment sources. The objectives of the study are as follows:

(1) To construct a record of sediment yield over the period 10 000–0 BP;
(2) To construct a record of sediment source (topsoil and subsoil/channel material) over the period 10 000–0 BP;
(3) To compare the recent records of sediment yield and sources with contemporary data for slope processes on cultivated land;
(4) To compare the long records with data for land-use change in order to infer the controls on sediment movements;
(5) To analyse the combined sediment data for responses and recoveries to environmental change over different timescales.

Previous papers (Dearing *et al.*, 1987; Dearing, in press) described some of the results presented here in terms of the influence of past agricultural societies and erosion within the region. This chapter utilizes an improved chronology which has been both extended and refined, and includes new information about the history of land management and drainage practices.

STUDY AREA

The catchment of Lake Bussjösjö is situated in southern Skåne, in the south of Sweden (Figure 12.1). The topography of southern Skåne is undulating and hummocky, with Quaternary deposits ranging from 20 m to 60 m in thickness. The stratigraphy of the area is highly complex, made up of several till units deposited from different ice-streams interbedded with fluvial and glacio-fluvial deposits. The topography has taken its form partly from these older deposits, but is also the result of dead-ice left during the retreat of the last ice-stream. The surficial till is made up of Cretaceous and Cambro-Silurian erratics, reflecting the local bedrocks of the area. The till also contains igneous and metamorphic erratics representing the last ice-stream which moved from the north. The till is largely calcareous with a high clay content (clayey to clay tills), with organic deposits lying in depressions.

Records of Soil Erosion from Southern Sweden 175

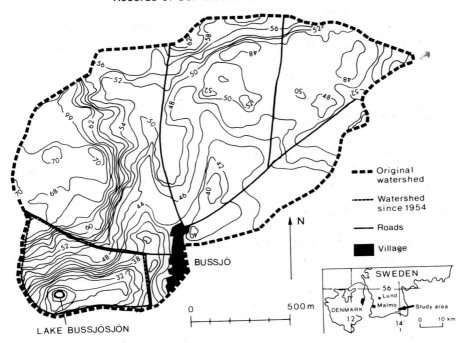

Figure 12.1 Bussjösjön; site and location (inset), showing the lake and catchment boundaries (since 1954 and before 1954) and topographic contours

Overall, the soils are fine-textured (loams to clay loams) brown earths or calcareous brown earths with a high natural fertility. Today they are intensively cultivated, a practice promoted by a long history of undersoil drainage which recently has included the burial of natural streams into wide-diameter concrete pipes.

The catchment which drains into Lake Bussjösjön (Figure 12.1), as defined by the natural topography, covers an area of $c.198$ ha. The lake has a present-day surface area of 0.2 ha, though in the last century the area was $c.0.7$ ha and the lake area in early post-glacial times may have been over 2 ha (Thuning and Linderson, 1986). Changes in the drainage system have reduced the effective surface drainage area to the lake to $c.24$ ha and, today, water enters the lake from two short pipes, from groundwater and by overland flow from the surrounding slopes.

The relative relief of the catchment (198 ha) above lake level is $c.46$ m, but the steepest slopes are found close to the northern edge of the lake. There, mean slope angles are 6.5° over a distance of 300 m, with the steepest slopes over shorter distances reaching 9–10°. Today the catchment (198 ha) is dominated by arable farming with over 90 per cent of the farmland put down to cereals, oilseed rape, sugar beet and leguminous fodder crops. Less than 10 per cent of farmland is put down to ley. There is one small settlement in the catchment, the hamlet of Bussjö.

LAND USE AND DRAINAGE DURING THE NINETEENTH AND TWENTIETH CENTURIES

Land-use records (Figure 12.2) show that the area of arable farming was relatively low prior to 1800, but increased rapidly during the nineteenth century to reach present-day levels by 1900. The results of a more detailed study of the changes in drainage and land use, using maps from 1798 and aerial photographs from 1939 and 1985, are shown in Figures 12.3 and 12.4. In the beginning of the nineteenth century more than 20 per cent of the catchment area was meadow or wetland (Figure 12.3); all the streams, a total length of $c.2$ km (Figure 12.4), were surrounded by meadows. A major change of the drainage system started in about 1870, when some of the streams were straightened and deepened. In 1900 most of the wetlands were drained and the area of wetland and meadows decreased to less than 8 per cent of the catchment area. New ditches were dug to drain the wetlands, and the total length of streams increased to its maximum ($c.3.5$ km), where a large proportion of the stream length was adjacent to arable land. The meandering part of the stream near the lake was straightened and deepened $c.1900$.

During the twentieth century the drainage system passed through another major change when almost all the streams and ditches were piped. In 1939 the total stream length had decreased to $c.1.5$ km with 50 per cent of the length in arable land. In 1954 the main stream was rerouted past the lake around its southern edge in covered pipes. This action, coupled with road-building upstream, reduced the effective surface drainage area to the lake to $c.24$ ha.

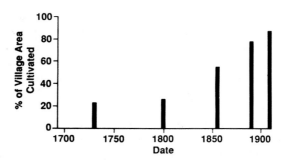

Figure 12.2 Arable land in the Bussjösjön catchment, 1700–1910, expressed as a percentage of total agricultural land (based on land-use documents)

Records of Soil Erosion from Southern Sweden 177

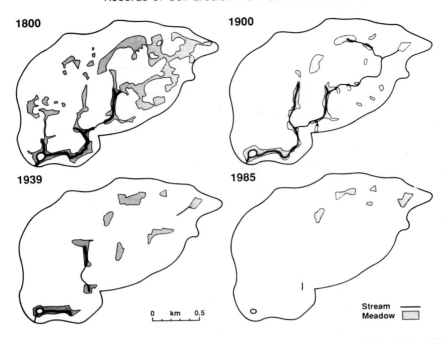

Figure 12.3 Streams, ditches, wetlands and meadows in the catchment, 1800, 1900, 1939 and 1985 (based on maps from 1798 onwards, aerial photographs from 1939 and 1985, and interviews with farmers)

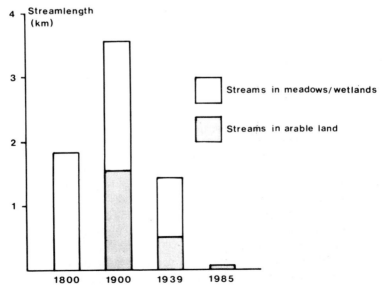

Figure 12.4 Total length of surface streams in arable land, meadows and wetlands, 1800, 1900 1939 and 1985 (sources of data as for Figure 12.3)

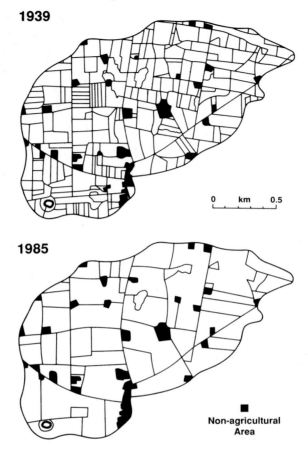

Figure 12.5 Field boundaries and settlement in the catchment, 1939 and 1985 (based on aerial photographic interpretations)

In 1985 the meadows and wetland had declined to only 2 per cent of the original catchment area. Today there is only 100 m of open stream, and this has no connection with the lake.

In the twentieth century the intensification of agriculture has led to changes in field size. Over the catchment there has been a reduction in field units from over 200 in 1939 to c.50 in 1985 (Figure 12.5), thus significantly increasing the maximum slope length within one field (normal to contours) from less than 100 m to more than 300 m. According to local farmers, the major changes in field size took place in the early 1960s. This implies that the slopes surrounding the lake are more susceptible to erosion today than ever before, but that the area contributing sediment to the lake is very small because of the absence of surface streams.

In contrast, the maximum total length of streams within the catchment was reached at the beginning of the twentieth century. This implies that there was an enlarged potential contributing area for slope-derived material, and potentially high channel and stream bank erosion at that time.

CONTEMPORARY EROSION

A partial assessment of erosion on slopes was made on the relatively steep slopes to the north and west of the lake, and measurements were made through two winter/spring seasons, 1986/7 and 1987/8. Details of the methods are presented in Alström and Bergman (1988a).

The sheet erosion was measured by trapping the eroded soil in metal troughs. Rill erosion was quantified by surveying the volumes of rills and sedimentation fans. The total mass of soil transported to the edge of the lake is calculated as the difference between the total volume of eroded soil and the volume of material in the fans, corrected for density. The results are presented in Table 12.1. The erosion was very severe during the first monitored season, with c.218 t of material transported to the edge of the lake. During the subsequent season the erosion rate was considerably lower (c.4 t). Rill erosion was the most significant process during both years.

The fields were ploughed and left bare in the same way during both periods, which suggests that the difference in soil losses was a result of climatic factors. The winter 1986/7 was cold with snow, and the soil was frozen to a depth of more than 1 m. The winter 1987/8 was milder with no snow or frozen ground. In 1986/7, 99 per cent of the sediment was trapped during two periods of a week each in the spring. On both these occasions there were similar climatic

Table 12.1 Sediment losses by water 1986/7 and 1987/8 (after Alström and Bergman, 1988)

		1986/7 (kg)	1987/8 (kg)
Sheet erosion	November	16	18
	December	16	12
	January	0	20
	February	889	16
	March	14 754	1
	April	3	0
	Total	15 678	67
Rill erosion	Nov.–April	208 215	4 950
Sedimentation in the field		6 150	600
Total transport of soil to the edge of the lake		217 743	4 417

conditions; a sudden thaw had led to a few centimetres of saturated soil overlying a frozen subsurface, which was then exposed to meltwater flow and a moderate rainfall. The total rill erosion caused by these events was not measured, but discharges were observed to reach $10 \, \text{l s}^{-1}$ transporting $c.0.06 \, \text{kg s}^{-1}$. On these slopes the eroded sediment was transported to the flatter slopes adjacent to the lake. In contrast, the total sediment losses on the same slopes in the period January to April 1988 were $c.0.2$ per cent of the losses during the same period in 1987, even though the precipitation was twice as high.

LONG RECORDS OF SEDIMENT YIELDS AND SEDIMENT QUALITY
Sediment Yields 1700–1985

Recent sedimentation rates in Bussjösjön were calculated from dated changes in the pollen record and from stratigraphic changes, whose dates were assumed to match those of recent drainage events in the catchment. The preferred means for dating recent sediments, through lead-210 and caesium-137 (^{210}Pb and ^{137}Cs analyses), has not been successful yet because of the high sedimentation rate. The level where Brassica pollen increases is dated to $c.1945$, the first time oilseed rape was locally grown, and two clay layers are assumed to correspond to the excavation of drains in 1955 and to a large flood caused by a blocked drain in 1960. A date of 1900 is given to the rise in organic matter and $CaCO_3$ at a depth of 80 cm, a feature which is recorded in other lakes and dated by ^{210}Pb analyses (Regnell, 1989). Intermediate and older dates were calculated by linear interpolation between dated points in the pollen record, with the exception of clay layers which were assumed to represent one year (maximum) of accumulation. This chronology is an improved version of that described by Dearing (in press), and consequently the sediment yields presented here are different in their absolute values, though the trends remain much the same.

Accumulation rates in the central core BU4 were transformed into values of minerogenic accumulation ($\text{g cm}^{-2} \text{yr}^{-1}$) by adjustment for dry bulk density and the content of organic matter (% LOI at 550°C) and carbonates (% LOI between 550°C and 925°C). The accumulation rate of minerogenic sediment was multipled by the lake area and lake:catchment ratio to produce figures for sediment loading to the lake (t yr^{-1}) and sediment yield from the catchment ($\text{t ha}^{-1} \text{yr}^{-1}$). Such calculations are valid if the accumulation rates used are the mean values across the bed of the lake. Correlation of 7×1 m sequences to core BU4, using magnetic measurements, showed that the accumulation rates over the past century have not differed greatly within the area of open water: BU4 has accumulated at a slightly higher rate than the mean, and sediment yields are therefore maximum estimates. Other sediment parameters were expressed on influx bases by multiplying the minerogenic mass–specific property by the minerogenic accumulation rate.

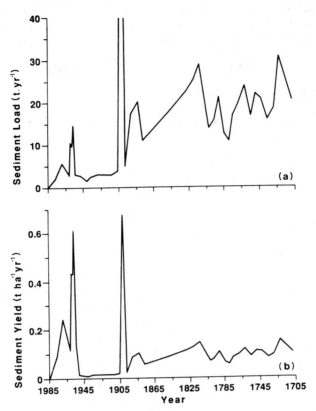

Figure 12.6 (a) Loading of minerogenic sediment to the lake (t yr^{-1}), 1700–1985; (b) record of minerogenic sediment yield from the catchment (t ha^{-1} yr^{-1}), 1700–1985

Figure 12.6 shows the results of sediment load and sediment yield. The record of sediment load (Figure 12.6(a)) shows relatively high though fluctuating values, of between 10 and 32 t yr^{-1} before 1900. In the twentieth century, the values are lower (around 2–3 t yr^{-1}), apart from two influx events c.1900 and c.1955; there is a peak dated to c.1970 and the trend since then has been downwards. Both major influx events are assumed to have been caused by drainage operations.

These values represent the loading of sediment to the lake from the catchment but do not take into account the area of the contributing catchment. The Bussjösjön catchment is difficult to define today because of the much-reduced surface drainage network. In calculating sediment yield, areas of 24 ha and 198 ha have been used for post-1954 and pre-1954 catchments, respectively. Figure 12.6(b) shows that yields before 1880 lie between values of 0.05 and 0.15 t ha^{-1} yr^{-1}; the values drop to c.0.02 t ha^{-1} yr^{-1} during the period

1900–55 (except for the massive influx c.1900), and rise to a peak value of 0.24 t ha^{-1} yr^{-1} in 1965–70, with a major influx event of c. 0.8 t ha^{-1} yr^{-1} in 1955. Since 1970, sediment yields have fallen steadily to values of less than 0.05 t ha^{-1} yr^{-1} in the 1980s, lower than yields in the eighteenth century.

Sediment quality 1700–1985

Identification of both the sources of eroded sediment and the type of erosional processes may be inferred through comparisons of the mineral magnetic properties of lake sediment and soil. Table 12.2 shows selected mean magnetic data for soils in the catchment. Measurements on 35 samples, comprising different levels in 20 soil profiles, indicate the widespread occurrence of secondary ferrimagnetic minerals (cf. Dearing et al., 1986). These minerals (ultrafine magnetite and maghemite) may be detected by the measurement of frequency-dependent magnetic susceptibility (X_{fd}) and, as Table 12.2 shows, their distribution varies with both depth and particle size. Values are marginally higher in bulk topsoils (0–10 cm) than subsoils (40–50 cm), but analysis of different fractions shows a high concentration in the clay fraction. Sediment trapped from flowing rills has values even greater than those found in the clay fractions of topsoils, perhaps indicating the transport of particles finer than 2 μm. Samples of unweathered parent material from below 100 cm in the profiles show magnetic properties typical of antiferromagnetic minerals (goethite and hematite) with high levels of magnetic remanence in a back field of 100 mT after forward saturation (HIRM). The manner in which magnetic properties distinguish between topsoil and the original glacial deposits is most clearly shown by the ratio HIRM/X_{fd} (Table 12.2): values for bulk samples increase with depth, and are lower in the clay fractions.

Figure 12.7 displays the record of X_{fd} and HIRM/X_{fd} values in the sediment record. Before c.1850 the X_{fd} values, expressed per gram of minerogenic material (Figure 12.7(a)), fluctuate but lie within a range of between 15 and 50 nm^3 kg^{-1}. After this time the values fluctuate more widely, with extreme values of 2 and 60 nm^3 kg^{-1} within the last 80 years. It is noticeable that values

Table 12.2 Magnetic properties of soils and rill sediments (after Dearing, in press)

Soil depth (cm)	X_{fd} nm^3 kg^{-1}			HIRM/X_{fd} kA m^{-1}	
	Clay <2 μm	Coarse silt 32–63 μm	Bulk <2000 μm	Clay <2 μm	Bulk <2000 μm
0–10	43.4	15.2	12.7	19	77
40–50	21.7	9.2	10.6	46	82
100–125	nd	nd	nd	nd	381
Rill sediment	nd	nd	>60	nd	<21

nd = no data.

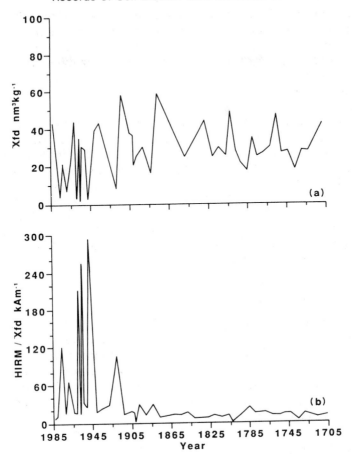

Figure 12.7 (a) Record of mass-specific frequency-dependent magnetic susceptibility (X_{fd}: $nm^3 kg^{-1}$), 1700–1985, corrected for masses of organic and carbonate materials, and used to infer the contribution made by clay particles of topsoil material to the sediment load; (b) record of HIRM/X_{fd} ratio (kA m^{-1}), 1700–1985, used to infer relative contributions to topsoil (low values) and subsoil (high values) material to the sediment load

are frequently very low ($<5 \, nm^3 \, kg^{-1}$) in the period 1945–80, though the most recent values (c.1980–85) are significantly higher ($>30 \, nm^3 \, kg^{-1}$). By comparing Figure 12.7(a) with the data in Table 12.2 it is possible to infer changes in the accumulation of topsoil. If a value of $40 \, nm^3 \, kg^{-1}$ is taken to represent an accumulation of clay derived entirely from topsoil then it may be inferred that topsoil (transported as clay) has not dominated the sediment source; many more values in the sediments lie below $40 \, nm^3 \, kg^{-1}$ than above it.

As with all mass-specific parameters, changes in X_{fd} values may be caused by changes in the mass of sediment constituents other than those of secondary minerals from topsoils; values of X_{fd} may not always be directly compared to the values in soils. One way of making a more direct comparison between sediments and sources is to use ratios which are independent of sediment mass. Table 12.2 shows that the HIRM/X_{fd} ratio distinguishes between topsoils and unweathered parent materials, and that the ratio relates to the relative proportions of these two different types of material within the bulk sample.

The record of HIRM/X_{fd} (Figure 12.7(b)) in the sediments is one of contrast between the twentieth century and earlier times. Before $c.$1900 the values are fairly constant, with small variations about a mean value of $c.10\,\text{kA m}^{-1}$. These values are typical of fine ($<2\,\mu\text{m}$) particles from topsoil zones or rill sediments (Table 12.2). Since then, the record shows values between $7\,\text{kA m}^{-1}$ (1985) and $295\,\text{kA m}^{-1}$ (1945); the high values appear to register the influx of almost pure unweathered subsoil, and major erosional events involving the transport of subsoil through channel scour or coarse particles (Table 12.2) are dated to $c.$1920, 1945, 1950–60, 1970 and 1975. During the twentieth century, topsoil material has only dominated the sediment composition in the periods 1925–40, 1965–60 and 1980–85.

Sediment Yields 10 000–0 BP

Long-term records of sediment accumulation were obtained from analyses of two complete sequences of sediment covering the last 10 000 years. A methodology similar to that used for the recent sediments was adopted, with sediment ages derived from pollen records correlated to local radiocarbon-dated pollen diagrams. The sediment sequences showed good depth-by-depth correlations between records of loss-on-ignition, an observation indicating fairly constant accumulation rates in the central part of the lake. As for recent sediments, dry mass, organic matter, minerogenic matter and carbonate contents were plotted against sediment age by using interpolated dates from the pollen-based depth–age curve, and expressed as influxes by multiplying mass-specific sediment properties by the accumulation rate.

Figure 12.8(a) shows the record of sediment loading to the lake (t yr^{-1}), calculated by multiplying the accumulation rate of minerogenic material by a lake area of 1 ha. The use of this constant lake area will have the effect of slightly overestimating the loading values in the youngest part of the record (by $c.$30 per cent) and underestimating the oldest records (by $c.$100 per cent). The underestimations are, to some extent, offset by changes in the pattern of accumulating sediment: it may be assumed that accumulation rates in the central part of the lake are not always mean rates. They are more likely to be higher than average, especially in the past, when sediment may well have been focused into the deep central parts of the lake. Consequently, the absolute load values represent estimates with perhaps significant errors, and more reliance should be placed on the trends as an indication of changing load.

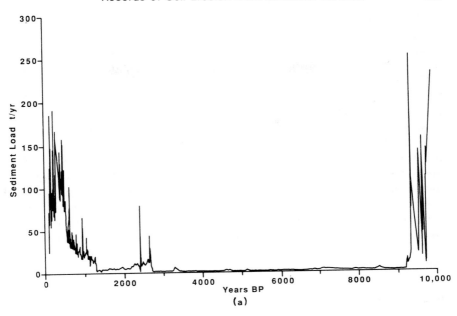

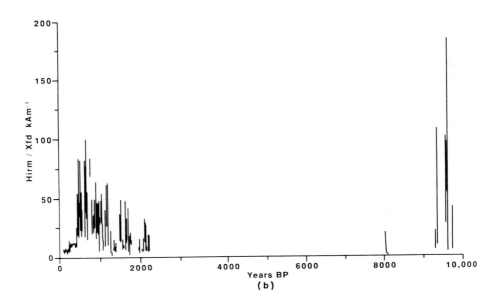

Figure 12.8 (a) Loading of minerogenic sediment to the lake (t yr^{-1}), 10 000–250 BP; (b) record of HIRM/X_{fd} ratio (kA m^{-1}), 10 000–250 BP. See also Figure 12.7

The sediment load record ($t\,yr^{-1}$) is characterized by high values before $c.9250$ BP, followed by low values until $c.2750$ BP, when there is the beginning of an overall rising trend which peaks $c.500–250$ BP. Note that the recent load record (Figure 12.6(a)) is hardly represented in Figure 12.8(a), which terminates $c.200$ BP. It is difficult to express the load figures in terms of sediment yield per area of catchment because of the difficulties in estimating lake and catchment sizes in the past. However, assuming a maximum catchment size of 198 ha, the sediment yield values ($t\,ha^{-1}\,yr^{-1}$) are estimated to be up to $1.3\,t\,ha^{-1}\,yr^{-1}$ before $c.9250$ BP, less than $0.5\,t\,ha^{-1}\,yr^{-1}$ between $c.9250$ BP and $c.2750$ BP, and rising to peak values $c.700–250$ BP of over $0.8\,t\,ha^{-1}\,yr^{-1}$. After 250 BP there appears to be a decline to values of less than $0.2\,t\,ha^{-1}\,yr^{-1}$, though these figures are not comparable with those for the recent period (Figure 12.6(b)), which utilized a smaller catchment size in the calculations. Although, as with the short record, the record of sediment load in Figure 12.8(a) should be used cautiously, there is substantial evidence to argue that the highest sediment yields during the post-glacial were reached during the period 700–250 BP (AD 1285–1735).

Sediment Quality 10 000–0 BP

Figure 12.8(b) shows the long record of $HIRM/X_{fd}$ which describes a data distribution similar to that for sediment load (Figure 12.8(a)). As with the recent record (Figure 12.7(b)) high values indicate a dominant subsoil source in stream channels and low ones a dominant topsoil source, or fine clay particles. Data for the period $c.9000–2500$ BP were not obtained easily because the concentration of ferrimagnetic minerals in these sediments is extremely low. There is no doubt, though, that during this period the ratio is also low, indicative of topsoil sources or clay particles. Before this long period of low values, subsoil appears to be the major source of sediment, although it is possible that immature soils at this time had low concentrations of secondary ferrimagnetic minerals which would result in enhanced $HIRM/X_{fd}$ ratio values. After $c.2500$ BP there are peaks in subsoil sources during the periods $c.2500–2200$ BP, $c.1800–1500$ BP and generally high contributions thorughout the period $c.1250–500$ BP: topsoil makes a significant contribution during the periods $c.2000–1800$ BP, $c.1500–1250$ BP and after 500 BP. The sediments deposited within the last 100 years which have high values (Figure 12.7(b)) are not represented in Figure 12.8(b).

It is interesting to compare the records of sediment load and $HIRM/X_{fd}$ ratio. They are completely independent records, except for their being based on the same sediment chronology, and yet there are some striking correlations. Before $c.500$ BP there is an apparent positive correlation between the records; high sediment accumulation is related to subsoil from channels. Between $c.500$ and 250 BP the relationship changes to one where high-sediment accumulations are related to topsoil, and after $c.250$ BP there appears to be a return to the positive link, with low accumulation rates correlating with topsoil material. The period around 500 BP (AD 1450) appears to have witnessed a radical change in sediment source, from predominantly channels to topsoil, while the sediment yield remained constant and very high.

DISCUSSION

At a broad level, the complete record of sediment load and sediment source for the last 10 000 years (Figure 12.8) describes sediment responses to four different sets of environmental conditions. The earliest part of the record indicates that high levels of sediment movement, initiated in periglacial conditions, continued into the post-glacial until $c.9250$ BP. This suggests that even with an ameliorating climate and rapid growth of vegetation, sediment continued to move in relatively large amounts for several hundreds of years into the post-glacial period; sediment which was of a non-topsoil origin (Figure 12.8(b)). At this time the system was still adjusting to previous meltwater flows, and the presence of non-topsoil material suggests that previously deposited unweathered deposits were moving through the system.

The middle part of the record, 9250–2750 BP, is a 6500-year period of catchment stability, or at least of constantly low sediment loadings to the lake, under a well-established forest cover (Regnell, 1989). Sediment sources were of topsoil origin, but the low energy conditions which seem to have prevailed suggest that sediment mobilization was through periodic bank collapse (debris dams, uprooted trees) which brought topsoil from the stream margins to the channel.

The period 2500–500 BP registers the effects of human impact on the sediment record. Initially, the sediment load increased in response to major deforestation, but to levels that were much lower than those associated with later agriculture (see below and Dearing, in press). During this period, though, the dominant sediment source was subsoil, an indication that early anthropogenic impact caused changes to the hydrodynamics of the channels. It seems likely that the impact was such that runoff levels increased, altering the magnitude and frequency of channel-forming flows which stimulated channel adjustment. This adjustment may have been in terms of channel deepening, but also in terms of reworking floodplain deposits; present-day samples of floodplain material are indistinguishable from subsoil on the basis of $HIRM/X_{fd}$ measurements. Whether or not the anthropogenic impact also caused changes to the erodibility or mobility of soil on slopes is open to question. The data show that some topsoil may have been reaching the streams, but an interpretation of this information is highly problematic. It appears from a comparison of sediment records (Dearing, in press) at other lake sites in the region that substantial sediment storage on slopes exists in drainage catchments, like Bussjösjön, which have relatively large areas in relation to the drainage network and lake area.

In contrast, the period 500–300 BP is typified by a change to relatively greater losses of topsoil and finer particles. Since the sediment load remains high it seems that this period witnessed an absolute increase in the amount of eroded topsoil, rather than an absolute decrease in the amount of subsoil erosion. Agriculture is thought to have intensified through the transition between this and the previous period, and it is unlikely that there was a loss of erosive power in the channels brought about by reduced runoff. However, there is little evidence to link the accelerated topsoil erosion to specific land-management practices; the area of ploughed land in the catchment is thought to have been less than 20 per cent of today's figure, and the number of cattle was much lower than in the period 1850–1950. In addition, it is thought that the riparian wet meadows were not reclaimed for cultivation until after 1800 (Figure 12.3), and their presence would be expected to have created an effective buffer to sediment moving from hillslopes to channels. Given these environmental conditions, an increase in the amount of topsoil reaching the channels would have had to have been the consequence of a significant increase in the mobility of soil on existing cultivated slopes. One explanation may be the onset of colder winters and more frequent periods of frozen ground, associated with the Little Ice Age, giving rise to highly erosive conditions as seen in the contemporary erosion data for 1986/7.

Superimposed upon the effects of increased sediment delivery from the slopes is the possible reduction in subsoil material caused by a reduced availability of sediment within the channels. Since $c.300$ BP the channels have only eroded significantly at times of disturbance during drainage operations, as seen in the load and source records for the past 300 years (Figures 12.6 and 12.7). Taking a long-term view of channel dynamics, it is therefore possible to argue that before 500–300 BP subsoil material dominated the sediment load because channel material was constantly available for erosion, a situation altered after 300 BP. The previous period 2500–500 BP may therefore represent the timescale of channel reaction and recovery to deforestation, under agricultural conditions. Perhaps the stabilization of channel form during the period 500–300 BP led to a reduced availability of subsoil material, thus accentuating the change in the proportions of topsoil- and subsoil-derived sediment. It is not possible to make more certain explanations for the cause of the change in sediment source during the period 500–300 BP without further land-use and climatic data. However, it was apparently a time of considerable hydrological change, a period which should be focused upon in other studies.

Reduction in the overall surface drainage network (Figures 12.3 and 12.4) during the last 85 years has led to a loss of sediment pathways and, as a consequence, sediment yields in the mid-twentieth century were probably lower than when the catchment was forested. The intensification of agriculture in the last 40 years, involving the rationalization of field units (Figure 12.5) and improved drainage, caused an increase in sediment load and yield up to 1970 (Figure 12.6). The dominant source was subsoil, probably as a result of operations of undersoil drainage. Since 1970 these values have declined to mid-twentieth-century levels, even though field sizes are now at their largest and the fields are laid bare throughout the winter because of the recent reduction in the area of ley and permanent pasture. However, data for slope processes confirm that erosion through rilling and overland flow in certain climatic conditions are responsible for high levels of soil loss from slopes. The winters 1986/7 and 1987/8 were both abnormal; the first was exceptionally cold with an intense snowmelt, and the second was unusually mild. This suggests that the sediment losses amounting to 218 t and 4 t in 1986/7 and 1987/8 (Table 12.1), respectively, are both extreme values, and that the normal sediment loss is intermediate in value. The monitored losses may be compared with contemporary sediment loadings from the whole catchment to the lake of less than 5 t, a comparison which indicates that sediment delivery from the slope to the lake will normally be very low. This suggests that either the monitored slope is unrepresentative of the whole catchment or that substantial sediment storage exists between the middle slopes and the lake. Since the monitored slope is the steepest in the present catchment and has the highest sediment loss compared with a large number of fields in the region (Alström and Bergman, 1988b), the first of these alternatives is quite plausible. However, the monitored slopes are also likely to be the most important source areas for lake sediment, given that there are no other direct pathways for eroded material to the lake. Clearly, there must be substantial sediment storage, a fact which confirms that the lake sediment record will mask the high magnitude–low frequency erosion event in today's catchment. The absolute definition of the 1986/7 soil losses in terms of magnitude and frequency will only be possible through further monitoring, but logical consideration of historical farming practices suggests that the high measured soil losses have not occurred frequently in the past; if they had, the lake records would show an upturn in sediment loading.

Overall, the data reveal that over long timescales, the quantities and magnetic qualities of sediment have varied considerably. It has been shown that the use of sediment yield in isolation is of little benefit to an interpretation of sediment dynamics. More preferable are data for total sediment load (either to a lake or to a downstream gauging station) coupled with sediment source information. Further work is in progress to calibrate the lake sediment properties in terms of quantitative contributions from alternative sources, but already the data for Bussjösjön show that many, if not most, of the controls on sediment movement take place within and around the stream channels, and that there may have been times in the past when the channel system was effectively isolated from the slope system. When compared with slope processes, the changes to the runoff regime and long-term channel development, involving the exploitation of channel and floodplain deposits, probably exerted a greater control over sediment losses. In one sense, the changes in channel dynamics prior to drainage may represent the recovery of the catchment system after the loss of natural vegetation, over a timescale which was considerably longer than the stabilization following the cessation of periglacial conditions. However, as the monitored process data for the present show, considerable variations occur in the effectiveness of slope processes, and there is no reason to suggest that this situation was not the same in the past. Therefore, although evidence for channel stabilization exists the same may not hold true for the slopes. An agricultural history of continually improved techniques represents a sequence of environmental triggers and reaction at the soil level, with only short periods for recovery or stabilization. The pace and direction of these reactions are not easily detected in sediment data which are dominated by changes in the channel and its hydrology.

In terms of using lake sediment-based data to construct or to test sediment models, several points may be noted. It is unlikely that empirical models of erosion and sediment yield may be developed from these data, except for the recent past, where land use and other environmental data allow correlations to be made between apparent causes and effects. As the records become older, the resolution of environmental reconstruction becomes weaker, with much of the interpretation of historical erosional data reliant upon long-term and generalized changes in pollen and archaeological records. The relatively poor spatial and temporal definition of such data does not normally provide information which may be compared to monitored yield or source data, sediment data which may reflect localized process operations over short timescales. However, the historical data for yield and source should be accurate enough to allow process-based models of long-term sediment yield to be tested, and models for this purpose are being sought. In this respect, the data perhaps show the type of processes which may be included in any model, processes which can be calibrated and parameterized from historical data. Certainly, pollen diagrams may provide data on local and regional climatic conditions, but local pollen diagrams of high spatial and temporal resolution (Bradshaw, 1981) may also enable catchment biomass to be calculated and soil organic matter content to be estimated through calibration with contemporary environments. Local pollen diagrams may provide the means for setting the correct parameters in models of runoff generation and soil detachment. It would also be useful to create models which predict the sediment movements from both slopes (topsoil) and channels (subsoil), which may then be tested against magnetic or other 'fingerprint' data.

ACKNOWLEDGEMENTS

Most of the results summarized here were obtained within the programme of research 'The Cultural History of the Ystad region during the last 6000 Years', financed by the Riksbanken jublileumfond (Stockholm). We would like to thank Professor B. E. Berglund for his continued support for the soil erosion research, Dr I. D. L. Foster for discussions on a previous version of the paper and Mrs S. Addleton for much of the cartography.

REFERENCES

Alström, K. and Bergman, A. (1988a). Vattenerosion och närsaltförluster via ytavrinning i åkermark i Skåne, *Lunds Universitets Naturgeografiska Institution, Rapporter och Notiser* **71** (English summary).

Alström, K. and Bergman, A. (1988b). Sediment and nutrient losses by water erosion from arable land in the south of Sweden. A problem with non-point pollution? *Vatten* **44**, 193–204.

Bradshaw, R. H. W. (1981). Quantitative reconstruction of local woodland vegetation using pollen analysis from a small basin in Norfolk, England, *J. Ecol.* **69**, 941–55.

Dearing, J. A. (In press). Erosion and land use, In Berglund, B. E. *et al.* (eds), *The Cultural Landscape of the Ystad area, 0–6000 BP*, Ecobulletin.

Dearing, J. A., Håkansson, H., Liedberg-Jonsson, B., Persson, A., Skansjö, S., Widholm, D. and El-Daoushy, F. (1987). Lake sediments used to quantify the erosional response to land use change in southern Sweden, *Oikos* **50**, 60–78.

Dearing, J. A., Morton, R. I., Price, T. W. and Foster, I. D. L. (1986). Tracing movements of topsoil by magnetic measurements: two case studies, *Physics of the Earth and Planetary Interiors*, **42**, 93–104.

Kelsey, H. M. (1982). Influence of magnitude, frequency and persistence of various types of disturbance on geomorphic form and process. In Swanson, J. *et al.* (eds), *Sediment Budgets and Routing in Forested Drainage Basins*, USDA Forest Service Tech. Rep. PNW-141, pp. 86–96.

Regnell, J. (1989). *The Cultural Landscape of Southernmost Sweden 6000-0 BP: A Palaeoecological Study*, PhD thesis, Department of Geology, University of Lund.

Thuning, B. and Linderson, H. (1986). *Stratigrafi och överplöjning i Bussjösjöområdet. Ystad*, Geologiska Institutionen Lunds Universitet, No. 15 (English summary).

13 Valley Sedimentation at Slapton, South Devon, and its Implications for the Estimation of Lake Sediment-based Erosion Rates

PHILIP N. OWENS
Department of Geography, University of British Columbia

INTRODUCTION

It has been recognized that the estimation of sediment yields to infer rates of catchment erosion may form an important basis for studying and understanding contemporary and palaeohydrological processes in drainage basins (e.g. Slaymaker, 1977). Sediment yield studies are also important in that they provide a useful index of the impact of society on the environment (e.g. Wolman, 1967). Many studies have been based on contemporary stream monitoring over short timescales (usually 1–10 years) and have attempted to extrapolate current process rates by making assumptions regarding bulk density of sediment. These may be misleading and imply a constancy of geomorphic process rates over a long period of time, usually failing to identify the significant variability in hydrological regimes (Dearing *et al.*, 1982). It is also suggested that the present climatic conditions are 'normal', but we are unsure of what 'normal' climate is (Lamb, 1977).

In order to extrapolate sediment yields over longer periods of time with some confidence, an alternative source of information is required. One such alternative is the sedimentary record contained within lakes. Oldfield (1977) identified lake sediments as providing 'possibly the most powerful basis for integrated insight into ecosystems that the world presents' (Oldfield, 1977, p. 462).

Soil Erosion on Agricultural Land
Edited by J. Boardman, I. D. L. Foster and J. A. Dearing
©1990 John Wiley & Sons Ltd

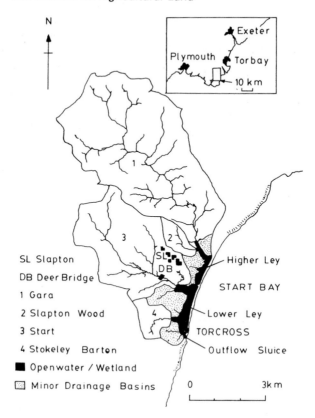

Figure 13.1 The catchments and location of Slapton Ley

SLAPTON LEY

Slapton Ley is a small (0.77 km²) freshwater coastal lake system in South Devon, England (Figure 13.1). The Ley is of recent origin as radiocarbon dating suggests that it is no more than 1000 years old (Morey, 1976). It is divided into two wetland areas: the Higher Ley (which is now largely reedswamp) and the Lower Ley (an open-water lake with a discontinuous reed fringe). The catchment of Slapton Ley occupies 46 km² and is largely composed of flat top ridges dissected by steep valleys, often covered by dense woodland. Van Vlymen (1979) superficially divided the catchment into four main basins; the Gara, Slapton Wood, Start and Stokeley Barton, of which the principal tributaries are the River Gara and Start Stream, which drain 59 per cent and 29 per cent of the total catchment, respectively. The underlying rocks are of Lower Devonian age with the coarser Meadset Beds overlying finer Dartmouth Slates. Climate is maritime with a mean annual rainfall of 1043 mm.

Outflow of the lake is controlled by a horizontal weir, culvert and gate system constructed in 1854 at the southern end of the Lower Ley, to stop winter flooding at Torcross (Morey, 1976).

LAKE-BASED SEDIMENT YIELDS

Within the last decade a variety of studies have used a multiple-core approach combined with an absolute chronology of sediments in the Lower Ley to establish sediment yields and infer rates of catchment erosion. Crick (1985) calculated mean rates of 53.5 t km^{-2} yr^{-1} for a 9 year period for the two sub-basins (Start and Stokeley Barton) that drain into the Lower Ley.

However, the relationship between erosion and sediment yield is a complex one, since not all the material detached from hillslopes and channels will reach the lake basin. A large proportion of sediment is often deposited at intermediate locations where entrainment and transport velocities fall below a critical threshold. Vanoni (1975) estimated that, for basins larger than 1 km^2, the amount of particulate material that arrives downstream will often be less than 25% of the amount eroded. Trimble (1981) in the 360 km^2 Coon Creek Basin, USA, found that sediment yields were only about 6 per cent (75 t km^{-2} yr^{-1}) of all upland erosion estimated to have occurred between 1853 and 1938.

THE 'DEER BRIDGE SEDIMENT SINK'

Upstream of one of the two feeder streams to the Lower Ley, the Start, exists a large area of recently deposited material. It extends upstream from Deer Bridge. In order to calculate the volume of deposited material a series of transects were taken across the valley floor. The location of the transects is shown in Figure 13.2, such that sites 7 and 1 represent the upper and lower limits of the sink, respectively. Between these extremities, five transects were taken so as to be fully representative of the variations in sediment thickness. The transects were surveyed with a Wild Heebrugg level (type NKOI), tripod and staff from known benchmarks. The depth of the recently deposited sediment was obtained using a 1.5 m ranging pole. The ranging pole was immersed in the deposited sediment, which was relatively soft, until a gravel base was reached. In the case where no gravel base was reached, the sediment depth was recorded as 1.5 m. Depths were defined along each of the seven surveyed transects (Figure 13.3).

The total volume of sediment in the sink, V, was calculated as 75 711 m^3, based on:

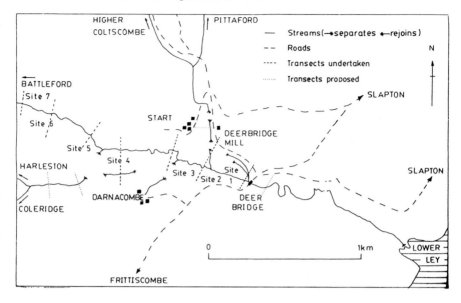

Figure 13.2 Map showing the location of survey transects

$$V = \sum_{n=1}^{6} \left[\frac{(W_n d_n + W_{n+1} d_{n+1})}{2} (L_{n+1} - L_n) \right]$$

where W = sediment width,
d = mean sediment depth, and
L = distance of transect from Deer Bridge.

To calculate the total mass of the sediment, the volume was multiplied by the dry density. This was based on the analysis of sediment cores taken at seven sites (Foster, pers. comm.), and was estimated at 34 070 t. Due to several limitations in the techniques used (Owens, 1988), the volume and mass calculated are underestimates of the total sediment in the sink. Figure 13.2 indicates the areas where future transects should be included to give a better estimate of total volume.

In order to obtain the date of the initiation of sedimentation, analytical techniques and documentary evidence were used. Analysis of historical maps from 1765 to 1965, aerial photographs from 1945 to 1980, local newspapers and interviews with farmers suggests that the age of the sink lies between the limits of 30–100 years. Sediment cores taken at the sites used for the surveying were also analysed for ^{137}Cs (Caesium-137). This artificial nuclide was first recorded in the Northern Hemisphere in 1954 due to nuclear weapons tests in the atmosphere (Petts and Foster, 1985). Cores were removed using 1 m long

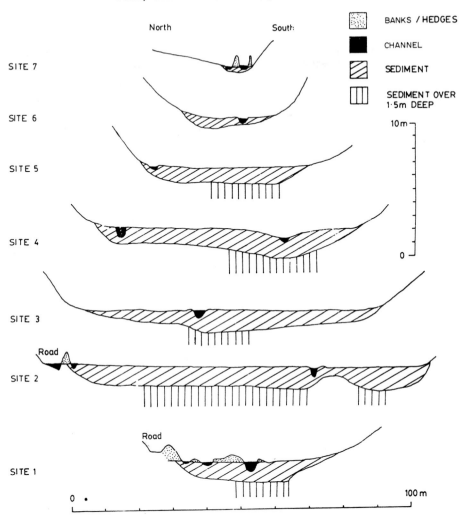

Figure 13.3 Survey transects showing depth of sediment

plastic drainpipe (10 cm diameter) that had been modified so that they could be drilled into the sediment using a Cobra Hammer Drill, and removed by a portable winch and frame. In order to provide a comparison of the ^{137}Cs in the deposited areas with that expected 'naturally', a core was taken nearby at an orchard known to have been undisturbed for at least 50 years. Due to the time required to perform ^{137}Cs on sediment cores, most of the cores have not yet been fully analysed. The results obtained so far, however, indicate that

198   Soil Erosion on Agricultural Land

sedimentation of the top 40 cm has occurred rapidly since 1954 (Walling, pers. comm.).

Documentary evidence (such as historical maps and aerial photographs) also suggest that the probable cause of sedimentation is lake-level change over the past 100 years (Owens, 1988). This has caused changes in channel gradient of the Start stream which in turn has led to a ponding of water upstream of Deer Bridge and a deposition of the sediment load. Superimposed on this may be an increased supply of sediment during the post-war period due to agricultural intensification.

REVISED EROSION RATES

From the above it is possible to calculate sedimentation rates in the sink of between 341 and 1136 t yr^{-1}. If the dimensions of the sink are correct, when combined with the sediment yield obtained from the lake cores, the erosion rate

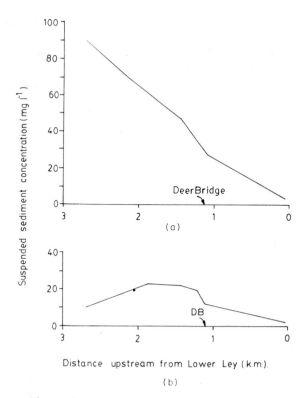

Figure 13.4 Variations in suspended sediment concentration along the Start Stream (a) Medium discharge; (b) low discharge

within the Start catchment probably lies between 82 and 148 t km^{-2} yr^{-1}, depending on the age of the sink. This suggests that a large proportion of eroded material does not reach the lake. This observation was reinforced by suspended sediment sampling above and below the sediment sink between November 1986 and June 1987. Water samples were taken from the stream at sites chosen to conform to the sites used for the surveying, using a hand-held sampler (Owens, 1988) and then filtered through fine-grade fibre-glass filter papers (Whatman type GF/C) by the use of an air vacuum pump. The results obtained (Figure 13.4) indicate that, at high to medium discharge, high suspended sediment loads entering the sink are quickly deposited, with loads entering the lake only being a fraction of upstream values. At low discharges little sediment is supplied to the sink and resuspension and redistribution within the sink itself seems dominant.

CONCLUSION

The results obtained underline the necessity of understanding fully contemporary processes operating within the catchment before lake sediment-based data can be correctly interpreted. When using lake sediments to infer catchment erosion rates, it is necessary to take account of intermediate sediment sinks to obtain more realistic estimates of catchment erosion. As shown above, much of the eroded sediment from the slopes and stream banks in the Start catchment is deposited in an intermediate sink before reaching the Lower Ley. Crick (1985) estimated catchment erosion rates of 53.5 t km^{-2} yr^{-1} for the catchments that drain into the Lower Ley based on lake cores. This is clearly an underestimate. The erosion rate within the Start catchment probably lies between 82 and 148 t km^{-2} yr^{-1}.

ACKNOWLEDGEMENTS

This study was undertaken while at Coventry Polytechnic and Slapton Ley Field Centre. The author is grateful to Dr I. D. L. Foster who helped with most aspects of this study and would like to thank Professor D. E. Walling for advice and help with the water sampling and ^{137}Cs analysis. Thanks are also due to D. A. Job and S. O. Whitfield for assistance in the field, and to Professor H. O. Slaymaker and Dr J. McManus for reviewing the manuscript.

REFERENCES

Crick, M. I. (1985). *Investigations into the Relationship between Sediment Accumulation in the Lower Ley, Slapton, and Spatial Patterns of Erosion within its Catchment, Using Magnetic Measurements.* Unpublished BSc dissertation, Plymouth Polytechnic.

Dearing, J. A., Foster, I. D. L. and Simpson, A. D. (1982). Timescales of denudation: the lake drainage basin approach. In *Recent Developments in the Explanation and Prediction of Erosion and Sediment Yield*, Proc. Exeter Symp., International Association of Hydrological Sciences, Publication No. 137, 351–360.

Lamb, H. H. (1977). *Climate: Present, Past, and Future. Vol. 2: Climatic History and the Future*, Methuen, London.

Morey, C. R. (1976). The natural history of Slapton Ley Nature Reserve IX: the morphology and history of the lake basins, *Field Studies*, **4**(3), 353–68.

Oldfield, F. (1977). Lakes and their drainage basins as units of ecological study, *Progress in Physical Geography*, **1**, 460–504.

Owens, P. N. (1988). *Valley Sedimentation at Slapton, South Devon, and its Implications for Management*, Unpublished BSc dissertation, Coventry Polytechnic.

Petts, G. E. and Foster, I. D. L. (1985). *Rivers and Landscape*, Edward Arnold, London.

Slaymaker, H. O. (1977). Estimation of sediment yield in temperate alpine environments. In *Erosion and Solid Matter Transport in Inland Waters*, Proc. Paris. Symp., International Association of Hydrological Sciences, Publication No. 122, 109–17.

Trimble, S. (1981). Changes in the sediment storage in the Coon Creek Basin, Drittley Area, Wisconsin, 1853–1975, *Science*, **214**, 181–3.

Vanoni, V. A. (ed.) (1975). *Sedimentation Engineering*, Report in Engineering Practice No. 54, *American Society of Civil Engineers*, New York.

van Vlymen, C. D. (1979). The natural history of Slapton Ley Nature Reserve XIII: the water balance of Slapton Ley, *Field Studies*, **5**, 59–84.

Wolman, G. R. (1967). A cycle of sedimentation and erosion in urban river channels, *Geografiska Annaler*, **49A**, 385–95.

14 Lake Sediment-based Studies of Soil Erosion

FRANK OLDFIELD
Geography Department, University of Liverpool

and

ROBIN L. CLARK
CSIRO, Australia

INTRODUCTION

We now have a wide range of techniques for dating and characterizing lake, reservoir and coastal sediments. Their successful application to a great variety of problems and environments should ensure that sediment-based research will remain a recognized aspect of erosion studies for the foreseeable future. Both the real and the perceived value of sediment-based studies for researchers and environmental managers will depend in part on the extent to which questions are addressed within a shared conceptual framework. Clearly, plot experiments will not tell us directly about long-term or large-scale catchment yields, nor can sediment-based historical reconstructions contribute to analyses of the processes of particle detachment and initial transport. Part of the value of a conference, or a symposium volume arising from it, should be a redefinition of problems and questions. Redefinition can sharpen research focus and identify goals more precisely, as well as enhance recognition of complementary themes and the prospects for exploiting them. The outcome should be a closer and more realistic integration of different kinds of evidence and of alternative approaches, whether they be empirical, conceptual, methodological, theoretical or managerial.

In this chapter we try to identify and address some of the themes that are crucial for fuller integration of sediment-based research into the whole field of erosion studies. We include a synopsis of the strengths of a sediment-based approach, introduce a simple conceptual model stressing linkages between different approaches to erosion and sedimentation, consider the question of temporal resolution, discuss the central question of sediment source identification

Soil Erosion on Agricultural Land
Edited by J. Boardman, I. D. L. Foster and J. A. Dearing
©1990 John Wiley & Sons Ltd

and conclude with some comments on the relationship between sediment-based research and current concerns with global changes. Questions that are just as important for sediment-based studies, such as the quantification of sediment yield to a lake of reservoir for any given time interval and the implications of varying catchment delivery ratios, are considered more fully in other contributions to this volume and earlier publications (Bloemendal, 1982; Bloemendal et al., 1979; Dearing and Foster, 1987; Dearing et al., 1981, 1987; Foster et al., 1985; Oldfield et al., 1985a).

THE POTENTIAL

Lake or reservoir sediments accumulating in efficient traps can be studied by a wide range of techniques appropriate to the questions being asked (Clark and Wasson, 1986). Methods are now available to date and correlate sediment cores and to quantify total sedimentation for each dated time interval, to establish sediment type and source from physical, chemical and biological constituents, and to reconstruct associated vegetation and land use changes. Some of the distinctive contributions sediment-based studies can make are outlined below.

(1) *Post-hoc* studies are possible once management problems have been recognized. This is important in a decision-making environment in which problems are increasingly addressed only when they are perceived as threatening to productivity, economic returns, health, livelihood or political survival. Sediment-based studies, unlike process-based observations and experiments, can be completed in a much shorter period of time than that spanning the period with which they are concerned. Thus, where issues of variability, recurrence and trend arise, sediment-based studies often have unique virtues in terms of both time- and cost-effectiveness.

(2) Sediment studies address longer time dimensions, providing a more secure empirical basis for evaluating the effects of secular environmental variation, for establishing the antecedents of present-day problems and for calculating trends and rates of change. These allow exploration of systems linkages subject to lead or lag times beyond the temporal range of direct, quantitative observations.

(3) Multidisciplinary studies of sediment, when sufficiently comprehensive, can yield *integrated* reconstructions of past changes. Such reconstructions can provide unique insights into the interacting effects of climatic, ecological, cultural and technological changes on land and water systems.

(4) Alongside the potential for integrating the complex expression of diverse processes, sediment studies also hold out the promise of discriminating between processes and interactions in terms of their effects and relative importance. In this context the limitations implicit in studying only the

terminal sink in large systems with low delivery ratios must be clearly recognized. In such systems, there is a clear need to broaden the sediment-based approach to an appraisal of temporarily stored material in streams, swamps and alluvial and colluvial deposits. In this way, analysis of the record of sedimentation and storage comes closer to linking up with direct studies of on-site erosion.

(5) Sediment-based studies can aid problem diagnosis to the point of contributing significantly to policy making. Recent studies of eutrophication (Battarbee, 1978, 1986) and acidification (Battarbee and Flower, 1983; Battarbee *et al.*, 1988) confirm this and challenge those involved in sediment-based studies of soil erosion to achieve comparable success at both a scientific and a practical level.

CONCEPTUAL MODELS

Palaeoecologists and geomorphologists have traditionally used sediments in lakes and swamps to study past climate, vegetation and landscape changes on relatively long timescales of millennia at the very least. With the development of appropriate dating techniques, the methods developed for long-term studies are now being applied to the recent past, on a timescale of a few hundred years or less and with a temporal resolution at the level of individual events to decades. Figure 14.1 summarizes the approach used, with the emphasis on linkages

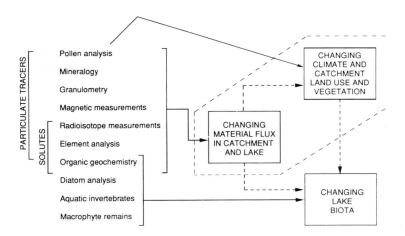

Figure 14.1 Lake sediment studies: methods and areas of interest. The concern of this chapter is with methods that contribute to the understanding of changing material flux in catchments and lakes, and the relationship between these changes and those in climate and catchment land use and vegetation (boxes enclosed by dashed line)

between a lake and its catchment. The three boxes on the right indicate the three main areas of interest of sediment-based studies. On the left are the most commonly used methods for extracting information about past environments from sediments. Some methods (e.g. magnetic measurements) are used only as particulate tracers, others (e.g. radioisotope measurements) can be used to trace either particulates or solutes. Pollen analysis can be used to reconstruct the vegetation history of the catchment or the lake and can also be used in some lakes to identify sediment sources.

For soil-erosion studies, information from the sediments is needed that sheds light on changing material fluxes in the catchment and lake and on changing climate, catchment land use and vegetation. Of the methods listed, in this chapter we shall concentrate on the application of magnetic and radioisotope measurements and of pollen analysis, as these appear to be the least-known, and yet in some ways most promising, of the techniques.

There is no mention in Figure 14.1 of dating techniques, which are essential for estimates of rates or frequencies. The most important of these for the timescales relevant to erosion studies, are carbon-14 (e.g. Wasson and Galloway, 1986), lead-210 (Appleby and Oldfield, 1978, 1983; Oldfield and Appleby, 1984a,b), caesium-137 (Miller and Heit, 1986), laminations (O'Sullivan, 1983), pollen and charcoal (Clark, 1986) and palaeomagnetic secular variation (Thompson and Oldfield, 1986). Instrumental and documentary records may also be useful where they can be linked to stages in sedimentation (Clark, 1986). Discussion of all these would require a separate chapter, so we wish only to stress that dating methods are now available for sediments deposited in the recent past and that fine time resolution right down the decade, year, season or even individual events is available at many sites.

Figure 14.2 shows a conceptual model of the linkages between events and processes relating to soil erosion, acting on a range of spatial and temporal scales. A single rainfall event may erode soil and transport it to one or more deposition sites. Sediment may be stored for very short or very long periods in a range of sites, or carried out of the terrestrial system to the seabed. Resuspension and redeposition can occur many times. The fate of the sediment affects the longer-term processes shaping the landscape, its vegetation and land use, and these in turn partly determine the antecedent conditions for subsequent rainfall events.

From Figure 14.2 it is evident that sediment studies complement other ways of investigating soil erosion, and vice versa. It also reinforces the need to apply the methodology to as many upstream sediment stores as possible.

With this general framework in mind, we propose to consider some methodological issues and to describe and illustrate some applications of sediment studies to soil-erosion questions.

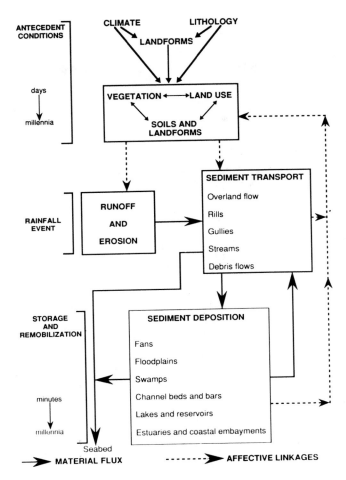

Figure 14.2 Conceptual model of linkages between events and processes over time in an eroding and sedimenting system (see text)

TEMPORAL RESOLUTION

Figure 14.3 is a block diagram, from Wasson and Galloway (1986), of the area around Umberumberka Reservoir near Broken Hill in western New South Wales, Australia. The Reservoir was created in 1913–15 by the damming of a creek which until then flowed out from its 420 km² catchment onto the Mundi Mundi Plain. Previously, drainage from the creek soaked into the plain, leaving the sediment in the form of an elongated alluvial fan. Paucity of organic matter within the deposited alluvium has limited to two (6000 BP to 3000 BP and 3000 BP

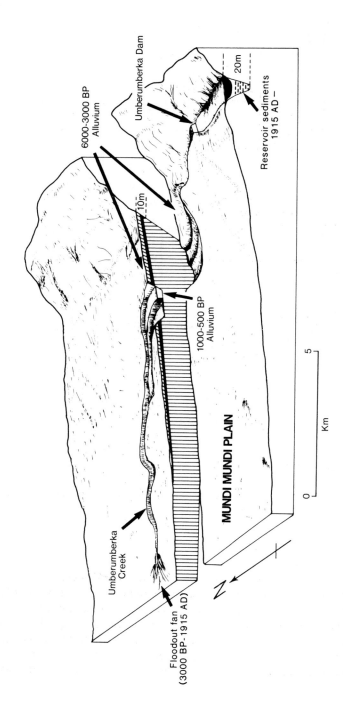

Figure 14.3 Block diagram of the area around Umberumberka Reservoir near Broken Hill, NSW, Australia. The 420 km² catchment (not shown) drained on to the Mundi Mundi Plain prior to 1915 when the dam was built. Between 6000 BP and 3000 BP, sediment was deposited in a large alluvial fan. Between 3000 BP and 1000 BP, this fan was gullied and the retransported sediment deposited in a floodout fan. Between 1000 BP and 500 BP, alluvium was deposited in the stream channel, then it was gullied and partly transported to the floodout fan. After AD 1915 sediment accumulated in the reservoir. (Adapted from Wasson and Galloway, 1986)

to AD 1915) the time intervals for which sedimentation on the Mundi Mundi plain can be calculated. The amount of material deposited on the floodout fan between first European settlement about AD 1850 and the completion of the dam in AD 1915 has been estimated (Wasson and Galloway, 1986). From the time of filling of the reservoir, estimates can be made from depth surveys of the reservoir sediments. Thus, rather diverse lines of evidence have been used for the estimates of sediment yield shown in Figure 14.4. Despite the lack of fine temporal resolution, this evidence points to major, presumably climatically modulated, shifts in yield from the catchment prior to European settlement, to a sharp contrast between sediment yields before and after European settlement and to trends during the 130 years of European impact.

The record of reservoir sedimentation since 1915 has yet to be studied in detail, but preliminary analyses illustrate well the level of temporal resolution such sites may offer. The Umberumberka site lies in the arid zone and receives a mean annual rainfall of about 200 mm, with a high degree of inter-annual variability and no regular or predictable seasonal pattern. During the period of existence of the reservoir, about 120 substantial rainfall events have been

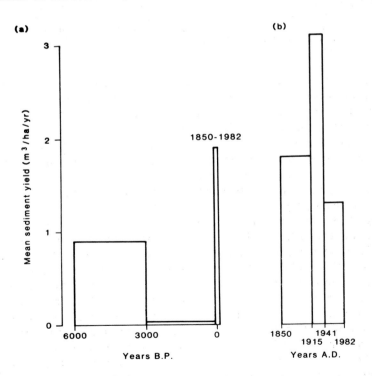

Figure 14.4 Mean sediment yield from the Umberumberka catchment reconstructed from calculations of rates of total sediment loss

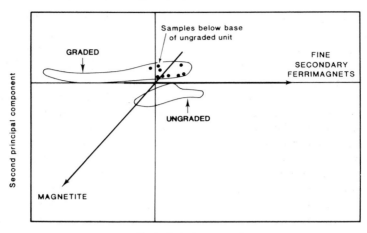

Figure 14.5 Ordination of the magnetic properties of two sediment sequences from a 14 m core of Umberumberka Reservoir sediments. The envelopes of values for the ungraded and graded units sampled do not overlap. Sediments immediately below the ungraded unit plot within the range of values for the earlier, graded one. The concentration of magnetite increases along the vector towards the lower left, and the horizontal vector indicates increasing fine secondary ferrimagnets towards the right

recorded. The 14 m sediment cores taken for initial study contain about 85 visually distinctive sediment units, which presumably relate to individual precipitation events delivering sediment to the reservoir. So far, only two such units have been studied in any detail. The earlier one, tentatively dated at 1928, is a graded deposit with visible upward fining over its 40 cm thickness, and the later one, possibly from 1937, is an upgraded unit 28 cm thick along with 9 cm of material from the unit below and 3 cm from immediately above.

Figure 14.5 shows that the two units can be clearly differentiated on the basis of their magnetic properties (cf. Thompson and Oldfield, 1986) and the samples underlying the later unit are magnetically indistinguishable from the finer sediments in the earlier graded unit. Even more remarkable is the consistency with which the magnetic properties mirror the upward fining trend within the graded unit (Figure 14.6). Magnetic property variations reflect mineralogy and magnetic grain size shifts, but the latter are not a simple and direct function of particle size although they are often strongly related (Oldfield *et al.*, 1985b; Maher, 1988). Figure 14.7 sets part of the magnetic evidence against pollen analyses from contiguous 1 cm samples through the graded unit. The two lines of evidence give a detailed record of sediment variations *within* a single major event, though no attempt has yet been made to interpret the full range of pollen-frequency variations, nor to distinguish temporal changes in sediment source from those due to different particle-settling velocities. Initial deposition of silt

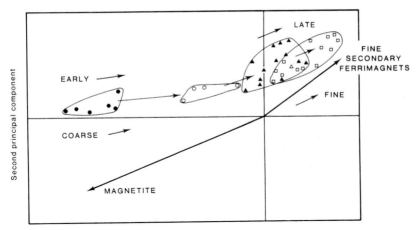

Figure 14.6 Ordination of the magnetic properties of the earlier, graded flood deposit. Upward fining within the unit is reflected in the magnetic measurements by a shift in properties away from those more typical of relatively coarse-grained primary magnetite associated with sandy channel infills to those derived from the silt-sized, clay-pellet dust mantle that covers much of the catchment and is rich in fine secondary ferrimagnets. The samples have been divided into four groups on the basis of their magnetic properties (envelopes and symbols)

is followed by clay derived from channel-side exposures of the mantle of fine 'soil' that covers much of the catchment. The sediment record from the Umberumberka catchment thus presents opportunities to resolve yields and erosion and sedimentation processes on timescales ranging from millennia to hours.

SEDIMENT SOURCES

Magnetic Measurements

Figure 14.8 links the magnetic properties of the Umberumberka Reservoir sediments to those of catchment soils. Most of the fine-grained material mantling the varied, predominantly igneous and often strongly mineralized bedrock of the catchment is derived from a Pleistocene dust mantle referred to as 'parna' (Chartres, 1982). The magnetic properties of this mantle are uniform and distinctive (unpublished data). Within the Umberumberka catchment the parna is mostly transported and redeposited through slope wash as colluvial and alluvial fill. Consequently, despite the very limited contribution to the regolith from *in situ* weathering, parna is often mixed with haematite-rich granitic material and, less frequently, with magnetic-enriched debris derived from both outcrops of basalt and zones of highly localized mineralization. The catchment is also

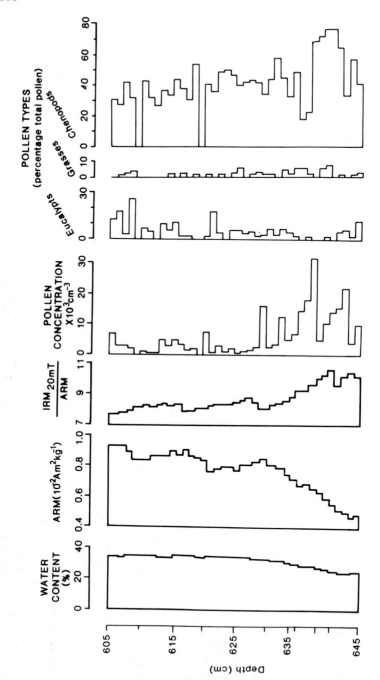

Figure 14.7 Selected pollen analytical and magnetic data from the graded Umberumberka flood deposit. Initial deposition of silt is indicated by water content, magnetic properties and pollen concentration. Variations in these and in abundances of the different pollen types in the clay deposited above may well reflect changes over time in sediment sources

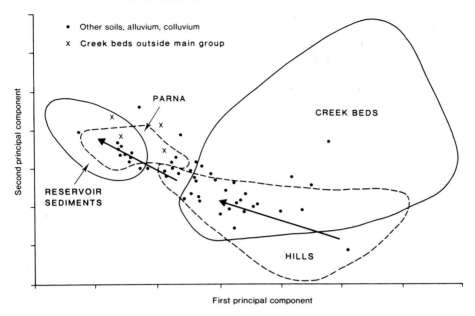

Figure 14.8 Ordination of the magnetic properties of samples from the Umberumberka Reservoir sediments and from the catchment soils. Solid and dashed lines are envelopes containing all samples in the locations indicated and for unmixed parna; individual points are not shown, except for four creek bed samples that fall outside the main envelope. Samples from other soils, alluvium and colluvium are shown as individual points. These samples are all from sites where the parna mantle has been mixed with material derived from local weathering or later aeolian accession. The long arrows indicate the trend in magnetic properties that reflects a fining of particle size, a decrease in the proportion of locally weathered material and a corresponding increase in the proportion of parna. Creek bed samples are largely sand and parna has been transported from the hills through lower slopes and widespread alluvial and colluvial deposits to the reservoir

laced by ephemeral creek beds choked with large volumes of locally heavy mineral-enriched sand. Small patches containing lag deposits of silt and clay also occur in the creek beds. The reservoir sediments fall within an envelope of values overlapping the parna values. The sum total of evidence from the site suggests that despite within- and between-event gradations in sediment types and sources, the bulk of the material yielded from the catchment is derived from the wind-deposited Pleistocene dust mantle.

Despite the lithological complexity of the Umberumberka catchment, the virtual ubiquity and relative homogeneity of both the reservoir sediments and the parna from which they have been largely derived makes the question of sediment source identification a comparatively simple one. No attempt has yet been made to derive quantitative estimates of the proportional contributions of different sources (cf. Stott, 1986; Smith, 1985). In the next example we summarize the results of an attempt to use magnetic measurements to provide quantitative estimates of contributions to sedimentation from distinctive source types.

In order to be fully reliable, magnetic sediment-source linkages should always be made on a particle-size related basis since, in most systems, magnetic properties and particle size are closely linked (Thompson and Oldfield, 1986).

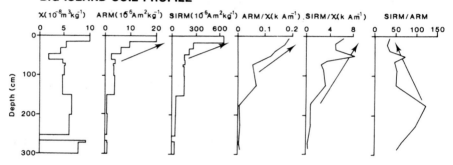

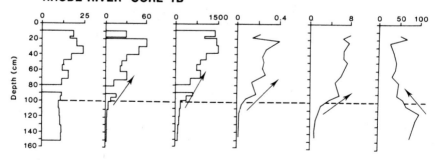

Figure 14.9 Soil profile (Big Island) and sediment core (1B) from the Rhode River, a tidal arm of Chesapeake Bay, USA. The changes in magnetic properties that distinguish weathered subsoil from surface soil in the deep profile, taken from a cliff section by the water's edge, parallel those in the sediment core at around 100 cm below the sediment/water interface. Changes in the sediment reflect an increase over time in the proportion of surface soil

Figure 14.9, illustrating results from the Rhode River, Maryland, a tidal arm of Chesapeake Bay, USA, shows how closely the changes in magnetic properties that occur upwards through the profile from unweathered parent material to surface soil in a deep cliff section are mirrored by changes within a sediment profile from the nearby estuary. This evidence, which is clearly indicative of a major shift in sediment sources over time (Oldfield et al., 1985b), can now be used to calculate and express source shifts on a quantitative basis using empirically derived multivariate mixing models and linear programming (Yu and Oldfield, 1989). Use of a wide range of magnetic measurements to characterize sources and sediments increases the confidence with which this approach can be used (Figure 14.10), but highlights the need for a more rigorous statistical approach to data processing and presentation. In this connection, the approach to cluster analysis and ordination, and the presentation of results, incorporated into the PATN package (Belbin, 1982), which was used to generate

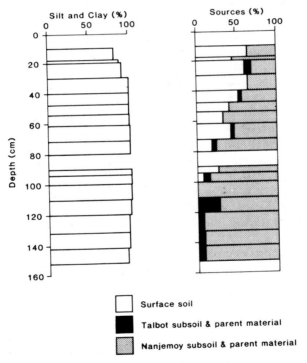

Figure 14.10 Sediment source proportions (percentages) for the silt and clay fraction of core 1B from the Rhode River, calculated by means of a multivariate mixing model derived from the magnetic properties shown in Figure 14.9 (from Yu and Oldfield, 1989). The diagram confirms a major shift in dominant sediment source types from parent material to surface soil (cf. Oldfield et al., 1985b). Nanjemoy and Talbot are two different Neogene sedimentary lithologies from which most of the sediment is derived

the ordinations shown in Figures 14.5–14.8, has proved especially useful. Used in combination with discriminant analysis (Hunt, 1986) and the multivariate modelling approach developed by Yu (1989), it promises to transform the basis for magnetically derived sediment–source linkage studies into a much more objective and quantitative research field than hitherto. The recent studies of Maher (1988) and of Maher and Taylor (1988) promise both to improve the physico-chemical basis for interpreting magnetic measurements and to increase the confidence with which, in appropriate systems, detrital, catchment-derived magnetic minerals can be used as a basis for reconstructing detailed soil-erosion histories.

Complementary evidence from magnetic and radiometric measurements for temporal and spatial shifts in sediment source in the Potomac River estuary, USA, is presented in Figures 14.11 and 14.12. The magnetic evidence (Figure 14.11) is similar to that from the Rhode River sediments (Figures 14.9 and 14.10) and shows an increase over time in input of sediment derived from surface soils, except in core 1, taken from near the mouth of the estuary. The radiometric evidence (Figure 14.12) from the same and other cores from the Potomac estuary shows that core 17, furthest from the mouth, contains far more unsupported ^{210}Pb from eroded surface soils than the downstream cores, while core 1, at the mouth, contains little more than the atmospheric flux of ^{210}Pb over the past 75 years. That the radiometric evidence supports the magnetic evidence in a system as large and complex as the Potomac River is especially encouraging, and reinforces the potential value of the magnetic approach (Oldfield *et al.*, in press). Australian studies on Lake Burley Griffin, ACT, (Caitcheon *et al.*, 1988) also confirm the potential economic effectiveness of the approach in evaluating soil-conservation strategies in rapidly eroding and sedimenting systems.

At the same time, it is necessary to point to important limitations in the magnetic approach if used in isolation. Magnetic measurements form a basis for sediment source tracing in erosion studies only when:

(1) The magnetic properties used are diagnostic of sources and relate to magnetic grains which form components of a fraction of the sediment that is significant in erosional terms;
(2) The magnetic properties of the sediments have not been subject to post-depositional changes.

Growing knowledge of the importance of atmospherically deposited magnetic particles confirms that in areas strongly affected by industrial deposition this source cannot be ignored. Even relatively remote sites in the UK (Battarbee *et al.*, 1988) and North America (Oldfield, 1985) can show a significant atmospheric input of magnetic minerals to the sediments. Many of the particulates derived from industrial combustion processes are more magnetic than most rocks and

Lake Sediment-based Studies of Soil Erosion

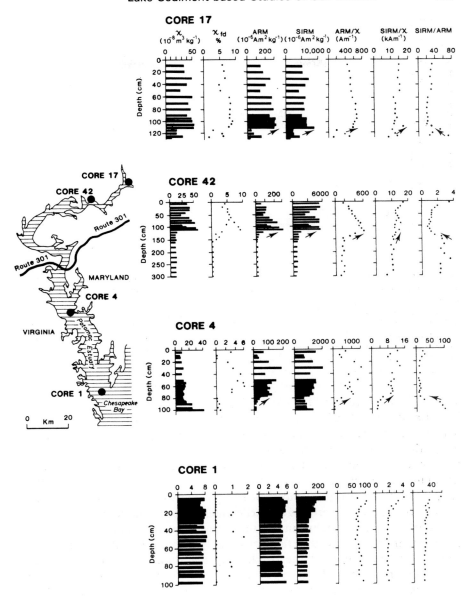

Figure 14.11 Magnetic measurements of four sediment cores from the Potomac Estuary, USA. In cores 17, 42 and 4 there are changes in magnetic properties that closely parallel those shown in Figures 14.9 and 14.10, which suggest a major shift in sediment source from subsoil to surface soil. In core 1, from the mouth of the estuary, close to the open waters of Chesapeake Bay, no such shift is apparent. Scales for each parameter change from core to core. (From Oldfield et al., in press)

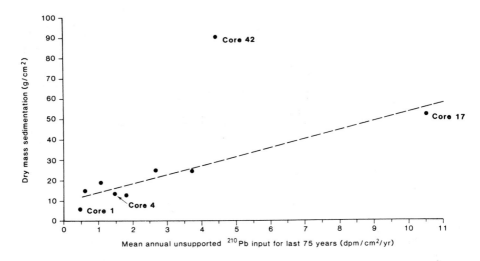

Figure 14.12 Unsupported ^{210}Pb inventories in cores from the Potomac Estuary (Figure 14.11) plotted against total dry mass sedimentation over the last 75 years (90% ^{210}Pb equilibrium depths). The total residual unsupported activity in core 1, from the mouth of the estuary, is comparable to the atmospheric flux, with no supplementary input via inwashed soils. For core 17, which is largely derived from surface soils, the value is some twenty times the atmospheric flux. The ^{210}Pb inventories thus parallel and reinforce the magnetic evidence for surface soil inwash to the upstream cores. In core 42 there is less ^{210}Pb per unit weight of sediment than at the other sites, suggesting a greater contribution from subsoils. (From Oldfield et al., in press.)

soils or the dusts (Oldfield et al., 1985c) and sediments (Scoullos et al., 1979; Puffer et al., 1980; Oldfield and Scoullos, 1984) derived from them. Atmospheric deposition can therefore make a disproportionately high contribution to magnetic concentrations in surface soils and recent sediments. Because atmospheric particulates are limited to catchment surfaces and recently accumulated material, they constitute potential tracers for surface-derived sediments. Their utility as tracers of soil particulates will be limited until we can determine the proportions of particulates deposited directly onto a lake or transported from the catchment. Further, they might act as indicators of surface processes affecting only superficial material lying on, rather than incorporated within, the upper part of the regolith. Under these circumstances, a more discriminating approach to separating out atmospheric from lithospheric components in the sedimentary magnetic record is required, as well as better information on the relative importance of atmospherically derived magnetic particulates in areas affected by industrial processes.

To complicate matters further, there have been several recent studies pointing to both the *in situ* formation of authigenic magnetic phases and the dissolution of detrital, catchment-derived magnetic minerals in Holocene sediments

(Anderson, 1986; Anderson and Rippey, 1988; Hilton and Lishman, 1985; Hilton et al., 1986; Karlin and Levi, 1983; Karlin et al., 1987; Lovley et al., 1987; Snowball and Thompson, 1988). The importance of authigenic and diagenetic effects remains a matter for further study, but in the majority of recent deposits, in all but the most organically polluted sites, the magnetic properties of the sediments are predominantly an expression of detrital input. As a general rule, the more recent and more rapidly accumulating the sediments, and the less organically productive the aquatic system with which they are associated, the more confident will be the sediment-source linkages based on magnetic measurements.

Radioisotopes

Most radiometric studies of sediment sources use only ^{137}Cs. It is worth reflecting briefly on its limitations and on the scope for using other radioisotopes alongside it. Radioactive tracers like ^{137}Cs will work effectively in erosion studies only insofar as their immediate and persistent adsorption onto fine particles can be relied on. Both theoretical considerations and empirical studies (e.g. Torgersen and Longmore, 1984; Oldfield et al., 1979) show that free-draining soils and acid organic soils and peats are unlikely to retain all deposited radiocaesium in particle-associated form. Moreover, the solubility and mobility of caesium in both acid and saline lakes and their sediments further limits its use in such environments (Longmore et al., 1986). Caesium-137 inventories and fluxes can be used in quantitative slope or catchment soil-erosion studies only where movement in solution is negligible. In these cases, its relative ease of measurement still makes it the most attractive option, an option that has been both enhanced and complicated over large areas of northern and western Europe by the deposition from the Chernobyl accident. Lake and reservoir sediment inventories and profiles of ^{137}Cs activity can complement catchment and stream-based measurements and reinforce other lines of evidence for sediment sources, erosion processes and rates of material flux.

The development of low-background gamma detectors opens up the possibility of measuring non-destructively, and of using as tracers, a wider range of radioisotopes and isotopic ratios in sediment source studies. Lead-210, one of the most useful radioisotopes from a chronological point of view, is especially attractive as a tracer too, since only exposed surfaces will receive the unsupported atmospherically derived ^{210}Pb that is out of equilibrium with, and hence in excess of, its parent lithospheric radioisotope radium-226 (^{226}Ra). Its half-life of 22.26 years limits its use as a tracer to sediments that have accumulated over the last few decades only, but within the timespan, high unsupported ^{210}Pb inventories in sediment cores will tend to indicate a soil source, low values a source deeper in the regolith, below the influence of atmospheric deposition (Figure 14.12). Moreover, lead would appear to be significantly less mobile than caesium in acid and organic soils. Wasson *et al.* (1987) elegantly illustrate the use of ^{210}Pb as a source tracer for the sediments of Burrinjuck Reservoir in eastern New South Wales, Australia, using the destructive alpha method of radioassay. Their results are discussed more fully in the next section and illustrated in Figure 14.15. The approach via non-destructive gamma measurement requires the most sophisticated of detection and screening systems (Appleby *et al.*, 1986, 1988) but it has the added advantages of small sample size and the detection, alongside radio-lead and radio-caesium, of a further range of isotopes in the thorium and uranium decay series (cf. Plater, *et al.*, 1988). This technology, which, like magnetic measurements, is non-destructive and totally compatible with virtually every other type of analysis, opens up exciting new prospects in sediment-source studies.

Burrinjuck Reservoir: Lead-210 and Pollen Analysis

Pollen analysis, the third and potentially one of the most sophisticated techniques for sediment-source tracing, is most conveniently introduced alongside a more detailed illustration of the ^{210}Pb-based approach to sediment-source tracing, through further consideration of results from the study of sediments accumulating in Burrinjuck Reservoir. The next section introduces the site and sediments preliminary to returning to our main concern with source tracing.

Burrinjuck Reservoir, at maximum water depth, covers an area of 57 km^2 and receives drainage from some 13 500 km^2 of land surface. There are three arms to the reservoir along the Yass, Goodradigbee and Murrumbidgee Rivers (Figure 14.13). The lands draining into the Yass are mostly pasture, those of the Goodradigbee mostly native *Eucalyptus* forests and those of the Murrumbidgee about half pasture and half native forests, with plantations of exotic conifers. The Canberra urban area lies within the Murrumbidgee catchment.

Figure 14.14 shows the cumulative dry weight of sediment at a site near the dam wall, plotted against a detailed timescale derived from a wide range of evidence summarized in Clark (1986). Over the 60 years since infilling began, the rate of sedimentation has tended to decrease overall. From this and other reservoir-based studies this is seen to be a widespread trend in south-eastern Australia and may be due to the effects of a vegetation response to increased rainfall after the mid-1940s as well as to massive reductions in rabbit populations in the 1950s. The graph also shows that some 57 per cent of the sediment was deposited in only 5 per cent of the years; indeed, most of it was deposited in three events. Only one of these *sediment yield* events, the first (in 1925) would have been predicted from rainfall or flood-frequency records. The 1945 and 1983 events represent extreme *local* erosion in summer storms at the end of droughts. In 1983, the high sediment yield at this site arose from heavy rainfall in overgrazed and almost totally devegetated pasture and from ploughed fields

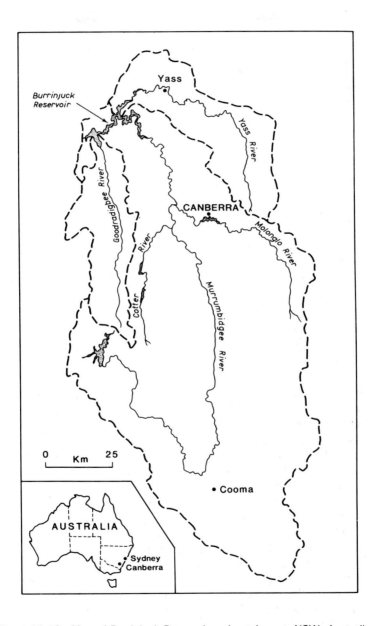

Figure 14.13 Map of Burrinjuck Reservoir and catchment, NSW, Australia

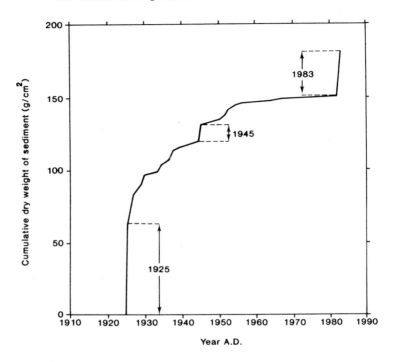

Figure 14.14 Cumulative dry weight of sediment in a core taken from 2 km upstream of the dam wall of Burrinjuck Reservoir plotted against a detailed timescale from Clark (1986). The three main sedimentation events of 1925, 1945 and 1983 are separately identified

left waiting for rain, as well as some resuspension of exposed upstream sediments. The sediment-based evidence thus highlights the need to understand the significance of antecedent conditions and the crucial role of land management in erosion within the region.

Figure 14.15 plots decay-corrected ^{210}Pb:^{226}Ra ratios for sediment columns from each of the arms of the reservoir (Wasson et al., 1987). Only in situations where a good independent chronology is available (Clark, 1986; Clark and Wasson, 1986) is it possible to correct the measured ^{210}Pb values for decay and so estimate the original ratios. ^{226}Ra is the parent radioisotope of ^{210}Pb, therefore in rocks and in the deeper layers of the regolith they will exist roughly in equilibrium with each other. Since atmospherically deposited unsupported (excess) ^{210}Pb will be limited to surface soils and plough layers, sediment with an initial or decay-corrected ratio of ^{210}Pb to ^{226}Ra in excess of the equilibrium ratio found below the atmospherically enriched layer must include a contribution from near-surface sources. Each of the sediment columns from the reservoir arms has a distinctive ^{210}Pb:^{226}Ra ratio trace. In the Murrumbidgee arm,

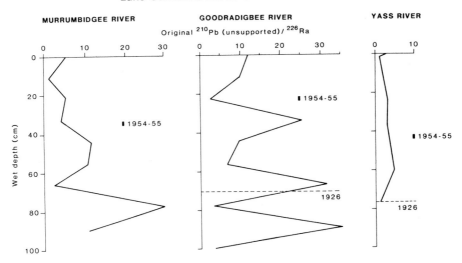

Figure 14.15 Decay-corrected ^{210}Pb:^{226}Ra ratios for sediment columns from each of the three arms (see Figure 14.13) of Burrinjuck Reservoir (from Wasson et al., 1986). The 1954-5 dates were determined from ^{137}Cs analyses; the dashed lines mark the stratigraphic boundary at the top of the sediment deposited in the 1925 flood in the Goodradigbee and Yass arms. For detailed interpretation, see text

decreasing topsoil contributions over time, suggested by the decreasing ratio, match historical records of earlier sheet erosion exacerbated by the effects of rabbits, followed by later subsoil erosion from gullying and channel widening. In the Goodradigbee catchment, which is relatively little disturbed or deforested and almost devoid of gullies, the ratio remains relatively high through to the present day, suggesting that the bulk of the sediment has come from topsoils. In the Yass arm, topsoils were probably eroded late last century before the Burrinjuck dam was built. The low ratio throughout suggests a dominance of subsoil erosion, and this is consistent with the high degree of active gully development within the catchment.

In turning from radioisotopes to pollen, we now consider evidence from the same core used for Figure 14.14 and taken closer to the dam wall of the reservoir where sedimentation receives contributions from all three catchment and river arms. Most palynologists have used their technique for vegetation and land-use reconstruction, proxy dating or palaeoclimatic inferences via a variety of more or less statistically sophisticated approaches. The requirements for using pollen analytical evidence for sediment-source tracing are that pollen types preserved in the sediment can be attributed to sources in identifiable vegetation and land-use categories within the catchment and that waterborne pollen predominates over airborne pollen deposited in the lake or reservoir. Failure to use pollen concentrations and relative frequencies hitherto in this way reflects

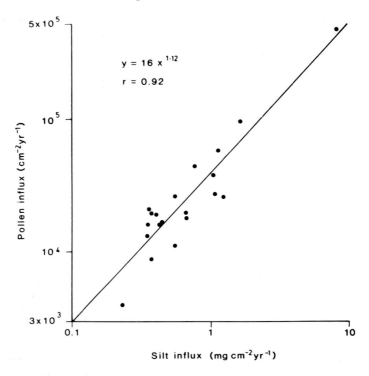

Figure 14.16 Annual pollen influx to the Burrinjuck sediments plotted against annual silt (2–63 µm) influx, both on logarithmic scales (from Clark, 1986)

the legacy of past tradition and disciplinary demarcation, rather than any rigorous evaluation of their sediment-source related responses. Large numbers of pollen grains are carried by streams (Crowder and Cuddy, 1973; Starling and Crowder, 1981; Brown, 1985) and several studies have shown a significant or dominant input to lake sediments of stream-borne pollen (Peck, 1973; Bonny, 1976, 1978; Pennington, 1978). Pollen grains range in size from 5 µm to 200 µm diameter, with most between 15 µm and 40 µm, but as their density is low they can be expected to move in air and water with smaller mineral particles. Clark (1986 and Figure 14.16) shows the excellent correlation between annual pollen and silt (2–63 µm) influxes to the sediments of Burrinjuck Reservoir between 1938 and 1957. If the pollen were predominantly air-borne, there would be no correlation.

In Figure 14.17 the three dominant pollen types are plotted as percentages of total annual pollen influx to the sediments in the years 1938–57 using the published chronology (Clark, 1986). Myrtaceae pollen comes mostly from native forests of *Eucalyptus* and related genera; these are widespread in the

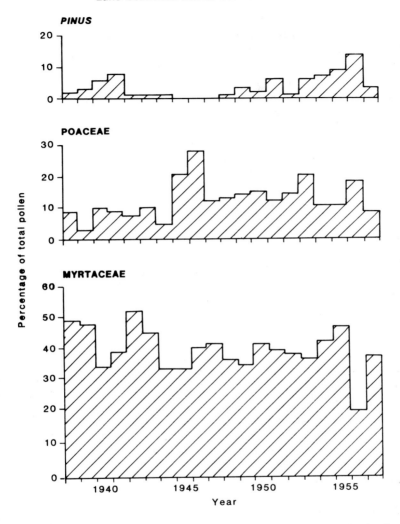

Figure 14.17 Percentages of the three major pollen types deposited in the Burrinjuck sediments in the years 1938–57 (from Clark, 1986). See text for interpretation

Goodradigbee and western Murrumbidgee catchments. Poaceae (grass) pollen comes mostly from pastures, especially in the Yass catchment and lower Murrumbidgee where they dominate the land use. *Pinus* pollen comes from exotic plantations largely in the Murrumbidgee catchment.

From these pollen-source linkages, deductions can be made regarding the areas, source-type and processes generating sediment at different times. For example, most of the 1942 sediment appears to come from the native forest areas of the Goodradigbee and Murrumbidgee catchments. From 1945 to 1947, a drought and post-drought period, there is no evidence for a significant sediment yield from the areas of pine forest in the Murrumbidgee, but in 1945 and 1946 most sediment came from the pasturelands of the Yass and Murrumbidgee. In the 1956 flood, the pollen content points to a sediment yield from both the grassland and pine forest areas of the Yass and Murrumbidgee, but not from the native woodlands of the Goodradigbee and Murrumbidgee. Such catchment- and land-use-related inferences have yet to be fully exploited, for they open up the possibility of establishing direct links with the instrumental record for specific areas and seasons. Indeed, this aspect of the methodology merits comprehensive development and evaluation in order to establish the spatial scales, land-use and vegetation contexts and hydrological systems within which its value can be maximized.

Synopsis

In the Burrunjuck project, results from magnetic, radiometric, pollen and other studies of sediments and soils are not yet complete. When all analyses have been made on the same cores and sediment samples then it will be possible to integrate the results and verify or contradict the interpretations given above. This highlights the importance of using independent lines of evidence for sediment sources *in conjunction*. Magnetic properties themselves can provide a range of independent quotients; non-destructive low-background gamma assay can contribute to both dating and sourcing through a whole series of independent concentrations, fluxes, inventories and ratios; and pollen can provide vital information on temporal changes in land use, on chronology over a wide range of timescales and, where appropriate, complementary evidence on sediment sources.

PROSPECTS

One of the most frequently expressed requirements for setting priorities for future studies of environmental change is the need to resolve change on timescales of decades and centuries. One of the great virtues of lake and reservoir sediment-based studies is their ability to do just this and, in so doing, provide an *empirical* basis for exploring links between the full range of interacting anthropogenic and non-anthropogenic processes. Such studies must be central to any research on global change and, more especially, on its local and regional environmental impacts. A concern with soil-erosion processes and rates must be an important component of this research. Linking our current concerns to the research issues that seem likely to command maximum attention in the near future will require, above all:

(1) Choice of environmental contexts where observation, experiments and historical reconstruction can not only be united but can also shed light on the response to secular environmental variations of rate and amplitude comparable to that envisaged for the near future;
(2) Collaboration on a multidisciplinary basis within these environmental contexts on a scale and with a coherence rarely achieved. In particular, there is an urgent need to communicate across the distance which still divides empirical sediment-based studies of the type summarized here and the development of models of environmental systems change;
(3) Broadening of our research priorities beyond agricultural field erosion into shared concerns with total material flux, of which field erosion is a part, with systems linkages and with coupled or nested scales of spatial and temporal resolution;
(4) Willingness to scale up spatially towards the global level by working together, for example, in near-shore sediment sequences that can form a bridge between the small-scale terrestrial fine time-resolution studies with which we are familiar and the coarse time-resolution records such as can be obtained from deep-sea sediments;
(5) Acknowledgement that the effectiveness with which we manage the changing conditions envisaged for the near future will depend in part on the extent to which we come to understand the interactions between, and consequences of, anthropogenic impacts and natural secular variation in the recent past.

REFERENCES

Anderson, N. J. (1986). *Recent Sediment Accumulation in a Small Lake Basin with Special Reference to Diatoms*, Unpublished PhD thesis. University of London.
Anderson, N. J. and Rippey, B. (1988). Diagenesis of magnetic minerals in recent sediments of a eutrophic lake, *Limnol. Oceanogr.*, **33**, 1476–92.
Appleby, P. G. and Oldfield, F. (1978). The calculation of lead 210 dates assuming a constant rate of supply of unsupported ^{210}Pb to the sediment, *Catena*, **5**, 1–8.
Appleby, P. G. and Oldfield, F. (1983). The assessment of ^{210}Pb data from sites with varying sediment accumulation rates, *Hydrobiol.*, **103**, 29–35.
Appleby, P. G., Nolan, P. J., Gifford, D. W., Godfrey, M. J., Oldfield, F., Anderson, N. J. and Battarbee, R. W. (1986). Lead-210 dating by low background gamma counting, *Hydrobiol.*, **143**, 21–7.
Appleby, P. G., Nolan, P. J., Oldfield, F., Richardson, N. and Higgitt, S. R. (1988). ^{210}Pb dating of lake sediments and ombrotrophic peats by gamma assay, *Sci. Tot. Env.*, **69**, 157–77.
Battarbee, R. W. (1978). Observations on the recent history of Lough Neagh and its drainage basin, *Phil. Trans. Roy. Soc. B*, **281**, 303–45.
Battarbee, R. W. (1986). The eutrophication of Lough Erne inferred from changes in the diatom assemblages of ^{210}Pb- and ^{137}Cs-dated sediment cores, *Proc. R. Irish Acad.*, **86B**, 141–68.

Battarbee, R. W. and Flower, R. J. (1983). Diatom evidence for recent acidification of two Scottish lochs, *Nature*, **305**, 130-33.

Battarbee, R. W., Anderson, N. J., Appleby, P. G., Flower, R. J., Fritz, S. C., Haworth, E. Y., Higgitt, S. R., Jones, V. J., Kreiser, A., Munro, M. A. R., Natkanski, J., Oldfield, F., Patrick, S. T., Richardson, N., Rippey, B. and Stevenson, A. C. (1988). *Lake Acidification in the United Kingdom 1800-1986. Evidence from Analysis of Lake Sediments*, Report to the Department of the Environment, Palaeoecology Research Unit, University College, London; Ensis Press.

Belbin, L. (1987). *PATN Pattern Analysis Package*, CSIRO, Division of Wildlife and Rangelands Research, Canberra.

Bloemendal, J. (1982). *The Quantification of Rates of Total Sediment Influx to Llyn Goddionduon, Gwynedd, North Wales*, Unpublished PhD thesis, University of Liverpool.

Bloemendal, J., Oldfield, F. and Thompson, R. (1979). Magnetic measurements used to assess sediment influx at Llyn Goddionduon, *Nature*, **280**, 50-53.

Bonny, A. P. (1976). Recruitment of pollen to the seston and sediment of some Lake District lakes, *J. Ecol.*, **64**, 859-87.

Bonny, A. P. (1978). The effect of pollen recruitment processes on pollen distribution over the sediment surface of a small lake in Cumbria, *J. Ecol.*, **66**, 385-416.

Brown, A. G. (1985). The potential use of pollen in the identification of suspended sediment sources, *Earth Surf. Proc. and Landf.*, **10**, 27-32.

Caitcheon, G. G., Hammond, R. P., Wasson, R. J., Wild, B. A. and Willett, I. R. (1988). The Lake Burley Griffin study: its implications for catchment management. In *Conference on Agricultural Engineering, 1988*, Institution of Engineers, Australia, publication No. 88/12, 216-20.

Chartres, C. J. (1982). Quaternary dust mantle soils in the Barrier Range, N.S.W., In Wasson, R. J. (ed.), *Quarternary Dust Mantles of China, New Zealand and Australia*, Department of Biogeography and Geomorphology, Australian National University, Canberra, pp. 153-60.

Clark, R. L. (1986). Pollen as a chronometer and sediment tracer, Burrinjuck Reservoir, Australia, *Hydrobiol*, **143**, 63-9.

Clark, R. L. and Wasson, R. J. (1986). Reservoir sediments. In De Deckker, P. and Williams, W. D. (eds), *Limnology in Australia*, CSIRO, Melbourne and Junk, Dordrecht, 497-507.

Crowder, A. A. and Cuddy, D. G. (1973). Pollen in a small river basin: Wilton Creek, Ontario, In Birks, H. J. B. and West, R. G. (eds), Blackwell, Oxford, pp. 61-77.

Dearing, J. A. and Foster, I. D. L. (1987). Limnic sediments used to reconstruct sediment yields and sources in the English Midlands since 1765. In Gardiner, V. (ed.), *International Geomorphology*, John Wiley, Chichester, pp. 853-68.

Dearing, J. A., Elner, J. K. and Happey-Wood, C. M. (1981). Recent sediment flux and erosional processes in a Welsh upland lake-catchment based on magnetic susceptibility measurements, *Quat. Res.*, **16**, 356-72.

Dearing, J. A., Hakansson, H., Liedberg-Jönsson, R., Perrson, A., Skansjö, S., Widholm, D. and El-Daoushy, F. (1987). Lake sediments used to quantify the erosional response to land use change in southern Sweden, *Oikos*, **50**, 60-78.

Foster, I. D. L., Dearing, J. A., Simpson, A., Carter, A. D. and Appleby, P. G. (1985). Lake catchment based studies of erosion and denudation in the Merevale Catchment, Warwickshire, UK, *Earth Surf. Proc. Landf.*, **10**, 45-68.

Hilton, J. and Lishman, J. P. (1985). The effect of redox change on the magnetic susceptibility of sediments from a seasonally anoxic lake, *Limnol. and Oceanogr.*, **30**, 907-9.

Hilton, J., Lishman, J. P. and Chapman, J. S. (1986). Magnetic and chemical characterization of a diagenetic magnetic mineral formed in the sediments of productive lakes, *Chem. Geol.*, **56**, 325–35.

Hunt, A. (1986). The application of mineral magnetic methods to atmospheric aerosol discrimination, *Phys. Earth Planet. Inter.*, **42**, 10–21.

Karlin, R. and Levi, S. (1983). Diagenesis of magnetic minerals in recent hemipelagic sediments, *Nature*, **303**, 327–30.

Karlin, R., Lyle, M. and Heath, G. R. (1987). Authigenic magnetite formation in suboxic marine sediments, *Nature*, **326**, 490–93.

Longmore, M. E., Torgersen, T., O'Leary, B. M. and Luly, J. S. (1986). Caesium-137 redistribution in the sediments of the playa, Lake Tyrrell, Northwestern Victoria. I. Stratigraphy and Caesium-137 mobility in the upper sediments, *Palaeogeog. Palaeolim. Palaeoecol.*, **54**, 181–95.

Lovley, D. R., Stolz, J. F., Nord, Jr., G. L. and Phillips, E. J. P. (1987). Anaerobic production of magnetite by a dissimilatory iron-reducing microorganism, *Nature*, **330**, 19.

Maher, B. A. (1988). Magnetic properties of some synthetic sub-micron magnetites, *Geophys. J. r. astr. Soc.*, **94**, 83–96.

Maher, B. A. and Taylor, R. G. (1988). Formation of ultrafine-grained magnetite in soils, *Nature*, **336**, 368–70.

Miller, K. M. and Heit, M. (1986). A time resolution methodology for assessing the quality of lake sediment cores that are dated by ^{137}Cs, *Limnol. Oceanogr.*, **31**, 1292–1300.

Oldfield, F. (1985). Magnetic studies of lake sediments and sources at Mirror Lake, New Hampshire, *National Geographic Society Research Reports*, **21**, 339–43.

Oldfield, F. and Appleby, P. G. (1984a). A combined radiometric and mineral magnetic approach to recent geochronology in lakes affected by catchment disturbance and sediment redistribution, *Chem. Geol.*, **44**, 67–83.

Oldfield, F. and Appleby, P. G. (1984b). Empirical testing of ^{210}Pb dating models for lake sediments. In Haworth, E. Y. and Lund, J. W. G. (eds), *Lake Sediments and Environmental History*, Leicester University Press, pp. 93–114.

Oldfield, F. and Scoullos, M. (1984). Particulate pollution monitoring in the Elefsis Gulf: The role of mineral magnetic studies, *Marine Polln. Bull.*, **15**, 6, 229–31.

Oldfield, F., Appleby, P. G., Cambray, R. S., Eakins, J. D., Barber, K. G., Battarbee, R. W., Pearson, G. R. and Williams, J. M. (1979). [210]Pb, ^{137}Cs and [239]Pu profiles in ombrotrophic peat, *Oikos*, **33**, 40–45.

Oldfield, F., Appleby, P. G. and Worsley, A. T. (1985a). Evidence from lake sediments for recent erosion rates in the Highlands of Papua New Guinea, In Douglas, I and Spencer, T. (eds), *Environmental Change and Tropical Geomorphology*, Allen and Unwin, London, pp. 185–95.

Oldfield, F., Maher, B. A., Donoghue, J. and Pierce, J. (1985b). Particle size related, mineral magnetic source sediment linkage in the Rhode River Catchment, Maryland, USA, *J. Geol. Soc.*, **142**, 1035–46.

Oldfield, F., Hunt, A., Jones, M. D. H., Chester, R., Dearing, J. A., Olsson, L. and Prospero, J. M. (1985c). Magnetic differentiation of atmospheric dusts, *Nature*, **317**, 516–18.

Oldfield, F., Appleby, P. G. and Maher, B. A. (in press). Sediment source variations and lead-210 inventories in recent Potomac Estuary sediment cores, *J. Quat. Sci.*

O'Sullivan, P. E. (1983). Annually-laminated lake sediments and the study of Quaternary environmental change—a review, *Quat. Sci. Rev.*, **1**, 245–313.

Peck, R. M. (1973). Pollen budget studies in a small Yorkshire catchment. In Birks, H. J. B. and West, R. G. (eds), *Quaternary Plant Ecology*, Blackwell, Oxford, pp. 43-60.

Pennington, W. (1979). The origin of pollen in lake sediments: an enclosed lake compared with one receiving inflowing streams. *New Phytol.*, **83**, 189-213.

Plater, A. J., Dugdale, R. E. and Ivanovich, M. (1988). The application of uranium disequilibrium concepts to sediment yield determination. *Earth Surf. Proc. and Landf.*, **13**, 171-82.

Puffer, J. H., Russell, E. W. B. and Rampino, M. R. (1980). Distribution and origin of magnetite spherules in air, water and sediments of the greater New York city area and the north Atlantic Ocean, *J. Sed. Pet.*, **50**, 247-65.

Scoullos, M., Oldfield, F. and Thompson, R. (1979). Magnetic monitoring of marine particulate pollution in the Elefsis Gulf, Greece, *Marine Polln. Bull.*, **10**, 287-91.

Smith, J. (1985). *Mineral Magnetic Studies of two Shropshire-Cheshire Meres*, Unpublished PhD thesis, University of Liverpool.

Snowball, I. and Thompson, R. (1988). The occurrence of greigite in freshwater sediments from Loch Lomond, *J. Quat. Sci.*, **3**, 121-5.

Starling, R. N. and Crowder, A. (1981). Pollen in the Salmon River System, Ontario, Canada, *Rev. Palaeobot. Palynol.*, **31**, 311-34.

Stott, A. P. (1986). Sediment tracing in a reservoir-catchment system using a magnetic mixing model, *Phys. Earth Planet. Inter.*, **42**, 105-15.

Thompson, R. and Oldfield, F. (1986). *Environmental Magnetism*, Allen and Unwin, London.

Torgersen, T. and Longmore, M. E. (1984). Cs-137 diffusion in highly organic sediments of Hidden Lake, *Aust. J. Mar. Freshwater Res.*, **35**, 537-48.

Wasson, R. J. and Galloway, R. W. (1986). Sediment yield in the Barrier Range before and after European settlement, *Australian Rangelands Journal*, **8**, 79-90.

Wasson, R. J., Clark, R. L., Nanninga, P. M. and Waters, J. (1987). ^{210}Pb as a chronometer and tracer, Burrinjuck Reservoir, Australia, *Earth Surf. Proc. and Landf.*, **12**, 399-414.

Yu, L. (1989). *Environmental Applications of Mineral Magnetic Measurements: towards the Quantitative Approach*, Unpublished PhD thesis, University of Liverpool.

Yu, L. and Oldfield, F. (1989). A multivariate mixing model for identifying sediment source from magnetic measurements, *Quat. Res.* **32**, 168-81.

Historical Surveys

15 Soil Erosion: Its Impact on the English and Welsh Landscape since Woodland Clearance

R. EVANS
Department of Geography, University of Cambridge

INTRODUCTION

At present, erosion is widespread in the lowlands and uplands of England and Wales (Evans and Cook, 1986). Water and wind act on bare soils of arable land in the lowlands; and water, wind, frost and animals affect the uplands where soil is exposed by animals, pollution and fire.

The impacts of water erosion on arable land have been recently monitored (Evans, 1988) but little is known, except in restricted areas, of either the impacts of wind erosion on arable land (e.g. Wilkinson *et al.*, 1969) or of erosion in the uplands (e.g. Tallis, 1985; Evans, in press). This chapter concentrates on the impact of water erosion in the past.

To assess the long-term impact of erosion on crop yield we need to know how much topsoil has been stripped off in the past. However, although there is much evidence for past erosion, there is little quantitative information. The approach used here is of necessity a broad one, because there is little information for individual localities about the impacts of past erosion. Also lacking for large areas are good records of land-use changes in the archaeological and historical past since the primeval woodland was cleared.

EVIDENCE OF EROSION IN THE PAST

There is much information on the occurrence of erosion in the past, both as physical features on the ground and in sedimentary sequences excavated and examined by, among others, palaeobotanists and archaeologists.

Soil Erosion on Agricultural Land
Edited by J. Boardman, I. D. L. Foster and J. A. Dearing
©1990 John Wiley & Sons Ltd

Visible Morphological Evidence

In the lowlands benches or steps are often found in valley floors where eroded soil has been held behind field boundaries. They are seen in the sandlands of the Midlands, on the Silurian dipslope in the Midlands, on chalk Downlands, in valleys cut into the till plateau of Essex, and in Norfolk, among other localities. The valley steps must have formed since the land was enclosed, and this can be quite recently, as on the Permo-Triassic sandstones of Nottinghamshire.

The distribution of soils may also indicate soil erosion in the recent past: for example, near Mansfield, Nottinghamshire. Here podzolic soils formed under the heath and woodland of Sherwood Forest. On slopes, the bleached and underlying iron- and humus-enriched horizons have been mostly eroded away since the land was enclosed for agriculture about 200 years ago, leaving shallow brown sands over rock.

Stratigraphic evidence

Evidence for water erosion is often in the form of mineral deposits within or on top of either lake muds (Tutin, 1969) or peat (Brown, 1982). The most extensive evidence is the alluvium deposited on floodplains and the colluvium on floors of dry valleys. These deposits, defined by Avery (1980), have been mapped by the Soil Survey throughout England and Wales.

TIME OF EROSION IN THE PAST

At present, erosion is more widespread and more frequent than two decades ago, because most erosion occurs in autumn sown crops and the area covered by these crops has increased by a factor of three (Evans and Cook, 1986). Land is under greater pressure now than previously, for a number of reasons (CPRE, 1985), and erosion probably occurs widely and frequently when the land is under pressure.

Can we identify when these periods of land pressure occurred? A crude attempt can be made by reconstructing in a generalized way the past land use of England and Wales based upon population pressures.

Documentary evidence of past erosion is very scarce. A few instances are known in Cornwall (Pickup, pers. comm.) and Devon (Clayden, 1964), when eroded soil deposited on valley floors was put back on the adjacent slopes. Jennings (pers. comm.) reported a tenancy agreement made in Sussex in the nineteenth century which stipulated that eroded soil should be replaced on slopes, presumably because erosion had been a problem. Mostly, however, erosion was not noted.

Most evidence for dating erosion comes from material contained in deposits, mineral or organic, in valley floors, lakes and moorland peat bogs. The sediments are dated by: pot sherds or other artefacts (Bell, 1981, 1982), pollen analysis (Pennington, 1969), radiocarbon dating (Hazelden and Jarvis, 1979; Shotton, 1978), molluscs (Evans, 1975) or combinations of these (Robinson and Lambrick, 1984).

Population in the Past

Prior to about 2500 BC the population of England was very low, and mostly concentrated in the south (Figure 15.1). It then increased rapidly, especially in Roman times, covering much of England and Wales, although thinly in the uplands. The population then dropped markedly, probably as a result of disease and weaker governance (Jones, 1979). Thereafter it slowly recovered with the onset of strong government after 1066 until the devastating impact of the Black death in 1349. Just prior to the Black Death population was similar to or greater than that in Roman times (Hallam, 1981). In the medieval period it was mainly south of the Humber in eastern England, and in southern and central lowland England (Butlin, 1982). Population increased rapidly in the sixteenth and early seventeenth centuries. For a century thereafter it was static or declined slightly, before increasing very rapidly after the 1750s.

Prior to the Roman period pressure on the land was generally slight, as there was always sufficient land to clear and use for arable. A land shortage would not be felt again until medieval times, and then in the seventeenth and eighteenth centuries.

Land Use and Erosion in the Past

Much has been written on man's impact on the vegetation of England and Wales and the survey below is not exhaustive. Until about 5000 years ago England and Wales was clothed by woodland (Evans, 1975; Pennington, 1969; Rackham, 1976), even at high altitudes (Turner, 1984). River channels were stable and small amounts of mostly organic sediment were deposited in streams and lakes (Bennett, 1983; Jones *et al.*, 1985; Pennington, 1980; Robinson and Lambrick, 1984; Rose *et al.*, 1980; Scaife and Burrin, 1983). Mesolithic man then began to clear the woods by axe and fire, but his clearings were often small and erosion was localized (Limbrey, 1983); for example in the North Yorkshire Moors (Simmons *et al.*, 1975). Only in Sussex is there evidence of widespread erosion (Scaife and Burrin, 1983).

Neolithic man cleared much of the chalk outcrop (Evans, 1972) and other parts of southern and eastern England of its woodland, and crops were widely grown. This sedentary agriculture was accompanied by water erosion on the chalk Downlands (Bell, 1982; Evans, 1972). Possibly in this period the topsoils

234

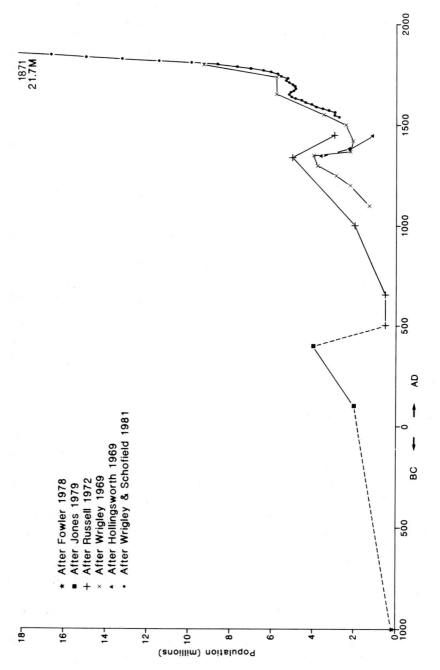

Figure 15.1 Population of England and Wales 1000 BC–AD 1871

were stripped from valley floors exposing argillic horizons upon which later colluvial deposits were laid (Smith, 1989).

There was much erosion on the chalk in Bronze and Iron Age times (Bell, 1982), probably associated with the use of better ploughs, a wider variety of crops (Jones, 1981), increasing population pressure and a change to a wetter climate about 500 BC (Lamb, 1982). Arable land use initiated the deposition of alluvium in river valleys in central and southern England (Brown and Barber, 1985; Hazelden and Jarvis, 1979; Limbrey, 1983; Shotton, 1978; Robinson and Lambrick, 1984). High moors were settled in the Bronze Age (Limbrey, 1983) and localized erosion took place when land was converted to arable (Simmons et al., 1975), as it did in the clearances within the wetter lowlands, such as Shropshire (Beales, 1980). Much land was deforested in Iron Age times (for instance, the lowland margins and uplands in Wales and the West Midlands (Brown and Barber, 1985), again leading to localized erosion, as in the Brecon Beacons (Crampton, 1969).

In Roman times the expanded population caused extensive erosion on the chalk, and in the lowlands alluvium was deposited along major valleys (Bell, 1982). Land was cleared in northern and western England and in Wales (Wiltshire and Moore, 1983). Winter wheat was widely grown (Jones, 1981), and in the Cotswolds land formerly under grass was used to grow cereals (Robinson and Lambrick, 1984).

Many alluvial sequences show a buried topsoil which may equate with the marked drop in population in the early Dark Ages (Figure 15.1), when much of the land reverted to woodland and scrub (Limbrey, 1983). Rates of alluviation in the Thames valley declined in early Saxon times (Robinson and Lambrick, 1984).

As the woodlands of central England were cleared by the Anglo-Saxons and much of this heavy land turned over with the mouldboard plough, so erosion was again initiated, and alluviation in some river valleys in central and eastern England began in this era (Bell, 1982; Evans, 1981a). The Norse invasion initiated woodland clearance in northern England, resulting in local erosion in the Lake District (Pennington et al., 1976), Howgill Fells (Harvey et al., 1981), Bowland Fells (Harvey and Renwick, 1987), the Yorkshire Dales (King, 1978; Smith, 1975) and probably the North York Moors (Richards et al., 1987). Tallis (1985) notes that erosion of peat in the southern Pennines was initiated about AD 900, and this could be partly attributable to clearance.

As population increased in the Medieval Period (Figure 15.1) so more woodland was chopped down to reach the level which survived until the 1960s (Rackham, 1986). The ridge and furrow of the open-field system was extensive throughout the lowlands and particularly the English Midlands (Baker and Butlin, 1973; Hall, 1982). In part of Warwickshire the arable was 94 per cent in 1300–13 compared to 57 per cent in about 1500 (Dyer, 1988). The medieval population was highest in East Anglia, central England and southern England on the chalk. Medieval colluvium is widespread on the South Downs (Bell, 1982; 1986) and has been found in the Craven lowland, North Yorkshire (Smith, 1975), and much of the colluvium found in valleys on Devonian rocks in Herefordshire and on drifts in Essex, for example, could well date from this period. Water flowing into Llangorse Lake, South Wales was red-brown with silt in the late twelfth century (Rhys, 1908, quoted in Jones *et al.*, 1985).

Because of population pressure, the Cistercians placed their abbeys and farms around the margins of settled England and Wales (Donkin, 1978). They (re-)introduced sheep and cattle to large areas of the English and Welsh uplands, removing scrub to improve the grazings (Donkin, 1978; Eyre, 1957). Wiltshire and Moore (1983) noted increased grazing pressures in the Cambrian Mountains, central Wales, at this time as a result of Cistercians founding an abbey at Strata Florida. The disturbance of the convex margins of the peat moors by clearance, fire and grazing, and the possible increase in runoff associated with these changes, may well have initiated or exacerbated erosion of the peat (Evans, in press).

With the Black Death and onset of a cooler climate, much arable land reverted to pasture (Parry, 1978) and has remained so ever since. By the eighteenth century the only land in lowland England not enclosed was heath and forest on poor sandy soils or thin soils on rock, but much of this was enclosed and ploughed at the onset of industrialization in the late eighteneth and mid-nineteenth centuries (Figure 15.1). The initiation of erosion on the sandlands in Nottinghamshire and the West Midlands probably dates from this period. In the uplands erosion was renewed because of changes in vegetation to heather and grass moorland caused mostly by industrial pollution in the southern Pennines (Tallis, 1985) or by final clearance of woodland to create heather moorland managed by burning in the North York Moors (Richards *et al.*, 1987). Erosion continues in these areas to the present.

The population pressure of the Georgian era was accompanied by a revolution in agricultural technology; not only was equipment improved and better animals bred, but crop rotation was introduced, especially the introduction of grazed grass leys and root crops. The fallow, previously used to restore fertility and cleanse the land of weeds, was superseded. These changes not only increased food production but may have curtailed the frequency of erosion.

Since the 1750s stress on arable land has often been relieved by importing food. However, during the Napoleonic Wars and the Victorian high farming period of the 1870s just prior to the opening up of the British markets to grain exports from the United States and elsewhere, the land was under pressure. Erosion may have been more widespread in these periods.

NEED FOR QUANTITATIVE DATA

There is much evidence therefore that water erosion has occurred widely in the past, but its quantitative impact on soils and landscape has hardly been considered (Bell, 1986; Evans, 1981a; Smith, 1975). To assess it, the volume of sediment deposited on valley floors or footslopes is divided by the area of the drainage basin from which the soil has been eroded to give a value of surface lowering of the land. However, there are two drawbacks to this approach. One is that deposits can only be related to the adjacent slope, and not to a wider area unless a number of assumptions are made. Nevertheless, Evans (1981a) and Bell (1986), examining these data for a number of sites, have calculated similar depths of topsoil stripped from slopes adjacent to valley floors. The other drawback has been the limited number of published descriptions of sections excavated by archaeologists. Detailed soil maps and the texts accompanying them can help overcome these drawbacks.

AMOUNTS OF SOIL ERODED IN THE PAST

The Soil Survey of England and Wales (now Soil Survey and Land Research Centre) has published 105 maps at a scale of 1:25 000 (Soil Survey, 1988), with field work based on 1:10 000 maps. Most maps cover 100 km^2 but a few up to 200 km^2. The maps were chosen to be representative of large areas of surrounding land and are widely scattered throughout England and Wales.

Method

Estimates were made of the volume of colluvium or alluvium stored in small and medium-sized drainage basins, in which all the colluvium or alluvium was included within the map area. The watersheds of some basins could extent slightly beyond the boundaries of the map.

The depth of colluvium and alluvium was estimated from soil profile descriptions given in the text accompanying the map. Mostly the depth given in the detailed profile description was used, as this is generally typical of the soil mapped. If the description suggested the soil was deeper than actually described, a further 5 cm was added to the depth. Estimates of depth of colluvium or alluvium are conservative because they take no account of sediment lost from the basins.

The area of colluvium or alluvium was measured to an accuracy of ±0.1 ha using a 1 ha grid inscribed on transparent non-deformable acetate sheet or plastic. The drainage basin area was estimated similarly to ±1 ha, unless it was very small (to ±0.1 ha). To obtain the minimum depth of topsoil removed from the adjacent slopes, the volume of deposit (area × depth) was divided by the

238 Soil Erosion on Agricultural Land

area of the drainage basin. Areas were measured of small basins and the larger basins to which they contributed.

Two problems arose. The first concerned the definition of colluvium; some soil series on limestones and cover loams were mapped as colluvium in some places but not in others. Only those soils on valley floors or footslopes were then considered colluvial. In one instance (map sheet TQ05) no colluvium was mapped on the chalk, although the text implied that loess had been redistributed by erosion. In this instance the upper part (0.3 m) of the profile of the soils mapped in valley floors was considered to be colluvium. The second problem was the difficulty of delineating drainage basins in drift-covered lowlands, such as the Vale of York and Solway Firth.

Of the 105 published maps, the first 60 in the series were examined. The task was time consuming and so only another 14 maps portraying colluvium were studied; areas of alluvium were also estimated on these. Of the 74 maps, five contained no small drainage basins with colluvium or alluvium. Fifty-seven maps portrayed alluvium and 33 colluvium in at least one watershed; 21 maps portrayed both colluvial and alluvial basins.

Results

For those areas in which colluvium occurred, two (6 per cent) of 33 maps had fewer than five drainage basins for which measurements of the area of colluvium could be made, and four maps had over 80 basins (Figure 15.2). Estimates of volumes of colluvium stored in valley floors were made for a total of 1090 basins.

For 17 (30 per cent) of 57 maps the area of alluvium was estimated for fewer than five drainage basins, and the maximum number of basins treated per map sheet was 35 (Figure 15.3). Estimates of volumes of alluvium were made for a total of 529 basins. For areas with both colluvium and alluvium, three (14 per cent) of 21 maps contained fewer than five basins for which estimates of volumes could be made.

Values of surface lowering in drainage basins of most areas are normally distributed, although values have a wide spread and distributions for alluvium are occasionally left-skewed. Those of drainage basin area in a locality are often very left-skewed and the best descriptor of the population may be the median value rather than the mean. However, the individual mean values are used for comparisons between areas.

Estimates of Surface Lowering—Colluvium

Averages of topsoil stripped off watersheds range from 9 mm (SK57) to 204 mm (SO74), but most mean values for individual map sheets (17) lie in the range 80–140 mm. Apart from two map sheets (SE36 and SP36) where single basins were examined (of 32 ha and 321 ha respectively), the mean areas of basins contributing runoff to erode topsoil are less than 150 ha.

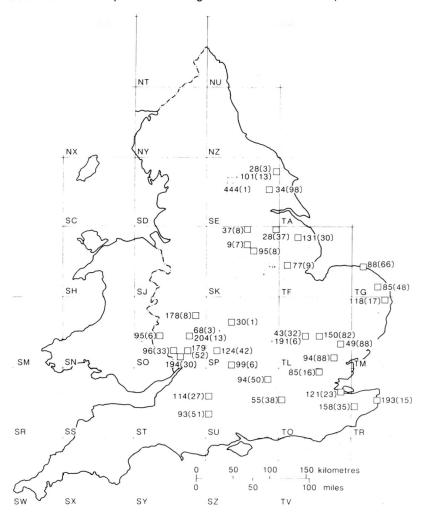

Figure 15.2 Colluvial basins: depth of topsoil stored and number of basins sampled per map

Estimates of Surface Lowering — Alluvium

Mean surface lowering for alluvial basins is usually between 20 and 30 mm (20 maps), but is quite frequently between 10 and 20 mm (11 maps) or between 30 and 40 mm (10 maps). It can be as low as 3 mm (TF04) or as high as 67 mm (SO53). Except for two values on one map (TQ64), all maps with alluvium have mean basin areas greater than 150 ha; most (35 maps) are between 150 and 600 ha; the largest is 5883 ha.

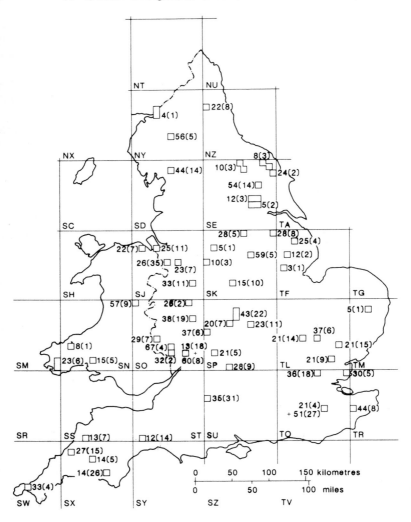

Figure 15.3 Alluvial basins: depth of topsoil stored and number of basins sampled per map

Estimates of Surface Lowering— Colluvium and Alluvium Derived from Similar Soils

Within mapped areas where colluvium and alluvium occur on similar substrata, the sum of the mean values of material deposited in valley floors is between 120 and 160 mm for nine of 18 maps, less than this for basins on six maps (as low as 52 mm) and more than 160 mm on three maps (as high as 216 mm) (Figures 15.2 and 15.3).

DISCUSSION

Colluvium — Distribution

The distribution of colluvium recorded on Soil Survey maps is very similar to that given by Bell (1981); much is mapped on chalk and other limestones. However, colluvium is also mapped on Devonian and Silurian rocks in the south-west Midlands, on medium- and heavy-textured drifts in southern East Anglia, on Permo-Triassic rocks (especially sandstones in the Midlands) and on cover loams and glaciofluvial deposits in north-east East Anglia. This distribution, except that on limestones, reflects the distribution of soils known to be susceptible to erosion (ENDS, 1984).

Colluvium is notably absent in the Midlands, where soils are heavier, and other areas in northern and south-western England, the Weald and in Wales, where land use is and probably has been largely pastoral. Erosion of grassland is rare at present, and when finer-textured soils erode much is transported out of the field into ditches and streams (Evans, 1988).

Colluvium — Depths of Soil Stripped from Slopes

Many drainage basins traverse different soil types even though the substrata may be similar (SSEW, 1983). It is most convenient therefore to describe the results in terms of broad substrata types. Seven main groupings are defined:

(1) On Cretaceous chalk (12 maps);
(2) On limestones of both Permo-Triassic and Jurassic age (seven maps);
(3) On Devonian and Permo-Triassic silty shales, siltstones, sandstones and mudstones (four maps);
(4) On heavy-textured drifts or bedrock in southern East Anglia (four maps);
(5) On light-textured cover loams and glaciofluvial deposits in East Anglia (three maps);
(6) On Permo-Triassic sandstones (two maps); and
(7) On Silurian siltstone and Devonian sandstones of Herefordshire and Gloucestershire (two maps).

Amounts of topsoil eroded on chalk are often considerable, (93–193 mm), but are less in three areas. In Surrey (TQ05) the chalk outcrop is narrow between presently and formerly well-wooded landscapes, and may in the past have been a grassed route rather than ploughed farmland. In Cambridgeshire the low value for map TL34 is associated with generally low relief, and many of the basins lie on argillaceous Lower Chalk. This is in contrast to the much higher value for map TL54, with its rolling chalk landscape and lighter-textured soils. The low value for map SE85 in the Yorkshire Wolds is anomalous. It is an area of high relief, and a value similar to that of the basins draining the north-facing escarpment of map SE79/89 (101 mm) was expected. There are several possible reasons for the low value. The soils in this area are mostly thin and stony over chalk, which is also harder than elsewhere (Furness and King, 1986). Although the Yorkshire Wolds have a long history of settlement, population was generally sparser than that elsewhere on the chalk (Butlin, 1982), and during medieval times the land was only intermittently under the plough (Sheppard, 1973).

More topsoil has been stripped from the chalk in Kent (TQ05, TQ35) and Cambridgeshire (TL54). In Kent silty loess was deposited towards the end of the Devensian Glaciation (Catt, 1985) and silty soils are very erodible; highest present erosion rates in the country are recorded on such soils (Evans, 1988).

On Jurassic and Permo-Triassic limestones amounts of surface lowering are generally less than those found on chalk. Usually this is because the limestone dipslopes are gentle, though in the Cotswolds, particularly in the high-relief area around Stow (SP12), rates are high. Soils are also generally of heavier texture on the limestones than on the chalk.

On siltstones and mudstones 68–96 mm of topsoil has been removed and on the long-settled light-textured soils of the coastal margins of East Anglia 85–118 mm. Relief is, in general, less pronounced than it is on the siltstones and mudstones, but soils on cover loams may be more erodible. On heavier-textured drifts in East Anglia amounts eroded are more variable. The very high mean value (191 mm) recorded in Cambridgeshire on map TL34 is associated with basins draining a low but steep south-facing escarpment flanking the margin of the chalky till plateau. Also a deserted medieval village lies close to the colluvial deposits. The low mean value (49 mm) of erosion around Halstead in Essex (TL83) is not easily explained, except that no silty loess deposits are mapped in this vicinity.

On the Permo-Triassic sandstones large amounts of topsoil have been stripped off and deposited in valley floors. Depths removed (178 and 179 mm) are almost as high as those in Kent. At present, erosion is more frequent and widespread on sandy soils than on any other soil type, although not as severe as on silty soils (Evans, 1988). Amounts of topsoil removed are also high on Lower Palaeozoic siltstones and sandstones in Herefordshire (204 mm) and Gloucestershire (179 mm).

The amounts of topsoil eroded from the individual basins in Warwickshire (SP36) and in North Yorkshire (SE65) relate well to the other findings. Sandy soils have eroded severely in the Vale of York, whereas the heavier-textured soils of the Midlands Plain have eroded only slightly.

Other estimates of surface lowering have been made. Evans (1981a) estimated rates of topsoil removal by erosion at seven sites on chalk in southern England, and they ranged from 58 to 216 mm. The values were higher in Kent than to the west and north-west. Bell (1986) has also estimated depths of topsoil stripped from nine sites on southern chalkland, and the range is from 25 to 240 mm. A further eight values derived from Soil Survey maps are given by Evans and

Cook (1986). These are higher on chalk and limestone than those given here and lower on sandy and light-textured soils. The values are for large blocks of land, not means derived from many small drainage basins. On chalk and limestone, large areas of colluvial soils flanking major through-flowing valley floors (or spreads of colluvium on footslopes whose provenance is unknown but probably in part came from the small drainage basins) were included in the estimates. In the other localities large areas of non-erodible interfluves were included. In the Craven district of North Yorkshire, Smith (1975) calculated that 250 mm of soil has been washed downslope from a lynchet on silty loessial soils. The values of topsoil stripping quoted elsewhere are therefore mostly similar to those found in this study.

Fewer instances of erosion have been recorded on limestones in the past two decades than on many other soil types, so soils on limestones are not considered to be at much risk (ENDS, 1984). This seems to be borne out by the smaller amounts of topsoil removed in the past over much of the limestone outcrops. Amounts of topsoil removed over archaeological and historical times reflect quite well relative present-day erosion rates.

Colluvium—Drainage Basin Area and Depth of Topsoil Stripping

Eroded drainage basins are often small (Figure 15.4), less than 40 ha in extent. There is no relationship between size of drainage basin and depth of topsoil

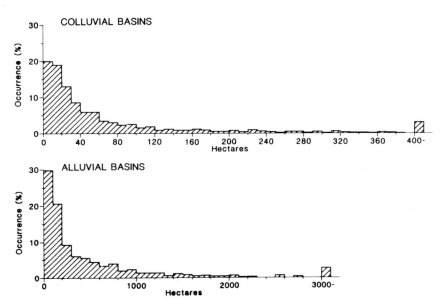

Figure 15.4 Size distributions of colluvial and alluvial basins

stripped. The amounts eroded are related instead to land use, length of time the land has been cropped, soil type, relief and storm intensity and frequency. A rare thunderstorm may cause much erosion or not, depending on the time of year it occurs, and it may be very localized in extent, so that even within the small area covered by an individual 1:25 000 scale map, only part may be affected.

Alluvium

In general, amounts of alluvium on valley floors are least, equivalent to less than 20 mm of topsoil stripped, in the uplands (North York Moors, Peak District, Mynydd Preseli, fringes of Dartmoor and Bodmin Moor), in pastoral areas in the wetter west of Wales and England, and where relief is low (Solway Firth, Vale of York, eastern England). Amounts stored are highest, equivalent to more than 40 mm surface lowering, in glaciated areas with drumlinoid relief (NY53, SD58, SO09); where soils are more erodible, as on light-textured soils of Jurassic sandstones (SE76) or silty unstable soils on siltstones/sandstones and mudstones (SK66, SO53 and 72, TQ64), where silty windblown deposits were widespread (TR04), or on heavy land where much of it was under medieval ridge and furrow (SP47 and 48).

In two localities substrate type is important in determining the amounts of soil transported. In Gloucestershire (SO72) alluvium on valley floors accounts for only 13 mm of topsoil eroded from adjacent slopes cut in Permo-Triassic sandstone, but 60 mm of topsoil from Permo-Triassic mudstones. The apparent anomaly of less alluvium deposited in a locality where erosion has been severe (179 mm of topsoil stored as colluvium) is easily explained, as coarse-textured soils contain only small amounts of fine silt and clay and it is those fractions which are transported out of fields (Evans, 1988).

In the second locality, in the Weald of Kent (TQ64), valley floors on low-lying gently sloping Weald Clays have less than half the amount of stripped topsoil of those on the higher relief of the erodible silty and sandy Hastings Beds.

The different values for topsoil stripping relate well to land use and soil differences. However, compared with the values for colluvium, there is less certainty of the accuracy of the estimates. This is because some sediment is transported out of the drainage basins, and not all sediment on floodplains is derived from farmland but may also be from eroding river banks.

Prior to about 5000 years ago, when the countryside was clothed with woodland, river channels were stable in lowland England and there was little alluviation (Rose *et al.*, 1980). Although woodland clearance followed by arable land use increased rates of sedimentation in river valleys (Brown, 1983; Hazelden and Jarvis, 1979; Robinson and Lambrick, 1984; Shotton, 1978), stream channels have often remained stable since Roman times (Ferguson, 1981). Where streams are currently eroding their banks, this is often related to an increased

Soil Erosion: Its Impact on the English and Welsh Landscape

area of urban land within the drainage basin (Mosley, 1975) or to man-induced changes in the stream course (Hooke, 1977). Even where banks are eroding rapidly, as in Devon (Hooke, 1979; 1980), the stream seems to be reworking a narrow portion of the floodplain; it is not removing the fine overbank deposits laid down away from the present channel.

The sources of sediment in streams does vary. In Devon the suspended sediment load comes mainly from splash erosion and surface runoff from arable land, not from stream banks (Walling et al., 1979), whereas in South Wales most suspended sediment comes from stream banks (Lawler, 1986), and in north Warwickshire from stream banks and adjacent land, and the balance of these may change over time (Foster et al., 1989). The balance between the amounts of soil removed by bank erosion and then redeposited downstream is therefore unknown, but if only small parts of the floodplain are being reworked, except in small narrow valleys, the differences in amounts of alluvium stored on valley floors will reflect the amount of erosion which has occurred within the drainage basin.

A serious problem in estimating amounts of alluvium derived from drainage basins is the transport of clay particles out of the basin into the estuary. The large amounts of alluvium, about 2.4 Mm^3 (Evans, 1981a), trapped since 1651 in the Ouse Washes which cross the Fenland, suggest how much soil could be carried out to sea.

Some of the alluvium must be trapped on coastal marshes. These gain sediment from two sources, and the ratio between that derived from coastal erosion and that from the land is not known. In the last 5000 years or so much sediment has accumulated on the coastal marshes of England and Wales (Devoy, 1977; Evans and Mostyn, 1979; Godwin and Clifford, 1938; Green, 1968; Greensmith and Tucker, 1971; Kidson and Heyworth, 1978; Rose, 1950; Tooley, 1978). If 1 m of this is attributed to soil erosion, this accounts for a further stripping of the topsoil from England and Wales of 33 mm, in addition to the figures given here. This figure is derived from the area of mapped coastal marsh (SSEW, 1983), estimating the volume for a depth of 1 m, and dividing this figure by the total land area of England and Wales.

The amount of soil eroded from slopes and stored in the floodplains of the major rivers of England and Wales is estimated to be 29 mm (Evans and Cook, 1986). Ths alluvium has been carried out of the small and medium-sized basins described in this study and is, therefore, additional to the figures quoted above. However, this depth of soil has not been stripped uniformly from the surface of England and Wales. The evidence from this study is that most alluvium is eroded from, and stored on floodplains in, lowland arable areas. The mean value of topsoil stored in floodplains for all the basins in this study is 32 mm, about the same as that stored in the larger basins of England and Wales. Doubling the figures derived in this study probably approximates to the amount of topsoil washed from the land in the various localities and then stored on the valley floors, and gives a range of 6–134 mm. This does not take into account the soil which may be stored in coastal marshes or that lost to the sea.

Alluvium — Drainage Basin Area and Depth of Topsoil Stripping

As drainage basin size increases, the amount of alluvium stored in the floodplain decreases exponentially (Figure 15.5). The reason for this is not clear. As basin area increases, relief should decrease and the land become less susceptible to erosion, but this is not always the case. Of the 13 larger basins ($>10\,km^2$) only four are in low-lying situations (NY36/37, SE63/73, TF04, TM28). The other nine often have much greater relief; except for the Cotswolds (SP12), they are dominantly down to grass, moorland or forestry. In these instances low storage of alluvium may be related to steep stream gradients or little erosion taking place on the vegetated slopes.

Colluvium and Alluvium — Depth of Topsoil Eroded

Colluvium and alluvium derived from similar lithologies are found on sandstones, chalk, interbedded siltstones, sandstones and mudstones, limestones, and drifts in East Anglia (Figures 15.2 and 15.3). Amounts of topsoil eroded and stored on valley floors are greatest on sandstones (207–216 mm), but the proportion stored as alluvium is small (7–18 per cent). On chalk amounts of topsoil eroded are slightly less than on sandstone where the relief is strong (149–187 mm), and much less where relief is less (64 mm); sediment storage as alluvium accounts for 16–33 per cent of the total. On interbedded siltstones, sandstones and mudstones the amounts of topsoil eroded range from 122 to 150 mm, and a greater proportion is stored as alluvium (23–44 per cent). Depths stripped in the past from limestones vary (52–145 mm), as do the proportions stored as alluvium (4–46 per cent), probably reflecting the variation in relief and soil textures in the five areas. On drifts in East Anglia amounts eroded range from 70 to 121 mm, and 18–30 per cent of this is stored as alluvium.

If the figures of eroded soil stored as alluvium are doubled, to account for soil stored in major valleys, this increases the amount of erosion which has occurred (sandstones 220–254 mm, chalk 85–224 mm, siltstones 153–199 mm, limestones 76–166 mm, drifts 91–157 mm); and also increases the proportion of eroded soil stored as alluvium (sandstones 12–30 per cent, chalk 23–49 per cent, siltstones 38–72 per cent, limestones 7–63 per cent, drifts 31–46 per cent). The last figures imply that small-scale erosion which preferentially removes fines from the topsoil at frequent intervals and deposits it in nearby valley floors is a more important process in England and Wales than has previously been considered (Evans and Cook, 1986).

The values given for slope lowering are averaged over complete drainage basins, but only parts of each basin are susceptible to erosion. For example,

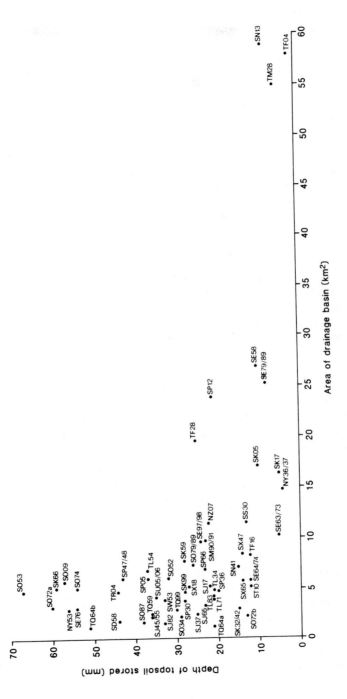

Figure 15.5 Mean area (km^2) of drainage basin: depth (m) of eroded topsoil stored in floodplain

Evans (1981b) showed that of an area of 7 km² in Bedfordshire 54 per cent was potentially at risk of water erosion, and for an area of 10 000 km² in East Anglia only 23 per cent was potentially erodible. Where erosion has occurred the values of surface lowering are therefore likely to be twice as high as the mean basin values given in this study.

Size of Colluvial and Alluvial Drainage Basins

Drainage basins in which colluvium is deposited are far smaller than those in which alluvium is laid down (Figure 15.4). Median values of drainage basin size are 32.1 and 197.6 ha for colluvial and alluvial basins, respectively.

On 1:25 000 soil maps alluvium is rarely portrayed in small drainage basins, where with high stream gradients overbank flows are probably too rapid to allow sedimentation. Alluvium is commonly mapped, however, in drainage basins of 40 ha to 200 ha (50.5 per cent of all occurrences). Deposits are not as extensive in large basins (>10 km²) as in smaller ones (Figure 15.5), possibly because the supply of sediment in the overbank flood flow is limited. In the lowlands the source of sediment, the eroded fields, may be so far from the river that much of the eroded soil is deposited before reaching the river, or the low-angled slopes adjacent to the stream are not eroding. In the uplands alluvium is mostly found well downstream, either because upstream gradients are too steep for sedimentation to occur or the supply of sediment is limiting in pastoral areas where erosion occurs infrequently.

THE IMPACT OF EROSION ON THE LANDSCAPE

Since the woodland was cut down man has been the major geomorphological agent acting in England and Wales, but the impact of erosion on the landscape has been slight. Over large parts of lowland England and Wales a thin layer of soil, often between 40–250 mm deep, has been removed and deposited in valley floors as colluvium and alluvium. Its most obvious manifestation in lowland England and Wales is the steps formed in valley floors and on slopes where colluvium has built up behind field boundaries.

EROSION RATES AND FREQUENCIES IN THE PAST

Present-day frequencies and rates of erosion do not account for the volumes of soil eroded and stored on valley floors since the land was cleared. In Nottinghamshire, for example, it is likely that erosion of the sandlands started when Sherwood Forest was converted to farmland about 200 years ago. Presently, water erosion removes about 0.4 mm per year of topsoil from slopes,

Soil Erosion: Its Impact on the English and Welsh Landscape

but this accounts for only about half the volume that appears, from soil-profile evidence, to have been removed (Evans, 1988). Elsewhere, similar discrepancies prevail. They may be partly explained by wind erosion but there are other possible reasons.

First, erosion may have been previously more frequent and/or rapid than at present. The widespread use of fallow prior to the eighteenth century may have exacerbated water erosion (see below). Second, particles may be carried by splash and sheet-wash which have not been measured in the monitoring project giving the present-day rates (Evans, 1988).

The third reason is that the present-day rates estimated from three years' data obtained in the mid-1980s are low. The data do not include the impact of a rare storm such as that reported by Boardman (1990) around Brighton in October 1987, which could increase mean annual erosion rates quite significantly over the short term.

A fourth possible reason is that erodible soils were once more widespread. Loessic soils at present have a discontinuous distribution, largely confined to wide ridge crests on calcareous rocks (Catt, 1985). However, windblown silts occur in many topsoils throughout England and Wales (Catt, 1985) and it is likely that a thicker veneer of loess has been stripped from slopes in many parts of the country.

FALLOW LAND

Land has been manured and marled since at least Roman times (Finberg, 1972), but until the agricultural revolution of the eighteenth century these were not the primary ways of restoring fertility (Mate, 1985). This was done by resting the land in bare fallow (Jones, 1981), usually for a growing season, but where the land was poor, for longer (Mate, 1985). In medieval times, and probably earlier, the land was often fallowed every two or three years (Baker and Butlin, 1973; Hall, 1982). Large areas of land were then at risk of erosion throughout the year, unless it become weed infested. Erosion could well have occurred twice in a season, when soils were saturated in winter and during summer thunderstorms.

The areas where thick deposits of colluvium occur are all areas of silty or sandy soils which are known to be structurally unstable and erodible at present. These soils are especially unstable when organic matter contents are low, as would be the case when land was under continuous arable. Slaking of the clods results in a smooth surface with low surface water storage and low infiltration rates, both of which encourage runoff. Moreover, capped soils inhibit seedling emergence so that weed cover may also have been delayed.

CONCLUSIONS

There is much evidence for water erosion in the past; in the recent past soil banked up against field boundaries, and in the ancient past colluvium and alluvium were deposited on valley floors.

Throughout England and Wales man cleared the land of wood. This happened earlier (in the Neolithic) on the southern and eastern chalklands and successively later to the west and north (in Bronze Age, Roman, Anglo-Saxon and Norse periods). Initially, erosion was local at the time of clearance and widespread later when clearings were used for arable. In the uplands and lowlands most of the woods were cleared by the time of the Black Death in the mid-fourteenth century.

Colluvium is widespread on soils which are known to erode today. These soils often have a high silt or sand content and are structurally unstable. The volumes of colluvium stored in valley floors are often equivalent to depths of 100–200 mm of topsoil stripped from the drainage basin.

Alluvium stored in small valleys is often equivalent to a surface lowering of 20–40 mm, similar in amount probably to that stored in large valleys. In total, therefore, in many parts of lowland England and Wales 40–80 mm of eroded topsoil is stored as alluvium. Where colluvium and alluvium have been eroded and deposited from similar parent materials the amounts stored in the valley floors are mostly equivalent to a topsoil stripping of 150–250 mm.

However, the figures given above are means for whole drainage basins; not all slopes are susceptible to erosion, so amounts of surface lowering were locally often greater than these figures, possibly by a factor of two or more. The estimates are also conservative, because soil carried out to sea or deposited in estuarine marshes is not accounted for.

Erosion has removed most topsoil where soils are sandy or silty, and probably where land has been settled longest. The present distribution of soils may well be different from that occurring when the land was first settled. Soils derived from erodible loess may then have been more widespread and thicker.

Most erosion probably took place in Bronze Age and Iron Age times, in the Roman period and in the medieval period prior to the Black Death. All these were times of rapid expansion of arable land, associated probably with agricultural innovation in the Bronze Age, and with population pressure in Roman and medieval times. Until the agricultural revolution of the eighteenth century the widespread use of bare fallow to restore fertility could have promoted erosion. Present-day rates and frequencies of erosion do not explain the depths of soil stored in valley floors. Erosion in the past may have been more rapid or frequent. During the Holocene, therefore, the impact of erosion on the landscape of England and Wales has been widespread but slight. A veneer of soil has been stripped off and deposited in valleys or in extensive coastal marshes.

ACKNOWLEDGEMENTS

John Boardman and John Catt made helpful comments on earlier drafts of this chapter.

REFERENCES

Avery, B. W. (1980). *Soil Classification for England and Wales*, Soil Survey Technical Monograph No. 14, Harpenden.
Baker, A. R. H. and Butlin, R. A. (1973). *Studies of Field Systems in the British Isles*, Cambridge University Press, Cambridge.
Beales, P. W. (1980). The late Devensian and Flandrian vegetational history of Crose Mere, Shropshire, *New Phytologist*, **85**, 133–61.
Bell, M. (1981). Valley sediments and environmental change. In Jones, M. and Dimbleby, G. W. (eds), *The Environment of Man: the Iron Age to the Anglo-Saxon Period*, British Archaeological Reports, British Series, **87**, 75–91.
Bell, M. (1982). The effects of land-use and climate on valley sedimentation. In Harding, A. F. (ed.), *Climatic Change in Later Prehistory*, Edinburgh University Press, Edinburgh, pp. 127–42.
Bell, M. G. (1986). Archaeological evidence for the date, cause and extent of soil erosion on the chalk, *SEESOIL*, **3**, 72–83.
Bennett, K. B. (1983). Devensian late-glacial and Flandrian vegetational history at Hockham Mere, Norfolk, England. I. Pollen percentages and concentrations, *New Phytologist*, **95**, 457–87.
Boardman, J. (1990). Soil Erosion in the South Downs: a review. In Boardman, J., Foster, I. D. L. and Dearing, J. A. (eds), *Soil Erosion on Agricultural Land*, John Wiley, Chichester, pp. 87–105.
Brown, A. G. (1982). Human impact on former floodplain woodlands of the Severn. In Bell, M. and Limbrey, S. (eds), *Archaeological Aspects of Woodland Ecology*, British Archaeological Report, International Series, **146**, 93–105.
Brown, A. G. (1983). Accelerated sedimentation and floodplain deposits in the Lower Severn valley. In Gregory, K. J. (ed.), *Background to Palaeohydrology*, John Wiley, Chichester, pp. 375–97.
Brown, A. G. and Barber, K. E. (1985). Holocene sediment history of a small lowland catchment in central England, *Quaternary Research*, **24**, 87–102.
Butlin, R. A. (1982). The historical geography of Britain. In *Atlas of Great Britain*, The Ordnance Survey, Southampton, pp. 144–57.
Catt, J. A. (1985). Soil particle size distribution and mineralogy as indicators of pedogenic and geomorphic history: examples from the loessial soils of England and Wales. In Richards, K. S., Arnett, R. R. and Ellis, S. (eds), *Geomorphology and Soils*, Allen and Unwin, London, pp. 202–18.
Clayden, B. (1964). *Soils of the Middle Teign Valley District of Devon*, Agricultural Research Council, Soil Survey of Great Britain, Bulletin No. 1, Harpenden.
CPRE (1985). *Soil Erosion in England and Wales. A Report from the Soil Survey of England and Wales to CPRE*, Council for the protection of Rural England, London.
Crampton, C. B. (1969). The chronology of certain terraced river deposits in the SE Wales area, *Zeitschrift fur Geomorphologie*, NF Bd **13**, 245–59.
Devoy, R. J. N. (1977). Flandrian sea level changes in the Thames estuary and the implications for land subsistence in England and Wales, *Nature*, **270**, 712–15.

Donkin R. B. (1978). *The Cistercians: Studies in the Geography of Mediaeval England and Wales*, Pontifical Institute of Mediaeval Studies, Toronto.

Dyer, C. (1988). Documentary evidence: Problems and Enquiries. In Astill, G. and Grant, A. (eds), *The Countryside of Medieval England*, Blackwell, Oxford, pp. 12-35.

ENDS (1984). Soil erosion: an unsustainable face of modern farming, *ENDS Report*, **115**, August, 9-10.

Evans, J. G. (1972). *Land Snails in Archaeology*, Seminar Press, London.

Evans, J. G. (1975). *The environment of Early Man in the British Isles*, Elek, London.

Evans, R. (1981a). Potential soil and crop losses in the UK. In *Soil and Crop Loss—Developments in Erosion Control*, Soil and Water Management Association/Royal Agricultural Society, Stoneleigh.

Evans, R. (1981b). Assessments of soil erosion and peat wastage for parts of East Anglia, England. In Morgan R. P. C. (ed.), *Soil Conservation: Problems and Prospects*, John Wiley, Chichester, pp. 521-30.

Evans, R. (1988). *Water Erosion in England and Wales 1982-1984*, Report for Soil Survey and Land Research Centre.

Evans, R. Erosion studies in the Dark Peak, *Journal Northern Soils Discussion Group* (in press).

Evans, R. and Cook, S. (1986). Soil erosion in Britain, *SEESOIL*, **3**, 28-58.

Evans, R. and Mostyn, E. . (1979). *Stratigraphy and Soils of a Fenland Gas Pipeline*, Agricultural Development and Advisory Service, Ministry of Agriculture, Fisheries and Food, Cambridge.

Eyre, S. R. (1957). The upward limit of enclosure on the East Moor of Derbyshire, *Transactions and Papers of the Institute of British Geographers*, Publication No. 23, 61-74.

Ferguson, R. I. (1981). Channel forms and channel changes. In Lewin, J. (ed.), *British Rivers*, Allen and Unwin, London, pp. 90-125.

Finberg, H. P. R. (ed.) (1972). *The Agrarian History of England and Wales, Vols I-II, AD43-1042*, Cambridge University Press, Cambridge.

Foster, I., Dearing, J. and Grew, R. (1989). The north Warwickshire experimental catchments; studies of historical sediment yield, sediment source and soil erosion. In Foster, I. (ed.), *Field Excursion Guide*, British Geomorphological Research Group, Coventry, pp. 1-30.

Fowler, P. J. (1978). Lowland landscapes: culture, time and personality. In *The Effect of Man on the Landscape: the Lowland Zone*, Council for British Archaeology, Research Report, **21**, 1-12.

Furness, R. R. and King, S. J. (1986). *Soils in Humberside II (Sheet SE85, Fridaythorpe)*, Soil Survey Record No. 97, Harpenden.

Godwin, H. and Clifford, M. H. (1938). Studies of the Post-Glacial history of British vegetation. I. Origin and stratigraphy of Fenland deposits near Woodwalton Fen. II. Origin and stratigraphy of deposits in southern Fenland, *Philosophical Transactions Royal Society B*, **229** (B526), 323-406.

Green, R. D. (1968). *Soils of Romney Marsh*, Agricultural Research Council, Soil Survey of Great Britain, England and Wales, Bulletin No. 4, Harpenden.

Greensmith, J. T. and Tucker, E. V. (1971). The effects of late-Pleistocene and Holocene sea-level changes in the vicinity of the River Crouch, east Essex, *Proceedings Geological Association*, **82**, 301-22.

Hall, D. (1982). *Medieval Fields*, Shire Archaeology, Princes Risborough.

Hallam, H. E. (1981). *Rural England (1066-1348)*, Fontana, Glasgow.

Harvey, A. M., Oldfield, F. and Baron, A. F. (1981). Dating of post-glacial landforms in the central Howgills, *Earth Surface Processes and Landforms*, **6**, 401-12.

Harvey, A. M. and Renwick, W. H. (1987). Holocene alluvial fan and terrace formation in the Bowland Fells of northwest England, *Earth Surface Processes and Landforms*, **12**, 249–57.

Hazelden, J. and Jarvis, M. G. (1979). Age and significance of alluvium in the Windrush valley, Oxfordshire, *Nature*, **282**, 291–2.

Hollingsworth, T. H. (1969). *Historical Demography*, Cambridge University Press, Cambridge.

Hooke, J. M. (1977). The distribution and nature of changes in river channel patterns: the example of Devon. In Gregory, K. J. (ed.), *River Channel Changes*, John Wiley, Chichester, pp. 265–80.

Hooke, J. M. (1979). An analysis of the processes of river bank erosion, *Journal of Hydrology*, **42**, 39–62.

Hooke, J. M. (1980). Magnitude and distribution of rates of river bank erosion, *Earth Surface Processes and Landforms*, **5**, 143–57.

Jones, M. (1981). The development of crop husbandry. In Jones, M. and Dimbleby, G. (eds), *The Environment of Man: the Iron Age to the Anglo-Saxon Period*, British Archaeological Report Series, **87**, 95–127.

Jones, M. E. (1979). Climate, nutrition and disease: an hypothesis of Romano-British population. In Casey, P. J. (ed.), *The End of Roman Britain*, British Archaeological Reports, British Series, **71**, 231–251.

Jones, R., Benson-Evans, K. and Chambers, F. M. (1985). Human influence upon sedimentation in Llangorse Lake, Wales, *Earth Surface Processes and Landforms*, **10**, 227–35.

Kidson, C. and Heyworth, A. (1978). Holocene eustatic sea level change, *Nature*, **273**, 748–50.

King, A. (1978). Ribblehead, *Current Archaeology*, **6**, 38–41.

Lamb, H. H. (1982). *Climate History and the Modern World*, Methuen, London.

Lawler, D. M. (1986). River bank erosion and the influence of frost: a statistical examination, *Transactions Institute of British Geographers New Series*, **11**, 227–42.

Limbrey, S. (1983). Archaeology and palaeohydrology. In Gregory, K. J. (ed.), *Background to Palaeohydrology*, John Wiley, Chichester, pp. 189–212.

Mate, M. (1985). Medieval agrarian practices: the determining factors, *Agricultural History Review*, **33**, 22–31.

Mosley, M. P. (1975). Channel changes on the River Bollin, Cheshire 1872–1973, *East Midland Geographer*, **42**, 185–99.

Parry, M. L. (1978). *Climatic Change Agriculture and Settlement*, Dawson/Archon Books, Folkestone.

Pennington, W. (1969). *The History of British Vegetation*, Unibooks, University Press, London.

Pennington, W. (1980). The origin of pollen in lake sediments: an enclosed lake compared with one receiving inflow streams, *New Phytologist*, **83**, 189–213.

Pennington, W., Cambrey, R. S., Eakins, J. D. and Harkness, D. D. (1976). Radiometric dating of the recent sediments of Blelham Tarn, *Freshwater Biology*, **6**, 317–31.

Rackham, O. (1976). *Trees and Woodland in the British Landscape*, Dent, London.

Rackham, O. (1986). *The History of the Countryside*, Dent, London.

Richards, K. S., Peters, N. A., Robertson-Rintoul, M. S. E. and Switsur, V. R. (1987). Recent valley sediments in the North York Moors: evidence and interpretation. In Gardiner, V. (ed.), *International Geomorphology 1986 Part I*, John Wiley, Chichester, pp. 869–83.

Robinson, M. A. and Lambrick, G. H. (1984). Holocene alluviation and hydrology in the upper Thames basin, *Nature* **308**, 809–14.

Rose, F. (1950). The east Kent fens, *Journal of Ecology*, **38**, 292–302.
Rose, J., Turner, C., Cooper, G. R. and Bryan, M. D. (1980). Channel changes in a lowland river catchment over the last 13,000 years. In Cullingford, D. A., Davidson, D. A. and Lewin, J. (eds), *Timescales in Geomorphology*, John Wiley, Chichester, pp. 159–75.
Russell, J. C. (1972). Population in Europe 500–1500. In Cipolla, C. M. (ed.), *The Middle Ages*, Collins/Fontana, London, pp. 25–70.
Scaife, R. G. and Burrin, P. J. (1983). Floodplain development in and the vegetational history of the Sussex High Weald and some archaeological implications, *Sussex Archaeological Collections*, **121**, 1–10.
Sheppard, J. A. (1973). Field systems of Yorkshire, In Baker, A. R. H. and Butlin, R. A. (eds), *Studies of Field Systems in the British Isles*, Cambridge University Press, Cambridge, pp. 145–87.
Shotton, F. W. (1978). Archaeological inferences from the study of alluvium in the lower Severn/Avon valleys. In Limbrey, S. and Evans, J. G. (eds), *The Effect of Man on the Landscape: the Lowland Zone*, Council for British Archaeology, Research Report, **21**, 27–32.
Simmons, I. G., Atherden, M. A., Cundill, P. R. and Jones, R. L. (1975). Inorganic layers in soligenous mires of the North Yorkshire Moors, *Journal of Biogeography*, **2**, 49–56.
Smith, R. (1989). A case for substantial unrecorded soil erosion in the prehistoric landscape, Paper presented at BGRC workshop, *Soil Erosion on Agricultural Land*, Coventry, January.
Smith, R. T. (1975). Early agriculture and soil degradation, In Evans, J. G., Limbrey, S. and Cleere, H. (eds), *The Effect of Man on the Landscape: the Highland Zone*, Council for British Archaeology, Research Report, **11**, 27–37.
Soil Survey (1988). *List of Publications*, Soil Survey and Land Research Centre, Cranfield.
SSEW (1983). *Soil Map of England and Wales*, Soil Survey of England and Wales, Harpenden.
Tallis, J. H. (1985). Erosion of blanket peat in the southern Pennines: new light on an old problem. In Johnson, R. H. (ed.), *The Geomorphology of Northwest England*, Manchester University Press, Manchester, pp. 313–36.
Tooley, M. J. (1978). *Sea-level Changes in North-west England During the Flandrian Stage*, Clarendon Press, Oxford.
Turner, J. (1984). Pollen diagrams from Cross Fell and their implication for former tree lines. In Haworth, E. Y. and Lund, J. W. G. (eds), *Lake Sediments and Environmental History: Studies in Palaeolimnology and Palaeoecology in Honour of Winfred Tutin*, Leicester University Press, Leicester, pp. 317–57.
Tutin, W. (1969). The usefulness of pollen analysis in interpretation of stratigraphic horizons, both Late-glacial and Post-glacial, *Mitteilungen Internationalen Vereinigung Limnologie*, **17**, 154–64.
Walling, D. E., Oldfield, F. and Thompson, R. (1979). Suspended sediment sources identified by magnetic measurements, *Nature*, **281**, 110–13.
Wilkinson, B., Broughton, W. and Parker-Sutton, J. (1969). Survey of wind erosion on sandy soils in the east Midlands, *Experimental Husbandry*, **18**, 53–9.
Wiltshire, P. E. J. and Moore, P. D. (1983). Palaeovegetation and palaeohydrology in upland Britain. In Gregory, K. J. (ed.), *Background to Palaeohydrology*, John Wiley, Chichester, pp. 433–51.
Wrigley, E. A. (1969). *Population and History*, Weidenfeld and Nicolson, London.
Wrigley, E. A. and Schofield, R. S. (1981). *The Population History of England 1541–1871*, Edward Arnold, London.

16 Some Magnetic and Geochemical Properties of Soils Developed on Triassic Substrates and their Use in the Characterization of Colluvium

J. P. SMITH, M. A. FULLEN and S. TAVNER
School of Applied Sciences, Wolverhampton Polytechnic

INTRODUCTION

Concern has grown about erosion on sandy agricultural soils in parts of the West Midlands which are underlain by Triassic red beds or their derived drift. Research to date has focused on spatial patterns and causes of erosion (Reed, 1979a; Fullen, 1985; Fullen and Reed, 1987) and erosion rates (Fullen and Reed, 1986). Estimates of contemporary erosion rates are available from plot studies (Fullen and Reed, 1986), observations of individual erosive events (Reed, 1979b) and catchment studies (Smith, 1986).

It is clear that much contemporary erosion is related to modern agricultural practices and that erosion rates have accelerated in recent years (Reed, 1979a). Estimates of contemporary erosion rates from experimental plots and from lake sediments differ markedly, even after allowance has been made for the percentage of sloping land within a catchment, land-use patterns and the capturing efficiency of lakes. For example, erosion rates for a range of experimental tilled and bare earth plots at Hilton in Shropshire range between 2 and 47 t ha^{-1} yr^{-1} (Fullen and Reed, 1986), while those for post-1965 topsoil erosion from the catchment of Bickley Brook (an area of about 14 km^2 in southern Cheshire), are of the order of 0.1–0.2 t ha^{-1} yr^{-1} (Smith, 1986).

One explanation for these differences is that much of the hillslope-derived sediment is stored as colluvium. During the mapping of the Claverley area (SO79E/89W), in areas underlain by Triassic substrates, Hollis (1978) recognized significant landscape features formed by colluvium. These he

Soil Erosion on Agricultural Land
Edited by J. Boardman, I. D. L. Foster and J. A. Dearing
©1990 John Wiley & Sons Ltd

identified as the 'colluvial variant' of the sandy phase of the Newport series.

The ages of colluvial deposits in the West Midlands are unknown, but there is evidence that some may have been initiated in Neolithic times (Twigger, 1983). It is known that at a number of sites significant colluviation has occurred during the last 40 years. Individual sites have been investigated where active build-up of colluvium at the base of slopes has been clearly, if not quantitatively, established (Reed, 1979a,b). In some locations buried fences and hedge lines attest to the very recent origin of these features. Similarly, active erosion on upper and middle slopes, with accumulation on basal segments, has been observed for particular erosional events.

There is a need to evaluate the role that colluviation plays in the modern agricultural landscape. In particular, four aspects merit further investigation:

(1) The identification of colluvial features and estimation of their areal extent and volume;
(2) Characterization of the sediment quality within and below colluvial features in terms of their topsoil content, entrapped fertility or resource sterilization;
(3) The age of colluvial features and rates of contemporary colluviation;
(4) The relationship between processes of colluvation and modern agricultural methods.

In practice, the differentiation between soil horizons in agricultural sandy soils on Triassic bedrock or Triassic derived drift is difficult, as is the identification of colluvium which is not unambiguously associated with terrace features. Brown Sands (cf. Bridgnorth, Newport series) under arable or temporary grass are characterized by relatively small differences in horizon appearances. These soils are sandy (>85 per cent sand content), and, except for some Ap horizons which exhibit weakly developed sub-angular blocks, they have a predominantly apedal single-grain structure. They often have a loose soil strength (as a result of the low silt and clay content) and are 'non-sticky' and 'non-plastic' (Crompton and Osmond, 1954). In some sites the horizons may show some colour differentiation (typically, Ap 7YR 3/2, Bw 5YR 5/7, Cu 6YR 5/6) but in others these differences are less evident. The organic content is often low (2 per cent by weight). Differentiation of topsoil, subsoil and parent materials, therefore, may rely on small variations in clay, silt or organic matter content. Nevertheless, the fertility (defined in terms of exchange capacity, soil biology, available nutrients and unused fertilizer) is primarily associated with the topsoil.

MAGNETIC AND GEOCHEMICAL METHODS

Mineral magnetic methods provide a rapid yet sensitive method of characterizing sediments (Thompson and Oldfield, 1986). Quite marked differences in magnetic signature can be attributed to assemblages of magnetic grains which form a very small mass fraction of the sediment, but which may be persistent through complete erosion–deposition cycles. It is the ability of magnetic measurements to contrast topsoils (Ap, Ah), subsoils (B, Bw, Eb) and parent materials (C, Cu, Cr) that is of particular significance in the study of colluvium.

The contrasts in magnetic properties between soil horizons results in many cases from the modification of primary magnetic minerals (present within the regolith as a result of its geological setting) by weathering and soil formation (Schwertmann and Taylor, 1977), secondary magnetic minerals formed during pedogenesis (Dearing et al., 1985; Mullins, 1977), burning of the topsoil and vegetation (Le Borgne, 1960), or the addition of magnetic grains as part of the atmospheric dustfall (Oldfield et al., 1978). In many soils the contrasting mineralogy, size, shape and stoichiometry of these magnetic minerals provides the opportunity to differentiate between soil types and between horizons.

Analysis of elemental geochemistry may also be carried out rapidly and non-destructively. The elemental geochemistry of a sediment is largely determined by its bulk mineralogy. However, trace constituents may reflect both provenance and pedogenic processes (Swaine and Mitchell, 1960), and provide a means of both sediment description and source identification. In the case of heavy metals the concentration in topsoils may reflect both their association with organic matter (Sposito, 1984) and post-industrial atmospheric deposition (Cawse, 1976). The use of these two independent methods, magnetic and geochemical, provides a convenient and quick technique of sediment characterization and also an internal check on the interpretation offered by either.

MAGNETIC MINERALOGY AND ELEMENTAL GEOCHEMISTRY OF TRIASSIC SOILS

Triassic soils (at sites away from localized surface contamination by industry or domestic activities) show distinctive magnetic characteristics both in terms of their bulk properties and on a particle size-specific basis. Table 16.1 shows typical magnetic values for a range of soil samples from the catchment of Bickley Brook in Cheshire, and Claverley Brook, a tributary of the River Worfe in east Shropshire. In both cases the landscape is underlain by Triassic strata with a variable thickness of drift cover. Four parameters are given: χ and IRM_{sat}, which respond principally to the concentration of ferrimagnetic minerals, and S (defined as $-IRM_{100mT}/IRM_{sat}$) and IRM_{sat}/χ, which relate to type of magnetic mineralogy, magnetic grain shape and size (Thompson and Oldfield, 1986). Figure 16.1 gives typical remanent coercivity spectra for Bickley Brook soils. The various substrate classes are clearly distinguished. When soils are examined in vertical section, well-ordered relationships between magnetic properties and depth can be seen. In Figure 16.2 data are shown from a soil

Table 16.1 (a) Mean and range values of magnetic parameters for bulk soil samples from the Bickley Brook catchment, Cheshire, England ($n = 58$)

		χ	IRM_{sat}	S	IRM_{sat}/χ
Topsoils					
Woodland (Ah)	Mean	1.48	229	0.59	15.2
	Max.	2.91	436	0.71	19.4
	Min.	0.48	73	0.41	13.7
Pasture/arable (Ap)	Mean	0.67	113	0.56	15.6
	Max.	0.95	171	0.69	17.3
	Min.	0.55	85	0.43	14.8
Subsoils					
Woodland (Bh)	Mean	0.05	6	0.23	13.1
	Max.	0.05	7	0.26	14.2
	Min.	0.04	6	0.20	12.4
Pasture/arable (B)	Mean	0.22	36	0.06	15.5
	Max.	0.27	49	0.26	17.5
	Min.	0.14	20	−0.11	13.2
Parent materials					
Fine-grained till	Mean	0.35	82	−0.47	22.7
	Max.	0.42	96	−0.37	24.7
	Min.	0.26	66	−0.58	19.9
Glacial Sands	Mean	0.18	49	−0.05	26.6
	Max.	0.21	52	0.10	27.8
	Min.	0.15	45	−0.22	24.0
Keuper Sandstone		0.21	77	−0.16	37.8

(b) Range values of magnetic parameters for bulk soil samples from the Claverley and Hilton Brook catchment, Shropshire, England ($n = 50$)

Topsoils				
(Ah + Ap)	0.85–1.87	166–314	0.10– 0.85	14–19
Subsoils	0.10–0.79	28–114	0.15– −0.35	19–23
Parent materials	0.21–0.53	74–142	−0.20– −0.55	28–39

$\chi = \mu m^3\ kg^{-1}$; $IRM_{sat} = \mu Am^2\ kg^{-1}$; $IRM_{sat}/\chi = kA\ m^{-1}$

pit in the catchment of Newton Mere in northern Shropshire. In addition to the parameters given in Table 16.1, the frequency-dependent susceptibility, χ_{fd}, is plotted. This parameter indicates the presence of very fine magnetic grains (Mullins and Tite, 1973), which are largely produced in the topsoil, during pedogenesis, forming secondary constituents of the magnetic assemblage (Mullins, 1977).

In general, then, topsoils, subsoils and parent materials can be differentiated to a greater or lesser extent by a series of magnetic measurements of bulk samples. Their respective characteristics can be described as follows:

(1) Topsoils have higher values of χ, χ_{fd}, IRm_{sat}; higher positive values of the S parameter and low values of IRm_{sat}/χ.

Magnetic and Geochemical Properties of Soils   259

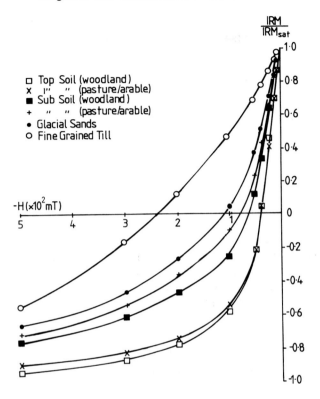

Figure 16.1 Mean remanent coercivity spectra for Bickley Brook catchment

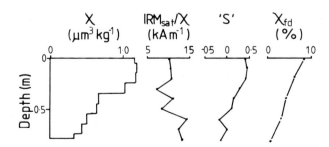

Figure 16.2 Bulk magnetic properties for soil pit B—Newton Mere

(2) Subsoils show a greater range of behaviour but generally have lower values of χ, χ_{fd}, IRM_{sat}, and intermediate values of S and IRm_{sat}/χ.
(3) Parent materials have intermediate values of χ and IRM_{sat}; lowest values of X_{fd}; lowest values of S, and the highest values of IRM_{sat}/χ.

Table 16.2 Mean and range values of elemental geochemistry for bulk soil samples from the catchment of Bickley Brook, Cheshire, England ($n = 58$)

		SiO_2	Al_2O_3	Fe_2O_3	MnO	CaO	K_2O	S	P_2O_5	MgO	Pb	Zn	Cu	Ni
Topsoils														
Woodland (Ah)	Mean	90.7	5.8	1.8	0.06	2.2	1.9	0.16	0.62	0.8	128	252	30	23
	Max.	96.5	7.1	2.2	0.13	4.1	2.2	0.30	1.04	1.1	280	512	38	39
	Min.	85.1	4.0	1.2	0.03	0.9	1.6	0.08	0.34	0.3	41	43	16	10
Pasture/arable (Ap)	Mean	89.9	7.1	2.1	0.06	0.4	2.2	0.07	0.24	0.8	32	62	10	17
	Max.	93.8	7.7	2.5	0.09	0.6	2.4	0.10	0.34	1.0	38	82	14	18
	Min.	87.1	6.3	1.8	0.03	0.3	2.0	0.04	0.11	0.7	27	51	4	15
Subsoils														
Woodland (Bh)	Mean	91.1	4.3	0.6	0.01	0.2	2.0	0.03	0.14	0.4	10	19	7	5
	Max.	99.0	4.9	0.7	0.01	0.2	2.0	0.04	0.17	0.5	13	28	12	7
	Min.	73.3	4.0	0.6	0.01	0.1	2.0	0.03	0.06	0.3	7	12	0	4
Pasture/arable (B)	Mean	86.9	6.3	1.8	0.03	0.5	2.3	0.08	0.19	0.9	26	28	15	17
	Max.	95.7	7.6	2.2	0.05	1.0	2.6	0.10	0.35	1.3	31	43	29	19
	Min.	73.2	5.1	1.6	0.01	0.1	2.1	0.04	0.11	0.4	18	7	4	15
Parent materials														
Fine-grained till	Mean	73.4	9.3	3.3	0.06	2.6	3.0	0.27	0.32	2.9	15	53	27	29
	Max.	88.3	11.7	4.4	0.09	6.2	3.6	0.87	0.74	5.8	19	63	32	40
	Min.	63.1	6.2	2.0	0.03	0.4	1.9	0.04	0.14	0.7	12	36	12	20
Glacial sands	Mean	88.0	5.7	1.1	0.01	0.3	2.1	0.05	0.18	0.6	17	23	11	10
	Max.	92.2	7.1	1.2	0.02	0.4	2.1	0.06	0.23	0.8	26	29	21	11
	Min.	80.0	4.5	1.1	0.01	0.3	2.1	0.04	0.08	0.5	12	15	0	9

All concentrations as dry weight of mineral matter. SiO_2 to MgO as weight %. Pb to Ni as $\mu g\,g^{-1}$.

Magnetic and Geochemical Properties of Soils

Differences can also be detected in the geochemistry of bulk soil samples (Table 16.2). Relationships are less clearly developed between horizons since the elemental geochemistry is dominated by gross mineralogy, but generally the following observations may be made:

(1) Topsoils have higher values of Fe, Mn, S, P, Pb, Zn, Cu and Ni than subsoils. Differences are particularly evident in Pb and Zn concentrations.
(2) The two parent materials shown have markedly different geochemical characteristics, as a result of contrasts in their bulk mineralogy. The glacial sands have low values of Fe, Mn, S, P, Pb, Zn, Cu and Ni. The fine-grained till is characterized, in terms of major elements, by the lowest values of Si and correspondingly higher values of Al, Fe, Ca, and Mg. Minor elements show a contrasting pattern to other substrates. Pb values are low, Zn and Cu values are intermediate and Ni values high.

Further insight into the mineral magnetic behaviour of Triassic soils can be gained by examining their behaviour on a particle size-specific basis. Figure 16.3 illustrates variations in magnetic parameters both with particle size and depth in a soil pit from the catchment of Newton Mere. Although the largest mass of sediment is in the 63-1000 μm size range, the highest values of χ and IRM_{sat} occur in the fines. The magnetic characteristics which provide the contrast between surface horizons and the subsoil are also most marked in the fine fractions. Table 16.3 illustrates the distribution of percentage frequency-dependent susceptibility (χ_{fd}). This indicates that the ultrafine magnetic grains predominantly occur in (1) the fine particle size fraction ($<4\,\mu$m) and (2) the surface horizons of the soil.

These magnetic signatures can be interpreted by modelling the soil magnetic assemblage as a mixture of primary and secondary magnetic minerals. The secondary magnetic minerals may occur as a very small proportion of the mixture, but strongly influence the magnetic properties where they are present.

On a particle size basis, the elemental geochemistry also displays ordered relationships. Figure 16.4 illustrates the particle size relationships for three selected substrates. In these substrates, the concentration of all elements except Si (and sometimes Zr) are enhanced in the fine fractions. The results illustrate the influence of quartz in the overall mineralogy of these substrates. There is a marked relationship between particle size and Pb, Zn and Ni concentrations.

FIELD STUDIES OF COLLUVIUM

Mineral magnetic and elemental geochemical analyses as described above can provide a rapid means of characterizing colluvium and help in understanding colluvial and slope processes. Two field examples are discussed below.

262 Soil Erosion on Agricultural Land

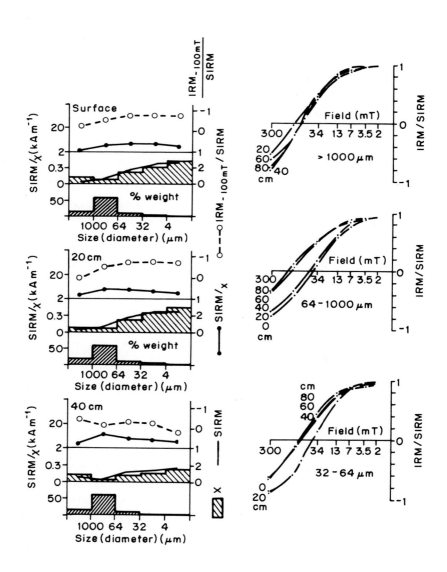

Figure 16.3 (caption opposite)

Magnetic and Geochemical Properties of Soils 263

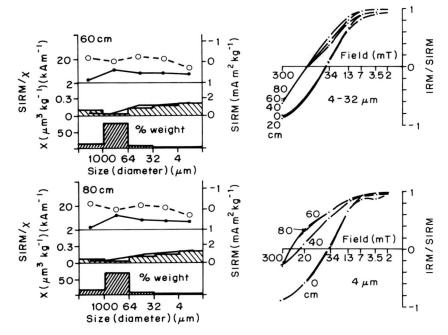

Figure 16.3 Mineral magnetic properties of particle size fractioned samples from a cultivated brown earth (Newton Mere)

Table 16.3 Percentage frequency-dependent susceptibility (χ_{fd}) for soil pit B from the catchment of Newton Mere, Shropshire

Depth (m)	Size fractions (ϕ)				
	0	0–4	4–5	5–8	>8
0	7	5	4	8	20
0.2	6	2	3	6	16
0.4	1	0	3	8	15
0.6	3	0	0	0	0
0.8	4	0	0	0	0

The 'Dell' (Figure 16.5) is a steeply sloping valley-side (maximum slope 24°) developed in brown sandy soils (the 'steep phase' of the Bridgnorth series on the upper slopes with a 'colluvial variant' of the Newport Series on the lower slopes (Hollis, 1978)). The site is too steep to be used for arable agriculture and is currently covered by a thin scrub of woodland and bracken. An obvious step at the base of slope divides the site from the washland of Hilton Brook. In section, the flat crest of the valley side grades into a convex upper segment, comprising about one third of the overall length of the slope. Below this convex

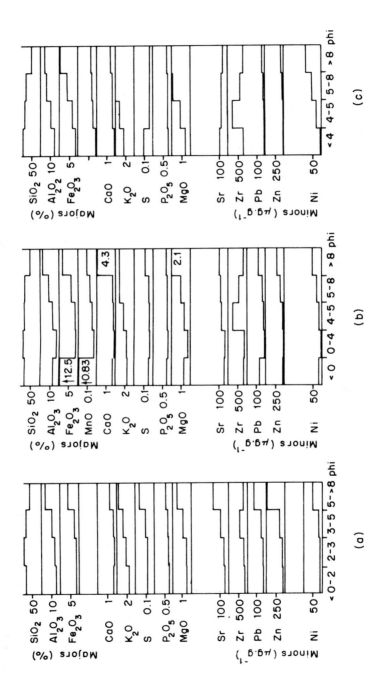

Figure 16.4 Elemental geochemistry of particle size fractioned samples of three selected Triassic substrates from Bar Mere catchment. (a) Topsoil; (b) subsoil; (c) fine-grained till

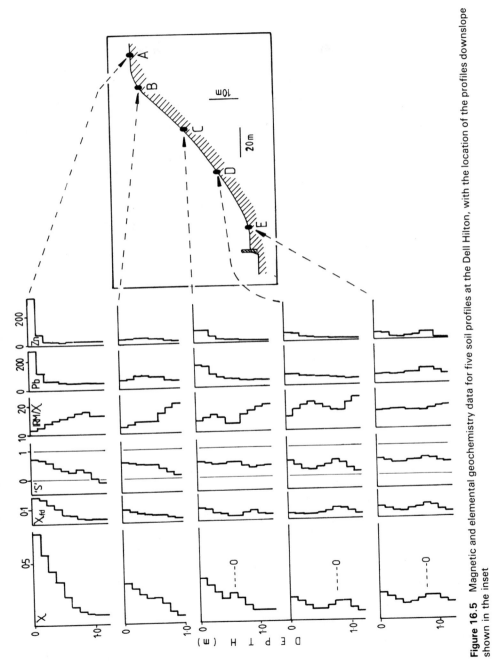

Figure 16.5 Magnetic and elemental geochemistry data for five soil profiles at the Dell Hilton, with the location of the profiles downslope shown in the inset

feature there is a short straight segment, leading into a long slightly concave section (Figure 16.5). Five soil pits were excavated and sampled. Throughout all the profiles, the soils are of a coarse/medium sand with an apedal single-grain structure. A fibrous raw vegetable mat is present at the surface. Below this only slight colour variations, with diffuse boundaries, are observed in the profiles. In pits C, D and E darker profile colours are associated with undecomposed coarse vegetable detritus. Pit A is at the crest of the valley-site and is not a receiving site. Pit B is situated on a moderate slope. A marked increase in slope angle occurs below this, and pits C and D are on steeply sloping ground. Pit E is on more gently sloping ground at the base of slope.

Figure 16.5 shows four magnetic and two geochemical parameters for each of the pits. Assuming pit A represents an undisturbed profile, it is possible to identify apparently ordered variations in these parameters with soil depth. These variations can be interpreted in the light of the magnetic and geochemical properties described earlier. Pit A displays marked increases in χ, χ_{fd}, Pb and Zn towards the surface. Pit B has a thin vegetable mat and also appears to be truncated. Compared to Pit A, it shows much less differentiation of its surface layers. (The surface sands of pit B are similar to those at a depth of about 30 cm in pit A.) Pit C shows a weak bimodal distribution of magnetic values down the profile. The lower peak in χ at 60 cm is associated with dispersed vegetable remains and a slightly darker soil colour. Pit D shows the bimodal character, but here the buried χ peak is at a depth of 90 cm. As with pit C, there is no associated peak in the heavy metal concentrations. Pit E has the same general characteristics but the lower χ peak has higher concentrations of Pb and Zn. The amplitude of the S and IRM_{sat}/χ variations are less marked than in pits C and D; but variation of χ_{fd} is greater. Overall, the slope sites do not reach the high values of χ, S, Pb or Zn displayed in pit A. The following interpretation is suggested:

(1) Pit A is a well-developed *in situ* soil profile. It is comparable with non-arable soils developed elsewhere on Triassic substrates.
(2) Pit B is truncated; it seems likely that this part of the slope is the source of the colluvium further downslope.
(3) Pit C shows the deposition of colluvium on a previously truncated soil.
(4) Soil pits D and E show either the deposition of colluvium over a previous *in situ* soil or a sequence of colluvial deposits in which the ratio of upslope-derived topsoil and subsoil material has varied.

Compared to the Dell, the 'Carrot Field' (Figures 16.6 and 16.7) is a more extensive slope system, of lower gradient. It has been under intensive arable cultivation for a number of years, although currently under grass. A marked step at the base of slope coincides with a hedge line. It is clear from accounts by local people that the terrace feature formed upslope of the hedge is entirely

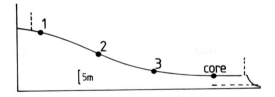

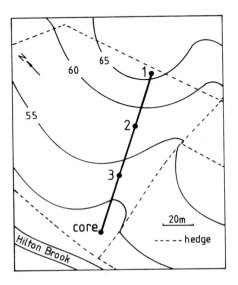

Figure 16.6 Profile and plan of the Carrot Field

the product of accelerated erosion since the 1940s. Reed (1979) noted recurrent erosion in the field, with the development of rills and gullies, while at the base of slope, deposition of sediment from slope wash was observed.

Three soil pits and a core were taken along a transect line from the highest point of the field to near the lower boundary of the terrace. Figure 16.7 shows selected magnetic and geochemical data for the three soil pits. Pit 1 has three identifiable horizons (Ap, Bw, B/C), based on field descriptions, including observations of distinct colour differences. These horizons are also clearly identifiable both magnetically and geochemically. From both pedogenic and magnetic data, pit 2 appears to be a shallower, truncated soil. Pit 3 displays three horizons. The surface horizon has higher values of χ, S and Zn than either of pits 1 and 2, and has a colour similar to the identified Ap horizon of pit 1. It may be suggested first, that this is a receiving site for sediment eroded upslope and, second, the material below this horizon is an *in situ* truncated soil.

268

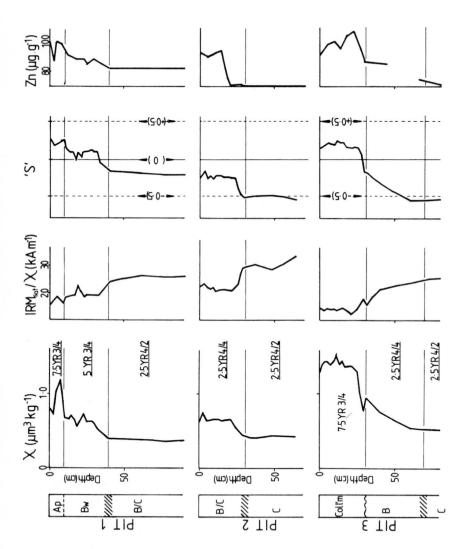

Figure 16.7 Selected magnetic and chemical measurements from soil pits 1 (top), 2 (middle) and 3 (bottom) for the Carrot Field, Hilton

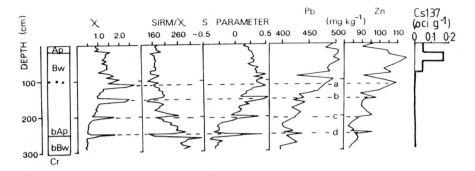

Figure 16.8 Magnetic and chemical data for the core from the Carrot Field, Hilton. Susceptibility peaks a, b, c and d are indicated

The surface horizon represents the upslope 'feathered' edge of the colluvium deposit which is displayed in the core (Figure 16.8).

The core taken through the terrace feature (Figure 16.6) consists of a remarkably uniform loamy coarse sand with an apedal grain structure. At 102 cm below the surface there is a stony layer and between 257 and 259 cm there is an horizon with weakly developed fine granular fragments, a higher organic matter content and fibrous roots. Bedrock is reached at a depth of 303 cm.

The magnetic and geochemical signatures from this core (Figure 16.8) are quite striking in comparison to those of the soil pits. χ values fluctuate markedly with depth but correlate strongly with S, IRM_{sat}/χ, Pb and Zn values. The peak values of χ, S, Pb and Zn are higher than found elsewhere along the transect.

The following interpretation of the core data is suggested:

(1) At a depth of >257 cm there is an *in situ* soil profile.
(2) Above 257 cm the sediment is colluvial in origin and consists of a variable mix of eroded topsoils and subsoils.
(3) The colluvium has been laid down as a number of pulsed events representing different regimes of upslope erosion and deposition. It seems likely that the differences in magnetic signature reflect both variations in contributing source area and the texture of material deposited during any one event.
(4) Much of the former, pre-arable topsoil has been stripped from the upslope area.

The ^{137}Cs data seem consistent with the overall (post-1945) age of the deposit. The absence of measured ^{137}Cs below 100 cm in the deposit could, however, be related to its dilution within the sediment mix rather than a pre-1960 age.

The two examples discussed above illustrate how magnetic and geochemical methods may provide a useful way of identifying colluvium and of discriminating between topsoil and subsoil components of the deposit. In suitable locations, field measurements of χ could provide a rapid method of surveying the extent and geometry of colluvium.

FUTURE RESEARCH AND MIXING MODELS

Investigations of the magnetic and geochemical properties of colluvium merit further attention. One aim is to interpret the measured properties using a sediment source mixing model, which would aid in the construction of a quantified, process-based, slope erosion model. Such a model will need to operate within an adequate dating framework and also link with other approaches, such as ^{137}Cs-dilution studies. However, any models should take account of the dynamic nature of the modern agricultural landscape. They thus need to include:

(1) The evidence for almost complete topsoil stripping from some slopes and even some interfluves. As a consequence, characterization of this lost material, and hence its inclusion in sediment-mixing models, is problematical and this may limit the straightforward application of types of particle size or geochemically based normalizing procedures similar to those developed for lake sediment-mixing models (Smith, 1986).
(2) The observations of active slope processes suggest that overland flow carries fines away from the site into the fluvial network. Maddox (1982), in Claverley Brook, identified both magnetically and geochemically the typical signatures of topsoil in-wash to the stream during storm events. Thus the colluvium must be seen as a variably incomplete fraction of the overall sediment flux from the slope.
(3) The evidence that some of the magnetic and heavy metal enhancement of topsoils has resulted from a progressive build-up of atmospheric fallout associated with industrialization. While this effect on the magnetic signature can be estimated by use of the χ and χ_{fd} parameters, separation of the atmospheric heavy metal contribution remains problematic. In the context of colluvium, buried topsoils identified by magnetic parameters which have no corresponding elevated heavy metal concentrations (cf. pits C and D at The Dell, above) may represent soils buried before the Industrial Revolution.

ACKNOWLEDGEMENTS

We would like to thank John Walden, David Mitchell and Gordon Whittaker for help with field work and sample analysis, Frank Oldfield and Des Walling for the ^{137}Cs data and the School of Construction Studies at the Polytechnic, Wolverhampton, for the loan of coring equipment.

REFERENCES

Crompton, E. and Osmond, D. A. (1954). *The Soils of the Wem District of Shropshire*, Mem. Soil Surv. Gt Br.

Cawse, P. A. (1976). *A Survey of Atmospheric Trace Elements in the UK*, HMSO, London.

Dearing, J. A., Maher, B. A. and Oldfield, F. (1985). Geomorphological linkages between soils and sediments: the role of magnetic measurements. In Richards, K. S., Arnett, R. R. and Ellis, S. (eds), *Geomorphology and Soils*, Allen and Unwin, London, pp. 245–68.

Fullen, M. A. (1985). Compaction, hydrological processes and soil erosion on loamy sands in east Shropshire, England, *Soil and Tillage Research*, 6, 17–29.

Fullen, M. A. and Reed, A. H. (1986). Rainfall, runoff and erosion on bare arable soils in east Shropshire, England, *Earth Surface Processes and Landforms*, 11, 413–25.

Fullen, M. A. and Reed, A. H., (1987). Rill erosion on arable loamy sands in the West Midlands of England. In Bryan, R. B. (ed.), *Rill Erosion Processes and their Significance*, Catena Supplement, 8, 85–96.

Hollis, J. M. (1978). *Soils in Salop 1*, Soil Survey Record, No. 49, Harpenden.

Le Borgne, E. (1960). Influence du feu sur les propiétés magnetiques du sol et sur celles du schiste et du granite, *Ann. Geophys.*, 16, 159–95.

Maddox, R. G. (1982). *Erosion Substrate Characterisation and Suspended Sediment in a Small East Shropshire Catchment*, Unpublished BA (Hons) thesis, The Polytechnic, Wolverhampton.

Mullins, C. E. (1977). Magnetic susceptibility of the soil and its significance in soil science—a review, *J. Soil Sci.*, 28, 223–46.

Mullins, C. E. and Tite, M. S. (1973). Magnetic viscosity, quadrature susceptibility and frequency dependence of susceptibility in single-domain assemblies of magnetite and maghemite, *J. Geophys. Res.*, 78, 804–9.

Oldfield, F., Dearing, J. A., Thompson, R. and Garret-Jones, S. E. (1978). Some magnetic properties of lake sediments and their possible links with erosion rates, *Polskie Archive. Hydrobiologia*, 25, 321–31.

Reed, A. H. (1979a). Accelerated erosion of arable soils in the United Kingdom by rainfall and runoff, *Outlook on Agriculture*, 10, 41–8.

Reed, A. H. (1979b). *Accelerated Erosion of Arable Soils with Special Reference to the West Midland*, Unpublished PhD thesis, University of Keele.

Schwertmann, U. and Taylor, R. M. (1977). Iron oxides. In Dixon, J. B. and Weed, S. B. (eds), *Minerals in Soil Environments*, Soil Sci. Soc. Am., 145–80.

Smith, J. P. (1986). *Mineral Magnetic Studies on two Shropshire–Cheshire Meres*, Unpublished PhD thesis, University of Liverpool.

Sposito, G. (1984). The chemical forms of trace metals in soils. In Thornton, I. (ed.), *Applied Environmental Geochemistry*, Academic Press, New York.

Swaine, D. J. and Mitchell, R. L. (1960). Trace element distribution in soil profiles, *J. Soil Science*, 11; 347–68.

Thompson, R. and Oldfield, F. (1986). *Environmental Magnetism*, Allen and Unwin, London.

Twigger, S. (1983). *Environmental Change in Lowland Cheshire*, Unpublished report. University of Liverpool.

17 Late Bronze Age–Iron Age Valley Sedimentation in East Sussex, Southern England

CHRISTINE SMYTH and SIMON JENNINGS
Department of Geography, The Polytechnic of North London

INTRODUCTION

Investigations into the palaeo-environments of the Combe Haven valley in East Sussex (Figure 17.1) have revealed extensive sequences of minerogenic estuarine and freshwater silty clays intercalated with biogenic freshwater deposits (Smyth, 1986; Jennings and Smyth, 1987; Smyth and Jennings, 1988a,b) (Figure 17.2). Pollen analysis has been the main technique used to examine the palaeo-environmental history of this valley from 6000 BP (base of the Combe Haven Peat—Figure 17.2) to the present day. One of the most significant events in the history of this valley was the change from biogenic to minerogenic floodplain sedimentation achieved largely during the Iron Age. It has been argued previously that this stratigraphic change was the consequence of industrial and agricultural activity (Jennings and Smyth, 1987; Smyth and Jennings, 1988b). In this chapter, new data are presented which provide a substantial body of additional evidence in support of this argument.

The Combe Haven lies at approximately + 2 m OD, which is below the height of the mean spring tide level. Flooding from the sea is prevented by a gravel barrier beach and a sluice at the coast. The valley is occupied by an underfit stream. The local geology consists of the Hastings Beds.

DATA ACQUISITION AND PRESENTATION

Hand auger and borehole data have permitted the construction of a detailed lithostratigraphy of the valley (Figures 17.1 and 17.2). The lithostratigraphic units of the valley are discussed in Smyth (1986), Jennings and Smyth (1987),

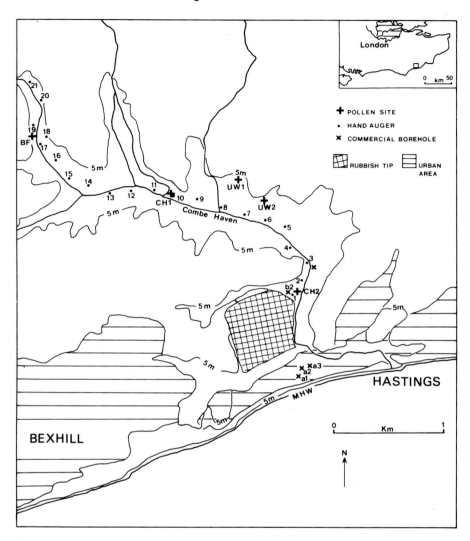

Figure 17.1 Location of pollen sites and augerhole and borehole investigations. Inset; regional location map

Smyth and Jennings (1988a,b), but the emphasis in this chapter is placed upon floodplain sedimentation associated with the switch from biogenic (Combe Haven Peat) to minerogenic sediments (Upper Silty Clay) that comprise the present floodplain. The Upper Silty Clay attains depths of approximately 0.5–1.0 m below the present ground surface (Figure 17.2), is structureless, has a low organic content, is stiff and with grey-blue to orange mottling.

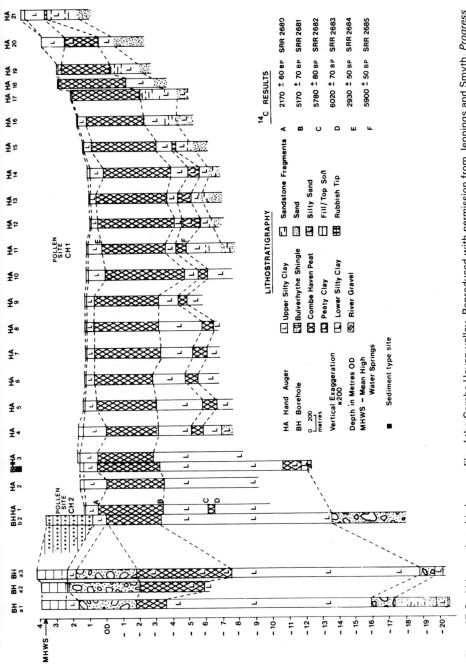

Figure 17.2 Lithostratigraphic long-profile of the Combe Haven valley. Reproduced with permission from Jennings and Smyth, *Progress in Oceanography*, **18**, 205–241. Copyright 1987 Pergamon Journals Ltd

Table 17.1 Radiocarbon dates for the contact between the Combe Haven Peat and the Upper Silty Clay

Site	^{14}C dating	Laboratory code
Combe Haven (CH1)	2930 ± 50 BP	SRR-2684
Buckholt Farm (BF)	2730 ± 70 BP	SRR-3214
Upper Wilting (UW1)	2180 ± 70 BP	SRR-3212
Combe Haven (CH2)	2170 ± 60 BP	SRR-2680
Upper Wilting (UW2)	2160 ± 60 BP	SRR-3213

Five sites from the floodplain have been investigated (Figure 17.1). Two are located centrally on the floodplain; CH1 is a mid-valley site and CH2 is situated down-valley. The pollen record from these two sites is discussed in Jennings and Smyth (1987) and Smyth and Jennings (1988b). In addition, three other sites have now been examined (UW1, UW2, BF) and these are located close to the junction of the floodplain with the valley-side (Figure 17.1). The lithostratigraphy and the pollen record have been established for each site from the upper part of the Combe Haven Peat, through the Upper Silty Clay to the present land surface. In addition, the contact between the peat and the silty clay has been radiocarbon dated at each site (Table 17.1). Together, the five floodplain sites provide a substantial database for the examination of floodplain development in the valley.

Relative pollen diagrams have been constructed for these five sites (Figures 17.3–17.7). For UW1, UW2 and BF the tree and main herb taxa are shown. The complete pollen record for CH1 and CH2 can be found in Smyth and Jennings (1988b). Summary diagrams for these two sites are presented here. The pollen sum for the sites is 300 dry land and marsh pollen excluding *Alnus*. The diagrams have been zoned into local pollen assemblage zones. The CH1 diagram provides the most complete record of vegetational change from c.6000 BP to the present day. In addition to the pollen assemblage zones, the diagrams for CH1 and CH2 have been divided into local forest composition zones based specifically upon arboreal pollen trends. This is elaborated upon in Smyth and Jennings (1988b).

Vegetation History

The pollen record from CH1 (Figure 17.3) establishes that the catchment was well forested from at least 5900 ± 50 BP (SRR-2685) up to the onset of Upper Silty Clay sedimentation. Two distinct forest components are recorded on the pollen diagram for CH1 (Smyth and Jennings, 1988b). First, a valley side component is represented by *Quercus*, *Tilia* and *Ulmus*, and second, pollen and macrofossil evidence indicates that *Alnus* and *Salix* largely comprised the floodplain component. All these taxa flourished, with some fluctuations,

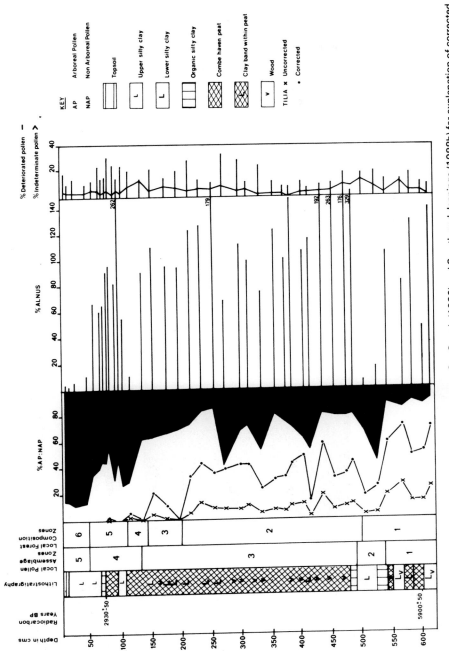

Figure 17.3 Combe Haven (CH1) summary pollen diagram. See Smyth (1986) and Smyth and Jennings (1988b) for explanation of corrected *Tilia* frequencies. All frequencies are expressed as percentage pollen sum, except deteriorated pollen (= percentage total pollen and spores). Reproduced by permission of Sussex Archaeological Collections

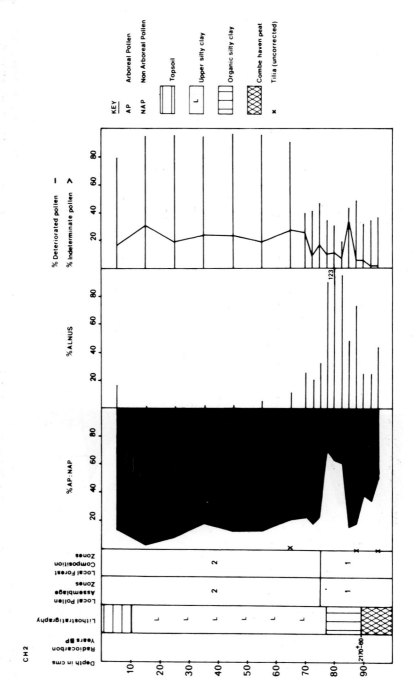

Figure 17.4 Combe Haven (CH2) summary pollen diagram. All frequencies are expressed as percentage pollen sum, except deteriorated pollen (= percentage total pollen and spores). Reproduced by permission of Sussex Archaeological Collections

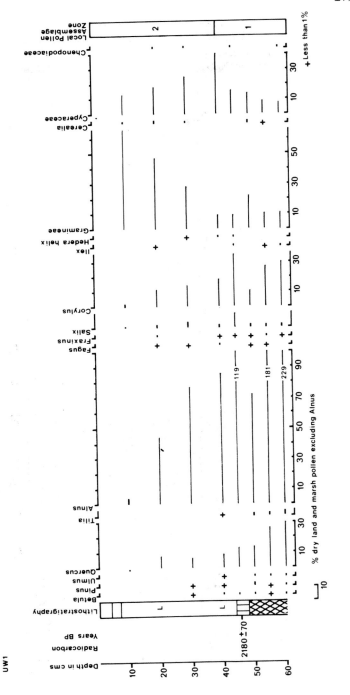

Figure 17.5 Upper Wilting (UW1) pollen diagram; tree and main herb taxa

280

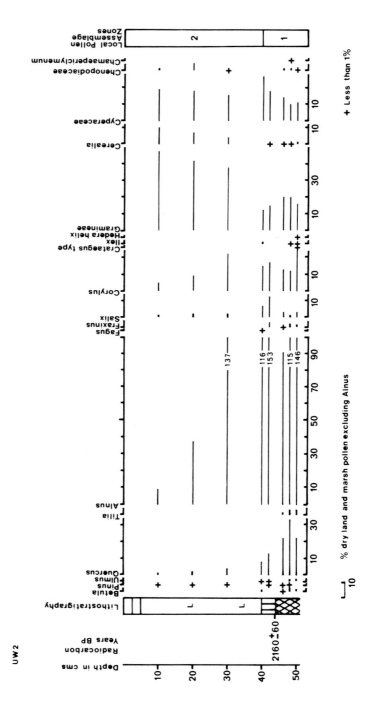

Figure 17.6 Upper Wilting (UW2) pollen diagram; tree and main herb taxa

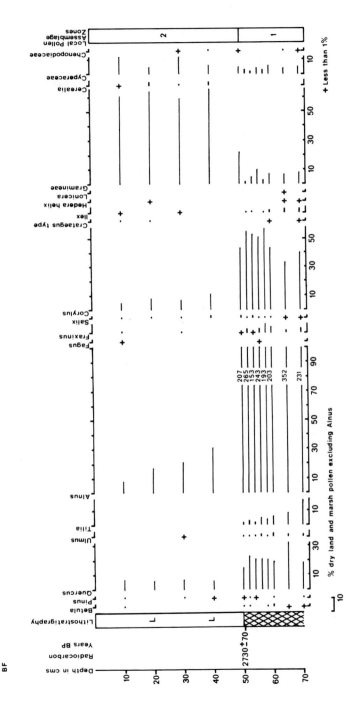

Figure 17.7 Buckholt Farm (BF) pollen diagram; tree and main herb taxa

until 2930 ± 50 BP, when the valley side taxa *Quercus*, *Ulmus* and *Tilia* decline or disappear, while the floodplain taxa, although experiencing a decline, maintain significant values. This pattern is reflected by the other diagrams, with the exception of CH2, where the floodplain taxa also decline to very low frequencies. Jennings and Smyth (1987) have argued that this was due to a marine incursion into the lower reaches of the valley at 2170 ± 60 BP, probably initiated by the decreased forest cover which resulted in a reduction in evapotranspiration and an increase in runoff. The consequential increase in river discharge may have widened the estuary, allowing an extension of the tidal limit up-valley. This would account for the significant reduction of *Alnus* at the onset of Upper Silty Clay sedimentation. Therefore, the Upper Silty Clay at CH2 has a lower estuarine facies and an upper freshwater facies, the boundary being 25 cm below ground surface (Jennings and Smyth, 1987; Smyth and Jennings, 1988b).

In detail, local pollen assemblage zone (lpaz) 4 of CH1 and lpaz 1 of the other diagrams correspond to the episode of vegetational history in the valley just prior to the substantial decline in arboreal pollen (AP), while lpaz 5 of CH1 and lpaz 2 of the other diagrams cover the period of low AP. Importantly, the transition from lpaz's 4 to 5 of CH1 and from 1 to 2 at the other sites is broadly contemporaneous with the change from biogenic (Combe Haven Peat) to minerogenic (Upper Silty Clay) sedimentation. In the pre-AP decline lpaz's, the valley side taxa attain high values. In general, *Quercus* reaches approximately 30 per cent, *Ulmus* 3 per cent and *Tilia* up to 17 per cent. On the floodplain, *Alnus* reaches a maximum of 352 per cent. In the subsequent non-arboreal pollen (NAP) dominated zone, all valley side arboreal taxa decline immediately or disappear; *Quercus* to 10 per cent, *Tilia* and *Ulmus* are no longer recorded. With the exception of CH2, floodplain taxa experience substantial reductions but do not decline to low values until near the top of the Upper Silty Clay. Radiocarbon dating indicates that the reduction in AP occurred between 2930 ± 50 BP (CH1) and 2160 ± 60 BP (UW2).

Due to the scale of the decline in AP and because it is recorded at all sites in the valley, the AP reduction is considered to be the consequence of anthropogenic forest clearance and is not related to either differences in pollen dispersal/preservation or to climatic change (Smyth and Jennings, 1988b). The pollen record indicates that the valley sides were cleared first, while the floodplain was not utilized until a later date. Also, closer inspection of the radiocarbon dates for the stratigraphic change from Combe Haven Peat to Upper Silty Clay and concurrent reduction in AP suggests that the dates cluster around two groups. For the up-valley sites, earlier dates are recorded; 2930 ± 50 BP from CH1 and 2730 ± 70 BP at BF. Down-valley, the stratigraphic change occurred later; 2180 ± 70 BP at UW1, 2170 ± 60 BP at CH2 and 2160 ± 60 BP at UW2. No hiatus is believed to exist between the Combe Haven Peat and Upper Silty Clay units for two reasons. First, with the exception of the BF site, there is a lithostratigraphic transitional phase represented by an organic silty clay layer. Second, all sites, including BF, exhibit a transitional sequence in the pollen record across the lithostratigraphic boundary. This pattern suggests that the Upper Silty Clay capped the Combe Haven Peat without any prior erosion.

The pollen data show that following clearance there was no regeneration of trees on either the valley side or floodplain, although a higher resolution sampling interval may be required to confirm this. Thus, once the trees had been cleared, there was continuous utilization of the valley, an assertion supported by the high Gramineae pollen frequencies and presence of anthropogenic indicator pollen types such as *Plantago* and Compositae. Additionally, high values of cereal pollen are recorded within the Upper Silty Clay, reaching a maximum of 12 per cent at UW2.

The stratigraphic data presented in this chapter show that the formation of the Upper Silty Clay was contemporaneous with forest clearance, especially on the valley sides. Anthropogenic activity may therefore have resulted in the colluviation of sediments from the valley sides and onto the floodplain. This activity certainly included agriculture, as evidenced from the pollen record, and may also have been associated with iron working.

CORRELATION OF ENVIRONMENTAL AND ARCHAEOLOGICAL DATA

The earlier up-valley dates for forest clearance (at CH1 and BF) are late Bronze Age, while the later down-valley dates are Iron Age. Thus Bronze Age and Iron Age activities had an identical impact upon the stratigraphic record of the valley.

From archaeological evidence it is known that the Combe Haven catchment and surrounding region was important for iron working during the Roman Period (summarized in Smyth and Jennings, 1988b; Drewett *et al.*, 1988). An analysis of the distribution of Iron Age sites in south-east England is given in Drewett *et al.* (1988, Figure 5.2, p. 128). This map reveals no known sites in the Combe Haven area. This lack of archaeological evidence may be attributable to insufficient archaeological research. However, the idea expressed by Drewett *et al.* (1988) that this region probably represents a 'true gap' (p. 127) in Iron Age activity and that the 'heavy clay areas were probably still forest [in the Iron Age]' (p. 127) is true if based on the distribution of archaeological finds, but is not supported by the stratigraphic evidence presented in this chapter, which demonstrates clearly intensive utilization of the Combe Haven valley from late Bronze Age–Iron Age times.

CONCLUSION

Forest clearance in the Combe Haven valley, undertaken largely during the Iron Age, was responsible for the erosion of sediment from the valley sides and onto the floodplain. This major anthropogenic environmental impact was the consequence of an expanding agricultural, and probably industrial (iron working) economy. The colluviation of sediments may have been encouraged further by increasing climatic wetness during the Iron Age (Barber, 1982). This has resulted in the floodplain of the Combe Haven valley acting as a substantial store of sediments. Subsequent to colluviation, the Upper Silty Clay sediments have experienced gleying with periodic oxidation, giving them their present characteristics.

Whether the colluviation took place as a single, catastrophic episode or in a sequence of lower-magnitude higher-frequency events is unclear. However, this episode of sedimentation in the Combe Haven valley provides an historical context in which to view present soil erosion problems. The volume of colluviated sediment stored on the floodplain of the Combe Haven approximates to 1.16 million m^3.

ACKNOWLEDGEMENTS

We gratefully acknowledge NERC (Dr D. D. Harkness) Radiocarbon Laboratories for the ^{14}C age measurements and the landowners of the Combe Haven valley who allowed access to their land.

REFERENCES

Barber, K. E. (1982). Peat-bog stratigraphy as a proxy climate record. In Harding, A. F. (ed.), *Climatic Change in Later Pre-History*, Edinburgh University Press, 103–13.

Drewett, P., Rudling, D. and Gardiner, M. (1988). *The South-East to AD 1000*, Longman, London.

Jennings, S. and Smyth, C. (1987). Coastal sedimentation in East Sussex during the Holocene, *Progress in Oceanography*, **18**, 205–41.

Smyth, C. (1986). A palaeoenvironmental investigation of Flandrian valley deposits from the Combe Haven valley, East Sussex, *Quaternary Studies*, **2**, 22–33.

Smyth, C. and Jennings, S. (1988a). Coastline changes and land management in East Sussex, southern England, *Journal of Ocean and Shoreline Management*, **11**, 375–94.

Smyth, C. and Jennings, S. (1988b). Mid- to late-Holocene forest composition and the effects of clearance in the Combe Haven valley, East Sussex, *Sussex Archaeological Collections*, **126**, 1–20.

18 Relationships between Catchment Characteristics, Land Use and Sediment Yield in the Midland Valley of Scotland

R. W. DUCK and J. McMANUS
Department of Geography and Geology, The University, St Andrews

INTRODUCTION

Sediment yield from a catchment may be determined by two principal techniques: the estimation of fluvial sediment transport and the computation of quantities of materials entrapped in reservoirs. In few cases has sampling of mobile sediments in rivers been carried out continuously over periods exceeding three years. As a result, figures derived from such studies are highly dependent on the precipitation characteristics of those specific years, which may be atypical. However, because they have accumulated sediments continuously over protracted time periods, often exceeding a hundred years, reservoir basins can be used to provide good long-term average values of sediment yield (Rausch and Heinemann, 1984). Beyond the 'bulk' sediment yields obtained there may be internal, dateable markers within reservoir sediment columns (geochemical or granulometric) which can enable subdivision of the record and thereby medium-term yields to be estimated. In this chapter we are principally concerned with 'bulk' sediment yields derived from reservoir surveys in the Midland Valley of Scotland.

SETTING

Some 15 000 km² in area, the physiographic province of the Midland Valley or Central Lowlands of Scotland is delimited to the north by the Highland Boundary Fault and to the south by the Southern Uplands Fault. The major underlying rock formations are mainly of Devonian (Old Red Sandstone) and

---

Soil Erosion on Agricultural Land
Edited by J. Boardman, I. D. L. Foster and J. A. Dearing
©1990 John Wiley & Sons Ltd

Carboniferous ages. These comprise varied sedimentary (including sandstones, limestones and shales) and contemporaneous igneous rock assemblages (intermediate to basic lavas and pyroclastics) with numerous intrusive bodies in the form of dykes, sills, vents and plugs. Importantly, the region is blanketed with varying thicknesses and types of Pleistocene till and fluvio-glacial deposits. In addition, post-glacial peat and alluvium are locally of significance.

Although over 80 per cent of the population of Scotland reside in the Midland Valley, it contains some of the country's most important areas of arable farming, together with local afforestation and upland moorlands, rising to over 700 m in height, which are largely devoted to rough grazing. In broad terms, the average annual precipitation varies from between 600 and 1000 mm in the relatively low-lying eastern part of the region to between 1000 and 2000 mm in the generally higher relief central and western sectors (see Meteorological Office, 1977, for details).

PREVIOUS STUDIES

In earlier publications (McManus and Duck, 1985; Duck and McManus, 1987), the authors reported the results of sediment surveys of six fully drawndown reservoirs in the Midland Valley in comparison with previous work on three other basins by Lovell *et al.* (1973) and Ledger *et al.* (1974, 1980). Of the nine estimates of sediment yield derived from these studies, four were from catchments in Fife (Lambieletham, Harperleas, Drumain, Cullaloe), two from south-east Perthshire (Glenfarg, Glenquey) and two from the Lothians (North Esk, Hopes). Hence there was a bias of data to the eastern half of the Midland Valley with one reservoir (Kelly) surveyed in the west (Figure 18.1).

PRESENT STUDIES

During the period 1987–8 five further surveys of reservoir sedimentation have been carried out by the authors on water bodies in three quite separate areas of the Midland Valley. Pinmacher Reservoir is situated near the west coast, 4 km south of Girvan in southern Ayrshire. Holl Reservoir lies in the Lomond Hills of Fife, some 2 km downstream of Harperleas Reservoir. An unsurveyed water body, Ballo Reservoir, forms the central member of this three-basin cascade. The remaining three reservoirs, Earlsburn No. 1, North Third and Carron Valley, are situated in the thickly peat-mantled uplands between the Gargunnock and Kilsyth Hills of Stirlingshire. Although there is still some imbalance of data points towards the east, the addition of these new sites to the inventory of reservoir surveys (Figure 18.1) affords a much better spread of information along the length of the Midland Valley.

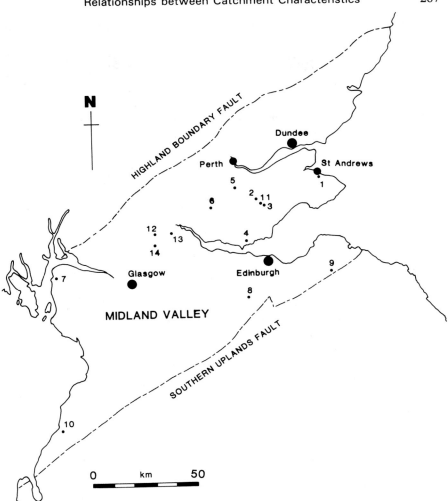

Figure 18.1 Location map of the reservoirs. 1 Lambieletham (Nat. Grid. Ref. NO 502134); 2 Harperleas (NO 212053); 3 Drumain (NO 223043); 4 Cullaloe (NT 188875); 5 Glenfarg (NO 016110); 6 Glenquey (NN 980027); 7 Kelly (NS 223685); 8 North Esk (NT 155582); 9 Hopes (NT 546621); 10 Pinmacher (NX 194937); 11 Holl (NO 228035); 12 Earlsburn No. 1 (NS 700895); 13 North Third (NS 756890); 14 Carron Valley (NS 680850)

An aim of this contribution is to collate and compare catchment sediment yields derived from all fourteen surveys in relation to catchment characteristics and land use. Some background data on the reservoirs and their catchment areas are presented in Table 18.1. Drainage densities have been determined from 1:10 000 or 1:10 560 scale maps of various editions as available.

Table 18.1 Background data on the reservoirs and their catchment areas

Reservoir	Construction date	Top water level (m above OD)	Water surface area (ha)	Original capacity (m^3)	Maximum depth (m)	Catchment characteristics Area (km^2)	Peak altitude (m above OD)
Lambieletham	1900	102.1	1.21	6.36×10^4	12.2	2.29	153
Harperleas	1875	259.1	16.2	8.22×10^5	10.1	3.44	522
Drumain	1867	231.3	2.02	7.57×10^4	7.8	1.53	318
Cullaloe	S. Bay 1876 N. Bay 1908	88.55	S. Bay 9.7 N. Bay 6.5	S. Bay 3.41×10^5 N. Bay 1.14×10^5	S. Bay 15.85 N. Bay 4.5	5.84, to 1925 4.13, 1925 to date	213
Glenfarg	1926	183.6	40.5	2.586×10^6	16.8	10.54, to 1972 23.5, 1972 to date	497
Glenquey	1909	287.8	16.7	1.191×10^6	18.0	5.58	643
Kelly	1895	c.200	5.4	2.4×10^5	c.5	3.4	290
North Esk	1850	c.335	10.0	5.76×10^5	c.12	7.0	564
Hopes	1935	c.275	13.0	3.5×10^5	c.6	5.0	528
Pinmacher	1903	136.9	1.42	6.82×10^4	10.7	0.425	263
Holl	1901	204.2	17.8	1.15×10^6	13.4	3.99	440
Earlsburn No. 1	1890	367.4	26.0	6.52×10^5	4.8	2.85	460
North Third	1911	171.1	46.0	3.39×10^6	13.3	9.31	441
Carron Valley	1939	225.4	3.9×10^2	2.14×10^7	11.2	38.7	570

Table 18.1 *(continued)*

		Catchment characteristics		
Relief ratio (m m^{-1})	Drainage density (km km^{-2})	Solid geology	Land use	Long-term average precipitation (mm yr^{-1})
0.017	1.40	Mainly Lower Carboniferous sandstones and shales	Mixed arable farming	820
0.142	1.80	Lower Carboniferous limestones with Carboniferous-Permian quartz dolerites	Rough moorland grazing with some afforestation	1000
0.041	1.83	Carboniferous-Permian quartz dolerite sills	Rough moorland grazing	1000
0.059	0.48	Lower Carboniferous sandstones with Carboniferous-Permian quartz dolerite and agglomerate	Mixed arable farming with afforestation	750
0.112	1.42	Lower Devonian andesite lavas and subsidiary volcaniclastics	Mixed arable farming with afforestation	1070
0.183	1.43	Lower Devonian andesite lavas and volcaniclastics	Rough moorland grazing with some afforestation	1650
0.035	2.35	Upper Devonian sandstones and conglomerates with minor Carboniferous and Tertiary intrusives	Rough moorland grazing	1650
0.135	1.26	Mainly Lower Silurian mudstones and shales of the North Esk inlier	Rough moorland grazing with minor afforestation	1150
0.127	1.40	Mixed Ordovician greywackes and agillaceous sediments	Rough moorland grazing	1000
0.103	2.94	Mixed Ordovician sediments and igneous rocks of the Ballantrae complex	Rough moorland grazing	1200
0.066	3.01	Lower Carboniferous limestones and sandstones with Carboniferous-Permian quartz dolerites	Rough moorland grazing with some afforestation	950
0.056	1.67	Carboniferous basaltic lavas	Rough moorland grazing	1600
0.069	1.61	Carboniferous basaltic lavas with some limestone and quartz dolerite intrusives	Improved and rough moorland grazing	1400
0.084	1.50	Carboniferous basaltic lavas with tuffs, shales and various intrusives	Rough moorland grazing with afforestation	1700

METHODOLOGY

The pre-1987 reservoir surveys were conducted by direct measurements of the thicknesses of the accumulated deposits during conditions of full drawdown as detailed previously (McManus and Duck, 1985; Duck and McManus, 1987). The one exception was Hopes Reservoir, which was surveyed by sounding line (Ledger et al., 1974).

Of the present group of surveys, those of Pinmacher and Holl Reservoirs were also carried out when the basins had been emptied to facilitate civil engineering works. Sediment thickness was determined by digging through the deposits or by removing columns of 'mud' bounded by polygonal desiccation cracks. This revealed, in most locations, either underlying till or a prominent soil layer containing root fibres. However, in the Pinmacher basin the reservoir sediments were found to be lying directly on top of bedrock over much of the flat central part of the floor. Probing with survey rods enabled detection of till, the soil horizon or bedrock at additional sites, confirmed where necessary by excavation. Observations were made at measured intervals along traverse lines between known end points which covered as much of the deposits as safety would permit. In the small Pinmacher Reservoir traverses were at 10–30 m intervals with thickness determinations at 2–10 m spacings, while in the larger Holl Reservoir traverse intervals were of the order of 50 m with observations at 5–50 m intervals. Additional valuable information on sediment thicknesses was gained from the continuous exposures revealed along the banks of re-excavated streams. The data were plotted at original scales of 1:500 (Pinmacher) and 1:2000 (Holl) and isopachyte maps constructed for each reservoir (Figures 18.2 and 18.3). Planimetric analysis of these maps enabled the total volume of wet sediment entrapped in each basin and the reduction in water storage capacity due to sedimentation to be evaluated (Table 18.2).

Several box core samples of the reservoir deposits were collected both from central areas characterized by fine-grained materials and from accumulations of coarser sediments localized off influent streams. The weight loss on drying these known volumes of wet sediments allowed their water contents and bulk dry densities to be determined. Hence estimates of the dry weights of materials entrapped in the reservoirs could be made. Grain size analyses were carried out on surface sediment samples and sectioned core blocks by a combination of sieving and the pipette method as required, following Folk (1974). Weight loss on ignition was used to provide estimates of the organic matter content in dried materials (Dean, 1974).

The closely clustered central group of reservoirs (Earlsburn No. 1, North Third, Carron Valley) were surveyed by the authors on behalf of Central Regional Council while impounded to top water level. In each case a bathymetric chart was produced from echo-soundings run from a 5 m survey vessel. The soundings were made using a 200 kHz Lowrance echo-sounder with position

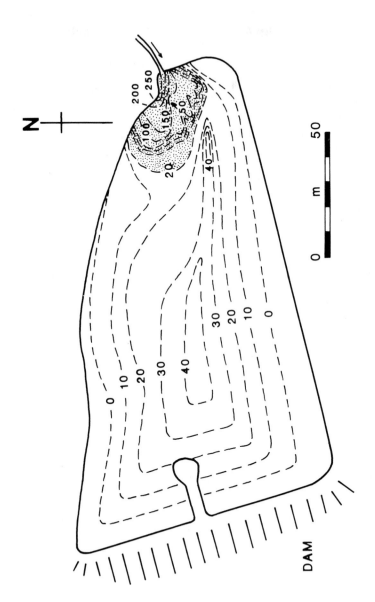

Figure 18.2 Isopachyte map of the deposits of Pinmacher Reservoir, contour values in centimetres. Note zero contour = 'mud line' above which no reservoir sediments are deposited. Stippling denotes area of sand and gravel accumulation below mouth of influent

Table 18.2 Reservoir sedimentation data

Reservoir	(years after impounding)	sediment volume (m³)	Reduction in capacity (%)	Reservoir trap efficiency (%)	Total dry sediment weight (t)[a]
Pinmacher	April 1988 (85)	2.4×10^{3b}	3.1	87–98	1.78×10^3
Holl	June 1987 (86)	5.28×10^{4c}	4.6	93–98	2.37×10^4
Earlsburn No. 1	April 1988 (98)	5.67×10^4	8.7	85–96	1.73×10^4
North Third	January 1988 (77)	4.85×10^5	14.3	90–98	1.40×10^5
Carron Valley	March 1988 (49)	8.57×10^5	4.0	92–99	2.42×10^5

[a] Increased according to minimum trap efficiency.
[b] Includes 504 m³ of sand and gravel.
[c] Includes 2.7×10^3 m³ of sand and gravel.

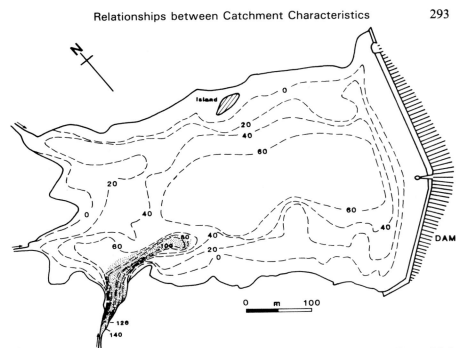

Figure 18.3 Isopachyte map of the deposits of Holl Reservoir, details as for Figure 18.2

fixing being achieved by means of a Motorola Mk III Mini-Ranger system, accurate to ±2 m, interfaced via a microcomputer with a chart plotter. In the relatively small Earlsburn and North Third basins traverse lines were run at 50 m intervals whereas a spacing of 100 m was adopted in the much larger Carron Valley Reservoir. The water depth data from the echo-sounder profiles were plotted at scales of either 1:2500 (Earlsburn and North Third) or 1:5000 (Carron Valley), then contoured at 0.5 m intervals to produce the bathymetric charts. Engineers from the Water and Drainage Department of Central Regional Council used these charts to construct hypsographic (depth–area) curves from which the volume of water in each reservoir was determined (cf. Hakanson, 1981). These figures were compared with the volumes derived from pre-impoundment surveys of the original river valleys in order to compute the losses of capacity through sedimentation (Table 18.2). As sampling of the bottom deposits was not carried out as part of this survey programme, determinations of reservoir sediment grain size, organic content and bulk dry density could not be made. However, analyses of the deposits of nearby water bodies which drain similar terrain, in particular Earlsburn No. 2 Reservoir, situated 1 km downstream from Earlsburn No. 1, indicate bulk dry densities in the range 200–320 kg m^{-3} for the characteristically peaty materials. This is below the values of 339 and 444 kg m^{-3} quoted by Labadz and Burt (1987) for the

Table 18.3 Long-term average sediment yields in the reservoir catchment areas

Catchment	Total sediment yield (t km^{-2} yr^{-1})	Inorganic sediment yield (t km^{-2} yr^{-1})
Lambieletham	2.1	1.8
Harperleas	13.8	11.5
Drumain	3.9	3.3
Cullaloe	30.8	26.2
Glenfarg	52.0	39.0
Glenquey	15.1	10.0
Kelly	41.0	n.r.
North Esk	26.0	n.r.
Hopes	25.0	n.r.
Pinmacher	50.9	45.6
Holl	72.3	61.7
Earlsburn No. 1	68.2	n.d.
North Third	205.4	n.d.
Carron Valley	141.9	n.d.

n.r. = not reported.
n.d. = not determined.

sediments of reservoirs draining peat-covered catchments in the southern Pennines. In the absence of further data, a mean bulk dry density value of 260 kg m^{-3} has been adopted for computing the dry weight of materials entrapped in the Earlsburn No. 1, North Third and Carron Valley basins.

The extent to which the quantity of sediment entrapped in a reservoir represents the total sediment yield from the contributing catchment during the period of impoundment depends on the trap efficiency of the water body. In all cases, reservoir trap efficiency (Table 18.2) was estimated from the ratio of its capacity/mean annual inflow using the curves established by Brune (1953) and Heinemann (1981, 1984). As only limited hydrological data were available for the water bodies and their influents it was necessary (cf. Ledger et al., 1974, 1980; Lovell et al., 1973) to compute mean annual inflows from estimated mean annual runoff values determined as mean annual precipitation minus estimated evapotranspiration loss (Meteorological Office, 1977 and pers. comm.; Fife and Strathclyde Regional Councils, pers. comm.).

The dry weights of sediment in each reservoir have been increased according to their respective estimated minimum trap efficiency (Table 18.2). For the computation of total sediment yields (i.e. incorporating *both* mineral and organic matter) these 'increased' weights have been divided by the reservoir impounding period and then by the catchment area minus the water surface area (Table 18.3). For Pinmacher and Holl Reservoirs inorganic sediment yields have also been determined after deducting the average concentrations of organic matter from the dry sediment weights (Table 18.3).

OBSERVATIONS AND RESULTS

The uppermost 2–6 m of the shores below top water level in each drawndown reservoir lacked materials deposited under lacustrine conditions. These upper slopes were characterized by wave-reworked till or rip-rap armouring to prevent wave erosion. Below wave base, effectively defined by the lower limits of these zones (the 'mud line' of Figures 18.2 and 18.3), the reservoir sediments thickened downslope from less than 0.5 cm to of the order of 40–60 cm in the deepest parts of the basins near the dams (Figures 18.2 and 18.3). Localized, relatively thick accumulations occupied former hollows, and also occurred against obstructions such as tracks and an old bridge as observed in Holl Reservoir. These sediments are dominantly of sandy-silts with subordinate silts (using the textural classification of Folk, 1974). Detailed textural analyses will not be included in this chapter. Average organic matter concentrations by weight were determined as 19.1 per cent (25 analyses) for the Pinmacher Reservoir sediments and 16.9 per cent (50 analyses) for those of Holl Reservoir. Bulk dry densities of these materials averaged 515 kg m^{-3} and 390 kg m^{-3} for Pinmacher and Holl Reservoir, respectively.

Deposits of cross-bedded fine, medium and coarse sands to fine gravels were observed as delta accumulations localized below the mouth of the sole surface influent to Pinmacher Reservoir and the main influent to Holl Reservoir (Figures 18.2 and 18.3). These contained some organic materials in the form of leaves and twigs. The two other influents to Holl Reservoir lack such accumulations because they are regulated by upstream reservoirs. The bulk dry densities of these materials averaged 1.6 t m^{-3} in the Pinmacher deposits and 1.2 t m^{-3} in Holl Reservoir.

The sediment yields derived, as described above, from the suite of 1987–8 reservoir surveys are compared with those previously published (Duck and McManus, 1987) for the pre-1987 surveys in Table 18.3. With the exception of that determined for the Pinmacher catchment, the new values of total sediment yield are considerably higher than those reported earlier. The highest figures are derived from the Carron Valley and North Third Reservoir echo-sounding surveys. Hence the full range of currently available sediment yield values, determined on the basis of reservoir sedimentation surveys, from the Midland Valley of Scotland, is between 2.1 t km^{-2} yr^{-1} and 205.4 t km^{-2} yr^{-1}.

DISCUSSION

Prior to a comparison and discussion of the data of Table 18.3 it is relevant to consider several of the sediment yield values individually. We have suggested (Duck and McManus, 1987) that the yield from the Lambieletham catchment (2.1 t km^{-2} yr^{-1}) is exceptionally low because of the operational management of the reservoir leading to an overestimation of the trap efficiency. A bypass channel is known to have been used to divert flows from the basin during much of its active life (Fife Regional Council, pers. comm.), thus preventing water and sediment from entering the reservoir. In contrast, we suggest that the yields from both the Holl and Carron Valley catchments should be considered as maximal values. In addition to receiving drainage from their direct catchment areas, these water bodies also collect waters diverted from indirect catchments. While the bed load sediments are trapped before entering the basins, some suspended sediments enter these reservoirs via the diverted waters. In the absence of data to quantify the suspended sediment loads, and thus the sediment yield contributions from the indirect catchment areas, the yield values for the Holl and Carron Valley catchments so presented (Table 18.3) must be regarded as overestimates.

There is a strong indication that the highest sediment yields recorded are associated with the dominantly peat-covered catchments of central Scotland (Earlsburn, North Third, Carron Valley). Reservoirs draining such catchments were not included in our pre-1987 surveys. However, at this stage it should be re-emphasized that the surveys of this group of reservoirs were undertaken by echo-sounding, not by direct observation of sediment thicknesses in drawndown conditions. Moreover, the results obtained from the two methods of survey are unlikely to be directly comparable. The use of the same 200 kHz echo-sounder in the three investigations does, however, permit intercomparability of the bathymetric data. This high-frequency transmission signal receives echoes from low-density surficial fluid muds at the top of the reservoir 'bed' (Duck and McManus, 1986) and so gives a worst-case estimate of capacity loss. We suggest that a potentially significant source of error lies in the lack of indigenous sediment bulk dry density determinations in these surveys. Indeed, errors may arise in all surveys from the use of average density values, as this parameter will not only vary between reservoirs but also within the deposits of individual basins (Rausch and Heinemann, 1984). A further source of error in the estimation of sediment volumes from bathymetric data stems from the level of accuracy and reliability of the pre-impoundment surveys, which cannot be assessed.

The argument for the computation of total sediment yields (i.e. incorporating *both* mineral and organic matter) assumes that all the sediment deposited in the reservoirs is the product of catchment erosion. This is true for the inorganic fractions and some organic materials (especially from peat-covered catchments). However, organic matter may also be of autochthonous origin, produced by phytoplanktonic micro-organisms and microbiological decay processes in the water bodies themselves. This fraction is particularly significant in eutrophic lakes and reservoirs in lowland agricultural areas which receive plentiful supplies of nutrients leached from fertilizers applied to the land. The problem of differentiating between allochthonous and autochthonous organic matter in sediment yield studies in one which remains to be addressed, having received only little attention hitherto (Foster et al., 1985; McManus and Duck, 1985). Some authors (e.g. Al-Bayati and McManus, 1984) have chosen to express sediment yields in terms of mineral matter alone, hence the inclusion of similar values for comparative purposes in Table 18.3. Undoubtedly, true sediment yields lie somewhere between the two extremes.

Despite the numerous sources of errors referred to above, a number of conclusions can be made regarding the sediment yield data. As we have previously concluded (McManus and Duck, 1985; Duck and McManus, 1987), although derived for catchments underlain by varying assemblages of rock types (Table 18.1), the sediment yields appear unrelated to the underlying bedrock. The drift deposits serve as the principal suppliers of debris.

Previously we drew attention to the relationships between increasing sediment yield with catchment area and with mean annual inflow (Duck and McManus, 1987). While precipitation, relief and drainage density undoubtedly exert local controls, no clear relationships between these parameters and sediment yield are apparent. It should be stressed that there are considerable difficulties in measuring drainage densities from maps of different scales and ages, especially in areas of peat moorland (Burt and Oldman, 1986). The values quoted in Table 18.1 could, in some cases, be substantial underestimates. Accurate drainage density determinations require combined aerial photography and field surveys. Another important feature of discussion when comparing sediment yields is the role of channel storage. We have no means of estimating the amounts of sediment stored within the various drainage networks. Thus the extent to which the reservoir-derived yields equate with soil erosion in the catchments is difficult to assess. We aim to address this important problem in the future.

On the basis of our present level of knowledge we believe that the major contributory factor in determining rates of sediment yield is the combination of the character of the drift and land use in the catchment. The pre-1987 results revealed that catchments dominated by arable farming generally show higher yields (31–52 t km^{-2} yr^{-1}) than those of open moorlands (4–41 t km^{-2} yr^{-1}), the greatest yields often being associated with catchments having substantial afforestation. This conclusion requires refinement on analysis of the enlarged set of sediment yield data. We suggest that it is necessary to distinguish between types of rough moorland, which apparently now show a very wide range of sediment yields from low values (4–15 t km^{-2} yr^{-1}) to the highest values so far computed (68–205 t km^{-2} yr^{-1}). Within this range a broad, twofold subdivision of catchments may be made. The catchments dominated by extensive, thick accumulations of peat (North Esk, Hopes, Earlsburn, North Third, Carron Valley) generally show much higher yields than those dominated by till soils with subordinate peat cover (Harperleas, Drumain, Glenquey, Holl). The yields derived from the Earlsburn, North Third and Carron Valley studies are directly comparable with the range of values of 120–213 t km^{-2} yr^{-1}

computed for peat-covered catchments in the southern Pennines (Labadz and Burt, 1987; J. Labadz, pers. comm.).

The yield derived for the Pinmacher catchment appears anomalously high for a till-veneered, upland moorland essentially devoid of peat. However, the influent stream to Pinmacher Reservoir flows in a narrow valley of steep gradient, deeply dissected into bare till which is prone to landslipping and gullying. Thus considerable quantities of coarse sediment (Figure 18.2, Table 18.2) are delivered to this small reservoir. This makes a substantial contribution to the sediment yield, which is of the same order as that determined for the similar Kelly catchment.

Elsewhere in Britain, surveys of reservoir sediment accretion have, for example, revealed sediment yields of 5–20 t $km^{-2} yr^{-1}$ (Merevale Lake, North Warwickshire; Foster et al., 1985), 25.4 t $km^{-2} yr^{-1}$ (Cropston Reservoir, Leicestershire; Cummins and Potter, 1967) and 39.4 t $km^{-2} yr^{-1}$ (Strines Reservoir, Yorkshire; McManus and Duck, 1985, after Young, 1958). These values lie within the range presented in this study.

Accepting the data of Table 18.3 as they stand, 43 per cent of the sediment yield values fall within the 20–60 t $km^{-2} yr^{-1}$ range suggested as characterizing small, well-vegetated catchments in upland Britain (Ledger et al., 1980). For reasons described earlier, it is quite possible that the true Lambieletham and Holl catchment yields also lie in this range. However, the newly acquired data support the findings of Labadz and Burt (1987) and J. Labadz (pers. comm.), and suggest that the upper limit to this range is much higher.

ACKNOWLEDGEMENTS

We wish to thank Fife and Strathclyde Regional Councils for providing access to, and information on, Holl and Pinmacher Reservoirs, respectively. We are very grateful to Central Regional Council for allowing us to publish information based on the bathymetric surveys carried out on their behalf. The data of Tables 18.1 and 18.3 which relate to the pre-1987 reservoir surveys are reproduced with the permission of the Royal Society of Edinburgh, which is gratefully acknowledged.

REFERENCES

Al-Bayati, K. M. and McManus, J. (1984). The bathymetry and sedimentology of Loch Benachally, Perthshire, *Scott. J. Geol.*, **20**, 65–71.

Brune, G. M. (1953). Trap efficiency of reservoirs, *Trans. Am. Geophys. Un.*, **34**, 407–18.

Burt, T. P. and Oldman, J. C. (1986). The permanence of stream networks in Britain: further comments, *Earth Surf. Proc. Landf.*, **11**, 111–13.

Cummins, W. A. and Potter, H. R. (1967). Rate of sedimentation in Cropston Reservoir, Charnwood Forest, Leicestershire, *Mercian Geol.*, **2**, 31–9.

Dean, W. E. (1974). Determination of carbonate and organic matter in calcareous sediments and sedimentary rocks by loss on ignition: comparison with other methods, *J. Sediment. Petrol.*, **44**, 242–8.

Duck, R. W. and McManus, J. (1986). Applications of geophysical techniques in the understanding of reservoir siltation, *J. Wat. Resour.*, **5**, 267–78.

Duck, R. W. and McManus, J. (1987). Sediment yields in lowland Scotland derived from reservoir surveys, *Trans. Roy. Soc. Edin: Earth Sci*, **78**, 369–77.

Folk, R. L. (1974). *Petrology of Sedimentary Rocks*, Hemphill, Austin, Texas.

Foster, I. D. L., Dearing, J. A., Simpson, A., Carter, A. D. and Appleby, P. G. (1985). Lake catchment based studies of erosion and denudation in the Merevale catchment, Warwickshire, UK., *Earth Surf. Proc. Landf.*, **10**, 45–68.

Håkanson, L. (1981). *A Manual of Lake Morphometry*, Springer-Verlag, Berlin.

Heinemann, H. G. (1981). A new sediment trap efficiency curve for small reservoirs, *Wat. Resour. Bull.*, **17**, 825–30.

Heinemann, H. G. (1984). Reservoir trap efficiency. In Hadley, R. F. and Walling, D. E. (eds), *Erosion and Sediment Yield: Some Methods of Measurement and Modelling*, Geo Books, Norwich, pp. 201–18.

Labadz, J. C. and Burt, T. P. (1987). Rates of moorland erosion and reservoir sedimentation in peat covered catchments in the southern Pennines, Paper presented at the BGRG annual conference, University of Oxford, September.

Ledger, D. C., Lovell, J. P. B. and McDonald, A. T. (1974). Sediment yield studies in upland catchment areas in south-east Scotland, *J. Appl. Ecol.*, **11**, 201–6.

Ledger, D. C., Lovell, J. P. B. and Cuttle, S. P. (1980). Rate of sedimentation in Kelly Reservoir, Strathclyde, *Scott. J. Geol.*, **16**, 281–5.

Lovell, J. P. B., Ledger, D. C., Davies, I. M. and Tipper, J. C. (1973). Rate of sedimentation in North Esk Reservoir, Midlothian, *Scott. J. Geol.*, **9**, 57–61.

McManus, J. and Duck, R. W. (1985). Sediment yield estimated from reservoir siltation in the Ochil Hills, Scotland, *Earth Surf. Proc. Landf.*, **10**, 193–200.

Meterological Office (1977). *Average Annual Rainfall (millimetres), International Standard Period 1941–1970, Northern Britain* map, Meteorological Office, Bracknell, Berkshire.

Rausch, D. L. and Heinemann, H. G. (1984). Measurement of reservoir sedimentation. In Hadley, R. F. and Walling, D. E. (eds), *Erosion and Sediment Yield: Some Methods of Measurement and Modelling*, Geo Books, Norwich, pp. 179–200.

Young, A. (1958). A record of the rate of erosion on Millstone Grit, *Proc. Yorks. Geol. Soc.*, **31**, 149–56.

19 Soil Erosion in Cavernous Limestone Catchments

PAUL HARDWICK and JOHN GUNN
Limestone Research Group, Department of Environmental and Geographical Studies, Manchester Polytechnic

INTRODUCTION

Caves form the accessible part of active and relict limestone drainage systems. The sediments which they contain are often well preserved and their study has revealed much valuable palaeoclimatic, palaeontological and geomorphological information (e.g. Bull, 1980; Noel, 1988). In contrast, there have been few studies of contemporary sediment accretion or erosion rates in caves, probably because of logistical problems and the general neglect of mechanical processes due to the assumed dominance of solutional erosion in limestone terrains (Newson, 1971). Moreover, those papers which have been published are often in somewhat obscure journals or reports. This is unfortunate, because the formulation of land-use strategies for karst terrain requires consideration of soil erosion and its impact on cave systems and limestone aquifers. Hence, one of the aims of this chapter is to provide a brief review of the literature on soil erosion in limestone terrains.

There are over 2700 caves in Britain (Hardwick and Gunn, 1989) and their value as scientific resources has been recognized by the Nature Conservancy Council's (NCC) designation of 48 sites as Cave Sites of Special Scientific Interest (SSSI) under the 1981 Wildlife and Countryside Act. The 48 sites include within their boundaries over 32 per cent of known caves and 80 per cent of cave passage. Landowners farming the SSSI are required by law to notify the NCC of their intention to carry out any of a series of potentially damaging operations (PDOs) prescribed for each site. The NCC may then object, in which case the farmer may be offered compensation. The PDOs should allow the NCC to protect the caves by exerting some control on land-use practices. However, problems have arisen because the impact on cave systems of the various agricultural operations which are covered by PDOs has not been adequately quantified. Hence, a second aim of this chapter is to address one aspect of this

---
Soil Erosion on Agricultural Land
Edited by J. Boardman, I. D. L. Foster and J. A. Dearing
©1990 John Wiley & Sons Ltd

problem by attempting to assess the impact of agriculturally induced soil erosion on cave systems. Consequently, it is necessary to consider first soil erosion on limestones overlying caves and the transport of derived sediments underground by autogenic waters and, second, inputs of sediments to caves from the catchments of allogenic feeder streams.

EVIDENCE FOR SOIL EROSION ON CAVERNOUS LIMESTONES

Limestones are frequently of high chemical purity with a relatively small insoluble content. Consequently, limestone terrain in general and karst in particular is often envisaged as consisting primarily of bare rock outcrops with little or no soil and vegetation cover. It has also been suggested that since those soils which do develop are largely organic, removal of vegetation cover will result in drying and oxidation rather than erosion of mineral particles (Trudgill, 1986). However, a more detailed examination reveals, first, that in many areas limestones are overlain by superficial deposits of allogenic origin on which mineral soils have developed and, second, that even where there are extensive areas of bare limestone the intervening fissures and underlying caves frequently contain substantial amounts of sediment, presumably the result of an earlier phase or phases of soil erosion. Furthermore, the bare appearance of much of the Mediterranean karstlands, including the 'classic' karst of Yugoslavia, is now thought to be a result of human-induced deforestation and subsequent soil erosion. It has also been argued that early agriculture in the northern Pennines, the classic karst of England, provoked large-scale soil loss and is at least in part responsible for the bare limestones of the area (Fleming, 1976). In Ireland, the thin soils, patchy vegetation and large areas of bare rock of the Burren plateau karst have been ascribed to glacial erosion, but Drew (1983) has argued that the area was once well populated and forested, with an extensive cover of mineral soil which was removed by erosion initiated by forest clearance during the late Bronze Age. Similar processes of erosion induced by deforestation can be observed in the developing countries of the Third World. For example, direct evidence is provided by Kiernan (1987), who undertook a limited study of soil erosion from opium swiddens in an area of the 'Golden Triangle' (formed by the borders of Burma, Laos and Thailand) which is underlain by Permian limestones. He found that shifting cultivation and anthropogenically induced forest fires have resulted in the rapid erosion of terra rossa soils, with accretion of derived sediments in dolines and caves. Evidence from areas of modern intensive agriculture is more indirect and fragmentary, but suggests that erosion of soils overlying cavernous limestones may be more widespread than has previously been supposed. For example, Lewis (1981a) in a study of CO_2 levels in Coldwater Cave, Iowa, noted that large amounts of mud and fine organic debris were washed into the cave from the overlying farmland, following storm events. Similarly, Weirsma *et al.* (1986) in their study of surface and groundwater interactions in the Door County karst, Wisconsin, drew attention to the impact which loss of soils into dolines had on groundwater quality. They also noted that a reduction in agricultural productivity had occurred due to loss of tillable acreage.

Closed depressions (dolines/sinkholes) are the most common surface expression of karstification processes. Four main types are recognized (solution, collapse, subsidence and interstratal collapse), the most common in Britain being subsidence dolines, which are formed entirely in soils. These are manifest at the surface either by rapid, episodic collapses or by more gradual flow or slumping processes, both of which are the result of sediment removal from the base of the soil profile by downwashing into bedrock fissures. Hence, it is argued that soil erosion is the major process of subsidence doline formation, rather that solutional erosion of the underlying bedrock. Subsidence dolines occupy a significant percentage of the surface area of Britain's Carboniferous Limestone outcrops, and it can be argued that since the volume of subsidence dolines formed in mineral soils represents the volume of soil eroded into underlying joints, fissures and ultimately caves, its measurement may provide evidence of surface soil loss. However, as Trudgill (1985, 1986) has noted, the lowering of soils *in situ* as differential (solutional) erosion deepens joints and fissures does not represent soil erosion *sensu stricto*.

AUTOGENIC SOIL EROSION PROCESSES

If it is accepted that the soils above karstified limestones, whether derived from bedrock solution or from the weathering of superficial deposits, may be subject to active erosion then consideration must be given to how material is removed and to its destination. The ultimate destination of eroded material must be the underlying cave and conduit systems, but the speed and mode of transmission are less clear, and demand consideration of geomorphological and hydrological processes. The model for the hydrology of closed depressions (dolines/sinkholes) put forward by Gunn (1981) provides a useful starting point, as similar processes may also be envisaged on the slopes of limestone valleys. Overland flow is generally rare unless the overlying deposits are both thick and/or of low permeability, in which case similar considerations apply as to soils on non-carbonate bedrock. However, some overland flow, with consequent soil erosion, has been observed on saturated areas near the base of large dolines (Gunn and Turnpenny, 1986). In contrast to overland flow, throughflow is thought to be widespread in soils above karstified limestones and to provide an important mechanism for the concentration of recharge at the base of dolines (Gunn, 1981). While throughflow is generally slow and laminar, rapid movement of water through macropore structures in soils above karstified limestones has also been noted (e.g. Thomas and Phillips, 1979) and provides a mechanism for soil erosion. Of particular relevance is the work of Simpson and Cunningham (1982), who identified macropore structures, coincident with bedrock solution channels, in irrigated soils derived from a dolomitic limestone bedrock. Field observations showed that flow velocities through macropores were four to eighteen times greater than in the adjacent soil matrix, and that 25–65 per cent of runoff could occur through channels occupying less than 10 per cent of the soil. The bedrock solution channels lie within the upper, most weathered layer of limestone, the subcutaneous zone, which is also an important zone of lateral water movement (Gunn, 1981; Williams, 1985). Although there is no direct evidence of sediment transport by subcutaneous flow it is considered to be highly likely.

The model developed by Gunn (1981) envisages a spectrum of flow routes by which water moves vertically from the subcutaneous zone to the underlying aquifer. Three points are identified on the basis of physical size: shaft flow (thin films of water flowing down the walls of openings over 1 m in diameter), vadose flow (meteoric water flowing for the most part in solutionally enlarged fissures and joints 0.01–1.00 m in diameter) and vadose seepage (water percolating through tight fissures and fractures or moving as intergranular flow). Shaft flow and most vadose flows will be turbulent and therefore capable of transporting soil and rock particles underground (Williams, 1985). Nevertheless, some studies have suggested that eroded sediments will be removed only locally to the subcutaneous zone. For example, Watts (1984) postulates that the products of soil erosion in the Burren (Co. Clare, Ireland) might end up in fissures in the limestone. Similarly, Trudgill (1985) considers that there is little evidence of significant inputs from autogenic recharge to caves despite noting that soil and peats have been found running down the walls of shallow cave systems. In contrast, the work of Bull (1981) on fine-grained sedimentary structures on Ogof Agen Allwedd, South Wales, suggests that pulses of sediment are emitted from the base of soil-filled fissures by a translatory flow (shunting) mechanism in response to precipitation events. Bull proposed this rhythmic pulsing of sediment as the process whereby parallel accretion of fine-grained sediments in cave conduits occurs, and suggested that the process is continuing today, quoting examples of recently deposited sediment slopes in Otter Hole, Chepstow; Mangle Hole, Mendip; and Ogof Craig-ar-Ffynnon in South Wales. Bull (in Trudgill, 1985, p. 119) stated that there was little evidence of deposited sediment being younger in age than late glacial times. However, it can be suggested that this supports Bull's translatory model, since older sediment may be extruded from the base of the soil column as modern soils are deposited at its top. This type of model implies that soil material may take up to 10 000 years to be translated from the surface to cave passages by autogenic recharge.

Few empirical measurements of autogenic sedimentation rates have been undertaken. The only studies currently known to the authors are those of Kogovsek and Habik (1980) and Kogovsek (1982), who found that autogenic recharge may supply significant amounts of clastic material to the Postojna and Planina caves in the Dinaric karst of Yugoslavia. The suspended sediment concentrations of a number of cave 'trickles' (vadose flows) were measured and related to precipitation. Percolating pulses of water were shown to have almost equal capability to transport both dissolved and suspended material. In one event, 42 m^3 of water from a trickle with a mean discharge of 41.2 l min^{-1} transported 7 kg of dissolved and 6 kg of suspended material, the mean sediment concentration being 143 mg l^{-1}. Lag times in peak trickle discharge response ranged from 6 to 7.5 h, giving a linear velocity through the 100 m thick cave roof of 13.3–16.7 m h^{-1}. The trickle hydrographs are attenuated, leading Bonnacci (1988) to suggest that the routeways followed by drainage waters are neither vertical nor direct; hence, point velocities could be considerably higher. Unfortunately, no attempt was made to determine the age or source of the inwashed material, and consequently it is not known whether it is of recent (surface) origin or has been moved through a series of stores in the unsaturated zone over a longer period of time, as suggested by the translatory flow model.

SOIL EROSION IN ALLOGENIC CATCHMENTS

Many limestone areas receive substantial point sources of allogenic recharge from adjacent non-carbonate rocks. The sinking streams carry erosion products from their catchments, and these are input to the karst drainage systems. Where the karst conduits contain freely flowing streams the features of alluvial channels are to be expected, and deposition zones will be similar to their surface counterparts (e.g. Jones, 1971). In addition, major deposits of sediment are frequently found in cave systems where flow velocities are rapidly reduced, for example at a constriction or where a narrow vadose stream meets a wider, slower phreatic flow (Ford, 1975), and many such deposits are characterized by depositional cycles of bedload to suspended sediment load (Wolfe, 1972).

In some caves there is evidence of an increased level of aggradation due to anthropogenically accelerated erosion rates in allogenic catchments. Gillieson *et al.* (1986), studied allogenic slopewash sediments accumulating in limestone rock shelters in the Papua New Guinea highlands. They suggested that late Pleistocene erosion rates were minimal under stable primary forest but that significantly increasing rates of erosion occurred from 6000 BP, and especially from 300–400 BP due to horticultural intensification. In temperate regions, Tucker (1982) noted continuing accretion of up to several metres of agriculturally derived sediment in Kentucky caves. Several authors (e.g. Williams, 1975; Hawke, 1982) have documented rapidly increasing sediment levels in cave systems due to anthropogenic activities in the catchments of allogenic streams. These range from public road maintenance and deforestation activities in surface catchments, to the artificial raising of stream levels downstream of cave risings which reduces the hydraulic gradient of the system leading to further depositions of sediment. Moreover, blockage and disruption of underground drainage routes can also create problems, particularly flooding, within their surface catchments (e.g. Dougherty, 1983).

IMPACTS OF SOIL EROSION ON CAVE SYSTEMS

Sediment accretion may result in the complete or partial infilling of cave passages and the burial of older clastic deposits which may be of scientific importance. Backflooding may follow and access to caves or parts of cave systems may be permanently lost. In contrast, the restoration of an allogenic catchment to natural or semi-natural vegetation may cause downcutting of cave sediment infills due to an increase in stream competence (e.g. Kranjc, 1979; Dougherty, 1983). Long-term changes in cave passage morphology may also result from sediment infill. For example, Jones (1971) suggested that the perching of streams on bedload material which armours the bedrock of a solution conduit may prevent a stream from downcutting in response to the lowering of regional base levels and that the meandering of such streams may play an important part in shaping the cross-sectional form of stream passages.

The adverse impact of soil materials transported by autogenic recharge on speleothem growth and development has been demonstrated in a number of studies, most notably by Jakucs (1977), who found that speleothems beneath barren and deforested karst were inactive and usually yellow, brown or ochre in colour. When surface vegetation died off or was removed, there was a marked change in colour of speleothem growth layers, usually towards red, and this was attributed to the inwashing of terra rossa soils and the deposition of clay minerals on the speleothem surface. If afforestation recommenced, speleothems grew lighter with fresh deposition of white or translucent calcite.

Anthropogenically induced allogenic sedimentation may also impact on cave ecosystems. For example, silting within Mammoth Cave, Kentucky, is known to have affected the geographical distribution of hypogean *Asselidae* communities (Lewis, 1981b), since pool and riffle sequences, together with gravel areas essential for reproduction and survival, were lost. Similarly, sedimentation is known to have reduced the breeding sites of a species of glowworm at the Waitomo Glowworm Grotto, Waitomo New Zealand (Pugsley, 1981). The impacts of autogenic sedimentation on cave ecosystems remains unquantified, but autogenic recharge is considered to provide one of the major routeways for the ingress of food and energy for cave (hypogean) communities (e.g. Chapman, 1983). Moreover, there is evidence to suggest that the preferred habitat of many hypogean communities is the Superficial Underground Compartment (SUC) proposed by Juberthie and Delay (1981), which appears to correspond with the subcutaneous zone (Gunn, 1981; Williams, 1985). Thus localized movement of soils may impact on hypogean communities.

CASE STUDY: THE CASTLETON KARST

Soil erosion in the catchments of allogenic streams is generally assumed to be a significant source of sediment and also to be sensitive to acceleration as a result of changing agricultural practices. However, there is a dearth of quantitative data, and hence a study has recently commenced in the Castleton Caves SSSI to investigate whether particular agricultural operations significantly modify the input of clastic sediments to cave systems by allogenic and autogenic recharge.

The Castleton karst lies on the northern margin of the White Peak, and is the most comprehensively studied karst in the Peak District, although the geomorphic history and detailed hydrology of the area are still not clearly understood (Ford, 1986). Limestones crop out over an area of $c.13$ km^2 but the water catchment is larger, since a series of twelve insurgences (P0-P10,P12) in the north-west supply concentrated allogenic recharge from a further 3 km^2 of Namurian strata. Water from the insurgences drains through the Peak-Speedwell cave system to three risings, Russet Well, Slop Moll and Peak Cavern Rising, although the precise routes followed by the drainage are unknown (Gunn, 1985). The inputs of sediment and water from three of the allogenic catchments (P1, P6, P9) which are characterized by differing agricultural land uses are being monitored. The P1 catchment is largely rough pasture which has not been ploughed for at least 10 years. In contrast, the P6 catchment has seen several agricultural improvements including reseeding and drainage improvements. The P9 catchment is a mixture of improved and unimproved grassland together with some new drainage works.

In addition, the hydrological and sedimentological characteristics of autogenic inputs to P8 cave are under investigation. The cave receives autogenic recharge from several vadose flows, some of which are instrumented with 'V' notch weirs and stage recorders. The weirs also act as sediment traps, and collection of sediment from these together with a grab sampling programme will enable rates of sediment input to be calculated. The field overlying the cave is due to be ploughed during the study, enabling the hydrological and sedimentological impact to be assessed.

CONCLUSIONS

On the basis of an extensive literature search it is clear both that soil erosion in cavernous limestone catchments has been inadequately investigated and that the impacts of derived sediments on cave systems may be considerable. Modern agricultural practices may increase the material taken underground by allogenic streams and may also increase the rate of removal by autogenic recharge. For example, the dependence on agrochemicals to maintain soil fertility may promote long-term soil degradation due to the loss of organic material from soil matrices. The presence of underlying joints and fissures is likely to facilitate the evacuation of soils from the surface, and this may be exacerbated by soil disturbance and compaction during soil cultivation. In addition, the application of liquid sewage sludges as fertilizers using irrigation techniques may accelerate soil erosion due to rapid subsurface runoff via the subcutaneous zone.

An investigation has been commenced in the Castleton karst to determine sediment sources, to quantify the rates of sediment delivery and to assess underground impacts. As catchment disturbance is likely to lead to increased sediment loads in insurgent streams, consideration needs to be given to agricultural activities in allogenic catchments. This raises a further problem with the present system for the designation of cave SSSI in Britain, since, in general, only land overlying known or hypothesized caves, not the allogenic catchment, is included in the site.

ACKNOWLEDGEMENTS

This project has been funded by the Nature Conservancy Council's Commissioned Research Programme and the help of the NCC and particularly Dr Laurie Richards is gratefully acknowledged. The authors would like to thank Messrs Bagshaw, Verner, Ritter and Watson for providing access to field sites; Dr D. P. Butcher and M. Beasant at Huddersfield Polytechnic for providing various items of field equipment; and the anonymous referee for comments on an earlier draft of this chapter.

REFERENCES

Bonacci, O. (1988). *Karst Hydrology*, Springer-Verlag, London.
Bull, P. A. (1980). Towards a reconstruction of time-scales and palaeoenvironments from cave sediment studies. In Cullingford, R. A., Davidson, D. A. and Lewin, J. (eds), *Timescales in Geomorphology*, John Wiley, Chichester, pp. 177–87.
Bull, P. A. (1981). Some fine grained sedimentation phenomena in caves, *Earth Surface Processes & Landforms*, 6(1), 11–22.
Chapman, P. (1983). Cave Life part 9: Cave Communities and the Future, *Caves & Caving*, 22, 24–5.
Dougherty, P. H. (1983). Valley tides—land-use response floods in a Kentucky karst region. In Dougherty, P. H. (ed.), *Environmental Karst*, Geospeleo Publications, Cincinatti Ohio, pp. 3–15.
Drew, D. P. (1983). Accelerated soil erosion in a karst area: the Burren, western Ireland, *Journal of Hydrology*, 61(1–3), 113–24.

Fleming, A. (1976). Early settlement and the landscape in West Yorkshire. In Sieveking, G. (ed.), *Problems in Economic and Social Archaeology*, Duckworth, London, pp. 359–73.

Ford, T. D. (1975). Sediments in caves, *Transactions of the British Cave Research Association*, **2**, 41–6.

Ford, T. D. (1986). The evolution of the Castleton cave systems and related features, Derbyshire, *Mercian Geologist*, **10**(2), 91–114.

Gillieson, D., Oldfield, F. and Krawiecki, A. (1986). Records of Prehistoric Soil Erosion from Rock-Shelter Sites in Papua New Guinea, *Mountain Research and Development*, **6**(4), 315–24.

Gunn, J. (1981). Hydrological processes in karst depressions, *Zeitschrift für Geomorphologie*, **25**(3), 313–31.

Gunn, J. (1985). Hydrology of the Peak-Speedwell System, *Bulletin of the British Cave Research Association*, **29**, 10–12.

Gunn, J. and Turnpenny, B. (1986). Stormflow characteristics of three small drainage basins in North Island, New Zealand. In Paterson, K. and Sweeting, M. M. (eds), *New Directions in Karst*, Geobooks, Norwich, pp. 233–58.

Hardwick, P. and Gunn, J. (1989). The limestone cave resources of Great Britain. *Proc. 10th. International Speleological (UIS) Congress, Budapest 1989*, **1**, 194–195.

Hawke, D. V. (1982). *Fluvial Processes in the Upper Waitomo Catchment*, Unpublished PhD thesis, University of Auckland.

Jones, W. K. (1971). Characteristics of the underground floodplain, *Bulletin National Speleological Society*, **33**(3), 105–15.

Jakucs, L. (1977). *Morphogenetics of Karst Regions*, Adam Hilger, Bristol.

Juberthie, C. and Delay, B. (1981). Ecological and biological implications for the existence of a superficial underground compartment, *Proc. 8th. International Speleological Congress, Kentucky*, pp. 203–6.

Kiernan, K. (1987). Soil erosion from hilltribe opium swiddens in the Golden Triangle and the use of Karren as an erosion yardstick, *ENDINS*, **13**, 127–31, Publicacio d'Espeleologia Federacio Balear d'Espeleologia, Mallorca.

Kogovsek, J. (1982). Vertical percolation in the Planina Cave, *Acta Carsologica Krasoslovni Zbornik*, **10**, 107–25.

Kogovsek, J. and Habic, P. (1980). Study of vertical water percolation in the case of Planina and Postojna Caves, *Acta Carsologica Krasoslovni Zbornik*, **9**, 133–48.

Kranjc, A. (1979). The influence of man on cave sedimentation, *Actes du symposium international sur l'erosion karstique*, UIS, Aix en Provence- Marseilles- Nimes, pp. 117–123.

Lewis, W. C. (1981a). Carbon dioxide in Coldwater Cave, *Proc. 8th. International Speleological Congress, Kentucky*, pp. 91–2.

Lewis, J. J. (1981b). The subterranean Caecidotea (Asselids) of the interior low plateaus *Proc. 8th. International Speleological Congress, Kentucky*, pp. 234–235.

Newson, M. D. (1971). The role of abrasion in cavern development, *Transactions of the British Cave Research Association*, **2**, 41–6.

Noel, M. (1988). Palaeomagnetism of cave sediments from Mynydd Llangattwg, *Transactions of the British Cave Research Association*, **15**(1), 3–11.

Pugsley, C. (1981). Management of a biological resource—Waitomo Glowworm Cave, New Zealand, *Proc. 8th. International Speleological Congress, Kentucky*, pp. 489–492.

Simpson, T. W. and Cunningham, R. L. (1982). The occurrence of flow channels in soils, *Journal of Environmental Quality*, **11**(1), 29–30.

Thomas, G. W. and Phillips, R. E. (1979). Consequence of water movement in macropores, *Journal of Environmental Quality*, **8**(2), 149–52.

Trudgill, S. T. (1985). *Limestone Geomorphology*, Longman, London.

Trudgill, S. T. (1986). Limestone weathering under a soil cover and the evolution of limestone pavements, Malham District, north Yorkshire, U.K. In Paterson, K. and Sweeting, M. M. (eds), *New Directions in Karst*, Geobooks, Norwich, pp. 461–71.

Tucker, N. L. (1982). *Nonpoint Agricultural Pollution in a Karst Aquifer: Lost River Groundwater Drainage basin, Warren County, Kentucky*, Unpublished MS thesis, Western Kentucky University, Bowling Green, Kentucky.

Waltham, A. C. (1989). *Ground Subsidence*, Blackie, Glasgow.

Watts, W. A. (1984). The Holocene vegetation of the Burren, western Ireland. In Haworth, E. Y. and Lund, J. W. G. (eds), *Lake Sediments and Environmental History*, Leicester University Press, pp. 359–76.

Weirsma, J. H., Steiglitz, R. D., Dewayne, L. C. and Metzler, G. M. (1986). Characterisation of the shallow groundwater system in an area with thin soils and sinkholes, *Environment Geology Water Sciences*, **8**(1–2), 99–104.

Williams, P. W. (1975). Report on the conservation of Waitomo Caves, *New Zealand Speleological Bulletin*, **5**(93), 373–96.

Williams, P. W. (1985). Subcutaneous hydrology and the development of doline and cockpit karst, *Z. Geomorph. N. F.*, **29**(4), 463–82.

Wolfe, T. E. (1972). A classification of cave sediments, *International Geography*, **2**, 1332–3.

PART 2
ASSESSMENT AND PREDICTION

Soil Deterioration

20 Sediment-associated Phosphorus Transport From Two Intensively Farmed Catchment Areas

B. KRONVANG
National Environmental Research Institute, Silkeborg, Denmark

INTRODUCTION

Particulate phosphorus (PP) can play a major role in the eutrophication of lakes and estuarine systems. Moreover, PP compounds influence the flux of dissolved P via adsorption–desorption reactions (Sharpley and Syers, 1979). PP is thus of major importance for the P budget of drainage areas (Johnson *et al.*, 1976; Schuman *et al.*, 1976; Duffy *et al.*, 1978; Probst, 1985), yet our knowledge of the nature and dynamics of PP fractions is still extremely limited (GESAMP, 1987).

The effects of land use on the PP have been investigated only sporadically. Johnson *et al.* (1976) and Schuman *et al.* (1976) found that sediment losses from small contour-cropped watersheds accounted for 85 per cent and 78 per cent of the total phosphorus discharge. Alström and Bergmann (1988), in a study of cultivated fields in Southern Sweden, reported high losses of PP following soil erosion during surface runoff.

The processes underlying soil erosion, instream erosion, land-use characteristics and biological activity in the stream are all important factors governing the PP flux, but too little is known about these processes. This chapter provides detailed data on the organic and inorganic particulate phosphorus fluxes in two streams draining intensively farmed catchments in Denmark.

Soil Erosion on Agricultural Land
Edited by J. Boardman, I. D. L. Foster and J. A. Dearing
©1990 John Wiley & Sons Ltd

314 Soil Erosion on Agricultural Land

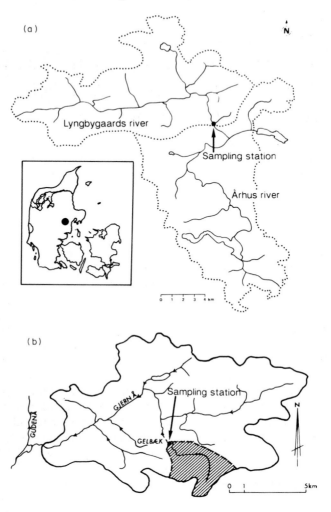

Figure 20.1 Catchment area and location of the sampling station in the Lyngbygaards river (a) and in the Gelbæk (b). Inset shows the location of the study area

Table 20.1 Land use in the study basins of the Lyngbygaards River and the Gelbæk

Land use	Arable (%)	Forest (%)	Meadow (%)	Lakes (%)	Uncultivated (%)
Lyngbygaards River	83	6	1	4	6
Gelbæk	89	4	3	—	4

STUDY AREAS

A medium-sized and a small basin located in Eastern Jutland, Denmark, were selected for this study (Figure 20.1). Both streams drain intensively farmed catchment areas (Table 20.1). The Lyngbygaards River drains 126 km² of fluvio-glacial sandy loams (Figure 20.1(a)). The main stream flows through a landscape of low relief. Along the tributaries slopes range from 3 per cent to 10 per cent. The Lyngbygaards River is fully regulated in its course. The Gelbæk drains approximately 8.5 km² and is a first-order tributary to the largest Danish river, the Gudenå (Figure 20.1(b)). The Gelbæk flows over glacial deposits, mainly of loamy composition. The stream is rather deeply incised in a narrow valley with slopes of 1–10 per cent and minor swampy areas are found along the course of the stream. The Gelbæk is regulated in its upper course.

Mean annual precipitation in both basins is approximately 650 mm. The discharge averages 255 mm yr^{-1} in the Lnygbygaards River and 215 mm yr^{-1} in the Gelbæk. Runoff is generally high in winter and low in summer. During the present study periods it was higher than average: 298 mm yr^{-1} in the Lyngbygaards River and 356 mm yr^{-1} in the Gelbæk.

METHODS

Field and Laboratory Procedures

Runoff, particulate organic matter (POM), particulate inorganic matter (PIM), particulate organic phosphorus (POP) and particulate inorganic phosphorus (PIP) concentrations were measured at two stations in the study basins. The study period was October 1986 to September 1987 in the Lyngbygaards River and June 1987 to June 1988 in the Gelbæk.

A basic sampling programme combined with more frequent sampling during storms was run at both stations. Basic sampling was weekly in the Lyngbygaards River and every 4 h in the Gelbæk. In the summer period basic sampling was increased to every 12–24 h in the Lyngbygaards river. Water samples were collected by simple dip sampling (weekly) and by automatic samplers installed at a fixed level above the stream bottom. During storm-runoff events, additional automatic samplers were started at the onset of the event to sample at fixed intervals. In the Lyngbygaards River the sampling station was placed at a rapid to ensure complete mixing.

Weekly samples and some of the samples taken in the Lyngbygaards River during storm runoff were analysed for POM, PIM, POP and PIP. All the water samples from the Gelbæk were analysed for POP and PIP.

On return to the laboratory, particulate matter (PM) concentrations were measured by filtering 100–500 ml water through 1.2 μm Whatman GF/C glass-fibre filters. Water samples were kept dark and cold (5°C) prior to analysis.

The filtration process was according to Dansk Standard (1985a). The filters were combusted at 500°C for 4 h to separate the organic and inorganic fractions. Concentrations of phosphorus (P) were obtained according to Dansk Standard (1985b), again using Whatman GF/C glass-fibre filters for the filtration procedure. The water samples were analysed for total P (*A*), total inorganic P (*B*), total dissolved P (*C*) and dissolved inorganic P (*D*). Particulate organic (POP) and inorganic phosphorus (PIP) concentrations were calculated as follows (Rebsdorf and Thyssen, 1986):

$$POP = (A - C) - (B - D); \quad PIP = B - D$$

Methods of Calculation

Instantaneous values and daily means of water discharge (*Q*) were obtained from a relationship between the measured values of *Q* and water level (*H*) of the

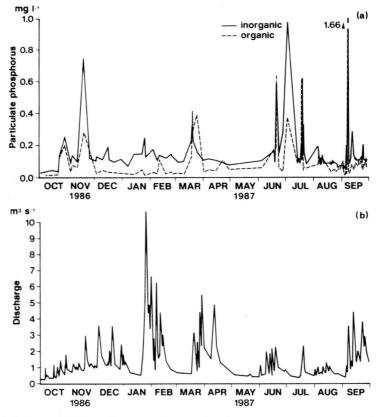

Figure 20.2 Seasonal fluctuations of particulate phosphorus concentration (a) and discharge (b) in the Lyngbygaards River

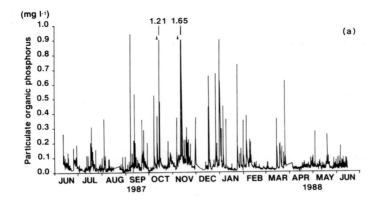
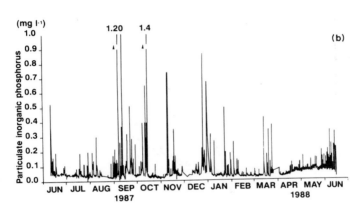
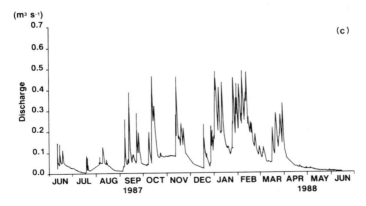

Figure 20.3 Seasonal fluctuations of particulate organic (a) and inorganic phosphorus concentrations (b) and discharge (c) in the Gelbæk

318   Soil Erosion on Agricultural Land

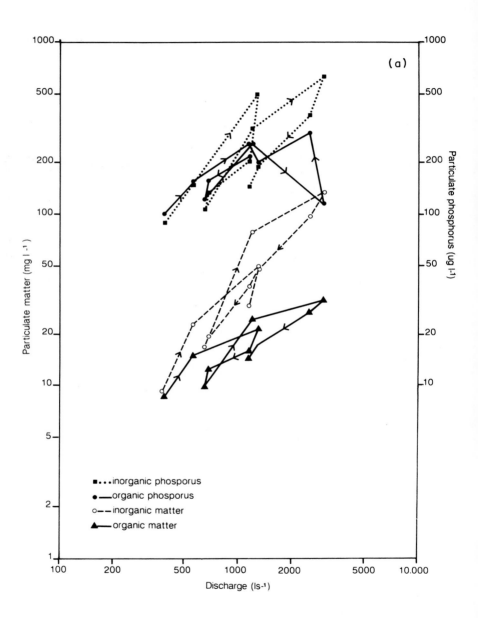

Figure 20.4 *(caption opposite)*

Sediment-associated Phosphorus Transport 319

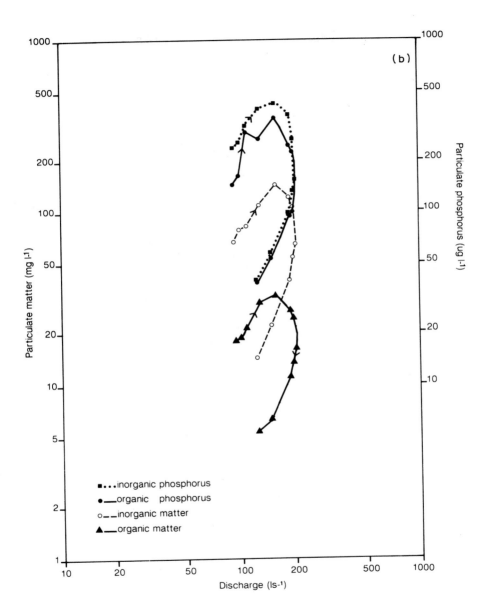

Figure 20.4 Trends of the particulate matter and particulate phosphorus rating relationships during two storm runoff events in the Lyngbygaards River (a) and a storm runoff event in the Gelbæk (b)

form $Q = aH^b$. The parameters were adjusted in accordance with the seasonal development of the submerged macrophytes.

In the case of the Lyngbygaards River the transport rates for the phosphorus fractions were calculated using PM−Q regression (Kronvang, 1988) and PP−PM regression. In the Gelbaek the transport rates for PP fractions were calculated using linear interpolation on the time−concentration curve. Because of the large number of samples (2300), the Gelbaek method is believed to give the 'true' PP transport, whereas the Lyngbygaards method is less reliable (Walling and Webb, 1981).

The hydrographs for the two streams were separated into periods of low flow and of storm runoff. Storm runoff is again separated into background flow and storm flow according to Fisk (1977). Low flow and background flow are hereafter denoted background flow. The hydrograph separation permitted the calculation of the PP load during background flow and storm flow, respectively.

RESULTS

Particulate Phosphorus Concentrations

High POP and PIP concentrations associated with storm runoff events were observed in both streams (Figures 20.2 ad 20.3). This was to be expected, since numerous authors have described this effect for suspended sediment (e.g. Walling, 1974). The relationship between the concentrations of particulate matter (PM) and particulate phosphorus (PP) fractions during two storms are shown in Figure 20.4. It is clear that PM and PP respond almost identically. They also show the usual hysteresis: for a given value of discharge, PM and PP concentrations are higher on the rising stage than on the falling one (Figure 20.4). Maximum concentrations precede peak discharge in the Gelbaek (Figure 20.4(b)) because of the exhaustion effect on PM and PP availability as the storm continues.

The mean annual concentration of PIP during periods of background flow is significantly ($P<0.01$) higher in the Lyngbygaards River (102 $\mu g\, l^{-1}$) than in the Gelbaek (45 $\mu g\, l^{-1}$) (Figure 20.5). This can be explained in terms of the more complex land-use pattern in the Lyngbygaards basin (Table 20.1), which facilitates a high inorganic phosphorus delivery. The PIP concentration increases slightly with increasing background flow in the Lyngbygaards River (Figure 20.5(a)). The opposite is true in the Gelbaek (Figure 20.5(b)). This indicates a more constant inorganic phosphorus delivery in the Gelbaek.

The concentration of POP during periods of background flow is only slightly higher in the Lyngbygaards River (49 $\mu g\, l^{-1}$) than in the Gelbaek (37 $\mu g\, l^{-1}$) (Figure 20.5). In both streams high POP concentrations due to the biological production in the river system are found in late spring−summer and, in the autumn, are due to the decay of the plant material.

Sediment-associated Phosphorus Transport 321

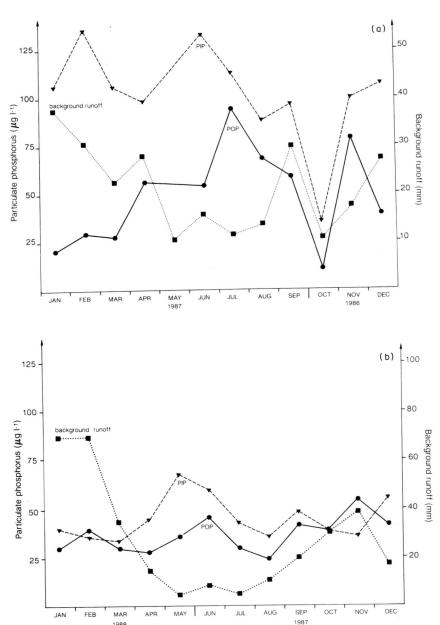

Figure 20.5 Seasonal fluctuations of the mean monthly particulate organic (POP) and inorganic phosphorus (PIP) concentrations during periods of background flow and monthly background runoff in the Lyngbygaards River (a) and the Gelbæk (b)

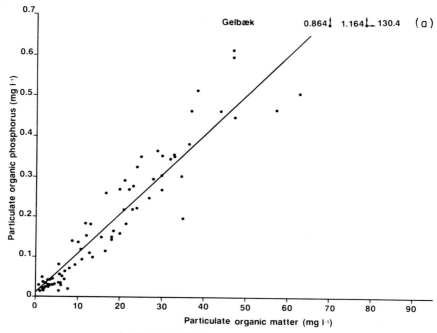
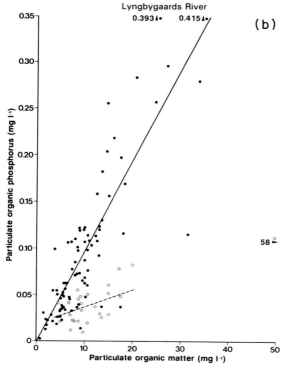

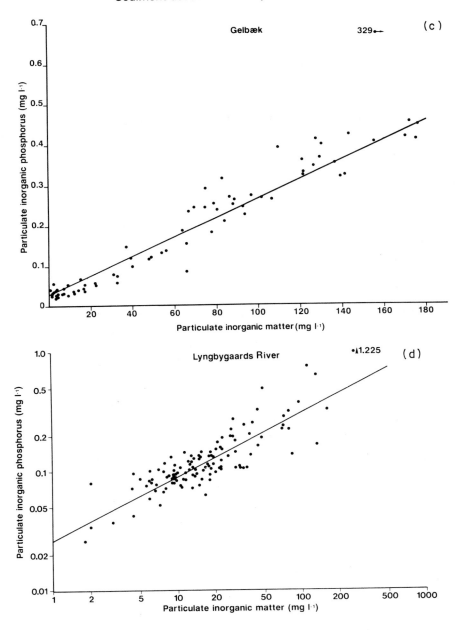

Figure 20.6 Relationships between particulate organic phosphorus and particulate organic matter (a,b) and particulate inorganic phosphorus and particulate inorganic matter (c,d) in the two study streams

Relationships between Particulate P and Particulate Matter

As illustrated in Figure 20.4, the organic and inorganic fractions of PM and PP show nearly the same concentration loops during storm runoff events. In fact, the POP versus POM concentrations illustrate almost the same linear relationship in the two streams (Figures 20.6 (a) and (b) and Table 20.2). The content of P in POM is almost constant at 1 per cent, except for a fortnight of low values in the Lyngbygaards River after weed cutting in the late summer (Figure 20.6(b). The deviation might be due to the resuspension of organic matter depleted in P because of mineralization.

A linear relationship is found between the PIP and PIM concentrations in the Gelbæk, whereas a log/log relationship is developed for the Lyngbygaards River (Figure 20.6 (c) and (d) and Table 20.2). Furthermore, it was observed that as PIM concentration increases, the P content falls: the average content at a low flow is 1.1 per cent but only 0.21 per cent at high flows. Perhaps this is because high flows can mobilize larger particles which are low in P content, as described by Viner (1988).

Particulate Phosphorus Transport

The PP flux has been calculated separately for storm flows and background flows. The annual transport of POP and PIP during storm flows is as large, or larger, than the transport during background flow, despite the fact that storm runoff volume comprises only 14 per cent and 18 per cent of the total runoff in the two streams (Figure 20.7).

The proportion of storm flow in the monthly runoff has a marked influence on the monthly POP and PIP transport (Figure 20.8 and 20.9). That overland flow generates a high PP flux is clearly seen in March 1987 in the Lyngbygaards River (Figure 20.8). A fortnight with high overland flow and low background flow led to an extremely high PP transport. The onset of autumn storms also

Table 20.2 Characteristics of the phosphorus versus particulate matter concentration relationships developed for the Lyngbygaards River and the Gelbæk

Period	n	Equation	r^2
		Lyngbygaards River (1986–7)	
Oct.–Sept.	117	$C_{PIP} = 0.0268 C_{PIM}^{0.532}$	0.72^a
Oct.–Aug.	85	$C_{POP} = -0.0015 + 0.0098 C_{POM}$	0.71^a
Sept.	28	$C_{POP} = 0.0190 + 0.0018 C_{POM}$	0.58^a
		Gelbæk (1987–8)	
June–May	82	$C_{PIP} = 0.0284 + 0.0024 C_{PIM}$	0.89^a
June–May	82	$C_{POP} = 0.0108 + 0.0098 C_{POM}$	0.93^a

[a] Significant at the 99.9 per cent level.

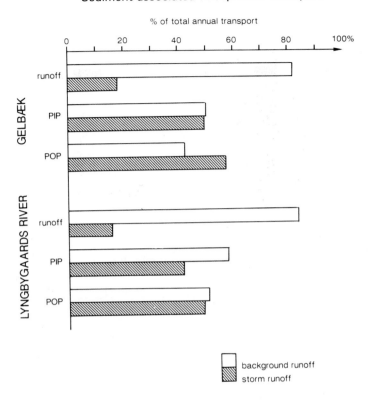

Figure 20.7 The proportion of annual runoff, particulate organic (POP) and particulate inorganic phosphorus (PIP) transport, respectively, during background flow and storm flow in the Lyngbygaards River and the Gelbæk

produces high PP transport rates, at least in part as a result of the resuspension of PM accumulated on the bed, during the previous summer drought. The availability of sediment also exerts a strong control on the PP transport. Thus, several events in January 1988 in the Gelbæk gave rise to low PP fluxes despite continued high proportions of storm flow (Figure 20.9).

In the background flow the POP and PIP fluxes are high in autumn and winter following the annual hydrological regimen (Figures 20.8 and 20.9). This implies that it is relatively simple to measure the transport of PP contributed by the background flow.

Sediment-associated P transport plays a very major role in the total P transport from the two study basins (Table 20.3). Equally high export coefficients for total P have recently been found in other intensively farmed Danish catchments (Hasholt, 1988). These results show that there is a great need to improve our understanding of, and the ability to measure, sediment-associated P transport.

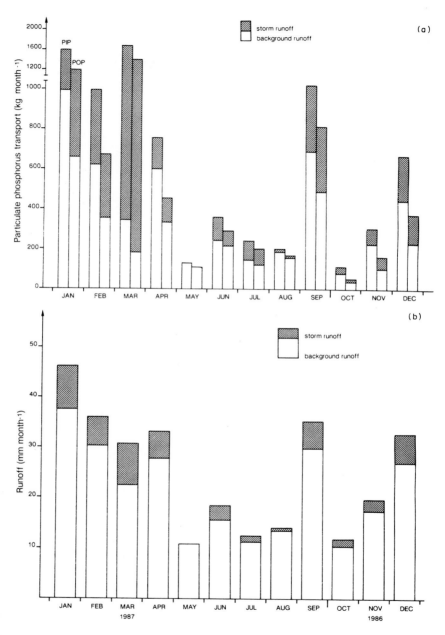

Figure 20.8 Monthly runoff, particulate organic phosphorus (POP) and particulate inorganic phosphorus (PIP) transport, respectively, during background flow and storm flow in the Lyngbygaards River

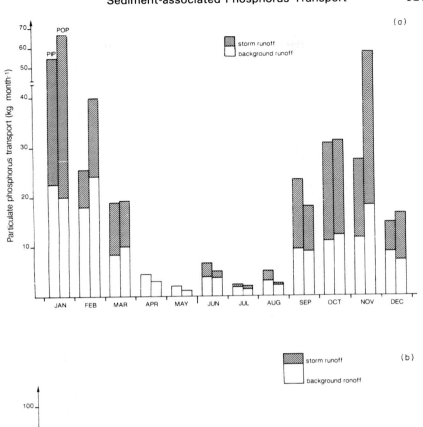
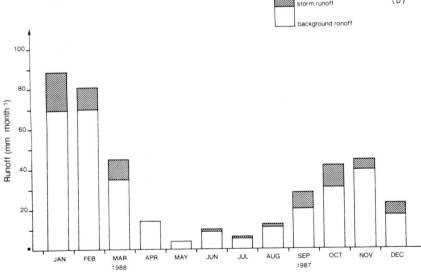

Figure 20.9 Monthly runoff, particulate organic phosphorus (POP) and particulate inorganic phosphorus (PIP) transport, respectively, during background flow and storm flow in the Gelbæk

Table 20.3 Annual runoff and export coefficients for the two catchments. Note that the two years are not identical

		Export coefficient	Percentage of total
Lyngbygaards River (Oct. 1986–Sept. 1987)			
	Area	126 km²	
	Total runoff	299 mm	
	Total P	1.69 kg ha^{-1}	100
	PIP	0.65 kg ha^{-1}	38
	POP	0.47 kg ha^{-1}	28
Gelbæk (June 1987–June 1988)			
	Area	8.5 km²	
	Total runoff	356 mm	
	Total P	1.03 kg ha^{-1}	100
	PIP	0.27 kg ha^{-1}	26
	POP	0.31 kg ha^{-1}	30

DISCUSSION AND CONCLUSION

This chapter has shown that sediment-associated P transport from two intensively farmed catchments is of major importance for the P budget. Moreover, the mean annual P content of particulate matter is appreciably higher (0.88 per cent and 0.62 per cent) than those reported by Johnson *et al.* (1976) and Probst (1985) for agricultural watersheds (0.11–0.21 per cent) and Rigler (1979) for grazed watersheds (0.2–0.25 per cent). The extremely high P content measured in this chapter shows the need for further investigation into sediment-associated P transport from agricultural watersheds.

Dividing the PP into the organic and inorganic constituents makes the relationship between PP and PM much clearer. Thus, POM and PIM are shown to exert a major control on POP and PIP concentrations, respectively. The P content of POM is high in the two streams and nearly constant at 1 per cent.

On the other hand, in both streams, there is a relative decrease in the P content as PIM concentrations increase. Walling and Kane (1982) and Probst (1985) reported an inverse relationship between P content of suspended sediment and sediment concentration or discharge. Probst (1985) also found that suspended sediment contained higher P-concentrations during low flow than during high flow. He assigned this to the higher organic matter content at low-flow periods. In order to improve our understanding of the variation in PIP it is necessary to investigate the precipitation–solution and the adsorption–desorption potential. Stevens and Stewart (1982) and Rebsdorf and Thyssen (1986) have described a positive relationship between PIP concentration and the concentration of iron.

More work is needed to determine how much of the POP and the PIP is directly accessible to biological production and how much is indirectly accessible in lakes and estuarine systems. Stevens and Stewart (1982) reported that only 20 per cent of the PP was directly accessible, partly depending on such processes as resuspension and sedimentation in the surface waters.

In the two study basins respectively 66 per cent and 56 per cent of the annual P fluxes consisted of PP. It is also shown that up to 70–90 per cent of the monthly PP fluxes were contributed by short-term storm events. It follows that the nature and the frequency of such events dominate the picture. Thus seasonal trends largely reflect changing storm frequencies. The exhaustion effects described by Walling (1974), and the resuspension of accumulated sediments as described by Rigler (1979), further complicate measurement of the P transport. An effective sampling strategy thus requires a clearer understanding of the sediment flux processes than we have at the moment.

The consequences can be remarkable when working with P budgets for whole river systems. Thus Walling and Webb (1981) found it very difficult to obtain reliable estimates of the suspended sediment transport, and Stevens and Smith (1978) found the same in estimating the P transport. Managers of river systems must therefore be made aware of the importance of sediment transport in nutrient fluxes and the need for monitoring to reflect this.

ACKNOWLEDGEMENT

Financial support for this work was given by the National Agency of Environmental Protection, the Aarhus City Council and the Aarhus County Council. The author is grateful to C. Aub-Robinson, Institute of Geology, University of Aarhus, for critically reviewing the manuscript. He would also like to express his gratitude to colleagues at the Freshwater Laboratory for assistance in the field, to Carsten Winther for drawing the figures and to Kirsten Thykjaer for typing the manuscript.

REFERENCES

Alström, K. and Bergmann, A. (1988). Sediment and nutrient losses by water erosion from arable land in South Sweden, a problem with nonpoint pollution, *Vatten*, **44**, 193–204.

Dansk Standard (1985a). *Vandundersøgelse, suspenderet stof og gløderest*, No. 207.

Dansk Standard (1985b). *Vandundersøgelse, total fosfor fotometrisk metode*, No. 292.

Duffy, P. D., Schreiber, J. D., McClurking, D. C. and McDowell, L. L. (1978). Aqueous- and sediment-phase phosphorus yields from five southern pine watersheds, *Journal of Environmental Quality*, **7**, 45–50.

Fisk, H. N. (1977). Magnitude and frequency of transport of solids by streams in the Mississippi basin, *American Journal of Science*, **277**, 862–75.

GESAMP (1987). Land/sea boundary flux of contaminants: Contributions from rivers, *Reports and Studies GESAMP*, No. 32.

Hasholt, B. (1988). *Transport af partikulært fosfor i danske vandløb, Nordisk Hydrologisk Conferens 1988*, Leena Ranta-jarvi (ed.), NHP-report No. 22, 243-51.

Johnson, A. H., Bouldin, D. R., Goyette, E. A. and Hedges, A. H. (1976). Phosphorus loss by stream transport from a rural watershed: Quantities, processes and sources, *Journal of Environmental Quality*, **5**, No. 2, 148-57.

Kronvang, B. (1988). The flux of particulate matter and phosphorus in a river draining an intensively farmed catchment area, Poster, presented at the Nordic Hydrologic Conference 'Simulering av strömning och vattenkvalitet i sjöar och alvar', 1-3 August, Rovaniemi, Finland, *National Agency of Environmental Protection, The Freshwater Laboratory*, Publ. No. 92.

Probst, J. L. (1985). Nitrogen and phosphorus exportation in the Garonne basin (France), *Journal of Hydrology*, **76**, 281-305.

Rebsdorf, Aa. and Thyssen, N. (1986). *Baggrundskoncentrationer af fosforog relationen mellem partikulært fosfor og jern i jyske kilder og kildebække, Foreløbige resultater, Partikulært bundet stoftransport i vand og jorderosion*, B. Hasholt, (ed.), NHP-Report No. 14, 135-46.

Rigler, F. H. (1979). The export of phosphorus from Dartmoor catchments: A model to explain variations of phosphorus concentrations in streamwaters, *Journal Marine Biological Association of the United Kingdom*, **59**, 659-87.

Schuman, G. E., Piest, R. F. and Spomer, R. G. (1976). Physical and chemical characteristics of sediments originating from Missouri Valley loess. In *Proceedings of the Third Federal Inter-Agency Sedimentation Conference*, Water Resources Council, Washington, DC., 5-28-3-40.

Sharpley, A. N. and Syers, J. K. (1979). Phosphorus inputs into a stream draining an agricultural watershed, II: Amounts contributed and relative significance of runoff types, *Water, Air and Soil Pollution*, **11**, 417-28.

Stevens, R. J. and Smith, R. V. (1978). A comparison of discrete and intensive sampling for measuring the loads of nitrogen and phosphorus in the river Main, county Antrim, *Water Research*, **12**, 823-30.

Stevens, R. J. and Stewart, B. M. (1982). Some components of particulate phosphorus in river water entering Lough Neagh, *Water Research*, **16**, 1591-6.

Viner, A. B. (1988). Phosphorus on suspensoids from the Tongariro river (North Island, New Zealand) and its potential availability for algal growth, *Archiv für Hydrobiologie*, **111**, No. 4, 481-9.

Walling, D. E. (1974). Suspended sediment and solute yields from a small catchment prior to urbanization, In Gregory, K. J. and Walling, D. E. (eds), *Fluvial Processes in Instrumented Watersheds, Institute of British Geographers*, Special Publication **6**, 169-92.

Walling, D. E. and Webb, B. W. (1981). The reliability of suspended sediment load data, *International Association of Hydrological Sciences publication* **No. 133**, 177-94.

Walling, D. E. and Kane, P. (1982). Temporal variations of suspended sediment properties, *International Association of Hydrological Sciences Publication*, **No. 137**, 409-419.

21 Soil Erosion and Organic Matter Losses on Fallow Land: A Case Study from South-east Spain

CAROLYN FRANCIS
Department of Geography, University of Bristol

INTRODUCTION

It is well established that soil erosion is, in part, influenced by vegetation which impedes erosion directly by protecting the soil surface and indirectly by ameliorating the soil conditions to improve both their structure and fertility. Thornes (1985) explored theoretically the competition between soil erosion and vegetation by considering the change in soil erosion rates for surface conditions varying from bare soil to optimum vegetation cover. This chapter studies empirically the interaction between erosion and vegetation by monitoring soil loss on fields in different stages of vegetation recolonization after having been abandoned for successively longer periods.

STUDY AREA

The area lies within the semi-arid belt of south-east Spain in the Province of Murcia, where the mean annual precipitation is below 300 mm and the mean annual temperature is about 18°C. Detailed descriptions of the climate and vegetation of the region are provided in López Bermúdez (1985) and Esteve Chueca (1972), and the site conditions are described in Francis (1986).

Four fields, which had been left fallow for 1, 2, 5 and approximately 20 years by November 1985 (according to local labourers) were chosen. These were located approximately 20 km west-north-west from Murcia (Figure 21.1), in an area of late Tertiary Andalucian gypsiferous marls with marly regosols developed on top (IGME, 1974, LUCDEME, 1986). Sites 1, 2 and 20 were roughly planar, while site 5 consisted of a depression flanked by two interfluves. South-facing

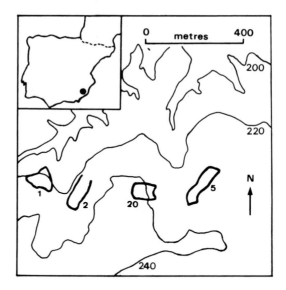

Figure 21.1 Location of field sites in the province of Murcia, Spain

slopes were avoided and the slope orientations were 0°N on site 1, 305° on site 2, 285° on site 5 and 90° on site 20. The land is terraced in broad fields with shallow slopes and sown for barley and corn or left fallow.

METHODOLOGY

Three Gerlach troughs draining an area 3 m² were installed on each of the four fields in January 1986 and maintained until July 1988, except on site 2, which was ploughed in May 1987. On site 5 one trough was placed on the spur, slope and hollow sections of the field while on the others the troughs were placed randomly. The mean slope angles of the runoff plots were 6.2°, 6.4°, 5.6° and 10.3° on sites 1, 2, 5 and 20, respectively. Each trough was connected to a 6 l bottle with a total capacity of about 9 l. Runoff was not measured due to overflowing problems but all the eroded soil in the troughs was collected after each storm, dried and weighed. The coarse organic matter was separated out by hand over a 2 mm sieve and the amount of fine organic matter was estimated for the fraction less than 2 mm by loss on ignition at 375°C for 48 h. This method ensures decomposition of the organic matter without the loss of carbonates (Davies, 1974) and was found to be highly and significantly correlated with organic carbon measured using the Walkley and Black method (Francis and Thornes, unpublished data). Rainfall was measured in each field using a simple rain gauge. Other field parameters such as vegetation cover, biomass, species composition, infiltration and bulk density were measured as part of a larger project using standard techniques (Francis, 1986).

RESULTS

Soil Erosion and Precipitation

Soil erosion occurred on only fourteen rain days out of 129 in the study period. Regression of 24 h precipitation on soil loss per storm was high and significant (Figure 21.2). A correlation of the site measures of precipitation with data for the nearest meteorological station at Alcantarilla (11 km away and about 170 m lower), undertaken as a check, was strong and highly significant ($r^2 = 0.864$; $\alpha = 0.001$) with lower rainfall at Alcantarilla. Soil loss was also correlated against the maximum 30-min intensity measured at Alcantarilla and found to be significantly correlated, and best fitted by a linear function (r^2 values were 0.579, 0.793, 0.583 and 0.709 on sites 1, 2, 5 and 20, respectively; $\alpha = 0.001$). The regressions suggest that erosion occurred when rainfall reached between 15.9 mm and 21.9 mm (using the linear model) or with 30-min intensities between 0.4 and 5.3 mm h^{-1}.

Annual rates of soil loss compared favourably with reported values for the province from Romero Díaz et al. (1988) on marl hillslopes of 2.5 t ha^{-1} yr^{-1} and from López Bermúdez and Gutierrez Escudero (1982) of 2–14 t ha^{-1} yr^{-1} using rates of reservoir sedimentation. The plot data show that overall there is no significant correlation between plot slope and soil loss for any of the storms, although plot slope may account for some of the within-site variability. On site 5 the plot slope is 7°, 6° and 3.8° on the spur, slope and hollow sections, respectively, while on site 1 plots a, b and c have slopes of 4.9°, 8.3° and 5.3°, respectively. However, on site 20 the plot slopes vary little (9.8–10.9°) but soil loss was markedly greater on plots b and c (Figure 21.1).

Soil Erosion and Vegetation

The amount of vegetation cover was measured monthly between January and August 1986, but showed little seasonal variation. By August the mean amount of vegetation cover for the three runoff plots was 26.2 per cent, 17.9 per cent, 21.7 per cent and 23.3 per cent on sites 1, 2, 5 and 20, respectively. While the cover did not distinguish between the more recently abandoned fields and the older ones, the species composition did, with a predominance of herbaceous plants and grasses on sites 1 and 2 (*Moricandia arvensis*, *Brachypodium distachyum* and *Lolium rigidum*) and mainly woody species on sites 5 and 20 (*Thymelaea hirsuta*, *Teucrium capitatum subsp gracillimum*, *Artemisia herba alba*, *Artemisia barrelieri* and *Thymus hyemalis*). This difference is reflected in the amount of biomass harvested from the runoff plots in June 1988 with

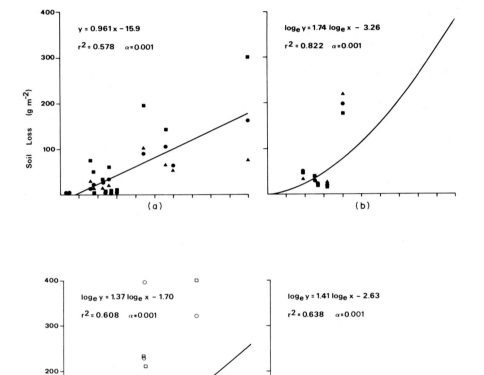

Figure 21.2 Regression of 24 h precipitation (mm) on soil loss (g m^{-2}) for the four fields. (a) Site 1; (b) site 2; (c) site 5; (d) site 20

mean values of 0.027, 0.203 and 0.128 kg m^{-2} on sites 1, 5 and 20. The high value on site 5 is influenced by the greater quantity of biomass in the hollow (0.367 kg m^{-2}) compared to the spur and slope positions (0.149 and 0.093 kg m^{-2}, respectively).

Soil loss was significantly correlated with the precentage vegetation cover (as measured in August 1986) for only four of the nine storms in 1986 and 1987 on 8 March and 4 October 1986 ($r^2=0.367$, $\alpha=0.05$; $r^2=0.601$, $\alpha=0.01$, respectively), and on 16 May and 4 October 1987 ($r^2=0.515$, $\alpha=0.05$; $r^2=0.586$, $\alpha=0.05$). For the storms in 1988 soil loss was not significantly correlated with biomass, and only once with percentage vegetation cover (both variables measured in June 1988) on 12 May ($r^2=0.525$, $\alpha=0.05$).

Organic Matter Losses in the Eroding Soil

The coarse organic matter in the eroding soil consisted of a wide range of petals, whole flowers, leaves, seeds, twigs and root fragments. Estimated annual losses varied little between sites and years, unlike those for the fine organic matter losses which did vary, mirroring differences in the amount of soil lost (Table 21.1).

The correlation of soil loss on coarse organic matter was only significant at site 1 ($r^2=0.799$; $\alpha=0.001$). However, this correlation was strongly influenced by one point and, when removed, the coefficient of determination (r^2) fell to 0.233. Biomass correlated significantly with the percentage of fine organic matter for only two storms in 1988 on 12 May ($r^2=0.705$; $\alpha=0.01$) and on 5 April ($r^2=0.557$; $\alpha=0.05$).

The amount of fine organic matter in the top 3 cm of soil on the plots increased with length of abandonment with mean values of 1.32 per cent at site 1, 1.60 per cent at site 2, 1.59 per cent at site 5-spur, 2.18 per cent at site 5-hollow and 1.95 per cent at site 20. However, the amount of fine organic matter was proportionally greater in the eroded material than in the remaining soil,

Table 21.1 Summary by year of soil loss, coarse organic matter and fine organic matter at sites 1, 2, 5 and 20

	Site 1	Site 2	Site 5	Site 20
Totals for 1986				
Soil loss (g m^{-2})	180.7	317.9	290.9	189.3
Coarse organic matter (g m^{-2})	0.36	1.01	0.55	0.69
Fine organic matter (g m^{-2})	5.4	9.0	13.9	9.0
Totals for 1987				
Soil loss (g m^{-2})	306.6	—	534.1	187.1
Coarse organic matter (g m^{-2})	0.47	—	1.02	0.90
Fine organic matter (g m^{-2})	6.3	—	23.6	13.7
January–July 1988				
Soil loss (g m^{-2})	137.2	—	195.1	86.0
Coarse organic matter (g m^{-2})	0.81	—	1.28	1.24
Fine organic matter (g m^{-2})	2.9	—	10.1	5.8

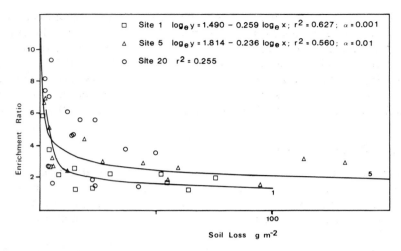

Figure 21.3 Regression of soil loss (g m^{-2}) on enrichment ratios for the storms occurring in 1988 on sites 1, 5 and 20

indicating enrichment ratios greater than unity. The enrichment ratios were poorly correlated with soil loss, but were generally higher for smaller soil losses and on fields abandoned longer (Figure 21.3). Within-site variations include relatively low enrichment ratios on site 20 plot 3 and relatively higher ratios for the hollow section on site 5.

CONCLUSIONS

Soil losses from these abandoned fields correlated well with precipitation characteristics but were poorly or not correlated with plot slope or vegetation parameters. While there was an evident decrease in soil erosion on fields abandoned for greater periods of time this did not appear to be due to the changes in vegetation. A decline in the soil erodibility would be a more likely factor. However, variations in the bulk density of the soil between sites do not readily support this, with mean values of 1.21 g cm^{-3} on site 1, 1.08 g cm^{-3} on site 2, 0.95 g cm^{-3} on site 5-spur, 1.13 g cm^{-3} on site 5-hollow and 1.17 g cm^{-3} on site 20. The mean infiltration rates, measured using a cylinder infiltrometer, of 199.6 mm h^{-1} at site 20 compared to 109.0 mm h^{-1} at site 5-spur, 92.5 mm h^{-1} at site 5-hollow, 104.3 mm h^{-1} at site 2 and 75.0 mm h^{-1} at site 1 (Francis, 1986) do point to greater infiltrability and hence less runoff and erosion on the older site. The effect of vegetation on increasing infiltrability of soils on natural slopes in the area is born out by comparing infiltration rates on weathered but unvegetated marls, which varied between 60 and

174.0 mm h^{-1}, and values between 138.0 and 894.0 mm h^{-1} on vegetated marls (López Bermúdez et al., 1984). The relatively lower rates of erosion on site 1 are probably due to the retaining effect of the still-prominent ridges and furrows of the plough and to a greater tendency to crusting where water had pooled in comparison to the other sites.

The amount of vegetation cover and biomass for much shrubland in lowland Murcia is low, in the order of 30–50 per cent (Esteve Chueca, 1972) or about 0.5 kg m^{-2} (Thornes and Francis, in press), which probably reflects the optimum, given the climatic conditions (Gandullo and Munoz, 1988). The small increase in vegetation between the sites may have been further suppressed by the effects of grazing, and more importantly, the severe drought in the late 1970s and early 1980s. Between 1978 and 1984 the mean annual rainfall of 172.4 mm in Murcia (excluding 1980, with 398.9 mm) was just over half the long-term mean.

The enrichment ratios for fine organic matter indicate a net loss of both nutrients and possibly the seed bank, suggesting that the rate of erosion currently occurring is sufficient to reduce soil fertility and productivity. The lack of correlation between soil loss and coarse organic matter suggests that transport of the latter material may be supply limited. Finally, the results suggest that where vegetation is low (cover values less than 30 per cent) it has little direct effect on erosion rates, which are affected more by variations in the soil condition for a given soil type.

ACKNOWLEDGEMENTS

The first year of this project was undertaken while the author was in receipt of a fellowship from the Royal Society as part of their European Exchange Programme in 1985/6. In the remaining 18 months the author was funded by NERC. She is indebted to the generous and enthusiastic support given by the Departamento de Geografía Física, Humana, y Analisis Regional, Universidad de Murcia, and in particular to Professor López Bermúdez, who proffered much advice and help. Finally, she would like to thank Professor J. B. Thornes for his advice during the project and for his comments on this chapter.

REFERENCES

Davies, B. E. (1974). Loss-on-ignition as an estimate of soil organic matter, *Soil Science Society America Proceedings*, **38**, 150–51.

Esteve Chueca, F. (1972). *Vegetación y flora de las regiones central y meridional de la Provincia de Murcia*, CEBAS, Murcia.

Francis, C. F. (1986). Soil erosion on fallow fields: An example from Murcia, *Papeles de Geografía Física*, **11**, 21–8.

Gandullo, J. M. and Munoz, L. A. (1988). Mapa de productividad potencial primaria neta de los ecosistemas españoles, *Boletin de la Estación Central de Ecología*, 3–17.
IGME (1974). *Mapa geologico de España*, 1:100 000, Mula-912.
López Bermúdez, F. (1985). *Sequía, aridez y desertificación en Murcia*, Academia Alfonso X El Sabio, Murcia.
López Bermúdez, F. and Gutierrez Escudero, J. D. (1982). Estimación de la erosión y aterramiento de embalses en la cuenca hidrográfica del río Segura, *Cuadernos de Investigación Geográfica*, VII, 3–18.
López Bermúdez, F., Romero Díaz, M. A., Ruiz García, A., Fisher, G. C., Francis, C. F. and Thornes, J. B. (1984). Erosión y ecologia en la España semiárida (Cuenca de Mula, Murcia), *Cuadernos de Investigación Geografica*, X, 113–126.
LUCDEME (1986). *Mapa de suelos*, 1:100 000, Mula-912.
Romero Díaz, M. A., López Bermúdez, F., Thornes, J. B., Francis, C. F. and Fisher, G. C. (1988). Variability of overland flow erosion rates in a semi-arid Mediterranean environment under matorral cover Murcia, Spain, *Catena Supplement*, 13, 1–11.
Thornes, J. B. (1985). The ecology of erosion, *Geography*, 70(3), 222–36.
Thornes, J. B. and Francis, C. F. (in press), Runoff hydrographs from three Mediterranean cover types. Paper to the BGRG Meeting, *Geomorphology and Vegetation*, Bristol, October, 1988.

22 The Influence of Forest-clearance Methods, Tillage and Slope Runoff on Soil Chemical Properties and Banana Plant Yields in the South Pacific

S. G. REYNOLDS
FAO, Rome*

INTRODUCTION

The utilization of nutrients from the natural organic cycle of tropical forests through systems of shifting cultivation is widely practised (Lal *et al.*, 1986). While these systems can be used for subsistence agriculture where land areas are large and population pressure relatively slight, cash crop production and increases in population pressure inevitably bring about increased forest clearance, more intensive land use, shortening of the fallow period and problems of maintaining soil fertility (Webster and Wilson, 1966).

Now that very large areas of tropical forest are being cleared for agricultural purposes (Lal, 1981, 1989; Lal *et al.*, 1986) it is important that information and guidelines should be available on appropriate systems so that use of improper methods of clearance, and ineffective erosion control following clearance, do not result in considerable damage to the soil and environment.

Lal (1981, 1982) and Lal *et al.* (1986) have reviewed many of the studies dealing with the effects of forest clearance on soil properties and crop growth and yield, and it is clear that some methods of mechanical land clearance may be detrimental to the yield of annual crops grown immediately after the clearance operation. The harmful effects of mechanical land clearing depend upon a number of

---

*The work reported here was carried out while the author was Lecturer in Agriculture at the South Pacific Regional College of Tropical Agriculture in Western Samoa. The views expressed are those of the author.

Soil Erosion on Agricultural Land
Edited by J. Boardman, I. D. L. Foster and J. A. Dearing
©1990 John Wiley & Sons Ltd

factors and, in particular, the type and weight of the crawler tractor used, the nature of the attachment and, above all, the efficiency of the operator. Scraping of the surface soil into windrows by inefficient and inexperienced operators can cause significant crop-yield reductions, unnecessary soil compaction and increased soil variability. Various authors (e.g. Street, 1980) have demonstrated that bush felling and exposure of soil to sunlight results in a rapid drop in organic matter levels and associated nutrients. Talineau (1968) has suggested that 'a fundamental rule for tropical soils is to leave them bare as shortly as possible'. Russell (1961) indicated that a basic principle for managing tropical soils is to devise a system of farming that, first, involves a minimum of clean cultivation (i.e. keeps the soil surface protected from the sun and rain for as long as possible) and, second, maintains a high organic matter content in the surface soil.

In Western Samoa in the South Pacific, traditional subsistence agriculture has been based on various shifting cultivation systems, and it is only in recent years that rapid population growth and increased cash crop production have resulted in widespread forest clearance and more intensive land use. This chapter examines the effect of forest-clearance methods, tillage and slope runoff on soil chemical properties and banana plant yield through three case studies. Soil physical properties are not discussed.

INFLUENCE OF DIFFERENT METHODS OF FOREST CLEARANCE ON SOIL CHEMICAL PROPERTIES

To establish arable land, large areas of forest were cleared at Togitogiga on the southern slopes of the island of Upolu (see Figure 22.1) in Western Samoa (latitude 14° S, longitude 171° 42′ W). Different methods of forest clearance were employed and the effects on soil chemical properties investigated.

Five areas with the following treatments were sampled:

(1) Forest, uncleared (control);
(2) Heavy bulldozing, one week before sampling;
(3) Heavy bulldozing, one year before sampling;
(4) Light bulldozing, one month before sampling; forest cut by hand one year before sampling;
(5) No bulldozing; forest cut by hand, piled and burnt four months before sampling.

(Heavy bulldozing—removal of debris and some topsoil; light bulldozing—efforts made to remove only debris and leave topsoil intact; hand clearing involved use of chainsaws and bushknives only.)

Soil samples were collected from three depths in each area (0–7.6 cm, 7.6–15.2 cm and 15.2–30.5 cm) with each sample being composited from five subsamples.

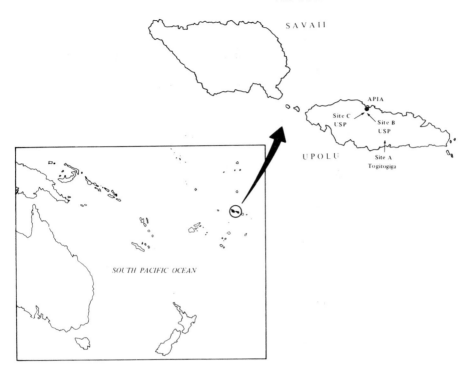

Figure 22.1 Western Samoa

Soil analysis data (Table 22.1) indicate that the method of land clearance has influenced soil fertility status. Probably between 7.6 and 15.2 cm of topsoil were removed along with forest debris by bulldozing in treatments (2)–(4). Some of this topsoil has been redistributed but much is probably concentrated in the debris piles bulldozed around trees left standing for shade. While surface organic debris accumulation was surprisingly light under natural forest (treatment 1), the data suggest that the soil was richer than that left on the bulldozed areas (treatments (2)–(4)).

However, the effect on soil properties of clearance by hand (without heavy machinery) was outstanding. Although the trunks and larger branches were pulled together into piles and burnt, much debris was left to rot back into the soil, which was quickly covered by weeds and served to protect the soil from exposure to sunlight as occurred where areas were bulldozed.

It is readily apparent that improved soil status under treatment (5) is related to the increased return of organic matter. Where soil was bulldozed the surface few centimetres with a high Org. C. content have been removed. Very close

Table 22.1 Soil analysis:[a] details from the five treatment areas at Togitogiga

Treatment no.	Depth (cm)	pH	Org C%	N%	CEC (m.e%)	TEB (m.e%)	B.S. (%)	Truog P (mg%)	P retention (%)	Ca (m.e%)	Mg (m.e%)	K (m.e%)
1	0–7.6	5.2	3.5	0.65	24.6	4.2	17	2.0	82	2.3	1.22	0.41
	7.6–15.2	5.3	1.8	0.34	24.3	1.0	4	4.0	85	0.3	0.23	0.13
	15.2–30.5	5.4	1.2	0.21	17.8	1.0	6	4.0	87	0.7	0.20	0.10
2	0–7.6	5.5	2.2	0.32	18.6	0.4	2	0.5	90	0.3	0.09	0.05
	7.6–15.2	5.5	1.3	0.18	18.3	0.6	3	0.5	92	0.4	0.09	0.04
	15.2–30.5	5.5	0.9	0.14	12.0	0.6	5	0.9	91	0.4	0.09	0.04
3	0–7.6	5.7	2.7	0.37	18.6	2.7	15	0.7	80	1.6	0.82	0.32
	7.6–15.2	5.6	1.2	0.15	11.1	0.6	5	0.9	87	0.4	0.12	0.07
	15.2–30.5	5.4	0.6	0.08	7.1	0.2	3	0.9	93	0.0	0.03	0.02
4	0–7.6	5.6	2.2	0.34	19.3	3.1	16	1.1	84	2.4	0.64	0.07
	7.6–15.2	5.5	1.4	0.22	14.7	0.5	3	0.7	85	0.3	0.20	0.02
	15.2–30.5	5.3	1.0	0.13	11.1	0.03	0	0.4	89	0.0	0.03	0.00
5	0–7.6	5.6	7.6	1.01	39.6	10.5	27	1.6	94	6.9	3.32	0.23
	7.6–15.2	5.7	5.8	0.81	44.5	7.0	16	0.9	94	4.4	2.22	0.13
	15.2–30.5	5.8	4.2	0.52	34.1	3.6	11	0.8	97	2.1	1.40	0.11

[a] Analyzed using methods described by Blakemore (1971b) and Reynolds (1970).

Table 22.2 Relationships between various soil properties and soil organic carbon content

Parameters correlated	n	r	r^2	Sig. level
r1.2	15	0.98	0.96	0.001
r1.3	15	0.93	0.86	0.001
r1.4	15	0.97	0.94	0.001
r1.5	15	0.96	0.92	0.001
r1.6	15	0.98	0.97	0.001
r1.7	15	0.52	0.28	0.05
r1.7*	6	0.99	0.98	0.001
r1.7**	9	0.97	0.93	0.001

1 = Organic carbon;
2 = Nitrogen;
3 = Cation exchange capacity;
4 = Total exchangeable bases;
5 = Exchangeable calcium;
6 = Exchangeable magnesium;
7 = Exchangeable potassium;
7* = Exchangeable potassium in soils 1 and 3;
7** = Exchangeable potassium in soils 2, 4 and 5.

relationships are shown to exist between Org. C. and N, CEC, TEB, exch. Ca and exch. Mg (Table 22.2).

INFLUENCE OF DIFFERENT DEGREES OF TILLAGE ON SOIL CHEMICAL PROPERTIES AND BANANA (*MUSA CAVENDISHII* L.) YIELDS

A large banana trial was established in Western Samoa at the South Pacific Regional College of Tropical Agriculture (now the Alafua Agricultural Campus of the University of the South Pacific (latitude 13° 52'S, longitude 171° 47'W) 5 km from the capital city of Apia (Figure 22.1). Although a reasonably flat, fairly uniform area was chosen for the trial, differences in tillage before trial establishment meant that artificial fertility gradients were established over the area. Effects of the different degrees of pre-plant tillage on soil properties and banana yields are examined below.

An area of secondary forest of some 3.6 ha in size was selected as the trial site. Workers with bush knives and chainsaws cut the trees and undergrowth, a bulldozer was used to drag away logs and remove large stumps but no bulldozing of debris as at Togitogiga was permitted. However, clearing was spread over some six months, and to keep regrowth of weeds and bushes under control a large tractor and heavy sixteen-disc plough was used on parts of the area as clearing progressed, resulting in three distinct tillage zones at the time of trial layout:

Zone I. Heavily cultivated (disc ploughed on six to seven occasions);
Zone II. Moderately cultivated (disc ploughed on two to three occasions);
Zone III. Non-cultivated (no disc ploughing).

On zone I surface organic debris was pulled to the sides by ploughing and topsoil and even subsoil exposed. On zone II the soil was similarly exposed but for a shorter period, and less debris was removed. Following cutting, much debris was left on the surface of zone III and the soil was not exposed. Composite soil samples were collected from zones I, II and III and banana blocks, using corms as planting material, were established on each of the three zones. Banana bunch weight data were collected over a twenty-month period.

Details of soil sample analyses are shown in Table 22.3. As well as a decrease in soil fertility with depth (0–7.6 to 7.6–22.9 cm), a number of soil properties are lowest on the heavily cultivated zone I and increase in zones II and III.

Details of banana yields on control (no fertilizer) plots in the three zones after 12 months are shown in Table 22.4. It is noted that in zone I only one plant has borne fruit and many plants were stunted and making poor growth whereas plants in zones II and III were making good growth and fruiting. Average bunch weight was greatest in zone III. On fertilized plots the number of bunches and total and average bunch weight considerably increased on zones I and II, and although zone III average bunch weight was similar on control and fertilized plots the increased number of bunches on the latter considerably increased the total weight of fruit harvested. Over a twenty-month period there is a clear trend of increase in number of bunches, average and total bunch weight from zone I to zone III. Average time to harvest decreased from zone I to zone III.

Excessive tillage appears to have considerably lowered soil fertility levels, as evidenced in particular by the low Org. C, nitrogen and CEC status of

Table 22.3 Soil analysis details from the three cultivation zones in the banana trial at Alafua[a]

Depth (cm)	pH	C (%)	N (%)	TruogP (mg%)	CEC (m.e.%)	TEB (m.e.%)	B.S. (%)	Ca (m.e.%)	Mg (m.e.%)	K (m.e.%)
Cultivation zone I										
0–7.6	5.9	2.6	0.34	0.8	15.5	7.4	48	4.5	2.6	0.25
7.6–22.9	5.8	1.5	0.18	1.2	11.1	2.6	24	2.0	0.5	0.04
Cultivation zone II										
0–7.6	6.7	4.9	0.62	1.2	32.8	27.4	84	19.1	7.8	0.40
7.6–22.9	6.4	2.0	0.26	1.2	20.3	7.1	35	4.8	1.9	0.11
Cultivation zone III										
0–7.6	6.3	5.7	0.68	1.4	29.4	23.7	81	15.0	8.1	0.46
7.6–22.9	6.0	2.6	0.28	1.4	16.8	5.9	35	4.1	1.7	0.10

[a]Analyzed using methods described by Blakemore (1971b) and Reynolds (1970).

Table 22.4 Banana yield details[a]

Cultivation zone	No. of bunches	Total bunch weight (kg)	Av. bunch weight (kg)	Description
Control plots after 12 months				
I	1	6.35	6.35	Poor growth; many plants only 60–90 cm in height
II	10	93.89	9.39	Good growth; fruiting plants
III	11	139.71	12.70	Good growth; fruiting plants
Fertilized plots after 12 months				
I	9	91.17	10.13	Reasonable growth; some fruiting plants
II	16	213.64	13.35	Good growth; fruiting plants
III	22	283.40	12.88	Good growth; fruiting plants
All plots after 20 months				Time to harvest (weeks)
I	486	5 458.14	11.23	77.0
II	744	8 575.12	11.53	68.8
III	984	11 944.75	12.14	63.9

[a] Receiving 112 kg N, 112 kg P_2O_5, 672 kg K_2O $ha^{-1} y^{-1}$.

zone I soils. A major reason for lower banana yields on zones I (particularly) and II is that cultivation and land preparation tended to smooth out microtopographical differences but probably increased soil variability by concentrating topsoil and exposing areas of subsoil, and also caused increased oxidation of organic matter, the most important parameter in Samoan soils because of its close relationship with other soil properties (Reynolds *et al.*, 1973).

In general, all yield parameters reflect a trend of increased yield with a decrease in the degree of pre-plant tillage. Relationships between soil Org.C, nitrogen and total bunch weight and mean time to harvest suggest that it is variations in these two important properties (and especially the soil Org. C) which have influenced soil productivity and banana yields.

INFLUENCE OF SLOPE GRADIENT AND POSITION ON SOIL FERTILITY AND BANANA PLANT YIELD

A second banana trial was established at site C near area B (Figure 22.1). After four months it was noticed that there were marked differences in banana plant size, with plants being much larger towards the lower part of the plantation. It was apparent that banana growth appeared to be related to slope gradient and position. Investigations were made to determine the likely causes of these height differences.

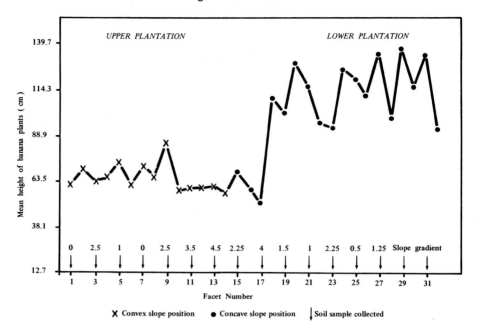

Figure 22.2 Relationship between mean height of banana plants and slope position

Banana suckers were planted out in rows of twenty with a spacing of 1.83 m within rows and 3.35 m between rows. A total of thirty-two rows were planted on an area with a gentle easterly slope and a maximum gradient of 4.5°. All rows were planted across the slope and each sucker was fertilized.

Average plant size was calculated from individual height measurements taken from the ground surface to the stem intersection of the youngest pair of leaves. Slope angles on a series of 3.35 × 6.10 m slope facets were measured (by abney level) along a transect line set out from slope top to bottom on the trial area. The slope was arbitrarily divided into convex and concave portions by the facet of steepest gradient. Soil samples from the surface 0–15.2 cm were collected from seven facets each on the convex and concave slopes, with ten subsamples being collected to produce one composite soil sample per facet.

Plant height was found to be more closely related to slope gradient on concave elements than on total slope or convex elements, with r^2 values of 0.56, 0.35 and 0.12, respectively. The relationship between plant height and slope position is shown in Figure 22.2. Plants on the concave slope positions were much larger and stronger than those on the convex slopes. Plants were fairly uniform on convex slope positions but size increased with decreasing slope gradient on concave slopes.

Table 22.5 Relationships between slope gradient, height of banana plants and various soil properties on concave and convex slopes

Soil property	Slope position	Slope gradient r	Banana plant height r
Org. C	CX	0.29	0.24
	CC	−0.15	0.35
pH	CX	−0.01	0.06
	CC	−0.60	0.75
Exch. K	CX	−0.72	−0.14
	CC	−0.89[a]	0.94[a]
Exch. Mg	CX	−0.70	0.06
	CC	−0.75[b]	0.83[b]
Exch. Ca	CX	−0.13	0.24
	CC	−0.86[b]	0.80[a]
Truog P	CX	−0.42	0.13
	CC	0.02	0.16
CEC	CX	0.91[a]	0.12
	CC	−0.34	−0.14
N	CX	0.27	−0.29
	CC	−0.60	0.46

CX = Convex.
CC = Concave.
[a] $P = 0.01$.
[b] $P = 0.05$.

Relationships between slope gradient (on convex and concave slopes) and eight soil properties are shown in Table 22.5, indicating that exchangeable K, Mg and Ca are more closely related to slope gradient on the concave slope positions with r^2 values of 0.79, 0.56 and 0.73, respectively. Relationships between soil properties and banana plant size (height) demonstrate the importance of the same three soil properties on the concave slope. Relationships on the convex slope element are generally low and non-significant.

Multiple-regression analysis demonstrated that although 86 per cent of the variation in plant height on the concave element was explained by exch. K, 69 per cent by exch. Mg and 65 per cent by exch. Ca, 97 per cent was explained by all three together. Coefficients of partial determination indicated that exch. K explained 86 per cent of the variation in height which the other two failed to explain, exch. Mg explained 75 per cent of the variation in height which the other two failed to explain, while exch. Ca explained only 12 per cent of the variation not explained by exch. K and Mg. The fact that soil potassium levels might be one of the main factors influencing plant growth is not surprising, because a widespread deficiency of potash on the volcanic soils of Western Samoa has been reported (Blakemore, 1971a; Reynolds and Lukes, 1971; Wright, 1963). On the convex slope very little of the variation of banana height was explained by the measured soil properties, the coefficient of multiple correlation showing that only 9 per cent of the variation was explained.

Probably long-term enrichment of soil on the concave slope and the immediate redistribution of fertilizer dressings by slope wash processes have caused the differences in size of banana plants. The early months of the trial coincided with heavy rains in the wet season (60 per cent of the average annual rainfall of 2892 mm falling in the period November to April) and no attempt was made to reduce runoff by mulching the ground. In fact, heavy use of herbicides exposed areas of bare soil. Removal of nutrient elements from the upper convex slope resulted in a uniformly low level of nutrients in these soils. On the concave slope, nutrient concentration was greater on the lower-lying more gentle slopes than on the upper steeper ones.

The soil–topography–plant size relationships reported above can be used to predict crop yields. It has been shown by Venkatesam *et al.* (1965) that plant height (and leaf area and circumference of the pseudostem) was positively correlated with fruit yield. Similar relationships have been found in Western Samoa (Reynolds, 1971), so that it is clear that the size of the plant at the four-month stage is a good indicator of expected fruit yield. The larger plants on the concave slopes will give an earlier and heavier fruit yield.

CONCLUSIONS

Although the three case studies reported here related to very small experimental areas and extrapolation of findings to much larger areas should always be made with caution, it is clear from other evidence not discussed here that similar problems and relationships have been found in various parts of Western Samoa.

The results from the first two case studies show clearly that soil organic matter content is of major importance in Western Samoan soils. Close relationships have been demonstrated between soil organic carbon and nitrogen, CEC, TEB, exchangeable Ca, Mg and K. Any practice which significantly decreases soil organic carbon will also diminish soil fertility and is likely to result in a decrease in crop yields. Excessive tillage or use of bulldozers for land clearance resulted in removal of the surface few centimetres of soil rich in organic carbon, exposure of subsoil and rapid oxidation of the organic fraction and a decline in soil fertility. Banana yields were shown to be related closely to degree of pre-plant tillage, yields being highest where the soil was not cultivated or disturbed except to dig holes for the banana plants.

In the third case study, excessive use of clean cultivation techniques and a failure to adequately cover the soil as protection against runoff during heavy tropical downpours has been shown to result in removal of surface soil from convex slope zones and deposition on concave areas. Banana plant growth was much more rapid on concave areas and was likely to result in much earlier and heavier fruit yields, which would have a direct impact on farmer income.

These findings add to the corpus of data from other tropical countries, and suggest that a few basic rules and principles, such as those proposed by Russell (1961), Talineau (1968) and von Uexkull (1989), be adopted whenever land is cleared or cultivated. There is a tendency in developing countries to adopt new 'Western' methods and use heavy machinery to replace traditional techniques which have evolved over many generations and which may often be more suited to the local environment. It would perhaps be unfortunate if Western Samoa simply abandoned many of its traditional minimum cultivation techniques for other methods which, while being quicker, may prove to have serious long-term effects on soil fertility and productivity. It is suggested that tropical farming practices have much to learn from the indigenous environment and technical knowledge of local farmers. It should be a two-way learning process. This is not to say that bulldozers, tractors and ploughs and other machinery should not be used, but to emphasize that clearance and cultivation should involve a minimum of soil exposure and disturbance, and farming systems should aim to maintain as high an organic matter content in the soil as is economically and practically feasible.

Perhaps, as indicated by Lal (1982):

> The adverse effects of mechanical clearing on soil is not a debatable issue. The fact remains, that for large-scale land development in the humid tropics, mechanical clearing is inevitable. It is here to stay. Manual labour, if available, is expensive and time consuming. Agronomists and soil scientists should design techniques of mechanical land clearing and soil management so as to minimize their adverse effects. It is more worthwhile to adopt a positive attitude towards this necessary evil than to ignore it.

What is most important is that the 'guidelines for clearing, development and protection of tropical lands for farming' proposed by the International Committee on Land Clearing and Development (ICLCD) (Lal et al., 1986, p. 427) at the First International Symposium on Land Clearing and Development in the Tropics held at IITA in 1982, and the appropriate technologies proposed by the International Board for Soil Research and Management (IBSRAM) (Lal, 1987) are widely disseminated and adopted.

ACKNOWLEDGEMENTS

The assistance of staff from the then South Pacific Regional College of Tropical Agriculture is gratefully acknowledged. Mrs D. Fabbri-Ruggeri and Mrs J. Brown-Moriani typed various drafts of the paper and Mr G. Beccaloni drew the figures.

REFERENCES

Blakemore, L. C. (1971a). *Potassium Status of Western Samoan Soils*, United Nations Development Programme/Food and Agriculture Organization, Agricultural Development Project, Western Samoa.

Blakemore, L. C. (1971b). *Recommended Methods of Soil Analysis for Use in Western Samoa*, United Nations Development Programme/Food and Agriculture Organization, Agricultural Development Project, Western Samoa.

Lal, R. (1981). Clearing a tropical forest. The effects on crop performance, *Field Crops Research*, **4**, 345–54.

Lal, R. (1982). Effective conservation farming systems for the humid tropics. In *Soil Erosion and Conservation in the Tropics*, American Society of Agronomy Special Publication No. 43.

Lal, R. (1987). Network on land clearing for sustainable agriculture in Tropical Asia. In Latham, M. (ed.), *Soil Management under Humid Conditions in Asia and Pacific*, International Board for Soil Research and Management Inc. (IBSRAM) Proceedings, No. 5, 35–44.

Lal, R. (1989). Postclearing soil management and sustainable land use. In *Soil Management and Smallholder Development in the Pacific Islands*, International Board for Soil Research and Management Inc. (IBSRAM) Proceedings, No. 8, 51–8.

Lal, R., Sanchez, P. A. and Cummings, R. W. (1986). *Land Clearing and Development in the Tropics*, Balkema, Rotterdam.

Reynolds, S. G. (1970). *A Manual of Introductory Soil Science and Simple Soil Analysis Methods*, South Pacific Regional College of Tropical Agriculture Monograph Series, No. 1.

Reynolds, S. G. (1971). Preliminary observations on a small banana fertilizer trial at Alafua, South Pacific Regional College of Tropical Agriculture Research Paper Series, No. 14.

Reynolds, S. G., Poleka, F., Peters, S., Moala, T., Laumoli, T. and Tiumalu, T. (1973). Soil property–plant growth relationships and relative productivity of soil horizons in a Samoan soil profile, South Pacific Regional College of Tropical Agriculture Research Paper Series, No. 21.

Reynolds, S. G. and Lukes, A. J. (1971). Nursery fertilisation of Sydney blue gum in Western Samoa. *Australian Forestry*, **35**(1), 27–35.

Russell, E. W. (1961). *Soil Conditions and Plant Growth*, Longman, London.

Street, J. M. (1980). Changes in carbon inventories in live biomass and detritus as a result of shifting agriculture and the conversion of forest to pasture, *Case studies in Peru, New Guinea, and Hawaii. International Seminar on Geography and the Third World*, Kuala Lumpur, May.

Talineau, J. C. (1968). Preliminary study on evolution of soil under some fodder and cover crops in lower Ivory Coast, *Cahiers ORSTOM Série, Biologie*, **5**, 49–64.

Venkatesam, C., Venkatareddy, K. and Rangacharlu, V. S. (1965). Studies on the effects of N, P_2O_5 and K_2O fertilization on the growth and yield of banana, *Indian Horticulture*, **22**, 175–84.

Von Uexkull, H. R. (1989). Nutrient recycling. In *Soil Management and Smallholder Development in the Pacific Islands*, International Board for Soil Research and Management Inc. (IBSRAM) Proceedings, No. 8, 121–32.

Webster, C. C. and Wilson, P. N. (1966). *Agriculture in the Tropics*, Longman, London.

Wright, A. C. S. (1963). Soil and land use in Western Samoa, *New Zealand Department of Scientific and Industrial Research Soil Bureau Bulletin* 22.

23 Phosphorus Transport in Agricultural Runoff: The Role of Soil Erosion

ANDREW N. SHARPLEY and S. J. SMITH
USDA-ARS, Durant, Oklahoma

INTRODUCTION

An increased awareness in the United States of non-point source pollution via erosion and associated agricultural chemical transport in runoff (USEPA, 1984) has stimulated an urgency in obtaining information on the impact of current and proposed agricultural management practices on surface water quality. Eutrophication of surface water leads to problems with its use for fisheries, recreation, industry or drinking, due to the increased growth of undesirable algae and aquatic weeds. Although nitrogen (N), carbon (C) and phosphorus (P) are associated with accelerated eutrophication, most attention has focused on P, due to the difficulty in controlling the exchange of N and C between the atmosphere and a water body and fixation of atmospheric N by some blue-green algae. Phosphorus is often the limiting element, therefore, and its control is of prime importance in reducing the accelerated eutrophication of surface water.

The transport of P to runoff occurs in soluble (SP) and particulate (PP) forms. In this chapter, PP includes inorganic P sorbed by soil particles and organic matter eroding during runoff. While SP is immediately available for biological uptake (Walton and Lee, 1972), PP can provide a long-term source of P for aquatic growth in a water body (Bjork, 1972; Carignan and Kalff, 1980; Wildung *et al.*, 1974). The amount of PP that is potentially available for uptake by algae (bioavailable P, BIOP) can be quantified by algal culture tests (EPA, 1971). These assays generally involve long-term incubations (100 days) and, thus, more rapid chemical extraction procedures, which simulate removal of P by algae, have been proposed for routine determination of the BIOP content of suspended and deposited sediment using NaOH, NH_4F and anion exchange resins (Dorich *et al.*, 1985; Huettl *et al.*, 1979; Logan *et al.*, 1979). A review

Soil Erosion on Agricultural Land
Edited by J. Boardman, I. D. L. Foster and J. A. Dearing
©1990 John Wiley & Sons Ltd

of limited field data presently available indicates BIOP to be a dynamic function of soil and runoff characteristics, constituting 0–95 per cent of PP transported (Sharpley and Menzel, 1987; Sonzogni et al., 1982; Wolf et al., 1985). Early research and predictive models assumed BIOP to be a constant proportion (20 per cent) of total P transported (Lee et al., 1979). Consequently, the measurement and simulation of not only PP transport but also its bioavailability will be necessary to determine the impact of agricultural management practices on the biological productivity of surface waters.

Soil erosion is a selective process, enriching runoff sediment in finer-sized particles and lighter organic matter. As P is strongly adsorbed on clay particles (Barrow, 1978; Parfitt, 1978; Syers et al., 1973) and organic matter contains relatively high levels of P, PP constitutes the major proportion of P transported in runoff from cultivated land (Burwell et al., 1977; Nelson et al., 1979). In runoff from grassland or forest with little erosion, SP is the dominant form transported (Burwell et al., 1975; Singer and Rust, 1975). Transformations between SP and PP forms can occur, however, during transport in runoff and stream flow. These transformations are accentuated by the selective transport of fine materials, which have a greater capacity to adsorb or desorb P and will thus be important in determining both short- and long-term bioavailability of P transported. The direction of the exchange between SP and PP will depend on the concentration of sediment and SP in stream flow and adsorption capacity of the sediment contacted, which includes suspended, stream bank and bottom material. The extent of these changes will depend on the labile or desorbable P content of the sediment material contacted and rate of stream flow. Consequently, the direction and extent to which transformation between SP and PP occur must be considered in modelling P transport and potential bioavailability in surface waters.

Most of these studies have been located in the cornbelt region of the mid-west United States and little information is available for the agriculturally important south-west United States, especially Oklahoma and Texas, where increasing rural and urban demands are being made on water supplies as population increases. As a result, the need exists to obtain information on and predict the transport of P in runoff associated with current and proposed agricultural management practices. Soluble P transport can be accurately simulated using equilibrium and kinetic approaches (Ahuja, 1986; Sharpley and Smith, 1989) and will, thus, not be discussed in this chapter.

The selective transport of clay-sized particles in runoff has led to the concept of enrichment ratios (ER) for P, defined as the ratio of the P content of sediment (eroded soil) to that of surface soil, which are used to predict PP transport (Menzel, 1980; Sharpley et al., 1985). Particulate P and BIOP concentrations of runoff are calculated from the total P and bioavailable P content of surface soil, respectively, using the enrichment ratio for each P form (PER and BIOER, respectively);

$$PP = \text{Soil total P. Sediment concentration. PER} \quad (23.1)$$
$$BIOP = \text{Soil bioavailable P. Sediment concentration. BIOER} \quad (23.2)$$

where the units of soil total P and bioavailable P are mg kg^{-1} and those for sediment concentrations in runoff are g l^{-1}. The enrichment ratios were predicted from soil loss (kg ha^{-1}) using the following equation developed by Sharpley (1985):

$$\ln(ER) = 1.21 - 0.16 \ln(\text{soil loss}) \quad (23.3)$$

This equation was developed due to earlier poor predictions of PP transport in individual runoff events (Sharpley *et al.*, 1985) and has had limited field testing.

This chapter presents the results of a study of P transport in runoff from 22 unfertilized and P-fertilized, grassed and cropped watersheds in Oklahoma and Texas, over periods of from 7 to 11 years. The relationship between soil erosion and bioavailability of P transport in runoff is investigated and predictions of PP and BIOP presented.

METHODS

Study Area

The location and management of the 22 watersheds used in this study are presented in Figure 23.1 and Table 23.1, respectively, and are representative of agricultural land use of the Southern Plains area of Oklahoma and Texas, USA. The major soil types at the El Reno and Fort Cobb, Oklahoma; Riesel, Texas; and Woodward, Oklahoma, watersheds are Kirkland silt loam (fine, mixed, thermic, Udertic Paleustolls), Cobb fine sandy loam (fine-loamy, mixed, thermic, Psammentic Paleustalfs), Houston Black clay (fine, montmorillonitic, thermic, Udic Pellusterts) and Woodward loam (coarse-silty, mixed thermic, Typic Ustochrepts), respectively. Watershed runoff was measured using precalibrated flumes or weirs equipped with FW-1 stage recorders and flow-weighted samples collected from each runoff event. Samples were refrigerated at 277 K until analysis.

Surface soil samples (0–50 mm depth) were collected at four sites in each watershed (near the flumes) at monthly intervals in 1981 and 1982 and in March of the other study years. Samples from each site were composited, then air dried and sieved (2 mm).

Figure 23.1 Location of the study watersheds in Oklahoma and Texas, USA

Chemical Analysis

Runoff

The concentration of SP and TP was determined on filtered (0.45 μm) and unfiltered (after digestion with perchloric acid; O'Connor and Syers, 1975), samples, respectively. Particulate P was calculated as the difference between SP and TP. Bioavailable P concentration was determined by extraction of 20 ml of unfiltered sample with 180 ml of 0.11 M NaOH (to give 0.1 M NaOH in 200 ml final volume) for 17 h on an end-over-end shaker (Dorich *et al.*, 1985). The amount of P in the filtered extract was determined and BIOP concentration obtained by subtraction of SP. Bioavailable P measurements were initiated in 1985 and are thus only available for runoff from watersheds at the El Reno, Fort Cobb and Woodward locations. Suspended sediment concentration of runoff was determined in duplicate as the difference in weights of 250 ml aliquots of unfiltered and filtered (0.45 μm) samples after evaporation to dryness.

Soil

The TP content of surface soil was determined by digestion with perchloric acid

Table 23.1 Watershed management characteristics

Watershed	Study period	Size (ha)	Crop type[a]	Fertilizer P applied (kg P ha^{-1} yr^{-1})
El Reno, Oklahoma				
FR1	1977–87	1.6	Native grass	0
FR2	1977–87	1.6	Native grass	2
FR3	1977–87	1.6	Native grass	0
FR4	1977–87	1.6	Native grass[b]	2
FR5	1978–87	1.6	Wheat[c]	16
FR6	1978–87	1.6	Wheat[c]	12
FR7	1978–87	1.6	Wheat[c]	13
FR8	1978–87	1.6	Wheat[c]	13
Fort Cobb, Oklahoma				
FC1	1982–7	2.56	Peanuts/grain	19
FC2	1982–7	2.08	sorghum rotation	18
Riesel, Texas				
Y	1976–82	122	Mixed[d]	4
Y2	1976–82	53	Mixed	1
Y6	1976–82	6.6	Cotton/oats/	14
Y8	1976–81	8.4	sorghum-	12
Y10	1976–82	7.5	rotation	15
Y14	1976–82	2.3	Klein grass	0
W10	1976–81	1.1	Coastal Bermuda grass	0
SW11	1976–82	8.0	Wintergreen Hardinggrass	8
Woodward, Oklahoma				
W1	1977–87	4.8	Native grass	0
W2	1977–87	5.6	Native grass	2
W3	1978–87	2.7	Wheat[c]	23
W4	1978–87	2.9	Wheat[c]	23

[a] Scientific names of crops as follows: wheat—*Triticum aestivum* L.; peanuts—*Arachis hypogena* L.; sorghum—*Sorghum bicolor* L.; cotton—*Gossypium hirsutum* L.; oats—*Avena sativa* L.; Klein grass—*Panicum Coloratum* L.; Coastal Bermuda grass—*Cynodon dactylon* L.; Wintergreen Hardinggrass—*Phalaris acquatica* L.
[b] Burned spring 1979, 1981, 1983 and 1987.
[c] Planted from native grass to wheat in September 1978.
[d] 60% bermuda grass and 40% a three-year rotation of cotton, oats, sorghum.

(Olsen and Sommers, 1982). Bioavailable P was determined by extraction of soil with 0.1 M NaOH for 17 h, at a solution soil ratio of 1000:1 (Dorich *et al.*, 1985). The extract was centrifuged (266 km s^{-1} for 5 min) and filtered (0.45 μm). Labile P was reported as anion exchange resin P, determined by shaking IRA-400 resin (bicarbonate form) at a water-to-soil ratio of 40:1 for 16 h and removal of P from the resin by 1.5 M NaCl (Sharpley *et al.*, 1984). The amount of P sorbed, x (mg P kg^{-1}), from one addition of 1.5 g P kg^{-1} soil (added as K_2HPO_4) was determined after end-over-end shaking for 40 h at a water-to-soil ratio of 100:1. The P adsorption index (PSI) was calculated using the quotient, $x \log C^{-1}$, where C is solution P concentration ($mg\,l^{-1}$) (Bache and Williams, 1971). The quotient was highly correlated with P adsorption maximum calculated from a Langmuir adsorption plot for a wide range of soils (Bache and Williams, 1971). Labile and PSI were determined on samples collected in March each year and values presented are averaged for the study period and each watershed at a given location.

For all samples, the concentration of P was determined colorimetrically on filtered samples by the molybdate-blue method (Murphy and Riley, 1962). Acid or alkali filtrates were neutralized prior to P determination.

Predictions were compared to measured values using linear regression analysis, analysis of variance for paired data and standard error of the y estimate. In the latter analysis, the measured value (x) was assumed to be correct and have no error, with the standard error in the predicted value (y) representing all variability associated with the predictive equations.

RESULTS AND DISCUSSION

Soil Erosion and Phosphorus Transport

Soluble P

Soluble P concentration decreased with an increase in suspended sediment concentration of individual runoff events from the unfertilized watersheds

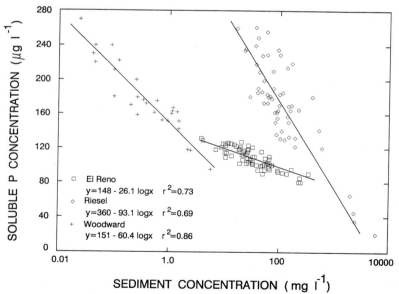

Figure 23.2 Relationship between sediment and soluble P concentration of runoff from the unfertilized native grassed watersheds at El Reno, Riesel and Woodward locations

(Figure 23.2). In fact, the inverse linear relationship between SP and the logarithm of sediment concentration was significant (at the 1 per cent level) over a wide range of sediment concentrations and was similar for different watersheds on the same major soil type at each location. Relationship slope and intercept values differed, however, between locations (Figure 23.2). Relationship slope increased ($r^2 = 0.99$, significant at the 1 per cent level) with an increase in the PSI of the dominant soil type at each location (755, 364 and 500 for Houston Black, Kirkland and Woodward soils, respectively). In other words, SP concentration can be modified to a greater extent during transport in runoff by suspended sediment from source soil having a higher PSI.

Intercept values of the SP-logarithm sediment concentration regression of Figure 23.2 represent the initial SP concentration of runoff before contact with surface soil. As the logarithm of sediment concentration is used, the intercept values of Figure 23.2 are actually at a sediment concentration of 0.01 mg l^{-1}, and thus negligible modification of SP occurs during transport. It is suggested that for these unfertilized watersheds the initial source of SP results from desorption of soil P and leaching of P from plant material by rainfall and runoff water. The intercept values were related ($r^2 = 0.99$, significant at the 1 per cent level) to the labile P content of surface soil (36, 14 and 16 mg kg^{-1} for Houston Black, Kirkland and Woodward soils, respectively) and were thus a direct function of soil P fertility which will also influence the P content of native grass.

Particulate P

Particulate P and BIOP content of runoff sediment decreased as sediment concentration of individual runoff events increased for all watersheds and locations (Figure 23.3). This can be attributed to an increased transport of silt-sized ($>2 \mu m$) particles, of lower P content than finer clay-sized ($<2 \mu m$) particles, as sediment concentration of runoff increases. The logarithmic relationship between sediment concentration and P content of runoff sediment shown in Figure 23.3 was similar for all watersheds and land-management practices. The decrease in BIOP content of runoff sediment, however, was greater than that of PP for a given increase in sediment concentration, as represented by regression slope values for BIOP (-0.63) and PP (-0.40). This may result from a decreasing bioavailability of P associated with increasing size of eroded soil particles, which contain less P sorbed and an increasing proportion of primary mineral P (i.e. apatite) of lower availability compared to finer clay-sized particles.

Concentration and Amounts Transported

Concentration and amounts of SP in runoff from wheat watersheds were greater (at the 5 per cent level) than from the other watersheds (Table 23.2). This resulted

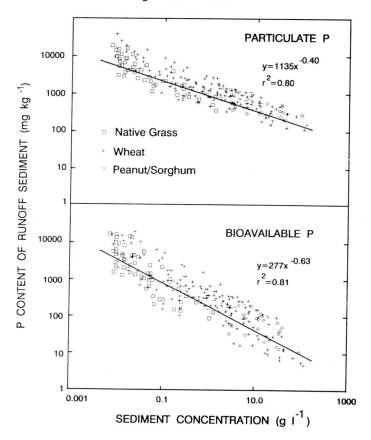

Figure 23.3 Relationship between sediment and particulate and bioavailable P concentration as a function of land use

in part from fertilizer application. This effect is also demonstrated by the greater SP transport from the fertilized (FR2, FR4, SW11 and W2) than unfertilized native grass watersheds (FR1, FR3, Y14, W10 and W1) at the same locations.

In the case of PP transport, significantly greater amounts (at the 5 per cent level) were transported from the cropped watersheds, particularly the clean-tilled peanut–sorghum watersheds (FC1 and FC2), than from native grass (Table 23.2). This difference is a result of the greater amount of erosion and associated P transport from management practices where lower amounts of crop residues are left on the soil surface. This is also reflected in appreciably lower amounts of sediment and PP transport in runoff from the conservation tilled wheat watersheds FR7 and W4 compared to the other wheat watersheds, which were

Table 23.2 Soil loss (kg ha^{-1} yr^{-1}), concentration (μg l^{-1}) and amount (g ha^{-1} yr^{-1}) of particulate and bioavailable P transported in runoff from the watersheds

Water-shed	Sediment loss	Soluble P Conc.	Soluble P Amount	Particulate P Conc.	Particulate P Amount	Bioavailable P Conc.	Bioavailable P Amount	%*
Native grass								
FR1	41	105	88	93	69	76	40	62
FR2	31	145	145	116	77	73	42	41
FR3	33	88	65	100	81	65	37	65
FR4	46	259	186	91	63	45	60	59
Y14	603	104	158	289	438	—	—	—
W10	74	106	118	95	106	—	—	—
SW11	791	215	227	685	724	—	—	—
W1	37	179	10	526	21	33	5	22
W2	617	245	208	1 266	45	90	29	22
Mean†	253^a	161^{ab}	134^a	362^a	180^a	64^a	36^a	45^a
Wheat								
FR5	4 602	236	261	2 191	2 381	457	608	36
FR6	4 874	266	157	1 233	2 874	801	1 674	18
FR7	287	392	436	733	408	135	133	49
FR8	2 899	270	210	1 432	2 225	724	1 288	26
W3	9 640	477	270	10 021	4 092	401	855	19
W4	634	910	373	1 345	553	343	360	43
Mean†	3 823^b	285^b	285^b	2 826^b	2 089^b	477^b	820^b	32^a
Mixed crop and grass								
Y	596	173	194	396	446	—	—	—
Y2	902	137	171	534	667	—	—	—
Y6	2 083	58	48	1 631	1 350	—	—	—
Y8	1 364	6	64	927	945	—	—	—
Y10	4 015	57	82	1 815	2 008	—	—	—
Mean†	1 792^c	86^a	112^a	1 061^b	1 203^b	—	—	—
Peanut-sorghum rotation								
FC1	19 944	192	226	4 354	4 835	924	2 046	34
FC2	20 647	216	143	6 897	4 601	1 460	1 700	24
Mean†	20 296^d	204^b	185^a	5 626^c	4 718^c	1 192^c	1 873^c	29^a

*Percentage of particulate P for corresponding years measured.
†Means followed by the same letter are not significantly different between land-use practices, as determined by analysis of variance for unpaired data.

conventionally tilled (Table 23.2). At the same time, however, concentrations and amounts of SP in runoff from conservation tilled wheat were slightly higher than from conventionally tilled wheat, due in part to an increased contribution of residue P leachate to SP transport.

Particulate P accounted for the major proportion of total P transported from the cropped watersheds (annual average of 81, 88 and 83 per cent for wheat, mixed and peanut-sorghum practices, respectively) (Table 23.2). In contrast no consistent trend in the proportion of total P transported as SP or PP was observed for the native grass watersheds, with PP constituting 48 per cent of total P averaged for all native grass watersheds.

The variation in concentration and amount of BIOP in runoff as a function of land use was similar to that of PP, with significantly greater (at the 5 per cent level) levels in runoff from the peanut–sorghum than wheat and native grass watersheds (Table 23.2). Additionally, conservation tillage of wheat (FR7 and W4) reduced the transport of BIOP in runoff compared to conventional tillage (FR5, 6 and 7, and W3) (Table 23.2). The proportion of PP that was bioavailable, however, was greater on average from conservation (46 per cent) compared to conventionally tilled wheat (25 per cent). This may be attributed to an increasing proportion of finer-sized particles transported in runoff from the former practice.

Mean annual flow-weighted SP concentrations of runoff were greater than the critical level of $10 \mu g \, l^{-1}$, above which the eutrophication of a water body receiving this runoff may be accelerated (Sawyer, 1947; Vollenweider and Kerekes, 1980). This was even the case for the unfertilized native grass watersheds, where native soil fertility was apparently high enough to enrich the SP concentration of runoff. However, soil total P contents of these watersheds ranged from 263 to 385 mg kg^{-1} soil, and are considered to be of 'low' soil P status compared with generally observed contents (100–2000 mg kg^{-1}) summarized by Brady (1984). It is apparent, therefore, that release of P from surface soil material can be great enough to support an immediate increase in growth of aquatic biota. As greater amounts of PP than SP were, in general, transported in runoff, there is an even greater potential for long-term P supply for aquatic growth from eroded soil. This emphasizes the need to predict the transport of PP in runoff and its bioavailability as a function of agricultural management practice in efforts to improve long-term water quality.

Prediction

The PP and BIOP concentration of individual runoff events was predicted using equations (23.1)–(23.3), soil loss and total P and BIOP content of surface soil before runoff. Measured and predicted annual PP concentrations calculated from each runoff event were not significantly different (at the 1 per cent level) over a wide range in measured values 31 to 14 074 $\mu g \, l^{-1}$ (Figure 23.4). For each management practice, slope values of the measured–predicted regression were close to 1.00 with a prediction error of 45, 244, 230 and 83 $\mu g \, l^{-1}$ for the native grass, wheat, mixed and peanut–sorghum watersheds, respectively. This represented a respective 18, 6, 13 and 0.4 per cent of mean predicted PP concentrations.

Measured and predicted mean annual BIOP concentrations of runoff were similar, covering a wide range in measured values (7 to 1765 $\mu g \, l^{-1}$)

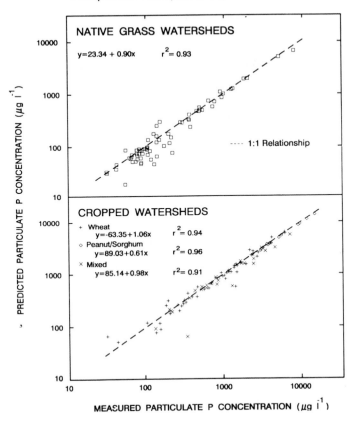

Figure 23.4 Relationship between measured and predicted particulate P concentration of runoff from the grassed and cropped watersheds

(Figure 23.5). The error in BIOP prediction was 33 µg l^{-1}, which represented 7 per cent of the mean predicted value for all watersheds and events (449 µg l^{-1}). As for PP, however, BIOP predictions were not as accurate for the grassed compared to the other watersheds, particularly at low concentrations (Figures 23.4 and 23.5).

The discrepancy between measured and predicted concentrations in runoff from the grassed watersheds may result from a lower soil loss from these watersheds (Table 23.2). As the predicted relationship between ER and soil loss (equation (23.3)) is logarithmic, predicted values of ER will be affected more by a unit quantity of soil loss at low values of loss (<50 kg ha^{-1} yr^{-1}) than higher values (>500 kg ha^{-1} yr^{-1}). Additionally, preliminary testing of equation (23.3) showed that this relationship varied between watersheds

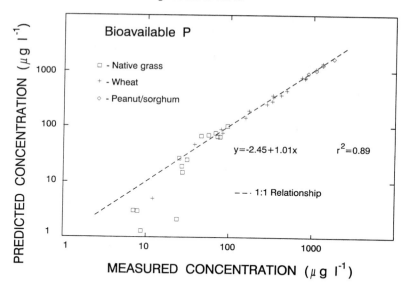

Figure 23.5 Relationship between measured and predicted bioavailable P concentration of runoff from the grassed and cropped watersheds

(Sharpley *et al.*, 1985). Consequently, making slope and intercept of equation (23.3) a function of factors affecting soil loss or runoff energy, such as rainfall intensity, vegetative cover and management practices, should improve PER and BIOER prediction and thus PP and BIOP transport in runoff.

At low runoff energy and subsequently low sediment concentrations, PP enrichment will be more a function of the transport of lighter organic matter than heavier soil particles. Although particulate organic P transport was not measured independently of inorganic P in this study, previous research has shown that enrichment of organic particles is greater than for inorganic particulates (Knoblauch *et al.*, 1942; Neal, 1944). Recently, Sharpley (1985) reported ERs of 2.00 and 1.56 for organic C and clay in runoff from several soils using simulated rainfall.

CONCLUSIONS

It is apparent that sediment plays a major role in controlling the partitioning and amounts of both SP and PP in agricultural runoff. The dynamics and direction of partitioning between SP and PP are a function of the complex rainfall–runoff–sediment interaction, involving rainfall intensity and duration; runoff volume; suspended sediment size, amount, P content and P sorption index; and SP concentration of runoff. Consequently, these processes must be considered in terms of quantifying and modelling P movement and the potential bioavailability of P in runoff from eroding agricultural soils.

In general, the major proportion of P transport in runoff from the watersheds studied was in the particulate form. Amounts of PP and BIOP transported increased with cultivation and associated reduction in surface soil vegetative or residue cover. The mean annual loss of total P in runoff from the P-fertilized watersheds is equivalent to an average 15 per cent, 12 per cent and 32 per cent of the annual fertilizer P applied to wheat, mixed crop and grass, and peanut–sorghum rotation practices, respectively. This P loss is not a direct estimate of fertilizer loss, due to unaccountable losses from native soil P. Although P losses, such as from the clean-tilled peanut–sorghum rotation may constitute a modest economic loss of P to the farmer, the data indicate that transport from all management practices can be of environmental concern.

Native soil fertility levels were high enough to increase SP concentrations in runoff above critical values associated with accelerated eutrophication ($10 \mu g\, l^{-1}$) even for unfertilized native grass watersheds. Consequently, reduction of fertilizer P application rates is not expected to decrease SP concentrations in runoff from the study watersheds to below the critical values. It is apparent, therefore, that control of sediment and associated P transport in runoff may be of potentially greater benefit in reducing P loads and improving long-term water quality of receiving water bodies than considerable reductions of fertilizer P applications.

Soil erosion and associated P loss in runoff may be controlled by increasing vegetative or residue groundcover through reduced tillage practices, although bioavailability of particulate P transported may increase. Badly eroded areas may be carefully fertilized to enhance plantcover and surface soil protection.

The accurate prediction of P transport in runoff will allow a more efficient evaluation and selection of management practices aimed at controlling and reducing P transport in runoff while maintaining soil productivity. The predictive equation used in this study (equations (23.1)–(23.3)) are being incorporated into models simulating agricultural chemical transport and long-term effects of erosion on soil fertility and crop productivity. Prediction of PP and BIOP transport may be improved, however, by use of a linear relationship between enrichment of P and clay or specific surface area of eroded material (Sharpley, 1985). These relationships are non-logarithmic and may thus provide a more sensitive means of predicting PP and BIOP transport than using soil loss. As clay and specific surface area enrichment are estimated in erosion models, more accurate estimates of the efficiency of agricultural management practices on long-term ($>50\,yr$) loss of sediment and associated P on both water quality and soil productivity may be afforded.

Buffer or riparian strips can effectively reduce sediment, PP and BIOP transport before runoff water enters a stream or river. A maximum of

85 per cent sediment removal can be achieved with a 2.5 m grassed strip during shallow overland flow (Barfield *et al.*, 1975). The removal of P is not as efficient due to the preferential transport of clay-sized material of higher P content than coarser silt and sand. In addition, small impoundments and reservoirs can be used as effective traps, reducing both SP and PP levels and improving downstream water quality.

Future research should be directed towards improvement of the partitioning models for SP and PP transport in runoff as a function of soil loss. This should focus on the mechanisms of exchange between sediment and solution P and standardization of routine methods to quantify BIOP. With the accumulation of fertilizer and residue P at the soil surface under conservation tillage practices, the relative importance of the partitioning processes outlined above may need to be re-evaluated. This information will improve the prediction of the amounts and forms of P transported into lakes and reservoirs, aiding agricultural management decisions to reduce P loss in runoff while maintaining crop yields.

ACKNOWLEDGEMENTS

The assistance of USDA-ARS personnel at Chickasha and Woodward, Oklahoma and Riesel, Texas, in collecting runoff and soil samples is appreciated.

REFERENCES

Ahuja, L. R. (1986). Characterization and modeling of chemical transfer to runoff, *Advances in Soil Science*, **4**, 149–88.

Bache, B. W. and Williams, E. G. (1971). A phosphate sorption index for soil, *Journal of Soil Science*, **22**, 289–301.

Barfield, B. J., Kao, D. T. Y. and Toller, E. W. (1975). Analysis of the sediment filtering action of grassed media, *Research Paper No. 90, Univ. Kentucky Water Research Inst.*, Lexington, KY.

Barrow, N. J. (1978). The description of phosphate adsorption curves, *Journal of Soil Science*, **29**, 447–62.

Bjork, S. (1972). Swedish lake restoration program gets results, *Ambio*, **1**, 153–65.

Brady, N. C. (1984). *The Nature and Properties of Soils, 9th edition*, Macmillan, New York.

Burwell, R. E., Timmons, D. R. and Holt, R. F. (1975). Nutrient transport in surface runoff as influenced by soil cover and seasonal periods, *Soil Science Society of America Proceedings*, **39**, 523–528.

Burwell, R. E., Schuman, G. E., Heinemann, H. G. and Spomer, R. G. (1977). Nitrogen and phosphorus movement from agricultural watersheds, *Journal of Soil and Water Conservation*, **32**, 226–30.

Carignan, R. and Kalff, J. (1980). Phosphorus sources for aquatic weeds: Water or sediments? *Science*, **207**, 987–9.

Dorich, R. A., Nelson, D. W. and Sommers, L. R. (1985). Estimating algal available phosphorus in suspended sediments by chemical extraction, *Journal of Environmental Quality*, **14**, 400–5.

Huettl, P. J., Wendt, R. C. and Corey, R. B. (1979). Prediction of algal available phosphorus in runoff sediment, *Journal of Environmental Quality*, **4**, 541–8.

Knoblauch, H. C., Kolosky, L. and Brill, G. D. (1942). Erosion losses of major nutrients and organic matter from Collington sandy loam, *Soil Science*, **53**, 369–89.

Lee, G. F., Jones, R. A. and Rast, W. (1979). Availability of phosphorus to phytoplankton and its implications for phosphorus management strategies, In Loehr, R. C., Martin, C. S., and Rast, W. (eds), *Phosphorus Management Strategies for Lakes*, Ann Arbor, New York, pp. 259–308.

Logan, T. J., Oloya, T. O. and Ytaksich, S. M. (1979). Phosphate characteristics and bioavailability of suspended sediments from streams draining into Lake Erie, *Journal of Great Lakes Research*, **5**, 112–23.

Menzel, R. G. (1980). Enrichment ratios for water quality modeling, In Knisel, W. (ed.), *CREAMS—A Field Scale Model for Chemicals Runoff and Erosion from Agricultural Management Systems*, Vol. III, Supporting Documentation, USDA Conservation Research Report, **26**, 486–92.

Murphy, J. and Riley, J. P. (1962). A modified single solution method for the determination of phosphate in natural waters, *Analytica Chimica Acta*, **27**, 31–6.

Neal, O. R. (1944). Removal of nutrients from the soil by crops and erosion, *Agronomy Journal*, **36**, 601–7.

Nelson, D. W., Monke, E. J., Bottcher, A. D. and Sommers, L. E. (1979). Sediment and nutrient contributions to the Maumee River from an agricultural watershed. In Loehr, R. G., Haith, D. A., Walter, M. F. and Martin, C. S. (eds), *Best Management Practices for Agriculture and Silviculture*, Ann Arbor, New York, pp. 491–505.

O'Connor, P. W. and Syers J. K. (1975). Comparison of methods for the determination of total phosphorus in waters containing particulate material, *Journal of Environmental Quality*, **4**, 347–50.

Olsen, S. R. and Sommers, L. E. (1982). Phosphorus. In Page, A. L. *et al.* (eds), *Methods of Soil Analysis*, Part 2, 2nd edition, *Agronomy*, **9**, 403–29.

Parfitt, R. L. (1978). Anion adsorption by soils and soil materials. *Advances in Agronomy*, **39**, 1–50.

Sawyer, C. N. (1947). Fertilization of lakes by agricultural and urban drainage, *Journal of New England Water Works Association*, **61**, 109–27.

Sharpley, A. N. (1985). The selective erosion of plant nutrients in runoff, *Soil Science Society of America Journal*, **49**, 1527–34.

Sharpley, A. N. and Menzel, R. G. (1987). The impact of soil and fertilizer phosphorus on the environment, *Advances in Agronomy*, **41**, 297–324.

Sharpley, A. N. and Smith, S. J. (1989). Prediction of soluble phosphorus transport in agricultural runoff, *Journal of Environmental Quality*, **18**, 313–316.

Sharpley, A. N., Jones, C. A., Gray, C. and Cole, C. V. (1984). A simplified soil and plant phosphorus model. II. Prediction of labile, organic, and sorbed phosphorus, *Soil Science Society of America Journal*, **48**, 805–9.

Sharpley, A. N., Smith, S. J., Berg, W. A. and Williams, J. R. (1985). Nutrient runoff losses as predicted by annual and monthly soil sampling, *Journal of Environmental Quality*, **14**, 354–60.

Singer, M. J., and Rust, R. H. (1975). Phosphorus in surface runoff from a deciduous forest, *Journal of Environmental Quality*, **4**, 307–11.

Sonzogni, W. C., Chapra, S. C., Armstrong, D. E., and Logan, T. J. (1982). Bioavailability of phosphorus inputs to lakes, *Journal of Environmental Quality*, **11**, 555–63.

Syers, J. K., Browman, M. G., Smillie, G. W. and Corey, R. B. (1973). Phosphate sorption by soils evaluated by the Langmuir adsorption equation, *Soil Science Society of America Proceedings*, **37**, 358–63.

US Environmental Protection Agency (1971). *Algal Assay Procedure—Bottle Test*, National Eutrophication Res. Program, Pacific North West Laboratory, Corvallis, Oregon.

US Environmental Protection Agency (1984). *Report to Congress: Nonpoint Source Pollution in the U.S.*, Washington, DC.

Vollenweider, R. A. and Kerekes, J. (1980). The loading concept as a basis for controlling eutrophication: Philosophy and preliminary results of the OECD program on eutrophication, *Progresses in Water Technology*, **12**, 5–38.

Walton, C. P. and Lee, G. F. (1972). A biological evaluation of the molybdenum blue method for orthophosphate analysis, *Verhandlungen Internationale Vereinigung Limnologie*, **18**, 676–84.

Wildung, R. E., Schmidt, R. L. and Gahler, A. R. (1974). The phosphorus status of eutrophic lake sediments as related to changes in limnological conditions—total, inorganic, and organic phosphorus, *Journal of Environmental Quality*, **3**, 133–8.

Wolf, A. M., Baker, D. F., Pionke, H. B. and Kunishi, H. M. (1985). Soil tests for estimating labile, soluble and algal-available phosphorus in agricultural soils, *Journal of Environmental Quality*, **14**, 341–8.

Risks and Costs

24 Some Costs and Consequences of Soil Erosion and Flooding around Brighton and Hove, Autumn 1987

D. A. ROBINSON and J. D. BLACKMAN
Geography Laboratory, University of Sussex

INTRODUCTION

In recent years, water erosion of arable soils in Britain has become a topic of increasing concern (Evans, 1971; Morgan 1977, 1980; Reed, 1979; Fullen, 1985a; Speirs and Frost, 1987; Boardman, 1986, 1988a; Robinson and Blackman, 1989). In a number of localities, erosion has caused not only soil loss from fields but also localized flooding of roads and houses (Stammers and Boardman, 1984; Evans and Skinner, 1987; Boardman and Spivey, 1987). Research has concentrated on estimating soil losses and on identifying the factors responsible for the apparent increase in the intensity and occurrence of erosion (Morgan, 1977; Evans and Nortcliff, 1978; Evans, 1980; Boardman, 1983; Reed, 1983; Frost and Speirs, 1984; Boardman and Robinson, 1985; Fullen, 1985b; Colborne and Staines, 1986; Fullen and Reed, 1986, 1987; Jackson, 1986). There has been little attempt to assess the economic costs of erosion and flooding to householders, farmers or local authorities (Boardman, 1986, 1988a).

An area affected by periodic severe erosion is the eastern South Downs, in Sussex (Boardman, 1984; Boardman and Robinson, 1985). Exceptional rainfall in October 1987 led to widespread erosion of arable soils on the Downs, and caused localized flooding of adjacent suburban areas of Brighton and Hove (Boardman 1988b; Robinson and Williams, 1987, 1988). Detailed information has been gathered on the costs incurred as a result of the erosion and flooding through follow-up surveys of the householders, farmers and local authorities.

THE STUDY AREA

Many of the outer suburbs of Brighton and Hove occupy the lower reaches of dry chalk valleys to the north and east of the town centres. The upper reaches of the valleys remain in agricultural use and over the past 40 years have been increasingly used to grow cereals. In recent years there have been a growing number of erosive events during autumn and winter which seem to be associated with the widespread adoption of autumn sowing (Boardman and Robinson, 1985). By far the most serious and widespread erosion to date occurred in the first two weeks of October 1987. The worst-affected area was Rottingdean (Figure 24.1), on the eastern outskirts of Brighton (Robinson and Williams, 1987, 1988), but other major areas affected included two localities at Mile Oak and one at Hangleton on the northern outskirts of Hove. In all four localities, homes were flooded, gardens inundated with mud, and roads and drains flooded or blocked. Emergency services had to be called out, and the local authorities

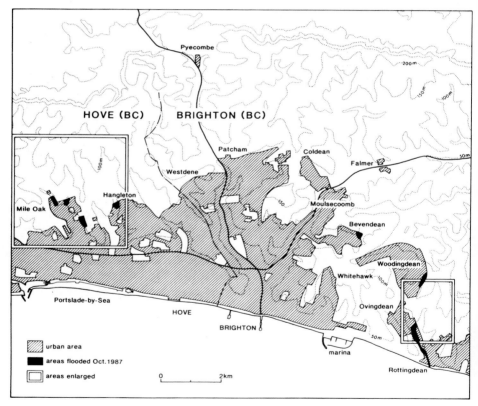

Figure 24.1 Suburban areas on the periphery of Brighton and Hove flooded by mud-laden water that originated as erosive runoff from arable land, October 1987

were involved both in emergency work and in longer-term protective measures designed to alleviate the risk of any repeat flooding in future years.

THE SURVEYS

All householders in the areas known to have been flooded were subject to a door-to-door questionnaire survey. They were asked about the nature and extent of the damage they suffered; the costs incurred; the extent to which costs were or were not covered by insurance; what they thought were the causes of the flooding; and any future changes they thought should occur to reduce the risk of further flooding. The response rate to the survey was good, with only one householder refusing to disclose the information requested, although there was some variation in the level of detail that households could provide on the breakdown of costs. Data could not be obtained from a few households because of a change of ownership since the flooding occurred, or because the owners were absent for the whole month of the survey. All households that suffered minor flooding at Mile Oak may not have been contacted because the properties affected were scattered across several streets. However, all houses that suffered internal flooding are believed to have been identified.

Farmers on whose land the floods originated were asked to quantify the extent of different forms of erosion suffered by their land; the costs they incurred in terms of direct crop and material losses; machinery and labour costs for cleaning up and for cultivations that had to be repeated; their views as to the cause of the flooding; and whether they intended modifying their agricultural practice in future, and if so, how, as a result of the erosion and flooding. Most of the floodwater originated as erosive runoff from just three large farms on the urban periphery.

The surveys were carried out in October 1988, one year after the flooding occurred, so that farmers could estimate any reductions in yield from eroded fields and for householders to have completed their repairs and insurance claims. Despite the lapse of time, one or two householders could still not give final costs because some damage had still not been fully rectified.

Local authority costs were obtained directly from council officers. Some of the figures supplied included costs budgeted for work still not completed at the date of interview.

RESULTS

Rottingdean

The Brighton suburb of Rottingdean has grown up around an old village centre located in the floor of a dry valley immediately adjacent to the coast. Flooding occurred on the evening of 7 October during an exceptionally intense rainstorm towards the end of a day of heavy, continuous rainfall, which itself had been preceded by several days of persistent rain (Robinson and Williams, 1987, 1988). The flood originated as runoff from some 68 ha of recently sown winter cereals in a normally dry, tributary valley to the north-east of the suburb. After ponding at the exit of the valley, where it flooded storage and machinery barns, the water broke through a field boundary, crossed a road and entered the housing area as a wave of water almost a metre deep. It then flowed down the main valley and through the centre of Rottingdean to the sea (Figure 24.2). Further flooding occurred on a more limited scale the next day, when more rain fell. Brighton Borough Council promptly installed a series of earth dams within the source valley and arranged pumps to try to remove water before it reached the housing area, and no further flooding occurred, despite further heavy rain.

The rapid runoff deeply rilled the valley-side slopes, and more than 5000 m^3 of soil was eroded from the fields. As a result, the floodwater was heavily loaded with mud and other debris, most of which was deposited in the flooded houses and gardens.

Household costs

A survey by the local authority in the days immediately following the flood indicated that some 66 properties were affected (Robinson and Williams 1987, 1988). In our follow-up survey, no information could be obtained about one badly flooded property that had been sold in the intervening year, a further property that was less extensively flooded and one or two that suffered flooding of part of their gardens. There were also two or three properties in Rottingdean High Street from which no replies were obtained because the owners were absent. One was probably flooded internally, the others only suffered minor flooding of their gardens.

The follow-up survey suggests that costs were incurred by only a little over half of the households initially identified as suffering some flooding, and internal damage was caused to just 25 of the properties in the sample interviewed, probably 29 in total. Because many of the properties were bungalows, 10 of the 25 houses that were flooded internally suffered damage to all rooms, probably 11 or 12 properties in total. Twelve households had to seek temporary accommodation while their homes were cleared, dried and repaired. Two stayed with friends or relations, but temporary accommodation for the remaining 10 averaged £1101, with costs for one household reaching £3400. The average cost of internal damage to those homes and garages flooded was £8351, the highest individual cost, £25 000. Most households successfully claimed the majority of these costs from their insurers, the overall rate of recovery for damage being 80 per cent and for accommodation 86 per cent. Nearly all shortfalls were because of

Some Costs and Consequences of Soil Erosion and Flooding 373

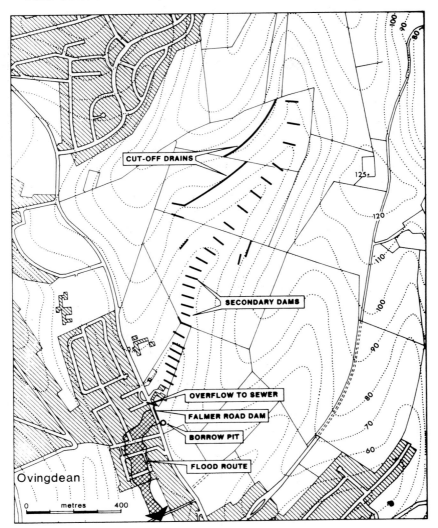

Figure 24.2 Flood route and emergency protection works installed at Rottingdean, October 1987

inadequate home contents insurance, with at least four households having little or no cover, which resulted in individual losses of as much as £9000.

Damage to gardens was more extensive, including all 32 properties with gardens in the sample, and probably around 40 in total. The costs of damage were much less overall, but nearly all the rectification costs had to be borne by the householders. Among the 33 respondents in the survey, only one had made a successful insurance claim, which was for the rebuilding costs of a boundary wall. Several respondents admitted that the cost estimates were rough approximations because rectification had often involved numerous small payments spread over the full year. Many of the householders were elderly and they had to pay both labour and material costs. Younger households mostly used their own labour, which lowered their financial losses. The most costly items were replacement garden buildings, fences, walls and the removal of mud and other debris. Average cost of damage and rectification for the surveyed households with flooded gardens was £1261, but several losses were in excess of £4000, while some were only £200–300. Total costs for the surveyed households was £37 815, and using estimates based on neighbouring properties for the seven additional gardens known to have been flooded, but from whom no data could be obtained, the total cost of damage to gardens and garden properties was approximately £41 000. Using similar estimates for house damage and temporary accommodation, the costs were £237 820 and £12 108, respectively, making the total cost of all damage to individual property a little less than £300 000. Of this total, 67 per cent was paid out by insurers, 33 per cent by the householders.

In addition to the financial costs, several householders complained of effects of the floods that could not be estimated in direct monetary terms. These included psychological problems associated with fears of repeat flooding; health and allergy problems attributed to agrochemicals contained within the mud and debris left behind by the floods; heart attacks brought on by worry and the hard work of clearing up; and distress over carefully tended gardens buried by mud and infested by weeds and cereal crops.

Despite many of the properties being sited in the floor of a valley, only one of the householders interviewed had ever considered that there might be a flood risk. Blame was mostly attributed to the exceptional rainfall (85 per cent of respondents) and farming practices (82 per cent). Sixty per cent also blamed inadequate drainage. Over 50 per cent of the households expected farming practices to change in order to lessen the risks of future flooding, and nearly 30 per cent expected drainage to be improved.

Local authority costs

Immediately after the flood, Brighton Borough Council constructed a series of 26 small protective earth dams and associated trenches across the floor of the tributary valley from which the flood emanated (Figure 24.2). Three further dams were built across a shallow side valley, two short-contour ditches and ridges across the north-facing slopes of the valley and two further ditches and ridges across the south-facing slopes, graded to drain to the upper end of the valley. An overflow drain and pump were installed to transfer water from behind the lowest dam into a surface-water sewer lying beneath the road running across the exit of the valley (Director of Technical Services, Brighton, 1987).

Householders were supplied with sandbags, and the Council had to clear mud and other flood debris from the roads and drains. The total cost of the emergency work, labour, materials and machinery was £83 000. This figure does not include the cost of direct labour. Other local authority costs included the fire brigade and police, who provided emergency help to residents on the night of the flood, pumping away large volumes of floodwater and redirecting traffic. The addition of these further costs raised total emergency costs to more than £100 000.

The emergency dams remained in place until after the 1988 cereal harvest, when, in accordance with an agreement by the farmer to reduce the area of autumn sown cereals in the valley by 50 per cent, the Council infilled and restored all but five of the dams, trenches and ridges (Figure 24.3). The remaining dams, one adjacent to each field boundary across the floor of the valley, were enlarged and strengthened, a water-storage area scraped out behind each, and it is proposed that both the dams and emergency water-storage areas will be grassed over during the next year. The entire floor of the valley between the lowest two dams was corrugated with gentle ridges and hollows aligned across the floor of the valley and the area will be sown to grass to increase surface water storage and infiltration. The drain to transfer water from the settling lagoon behind the lowest dam into the surface water sewer was left in place as a permanent feature. The budget for all these works was estimated at £20 000.

Mile Oak and Hangleton

Roads and houses occupying the floors of three separate dry valleys on the northern periphery of Hove were repeatedly flooded by erosive runoff from adjacent arable land during October 1987 (Figure 24.4). Minor flooding also occurred elsewhere. Flooding was never on as large a scale as at Rottingdean, with only a few houses flooded in each valley and maximum speeds and depth of flowing water were much less. Mile Oak Road has a history of minor flooding in the past and suffered flooding very similar to October 1987 in the late autumn of 1976, after which a stormwater storage dam was built in the floor of the valley above the residential area. This successfully protected the area in 1982 when flooding was widespread elsewhere on the Downs (Stammers and Boardman, 1984; Boardman and Robinson, 1985), but failed in 1987. The other two localities had never previously suffered serious flooding of this nature.

All stormwater drainage in these northern suburbs relies on soakaways into the underlying chalk, but the mud and silt in the floodwater choked many soakaways, which then ceased to function. Some attempts were made to dispose of water through the foul-water sewage system, but this was entirely inadequate.

Household costs

Tracing the precise route of the floodwater, and identifying exactly which houses were flooded, was more difficult than at Rottingdean because the shallower depth

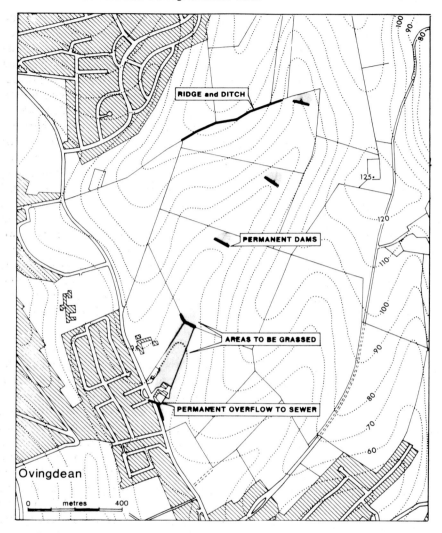

Figure 24.3 Permanent protective works installed on New Barn Farm, Rottingdean, summer 1988

of floodwater meant that slight variations in the height of door thresholds created great differences in the extent to which properties were flooded. Of the 31 households surveyed, muddy water entered 16 houses and 18 garages, causing £84 000 of damage. At least one other property at Mile Oak Road was reported by neighbours to have been badly flooded, but was standing empty at the time of the survey and was for sale for the second time since the flood. Assuming costs at this property to have been the same as the average for the others in the

Some Costs and Consequences of Soil Erosion and Flooding 377

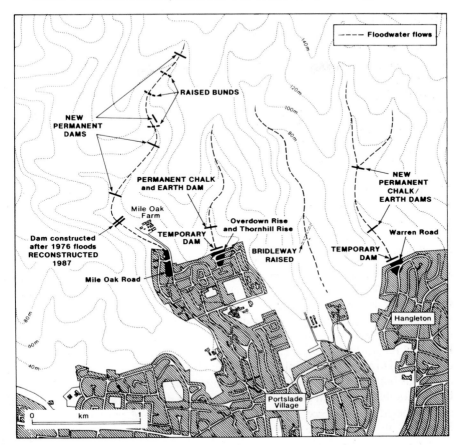

Figure 24.4 Surveyed areas of Portslade and Hangleton, Hove, flooded in October 1987 and the protective works installed by the Council

survey, the total cost increases to nearly £90 000. The highest costs were incurred by householders in Overdown Rise and Thornhill Rise, Mile Oak (Figure 24.4), where two properties each suffered £15 000 worth of damage, and six residents had to move into temporary accommodation at an additional cost of £1360 in total. Householders made successful insurance claims for over 91 per cent of these costs.

Damage to gardens was far more widespread. Gardens of all 31 properties surveyed had been inundated, and several additional properties that suffered minor flooding of parts of their gardens were undoubtedly omitted from the survey. The total cost of damage given by those surveyed was approximately £15 000, but several could not estimate their costs although some had been incurred. A total of £10 600 of these costs (71 per cent) were reclaimed from insurers, which is 11 per cent of the total off-farm costs, a much higher proportion than that at Rottingdean, where only 1 per cent was reclaimed. Several of these successful claims were for boundary walls, but the more youthful population who were physically able to undertake much of the necessary garden repair work themselves also helped to reduce the uninsured losses to the households.

Combining all the costs of damage to property and gardens, plus the costs of temporary accommodation, gives a minimum total householder cost of approximately £106 500 for the flooding at the three sites in Hove, of which 88 per cent was recouped from insurers.

Of the 31 households interviewed, only five had ever considered that they were at risk from flooding, three of whom lived at the Mile Oak Road location, which had suffered flooding before. Twenty-seven of the 31 blamed farming practices for the flooding, but most also recognized that exceptional rainfall and inadequate drains were major factors in this part of Hove.

Local authority costs

After the flooding at Mile Oak in 1976, Hove Borough Council commissioned a report (Furneaux, 1977) which suggested that either the agricultural land use in the valley above the residential area must change, or, alternatively, facilities for stormwater storage should be built in the floor of the valley above the residential area. The latter solution was adopted to avoid the reduction in financial return to the farmer, and to his landlord, that would otherwise have occurred (Director of Technical Services, Hove, 1987).

Similar but more widespread solutions have been sought to the problems posed by the floods of 1987. In the Mile Oak valley, the dam built after 1976 has been reconstructed and enlarged, a further four dams have been constructed higher up the valley, along with three small bunds across the valley floor (Figure 24.4). Two small emergency dams and one large permanent dam have been built up-valley of the second site in Mile Oak, and two dams up-valley of Hangleton. Old soakaways have had to be cleared and redrilled, new soakaways built and some road cambers realigned to direct runoff away from vulnerable houses and into low-lying storage areas free of housing. The total costs in 1987/8 were approximately £133 000, of which £65 000 was for emergency work (sandbags, garden clearance, pumping out and flushing of soakaways, etc.) and dam construction, and £68 000 for drainage improvements. A further £20 000 is budgeted for further soakaway improvements in the financial year 1989/90. These figures exclude police and fire brigade costs. Local authority costs for Hove were, as a proportion of total off-farm costs, more than twice those of Brighton for the Rottingdean flood.

On-farm Costs

Costs incurred by two of the three farmers were minimal. The two farmers at Hove each estimated that 20 ha of their land suffered noticeable erosion, at a cost of £700 to one and £350 to the other. In contrast, the farmer at Rottingdean, where both erosion and crop losses caused by the construction of emergency dams and trenches were far greater, estimated that his losses were much higher, amounting to £13 000 in total. The greatest component of this loss, £10 000 was for the loss of seed and crop from a total of 80 ha of land that he estimated had suffered from erosion. Of the combined area of 121 ha that suffered erosion on the three farms, 113 ha were recently sown seedbeds, all of which suffered rill erosion, and the remaining 8 ha were a grass ley which suffered from surface wash.

All three farmers believed the major cause of erosion to be the exceptional volume and intensity of rainfall in early October 1987. However, each acknowledged that their cultivation of the land was also a factor. In future years, two of the farmers intend to reduce the area they sow to winter cereals and to avoid autumn rolling. One of these, the farmer at Rottingdean, has agreed a 5-year management plan with his landlords, Brighton Borough Council, which includes a reduction of approximately 50 per cent in the annual acreage of autumn cereals (Robinson *et al.*, 1988). Both farmers have also lost a few hectares of productive land for the construction of protective dams.

The third farmer, most of whose land lies within the South Downs Environmentally Sensitive Area, has decided to benefit from the special grants available to farmers in these areas and convert the whole of his farm to a low-intensity beef- and sheep-rearing enterprise. However, this decision was made for reasons of personal preference prior to the erosion of October 1987, and not as a consequence of it.

DISCUSSION AND CONCLUSIONS

Flooding of suburban housing by erosive runoff from arable farmland on the periphery of Brighton and Hove in October 1987 caused substantial off-farm damage. Total off-farm costs attributable to flooding at the four major sites examined in this chapter are in excess of £660 000. These costs exclude those of the police and fire services. There were also several other sites where more minor flooding occurred (Figure 24.1). The majority of the costs were borne by the local authorities and insurance companies, but uninsured financial losses that had to be borne directly by the householders amounted to at least £112 500. This figure does not include several hidden costs such as free household labour and the loss of no-claims bonuses by people who had to claim for damage to vehicles (worth several hundred pounds over the next 2–4 years). The most common costs that had to be borne by householders were for reinstatement of their gardens, but several households also suffered large losses due to a lack of, or inadequate, home-contents insurance. The more elderly households tended to suffer the greater uninsured losses because they often had to pay labour charges for the reinstatement of their gardens, while younger, fitter households used their own, unpaid labour. Several householders at Rottingdean considered trying to claim their uninsured losses against the farmer, but no proceedings have yet commenced.

Local authority costs were split between emergency flood alleviation work, and longer-term protective measures designed to reduce future dangers of repeat flooding. Although the local authorities are aware that land use on the Downs is a major contributory factor to their flooding problems (Furneaux, 1977; Boardman et al., 1983; Robinson and Williams, 1987), they have adopted solutions that are essentially protective rather than preventative. Although some modification to arable practice has been voluntarily agreed at Rottingdean, the local authorities have chosen the engineering solution of building dams and improving drainage, rather than demanding changes in land use to reduce, or stop, runoff occurring. Essentially, they see the problem as one of controlling and disposing of the runoff, rather than of solving the problem of its origin. As a result, the costs of protection are to be levied, through the rates, on householders, some of whom have suffered from the flooding, and not borne by the farmers, who are largely responsible for causing the floods. In part this is because, under existing freehold or farm-tenancy agreements, it is difficult for landlords, or local authorities, to impose changes in land use or cultivation practice on farmers without their co-operation. Most farmers grow winter cereals because these are currently the most profitable crop, and they would expect to be financially compensated for any changes in cropping practice that lessen the risk of erosion but also give lower financial returns (Funnell and Blackman, 1987).

Financial losses by farmers were small. Even at Rottingdean, where soil erosion was extreme, the total loss by the farmer was considerably less than the individual cost of damage to several households, and only 52 per cent of the cost to the most severely flooded household. In Hove, the differential between the very small on-farm costs and the cost of damage to the worst-affected households was even greater. The losses to farmers exclude the value of the soil lost. Farmers do not seem to consider this a major loss, yet at Rottingdean, where the average depth of soil is only 200–300 mm, it is estimated that one field lost 10 per cent of its soil (Boardman, 1990), and this may feed back into lower yields, especially if erosion were to recur (Evans and Nortcliff, 1978; Evans, 1981).

Pressure for land-use changes to lessen the risk of flooding will undoubtedly have to come from the urban areas. It is here that the greatest costs were incurred, and press and publicity has convinced residents that farm practices were largely to blame for the flooding. However, although householders expect these changes, they also appear to expect the local authorities to translate their demands into action. The local authorities, meanwhile, appear to see their role restricted to their legal responsibilities as highway and drainage authorities, attempting to control and dispose of water entering the urban area so that roads and houses are not flooded.

Pressure for change will probably have to come from claims through the courts, by householders (or their insurers) against the farmers, for the costs of flood damage. The threat of this at Rottingdean appears to have faded, but repeat flooding would undoubtedly strengthen resolve and lessen the farmers' defence that erosion is the result of exceptional rainfall and is, as such, an 'act of God', for which they are deemed not liable. Ironically, a test case of this nature is currently going through the courts, not between an urban householder and a farmer but between two farmers, one of whom has a house and vineyard that was inundated several times in autumn 1987, and which has also been inundated in previous years by erosive runoff from an adjacent farmer's land.

ACKNOWLEDGEMENTS

The research was undertaken while Jonathan Blackman was supported by a NERC studentship awarded under the 'Agriculture and Environment' Special Topic. We are grateful to all the residents, farmers and council officers who willingly provided the information on which this chapter is based. The maps and diagrams were drawn by Sue Rowland.

REFERENCES

Boardman, J. (1983). Soil erosion at Albourne, West Sussex, England, *Applied Geography*, 3, 317–29.
Boardman, J. (1984). Erosion on the South Downs, *Soil and Water*, 12(1), 19–21.
Boardman, J. (1986). The context of soil erosion, *SEESOIL*, 3, 2–13.
Boardman, J. (1988a). Public policy and soil erosion in Britain. In Hooke, J. M. (ed.), *Geomorphology in Environmental Planning*, John Wiley, Chichester, pp. 33–50.
Boardman, J. (1988b). Severe erosion on agricultural land in East Sussex, U.K., October 1987, *Soil Technology*, 1, 333–348.
Boardman, J. (1990). Soil erosion on the South Downs: a review. In Boardman, J., Foster, I. D. L. and Dearing, J. A. (eds), *Soil Erosion on Agricultural Land*, John Wiley, Chichester, pp. 87–105.
Boardman, J. and Robinson, D. A. (1985). Soil erosion, climatic vagary and agricultural change on the Downs around Lewes and Brighton, autumn 1982, *Applied Geography*, 5, 243–58.
Boardman, J. and Spivey, D. (1987). Flooding and erosion in west Derbyshire, April 1983, *East Midland Geographer*, 10, 36–44.
Boardman, J., Stammers, R. L. and Chestney, D. (1983). Flooding problems at Highdown, Lewes: technical report, *Report to Lewes District Council*.
Colborne, G. J. N. and Staines, S. J. (1986). Soil erosion in Somerset and Dorset, *SEESOIL*, 3, 62–71.
Director of Technical Services, Brighton (1987). *Flooding and Soil Erosion at Rottingdean and Woodingdean—October 1987*, Brighton Borough Council, 6 pp. and map.
Director of Technical Services, Hove (1987). *Land Drainage Problems—Mile Oak/Hangleton Area*, Hove Borough Council, Highways and Works Committee, MIS261.220, 4 pp.
Evans, R. (1971). The need for soil conservation, *Area*, 3, 20–23.

Evans, R. (1980). Characteristics of water eroded fields in lowland England. In de Boodt, M. and Gabriels, D. (eds), *Assessment of Erosion*, John Wiley, Chichester, pp. 77–87.

Evans, R. (1981). Assessment of soil erosion and peat wastage for parts of East Anglia, England. A field visit. In Morgan, R. P. C. (ed.), *Soil Conservation: Problems and Prospects*, John Wiley, Chichester, pp. 521–30.

Evans, R. and Nortcliff, S. (1978). Soil erosion in north Norfolk, *Journal of Agricultural Science, Cambridge*, **90**, 185–92.

Evans, R. and Skinner, R. J. (1987) A survey of water erosion, *Soil and Water*, **15**(1/2), 28–31.

Frost, C. A. and Speirs, R. B. (1984). Water erosion in soils in south-east Scotland—a case study, *Research and Development in Agriculture*, **1**, 145–52.

Fullen, M. A. (1985a). Erosion of arable soils in Britain, *International Journal of Environmental Studies*, **25**, 55–69.

Fullen, M. A. (1985b). Compaction, hydrological processes and soil erosion on loamy sands in East Shropshire, England, *Soil and Tillage Research*, **6**, 17–29.

Fullen, M. A. and Reed, A. H. (1986). Rainfall, runoff and erosion on bare arable soils in East Shropshire, England, *Earth Surface Processes and Landforms*, **11**, 413–25.

Fullen, M. A. and Reed, A. H. (1987). Rill erosion on arable loamy sands in the West Midlands of England. In Bryan, R. B. (ed.), *Rill Erosion: Processes and Significance, Catena Supplement*, **8**, Catena Verlag, Cremlingen, pp. 85–96.

Funnell, D. C. and Blackman, J. D. (1987). An economic analysis of protection against flooding and erosion. In Robinson, D. A. and Williams, R. B. G. (eds), *Flooding and Soil Erosion at Rottingdean, October 1987*, Final Report to Brighton Borough Council, pp. 18–25.

Furneaux, A. A. S. (1977). *Report on the Flooding Problem at Mile Oak, near Portslade, Sussex*, Borough of Hove, A.R.364.

Jackson, S. J. (1986). Soil erosion survey and erosion risk in mid-Bedfordshire, *SEESOIL*, **3**, 95–105.

Morgan, R. P. C. (1977). Soil erosion in the United Kingdom: field studies in the Silsoe area, 1973–75, *National College of Agricultural Engineering, Silsoe Occasional Paper*, **4**.

Morgan, R. P. C. (1980). Soil erosion and conservation in Britain, *Progress in Physical Geography*, **4**, 24–47.

Reed, A. H. (1979). Accelerated erosion of arable soils in the United Kingdom by rainfall and run-off, *Outlook on Agriculture*, **10**, 41–8.

Reed, A. H. (1983). The erosion risk of compaction, *Soil and Water*, **11**(3), 29–33.

Robinson, D. A. and Blackman, J. D. (1989). Soil erosion, soil conservation and agricultural policy for arable land in the United Kingdom, *Geoforum*, **20**, 83–92.

Robinson, D. A. and Williams, R. B. G. (1987). Physical analysis of the flood event and likely recurrence. In Robinson, D. A. and Williams, R. G. B. (eds), *Flooding and Soil Erosion at Rottingdean, October 1987*, Final Report to Brighton Borough Council, pp. 1–17.

Robinson, D. A. and Williams, R. B. G. (1988) Making waves in downland Britain, *Geographical Magazine*, **60**(10), 40–45.

Robinson, D. A., Williams, R. B. G. and Blackman, J. D. (1988). *Estimates of Runoff and Comments on the Proposed Cropping Plan at New Barn Farm, Rottingdean*, Report to Brighton Borough Council, 12 pp.

Speirs, R. B. and Frost, C. A. (1987). Soil water erosion on arable land in the United Kingdom, *Research and Development in Agriculture*, **4**, 1–11.

Stammers, R. L. and Boardman, J. (1984). Soil erosion and flooding on downland areas, *The Surveyor*, **164**, 8–11.

25 An approach to the Assessment of Erosion Forms and Erosion Risk on Agricultural Land in the Northern Paris Basin, France

A. V. AUZET *CNRS, Strasbourg*, J. BOIFFIN *INRA, Laon*, F. PAPY *INRA, Thiverval-Grignon*, J. MAUCORPS *INRA, Olivet*, and J. F. OUVRY *AREAS, St Valery-en-Caux*

INTRODUCTION

As in other loamy regions of north-west Europe (De Ploey, 1986; Chisci and Morgan, 1986), there is an increasing awareness of erosional phenomena and associated muddy floods in the northern part of the Paris Basin in France (Lefevre, 1958; Maucorps, 1982; Monnier *et al.*, 1986; Auzet, 1987a,b). These phenomena occur despite the moderate topographic and climatic conditions (Pihan, 1979; Gabriels *et al.*, 1988).

An important proportion of the damage related to the erosion is the off-site effect. Here mudflows and muddy floods originating on agricultural fields enter urbanized areas which are situated in adjacent valley bottoms or at the outlet from usually dry catchments (Cemagref, 1986; Papy and Douyer, 1988). In the fields, the inconvenience caused to farming operations due to gullies and the burial of crops should now be considered as a more important factor than the soil loss itself (Papy and Boiffin, 1988b).

In spite of the apparent uniformity of agricultural landscapes, the sites with erosion risks, which are known in the Northern Paris Basin, illustrate a range of erosion forms and processes (Ouvry, 1982; Boiffin *et al.*, 1986; Auzet, 1987b; Maucorps, 1987; Wicherek, 1988).

Our objectives are to identify and to classify the sites in relation to the nature and importance of risks. This approach concerns catchments without a permanent channel and where the valley bottom is included in agricultural land.

The different erosional forms, which can take place in such catchments, imply different kinds of risks and the technical measures applicable to prevent these are not necessarily the same.

Of course, the nature and the importance of the risks also depends on the characteristics of receiving zones including houses, roads and rivers. Nevertheless, to restrict their almost unlimited variability, we confine the assessment of potential risk to characteristics of systems producing runoff and sediments. As a first step, we will concentrate our investigations on first-order basins, as defined by the dry valley network (and not by the stream network). Areas of the loamy parts of the Northern Paris Basin with such elementary basins vary in size, from a few hundred square metres to several square kilometres.

The concepts presented in this chapter must be considered as working hypotheses which seem reasonable in the present state of our knowledge. These aim at initiating a discussion on the following points:

(1) How to take account of the diversity of erosional forms.
(2) Which are the landscape features to take into account as relevant to an assessment of the risk of erosion.

TYPOLOGY OF EROSIONAL SYSTEMS

Observed erosional forms are distinguished by:

(1) The diffuse or concentrated aspect of the sediment sources;
(2) The density of rills and gullies whose spacing could be of the order of one metre in certain cases and of several hundred metres in others;
(3) The location of rills and gullies on hillslopes or in dry valley floors;
(4) The date of formation of these different erosional forms: there are two main periods in the Northern Paris Basin, one at the end of the autumn and the other during the spring (Boiffin et al., 1988b; Ouvry, 1987; Papy and Douyer, 1988).

These different forms reflect different combinations of basic processes of soil particle detachment and transport which are greatly influenced by their topographic situation.

In the initiation phase of erosional forms, detachment takes place through two kinds of agents:

(1) Raindrops through impact and also by causing the moistening of clods and aggregates (Poesen, 1985; Boiffin, 1984; Le Bissonnais, 1988); the spatial distribution of such effects is diffuse;

(2) Runoff through shear stress exerted on the soil surface; this stress depends on hydraulic characteristics of the flow and on sediment load deficit in relation to the transport capacity (Savat, 1979; Govers and Rauws, 1986; Rauws, 1987; Rauws and Govers, 1988); these effects of detachment by runoff are most frequently found as linear forms.

During the evolution of these linear erosional forms sediment production from the sidewalls of channels is also mainly dependent on the mechanical forces exerted by the flow, and on the mechanical and hydrodynamic properties of the soil profile itself, which can produce mass movement, soil fall or shallow sliding (Govers, 1987). In the absence of wind erosion, sediment transport over significant distances is due only to runoff.

Soil particles detached by raindrops on interrill areas can be transported to the rills and then evacuated downhill more or less efficiently. The contribution of interrill areas to the total soil loss is expected to decrease as rill spacing increases. From results on the Huldenberg experimental field, with a few metre rill spacing, Govers (1987) mentions that this contribution could correspond to 30 per cent of the sediment travelling in the rill, which seems a maximum in these kind of regions.

It is possible to classify the various forms of erosion with respect to soil particle detachment conditions on the slopes and in the valley floor (Figure 25.1). It is assumed that the plan form of slopes does not induce flow convergence.

We will consider the rain and runoff effects on both these land facets according to a limited number of possibilities:

(1) Two conditions of detachment (yes–no) resulting from raindrops; this effect only operates for the slopes; the valley floor area will be considered as negligible in comparison with the slope area;
(2) Three states concerning the runoff: no runoff, transport without detachment, transport and detachment (depending on the shear velocity).

Each compartment of the table corresponds to what we call an 'erosional system', since each of them relates to a specific combination of elementary processes with a specific result. These systems correspond to the different erosional forms described in the literature.

Sheet erosion: soil particle detachment is mainly due to the conditions of contact between soil and water (Le Bissonnais, 1988); runoff remains diffuse, able to transport the detached particles but unable to initiate incision. In the Northern Paris Basin such diffuse forms are observed in isolation and only on gentle slopes whose plan form does not favour flow convergence even in medium or high-intensity rain.

Rill–interrill erosion on slopes, described by Thornes (1980) among others: some flow lines on slopes are incised; the rill density and the contribution of

386   Soil Erosion on Agricultural Land

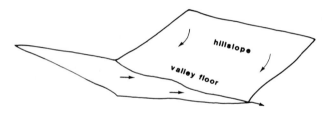

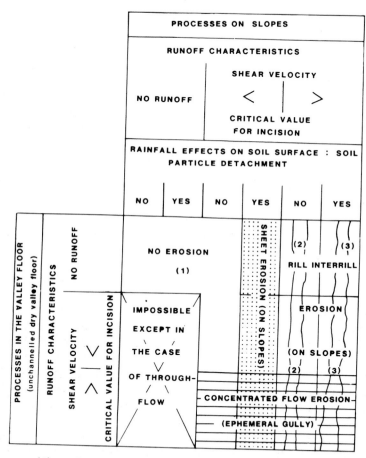

(1) crusting : soil particle displacement limited to ≃ 10 cm
(2) the interrill area is not a sediment source
(3) the interrill area is a sediment source

Figure 25.1 Erosional forms in relation to the nature of soil detachment processes and runoff characteristics

soil particle detachment on interrill areas vary. Transport to the rills is by splash and/or sheet runoff. In the rills, detachment is mainly due to the hydraulic shear stress exerted by the runoff on the bed and other mechanisms acting on sidewalls; transport is by concentrated flow. Such forms are common in Northern Europe as shown by photographs published by Boardman (1984), Fullen and Reed (1987) and Govers (1987), but do not seem dominant in the loamy parts of north-west France, the exception being where the loess cover is mixed with the tertiary sands, as, for example, on the hillslopes of Laonnois (Morand and Wicherek, 1985) and on the north-east side of Ile-de-France, when the slopes are fairly steep (Figure 25.2).

Concentrated flow erosion in the valley floor, also described by Foster *et al.* (1985) and by Spomer and Hjelmfelt (1986): detachment is mainly due to the shear stress exerted by runoff concentrated in the floor of valleys or hollows without a permanent channel and to various mechanisms affecting the sidewalls after the initiation of the incision. All the downhill transport is due to concentrated flow in the ephemeral channel whose location is determined by the topography. Such incisions in valley floors are particularly well represented on the loamy plateaus (Figure 25.3), even when the slope gradient remains less than the 3° mentioned by several authors as a critical threshold for rill initiation in the loamy lands of north-west Europe (Savat and De Ploey, 1982). Such erosional forms produce considerable off-site damage.

Figure 25.2 Rill–interrill erosion, Faucoucourt, Aisne, March 1981

Figure 25.3 Concentrated flow erosion, Saint-Rémy au Bois, Pas de Calais, April 1989

Erosion by breaking of hedgebanks was recently mentioned and accorded particular importance in Middle Belgium (Poesen, 1988). This fourth type of erosional form seems distinct from the three previous. Up to now, only a few cases were described in France, but information suggests that they are frequent enough and can attain impressive proportions (IGN, 1984). This particular form could be in addition to the others: we have not yet enough data to identify the main generating processes.

Assessment of Erosion Forms and Erosion Risks   389

The key to the erosional systems typology developed here is based on the presence or absence of soil detachment on slopes and in the valley floor. It remains to describe which landscape characteristics could determine such conditions.

CHARACTERISTICS OF AGRICULTURAL LANDSCAPES WHICH INFLUENCE EROSIONAL FORMS

The characteristics of a catchment liable to influence surficial water and sediment fluxes and the nature of erosional forms may be divided into three groups:

(1) Catchment morphology;
(2) Distribution of the soils and of their surface states (e.g. crusting);
(3) Land use on farmland related to the agricultural system in operation.

Morphological Characteristics

The morphological character of the landscape particularly influences the processes of soil particle detachment by runoff. It is essential for the initiation of rills to acquire critical shear velocity (Kirkby, 1980; Foster *et al.*, 1981; Rauws, 1987; Rauws and Govers, 1988) and this condition is dependent on several factors:

(1) The slope gradient, which increases the flow velocity;
(2) The runoff contributing area, in particular the slope length in the case of sheetflow, but also the variable area and character of the contributing zones in the case of concentrated runoff;
(3) The runoff route which influences the concentration time;
(4) The runoff convergence which increases the hydraulic radius and thus the flow velocity.

Considering these items, it is possible to assess the influence of the morphology through a limited number of criteria, as shown in Table 25.1.

Table 25.1 Some criteria related to the morphological parameters

	Sheet erosion	Rill–Interrill erosion	Concentrated flow erosion
Shear velocity >critical value for rill incision		Slope gradient	Valley floor slope gradient
			Runoff route
		Slope length	Contributing area/slope distribution along the valley floor
		Plan form	Topographic profile across valley floor

Morphology allows us to assess the possibility of occurrence of one erosional form rather than another in given climatic conditions. Morphological features give the main potential flow lines (convergence, divergence). In addition, these features make up the structure which is a guide for the distribution of influences of other types of parameters. This is particularly clear for concentrated flow erosion where channels, if they occur, are located each year more or less at the same place.

Pedological Characteristics

The influence of soil properties cannot be confined to the erodibility concept, in the sense that this notion is traditionally used in the USLE (Römkens *et al.*, 1987). In fact, the consequences of the presence of a given soil at a given point of a catchment could appear in a completely different part of the catchment. For example, runoff formation on crusted soils in the upper part of basins is often the main cause of gullying in valley floors. Therefore it is advisable to distinguish the influence of the soil properties on the different kinds of processes acting on erosion.

Towards Soil Particle Detachment

Under rainfall, particle detachment is the result of different mechanisms which break down soil and aggregates. This depends on their soil moisture status at the time of rain (Boiffin, 1984; Le Bissonnais, 1988). Slaking may result from air entrapped in soil pores escaping at the moment of contact between free water and unsaturated aggregates. Also, peripheral wear of clods may occur by raindrop impact, and microcraking may be due to swelling and shrinking caused by alternation of drying and wetting. Dispersion occurs for sodic soils which are uncommon in the Northern Paris Basin.

For soils with a clay content under 30 per cent, widely distributed in the regions under consideration, the first two mechanisms (slaking and peripheral wear) are dominant. Thus, the soil characteristics to take into account are clay content and, to a lesser extent, type and content of the organic matter (Monnier, 1965) and calcium carbonate status (Meriaux, 1958, 1961).

Through the action of runoff, the two stages of rill initiation and evolution must be distinguished. During the first the surface parameters have the main influence. During the second (mass-movement), soil fall and sliding affecting the sidewalls give more emphasis to the properties of the different layers of the soil profile below the plough layer. In both cases susceptibility to detachment depends on mechanical properties, relating to texture and water content. These characteristics greatly influence the relation between cohesion and soil moisture. Nevertheless, the role of the structural state in the plough layer is also very important (Lyle and Smerdon, 1965; Rauws, 1987; Rauws and Auzet, 1989). Considering the variability of compaction in relation to farming practices (e.g. wheeltracks), the role of texture is not dominant.

Towards Runoff Formation

Infiltration capacity and surface roughness are the main parameters to consider concerning the state of the soil surface with respect to runoff formation. These parameters depend on the structural state of surface layers as controlled by the interaction of rainfall and farming operations. Through soil crusting, rainfall leads to a decrease in the porosity, infiltration capacity (Table 25.2) and surface roughness of the soil (Onstad, 1984; Boiffin *et al.*, 1988a; Ouvry, 1987; King *et al.*, 1988).

In a given climatic situation, soil composition mainly influences the rate at which both parameters will reach the threshold values which generate the highest probability of runoff. For similar initial structural states, the cumulative rainfall at the moment when runoff occurs could differ by several hundreds of millimetres for soils with various compositions. Runoff risk is thus much higher in the case of soils with low aggregate stability which can quickly become capable of producing runoff.

The limiting values reached for infiltration capacity on crusted surfaces and within wheeltracks correspond approximately to the saturated hydraulic conductivity resulting from textural porosity (Fies and Stengel, 1981). These limiting values would seem to decrease when the clay content increases. For this reason, the role of the clay fraction is partly ambivalent.

Table 25.2 Probable range of wet soil infiltration capacity in relation to the facies during soil crusting process (after Boiffin *et al.*, 1988a)

Facies	Range of wet loamy soil-Infiltration capacity (mm/h)
Initial fragmentary structure	30–50
Altered fragmentary state with structural crusts	5–30
Transitional: local appearance of depositional crusts	2–5
Continuous state with depositional crusts	1–2
Disturbed depositional crusts (climatic and biological factors)	>5

Conclusion on the Effect of Soil Properties

In the range of soils whose ionic composition and organic status are not too variable, as is the case in the Northern Paris Basin, the assessment of erosion susceptibility could be based on textural characteristics. However, the knowledge of the links between relevant properties and characteristics of composition remains incomplete or at least insufficiently quantified. Particularly, the influence of the components other than clay still remains not properly defined. In any case, soil properties do not exert the same influence on different types of erosional forms (Table 25.3). The role that texture plays in influencing the

Table 25.3 Some criteria related to the soil properties

Soil properties	Sheet erosion	Rill–interrill erosion	Concentrated flow erosion
Susceptibility to raindrop detachment on the slopes	X	X	
Susceptibility to runoff incision on the slopes		X	
Susceptibility to runoff incision on the valley floor			X
Susceptibility of the soil profile to mass movement		X	X
Rates of change and limiting values of infiltration capacity on runoff-contributing area	X	X	X

soil's susceptibility to raindrop detachment, to runoff detachment and the risk of runoff production is probably not the same.

Agricultural systems

Characteristics to take into account

Agricultural systems, i.e. the way the rural environment is used by society, produce in the landscape some features whose effects on water and sediment transfer can be divided into three basic groups:

(1) The state of the soil surface depends on land use (e.g. woodland, grassland, arable land) and on the cropping systems applied to arable land. The changing character of the plough layer, and particularly its surface, in relation to the timing of farming operations and rainfall, should also be emphasized. Thus, infiltration capacity and water detention relate to surface crusting and roughness changes during crop rotation, as illustrated by Figure 25.4 and tables 25.4 and 25.5. The timing of rainfall and farming operations fix that of the maximum risk for runoff. The soil surface therefore acquires an important role for particle detachment in relation to the timing of particular farming operations (Imeson and Kwaad, 1990).

(2) The spatial distribution of the state of the soil surface. In the landscape the division into fields creates spatial discontinuities between states of the soil surface. Effects of upslope fields on downslope fields are more or less important and more or less frequent according to field shapes and sizes, location on hillslopes and, in the valley floor, kinds of boundaries (Papy and Boiffin, 1988b).

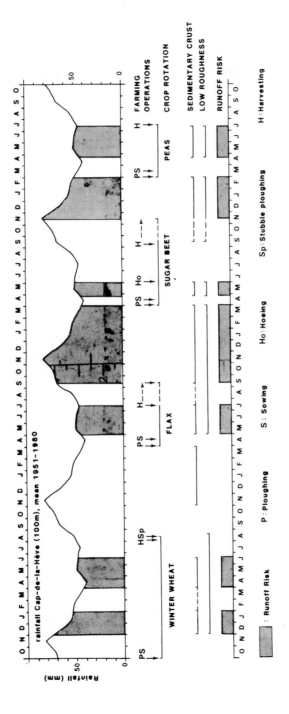

Figure 25.4 Risk of runoff in relation to crop rotation and climatic conditions

Table 25.4 Surface roughness in relation to the last farming operation (after Papy and Boiffin, 1988a)

Roughness grade	1	2	3	4
Situation	Strongly crusted sown fields; heavy compacted harvest	Slightly crusted sown fields; sown fields with wheeltracks on the soil surface	Sown fields with cloddy surface; stubble-ploughed without crop residues	Ploughed fields; stubble-ploughed with abundant residues

Table 25.5 Surface occupied by wheeltracks, in relation to the last farming operation (after Papy and Boiffin, 1988a)

	Percentage of surface	0%	5%	15–35%	70–80%
Wheel Tracks on soil surface	Situation	Tilled surface	Sowing of cereals, peas, flax	Sowing of sugarbeet, corn; planting potatoes	Sugarbeet and potato harvest

(3) Water and sediment pathways and storage. Man-made structures (hedges, banks, ponds, ditches, roads) can enhance or decrease the fluxes, depending on their location in relation to the slopes.

Thus the effects of agricultural systems appear quite complex. A given state of the soil surface could be considered as favourable or unfavourable, depending on its location in relation to the morphology. A hedge could have opposite effects according to its orientation either on a hillslope or in the valley floor. It is in relation to the erosional forms induced first by the morphology that we should assess the effects of characteristics due to the agricultural system (Table 25.6).

Characteristics of agricultural systems and erosional forms

(1) *Influence of agricultural systems on hillslope erosion (sheet erosion, rill-interrill erosion).* It is for such erosional forms that the USLE was designed (Wischmeier and Smith, 1965). The surface cover reduces the erosion risk: surface organic residues protect the soil against raindrop impact, stalks slow down and divide the flow, roots increase the resistance of surface layers. Crop residues could have partly the same role. The main criteria taken into account will be the percentage of the surface with a permanent cover (woodland, permanent grassland) and the percentage of arable land which will be affected by a rain-erosivity index (deduced from the rainfall depth and intensity) calculated for the time of uncovered or partially uncovered soil surface. The resistance to overland flow friction increases with increasing

Table 25.6 Some criteria related to the agricultural system. (ΣPmm is considered here as a first approximation for a climatic index)

Erosional forms Concerned domain	Sheet erosion Hillslopes	Rill-interrill erosion Hillslopes	Concentrated flow erosion Hillslopes	Concentrated flow erosion Valley floor
Properties				
Aptitude of runoff (infiltration capacity, roughness)	X	X	X	
Canopy protecting soil against raindrop impact	X	X		
Roots giving resistance to runoff incision to the soil		X	X	X
Criteria				
Forest	Areas of hillslopes (%)	Areas of hillslopes (%)	Areas of hillslopes (%)	Length of the valley floor (%)
Permanent grass	Areas of hillslopes (%)	Areas of hillslopes (%)	Areas of hillslopes (%)	Length of the valley floor (%)
Arable land	ΣPmm On a bare soil with low infiltration capacity and low roughness	ΣPmm On a bare soil with low infiltration capacity and low roughness	ΣPmm On a bare soil with low infiltration capacity and low roughness	ΣPmm On soil covered by crop with roots giving resistance to runoff incision

surface roughness: the contribution of form friction due to the presence of clods and residues seems particularly important to gentle slopes (Rauws, 1988). Therefore farming operations have an important impact on this parameter (Table 25.6).

(2) *The influence of agricultural systems on valley floor erosion.* The role of various catchment areas taking into account their function as runoff-contributing areas and/or sediment sources plus their position in space needs to be differentiated. The total assessment will result in a combination of criteria characterizing each of these main domains. On hillslopes, one should give priority to consideration of the evolution of infiltration capacity and surface detention resulting from the timing of farming operations under defined climatic conditions. Boiffin *et al.* (1988a, b) proposed a synthesizing approach to assessment of runoff risk in a field due to the cropping system. They used the depth of cumulative rain during the time of low infiltration

capacity and low roughness of the soil surface. In this way, periods with major risk of runoff should be very different from those with a protected soil surface (Figure 25.3). Therefore we will introduce an index of runoff risk distinct from the index expressing the resistance of the soil to particle detachment due to vegetation cover effects. On hillslopes the probable runoff pathways and the concentration times are influenced by the number and direction of linear landscape elements able to collect and route the flows in respect to the slope morphology. The main linear patterns are wheeltracks, and in certain cases the headlands, field boundaries, paths or roads.

In valley floors, the soil surface susceptibility to incision depends on two factors: the structural state of arable land and the presence of permanent cover (permanent grass, forest). Therefore, one should consider the influence of different land uses. For example, the risk decreases with the presence of permanent grass while it increases with arable land use. In this case, slight differences will be introduced in relation to the frequency of farming operations which break the soil surface and the time when the soil remains relatively uncovered. Moreover, field boundaries can constitute a factor able to diminish the risk of incision because they increase the chance of having various states of soil surface, infiltration capacities and roughness.

CONCLUSION AND PROSPECTS FOR RESEARCH

This approach is based on the following concept: in the context of regions such as the Northern Paris Basin erosion risk should be assessed at the scale of small spatial units appropriate to hydrological and sediment transfer principles: i.e.

(1) Small catchments at the level of lowest complexity;
(2) Composite catchments which are potential contributing areas for runoff and sediment sources for downhill zones.

Depending on the existence of several erosional forms, corresponding to various combinations of elementary processes of soil particle detachment and transfer, we propose to assess the susceptibility to erosion in relation to each of these erosional forms, first for small catchments and then for rather complex spatial units.

An initial sample of first-order basins (defined by the dry valley network) including only one segment of valley floor has been selected in different parts of the Northern Paris Basin. Working on this sample, we are now surveying various criteria and will initiate a systematic survey of erosion features and their locations in catchments.

Tables 25.1, 25.2 and 25.6 show how the different criteria will be combined for each of the three main erosional forms described. The catchment-classification system we are considering in terms of erosion risk is based on a score for each type of erosional form and then multi-criteria scoring for each catchment which will be compared with the occurrence, and importance, of observed erosional forms. Composite criteria will be used to characterize the spatial distribution in respect to the two main process responses, that of runoff formation and sediment production. For example:

(1) Percentage of soils in a state susceptible to rill incision on steep slopes (>3°);
(2) Percentage of valley floor occupied by cultivated fields;
(3) Percentage of the runoff contributing area, where the soil has a low aggregate stability.

The score will be referred to a climatic sequence from the end of winter and to a climatic sequence from the beginning of summer. Both climatic sequences correspond to the main risk time which clearly appears through observations of erosion phenomena in the Northern Paris Basin (Papy and Douyer, 1988; Cemagref, 1986; Ouvry, 1987).

This first classification will have implications for the conservation measures which may be implemented. For example, on steep slopes (>3° approx.), where small areas are able to produce enough runoff to attain critical shear velocities for rill incision, no-till or minimum tillage should be promoted (Wischmeier and Smith, 1965; De Ploey, 1988). In the situation where the incision risk concerns especially the valley floors, the runoff-contributing areas and the runoff concentration zone are sufficiently separate in space to justify different farming techniques which aim at limiting runoff formation on the contributing areas and at increasing soil compaction in the concentration zone (Monnier and Boiffin, 1986; Boiffin et al., 1988b; Ouvry, in press).

Nevertheless, we have still to define a combination which allows the building of a combined score for each site. As yet, little information exists to indicate a direction. There are two critical questions which must be considered:

(1) How do we combine in a score the interactions between the different types of erosional forms?
(2) How do we validate the classification in relation to different probable sequences of climatic conditions?

Such an empirical approach includes results of research concerning the deterministic relations between detachment and transport processes and erosional forms. If this approach is able to identify possible thresholds, direction of relationships and hierarchies between the main parameters, it would be possible to give a framework for physically based modelling, taking into account both hydrological and erosional processes.

An important aim of the classification is to recommend conservation measures and, especially, to define the measures suitable for the local conditions in the Northern Paris Basin. Often the validity of conservation measures is still not properly defined because they do not refer to the geography and timing of hydrological and erosional processes in the catchments. Such an approach could allow the identification of sensitive sites during land-consolidation operations, so defining the areas where farmers need to be aware of erosion problems and, if the occasion arises, to indicate means of taking land out of production.

ACKNOWLEDGEMENTS

The authors are grateful to the members of the Laboratory of Experimental Geomorphology from Leuven (Belgium) for helpful discussions. We also wish to thank the Laboratory of Biogeography and Ecology (Ecole Normale Supérieure de Fontenay-St Cloud) for collaborating in the programme in the Northern Paris Basin.

REFERENCES

Auzet, A. V. (1987a). L'érosion des sols cultivés en France sous l'action du ruissellement, *Annales de Géographie*, **537**, 529–56.
Auzet, A. V. (1987b). *L'érosion des sols par l'eau dans les régions de grande culture: aspects agronomiques*, Min. Env./Min. Agric., CEREG-UA 95 CNRS.
Boardman, J. (1984). Erosion on the South Downs, *Soil and Water*, **12**(1), 19–21.
Boiffin, J. (1984). *La dégradation structurale des couches superficielles du sol sous l'action des pluies*, Thesis Doct. Ing. INA-PG.
Boiffin, J., Papy, F. and Peyre, Y. (1986). *Systèmes de production, systèmes de culture et risques d'érosion dans le Pays de Caux*, Report INA-PG, INRA, Min. Agric.
Boiffin, J., Papy, F. and Eimberck, M. (1988a). Influence des systèmes de culture sur les risques d'érosion par ruissellement concentré. I.—Analyse des conditions de déclenchement de l'érosion, *Agronomie*, **8**(8), 663–73.
Boiffin, J., Papy, F. and Monnier, G. (1988b). Some reflections on the prospect of modelling the influence of cropping systems on soil erosion. In Morgan, R. P. C. and Rickson, R. J. (eds), *Erosion Assessment and Modelling*, Commission of European Communities, pp. 215–34.
CEMAGREF (1986). *Ruissellement, érosion, inondation dans le bassin du Croult (Val-d'Oise): recherches sur les causes d'aggravation*, Report CEMAGREF, Univ. Paris VII, Conseil Général du Val-d'Oise, Min. Env.
Chisci, G. and Morgan, R. P. C. (eds) (1986). *Soil Erosion in the European Community. Impact of Changing Agriculture*, Balkema, Rotterdam.
De Ploey, J. (1986). *Bodemerosie in de lage landen. Een Europees milieuprobleem*, Acco, Leuven.
De Ploey, J. (1988). No-tillage experiments in the central Belgian loess belt, *Soil Technology*, **1**, 181–4.
Fies, J. C. and Stengel, P. (1981). Densité texturale des sols naturels. II—Eléments d'interprétation. *Agronomie*, **1**(8), 659–66.
Foster, G. R., Lane, L. and Mildner, W. F. (1985). Seasonally ephemeral cropland gully erosion, *Proceedings of the ARS-SCS Natural Resources Modelling Workshop*, USDA-Agricultural Research Service, Washington, DC.
Foster, G. R., Lane, L., Nowlin, J. D., Laflen, J. M. and Young, R. A. (1981). Estimating erosion and sediment yield on field-size areas, *Transactions of the American Society of Agricultural Engineers*, **24**(5), 1253–62.

Fullen, M. A. and Reed, A. H. (1987). Rill erosion on arable loamy sands in the West Midlands of England. In Bryan, R. (ed.), *Rill Erosion*, Catena Supplement 8, pp. 85-96.

Gabriels, D., Cadron, W. and de Mey, P. (1988). Provisional rain erosivity maps of some European Community countries. In *Publication of the European Communities CD-NA-10860-EN-C*, Brussels, pp. 93-118.

Govers, G. (1987). Spatial and temporal variability in rill development processes at the Huldenberg experimental site. In Bryan, R. (ed.), *Rill Erosion*, Catena Supplement 8, 17-34.

Govers, G. and Rauws, G. (1986). Transporting capacity of overland flow on plane and on irregular beds, *Earth Surface Processes*, **11**, 515-24.

IGN (1984). *Erosion des terres agricoles. Ligescourt (Somme)*, Département de télédétection de l'Institut géographique National.

Imeson, A. C. and Kwaad, F. J. P. M. (1990). The response of tilled soils to wetting by rainfall and the dynamic character of soil erodibility. In Boardman, J., Foster, I. D. L. and Dearing, J. A. (eds), *Soil Erosion on Agricultural Land*, John Wiley, Chichester, pp. 3-13.

King, C., Maucorps, J., Aumonier, F., Renaux, B. and Lenotre, N. (1988). Indices d'érosion détectables par SPOT dans les sols limoneux du Nord/Pas-de-Calais, Report *ATP-CNRS* 88.DT010 TED, BRGM-INRA.

Kirkby, M. K. (1980). Modelling water erosion processes. In Kirkby, M. J. and Morgan, R. P. C. (eds), *Soil Erosion*, John Wiley, Chichester, 183-216.

Le Bissonnais, Y. (1988). *Analyse des mécanismes de désagrégation et de mobilisation des particules de terre sous l'action des pluies*, Thesis Université d'Orléans, INRA.

Lefevre, P. (1958). Quelques phénomènes d'érosion en Picardie, *Annales Agronomiques*, **1**, 91-129.

Lyle, W. M. and Smerdon, E. T. (1965). Relation of compaction and other properties to erosion resistance of soils, *Transactions of the American Society of Agricultural Engineers*, **8**(3), 419-22.

Maucorps, J. (1982). A note on farmland erosion in northern France. In *Soil Erosion* (abridged proceedings of the Workshop on Soil erosion and Conservation, Florence, October), Report EUR 8427, 50-52.

Maucorps, J. (1987). Estimation spatiale des risques de dégradation physique des sols dans le Nord/Pas-de-Calais, *Terres et Hommes du Nord*, **4**, 147-50.

Meriaux, S. (1958). Stabilité structurale et composition des sols, *Compte Rendu à l'Académie d'Agriculture*, 799-803.

Meriaux, S. (1961). Action du chaulage sur les composantes de la structure et sur l'état de saturation du sol, *Compte Rendu à l'Académie d'Agriculture*, 322-25.

Monnier, G. (1965). *Action des matières organiques sur la stabilité structurale des sols*, Thesis, Paris.

Monnier, G., Boiffin, J. and Papy, F. (1986). Réflexions sur l'érosion hydrique en conditions climatiques et topographiques modérées: cas des systèmes de grande culture de l'Europe de l'Ouest, *Cahiers ORSTOM, série Pédologie*, **12**(2), 123-31.

Morand, F. and Wicherek, S. (1985). Douze parcelles de mesures de l'érosion sur un versant de la france des plaines: l'exemple de Cessière (1977-1983). In Godard, A. and Rapp, A. (eds), *Processus et mesure de l'érosion*, CNRS, Paris, pp. 271-90.

Onstad, C. A. (1984). Depressional storage on tilled soil surfaces, *Transactions of the American Society of Agricultural Engineers*, **27**(3), 729-32.

Ouvry, J. F. (1982). *Localisation et description des sites d'érosion des sols agricoles du bassin inférieur de l'Yères (Seine-Maritime)*, Rapport DEA, ENSA et Université de Rennes, 72 pp. + appendices.

Ouvry, J. F. (1987). Bilan des travaux, Campagne 1986-1987, Report AREAS.

Ouvry, J. F. (in press). Effet des techniques culturales sur la susceptibilité des terrains à l'érosion par ruissellement concentré: expérience du Pays de Caux, *Cahiers ORSTOM, série Pédologie*.

Papy, F. and Boiffin, J. (1988a). Influence des systèmes de culture sur les risques d'érosion par ruissellement concentré. II.—Evaluation des possibilités de maîtrise eu phénomène dans les exploitations agricoles, *Agronomie*, **8**(9), 745–56.

Papy, F. and Boiffin, J. (1988b). The use of farming systems for the control of runoff and erosion. In Auerswald and Schwertman (eds), *Proceedings of the Workshop on Soil Erosion, Munich, 1988*, Soil Technology series 1, 29–38.

Papy, F. and Douyer, C. (1988). *Les déterminants des catastrophes liés au ruissellement des terres agricoles en Pays-de-Caux*, Report INRA-SERDA.

Pihan, J. (1979). Risques climatiques d'érosion hydrique des sols en France. In Vogt. H. and Vogt, T. (eds), *Seminar on Agricultural Soil Erosion in Temperate Non-Mediterranean Climate, Strasbourg-Colmar, September 1978*, Université Louis Pasteur, Strasbourg. pp. 13–18.

Poesen, J. (1985). An improved splash transport model, *Zeitschrift für Geomorphologie NF*, **29**(2), 193–211.

Poesen, J. (1988). Conditions for gully formation in the Belgian loam belt and some ways to control them. In Auerswald and Schwertman (eds), *Proceedings of the Workshop on soil erosion, Munich, 1988*, Soil Technology series 1, 39–52.

Rauws, G. (1987). The initiation of rills on plane beds of non-cohesive sediments. In Bryan, R. (ed.), *Rill Erosion*, Catena Supplement 8, pp. 107–18.

Rauws, G. (1988). Laboratory experiments on resistance to overland flow due to composite roughness, *Journal of Hydrology*, **103**, 37–52.

Rauws, G. and Auzet, A. V. (1989). Laboratory Experiments on the effects of simulated tractor wheelings on linear soil erosion, *Soil and Tillage Research*, **13**, 75–81.

Rauws, G. and Govers, G. (1988). Hydraulic and soil mechanical aspects of rill generation on agricultural soils, *Journal of Soil Science*, **39**, 111–24.

Römkens, M. J. M., Prasad, S. N. and Poesen, J. (1987). Soil erodibility and properties, *Transactions of the XIII. Congress of ISSC, Hamburg 1986*, 492–504.

Savat, J. (1979). Laboratory experiments on erosion and deposition of loess by laminar sheet flow and turbulent rill flow. In Vogt, H. and Vogt, T. (eds), *Seminar on Agricultural Soil Erosion in Temperate Non-Mediterranean Climate, Strasbourg-Colmar, September 1978*, Université Louis Pasteur, Strasbourg, pp. 39–43.

Savat, J. and De Ploey, J. (1982). Sheetwash and rill development by surface flow. In Bryan, R and Yair, A. (eds), *Badland Geomorphology and Piping*, Geo Books, Norwich, pp. 113–26.

Spomer, R. G. and Hjelmfelt, A. T. (1986). Concentrated flow erosion on conventional and conservation tilled watersheds, *Transactions of the American Society of Agricultural Engineers*, **29**(1), 124–7.

Thornes, J. (1980). Erosional Processes of running water and their spatial and temporal controls: a theoretical viewpoint. In Kirkby, M. J. and Morgan, R. P. C. (eds), *Soil Erosion*, John Wiley, Chichester, pp. 129–82.

Wicherek, S. (1988). Les relations entre le couvert végétal et l'érosion en climat tempéré de plaines. Ex: Cessières (Aisne, France), *Zeitschrift für Geomorphologie NF*, **32**, 339–50.

Wischmeier, W. H. and Smith, D. D. (1965). Predicting rainfall erosion losses from cropland east of Rocky Mountains, *USDA Agriculture Handbook*, 282.

26 A Rule-based Expert System Approach to Predicting Waterborne Soil Erosion

TREVOR HARRIS
Department of Geology and Geography, West Virginia University

and

JOHN BOARDMAN
Countryside Research Unit, Brighton Polytechnic

INTRODUCTION

The ability to predict the future impact of soil erosion is important for two reasons. First, erosion leads to the thinning of soils and a decline in their nutrient-holding capacity which results in declining crop yields or the necessity for increased inputs in order to maintain yields. The question of the timescale of such decline, and the problem of estimating the costs involved, are difficult to determine but are potentially amenable to a modelling approach. The on-farm impacts of erosion are clearly of greater concern in areas of already thin soils; for example, the South Downs of southern England, where soil thickness is rarely greater than 20 cm (Boardman, 1990). Second, the off-farm impacts of erosion are receiving extensive publicity in both North America and Europe. It is now widely recognized that these costs are far greater than those borne by the farmer, although precise estimates are lacking (e.g. Wall and Dickinson, 1978; Crosson, 1984; Boardman, 1986; Robinson and Blackman, 1989). The costs include property damage by soil-laden water, road clearance and watercourse pollution both by particulate matter and agricultural chemicals. In Britain, the evidence suggests that there has been a sharp increase in the number of erosion events since the mid-1970s (Evans and Cook, 1986) and that this has been accompanied by increased off-farm costs.

The increase in the erosion problem in Britain has so far generated little in the way of governmental response (Boardman, 1988). A soil-conservation policy

---

Soil Erosion on Agricultural Land
Edited by J. Boardman, I. D. L. Foster and J. A. Dearing
©1990 John Wiley & Sons Ltd

is notable by its absence. At the local level, district and county council plus water authorities and insurance companies are bearing the major costs of erosion. Thus limited resources are available to tackle the problem, and most of these are at present devoted to emergency engineering works which are treating the symptom, not the cause. In this atmosphere of short-term reaction to events there is a need for an ability to predict areas likely to erode or generate runoff. With limited resources available to the statutory authorities such predictive techniques have to be low-cost, applicable to many different areas and user-friendly.

Food surpluses in the European Community have led to the need to take agricultural land out of production. The 'Environmentally Sensitive Areas' and 'Set Aside' schemes are responses to this need. Both are suitable vehicles for erosion control in that erosion-prone land could be included in such schemes. However, for that to occur the concept of targeting would have to be introduced which requires an ability to predict land that is particularly prone to erosion. The justification for such an approach is that scarce public resources should be directed at areas where a need can be shown to exist rather than indiscriminately spread around. We suggest, therefore, that the ability to predict sites which are likely to erode, as well as the degree to which they will erode, is needed. Soil-erosion models are designed to meet this need.

Such models to date have largely been developed in the United States and have not achieved the quality of results that one would wish for. The most widely used, the Universal Soil Loss Equation (USLE), is claimed to be most accurate for medium-textured soils, where slope lengths are less than 122 m and mean slope angles are of the range 3–18 per cent (Wischmeier and Smith, 1978). The slope length constraint is especially restricting: for example, 92 per cent of the sites in the South Downs data set are greater than 122 m. The equation was used to predict long-term average soil loss for plots and these were compared with measured losses: about 53 per cent of the differences were less than 2.5 t ha^{-1} and 84 per cent were less than 12 t ha^{-1} (Wischmeier and Smith, 1978). Burwell and Kramer (1983) attempted to check the validity of USLE with long-term plot data and showed that average soil losses over 24 years for conventional and conservation tillage were 54 and 63 per cent, respectively, of those predicted by the USLE. The USLE is widely used for predicting erosion rates in fields (rather than on plots), but its accuracy in that situation is unknown. Burwell and Kramer (1983) further point out the difficulty of defining USLE factors for specific field areas; the implication being that field testing of the USLE would give differences between predicted and actual values far greater than those obtained on plots. The problems with the use of the USLE and the inadvisability of using it for purposes for which it was not designed and in areas outside the eastern United States are reviewed by Morgan (1986, p. 120).

A predictive model for the assessment of erosion risk developed by Morgan *et al.* (1984) was validated using data from 67 sites in 12 countries. Seven sites in the United Kingdom are included. For five of the sites the predicted soil loss is an order of magnitude too low; for one it is three orders too low and for the other it doubled the observed value.

Testing of the CREAMS model with British data gave better results (Morgan, 1986, p. 137). Of the seven sites, four gave mean annual observed and predicted soil loss values within the same order of magnitude; for two others the predicted value was an order of magnitude too high and for the remaining one it was an order of magnitude too low.

Williams (1985) reports limited testing of the EPIC model for three watersheds at Riesel, Texas. Mean annual sediment yield for the first watershed is accurately predicted by the model; for the second, the predicted value is 83 per cent and for the third it is 127 per cent of the measured value.

The limited testing that has been carried out on these models suggests that their use as reliable predictive tools is variable. There are also other disadvantages to their use which are relevant to a discussion of the alternative approach advocated in this chapter. In particular, the amount of data that many of the models require make them applicable only to small intensely investigated sites. Also, the demand for detailed accurate information is more easily satisfied on experimental plots than in the field. The model validation which has been carried out has largely used experimental sites and yet the results have not been impressive. A new generation of process-based models (Foster, 1990) is unlikely, in the foreseeable future, to overcome this problem.

With these problems in mind, our concern has been to develop a predictive tool which will be operative over large areas and may be used as an aid to management, for example, in identifying areas likely to suffer high rates of soil loss. Use of data gathered at the field scale means that very high levels of accuracy will not be attained but this is offset by certain advantages. This approach applies techniques from the field of artificial intelligence to an extensive data set of soil erosion sites in East Sussex.

THE AREA

The South Downs is an area which, in recent years, has been dominated by cereal farming and in particular the growing of autumn-planted ('winter') wheat and barley. Since 1982 soil erosion has become a significant environmental problem, especially as it has affected properties adjacent to agricultural fields. Rates of erosion have been monitored (Boardman, 1990) and sites of major erosion and property damage have been investigated in some detail (Stammers and Boardman, 1984; Boardman and Robinson, 1985; Boardman, 1989).

THE DATA

The data used in developing this approach were gathered during a six-year (1982-7) monitoring exercise on the South Downs. Data from sites on the Downs outside the monitored area were also included; details of the methods of data collection are described elsewhere (Boardman, 1990).

The data set used in this analysis comprised details of 334 erosion sites. For each erosion event information on twelve environmental variables or attributes were collected by field survey or from published topographic maps:

(1) Rainfall index: this characterizes the rainfall responsible for the measured soil loss. Rainfall events of 30 mm in two days are allotted a value of one; 30 mm in one day a value of two; 60 mm in two days a value of three; 60 mm in one day a value of four. Inspection of the rainfall record for the growing season gives the rainfall index for the year (Boardman, 1990);
(2) Catchment area: area (in hectares) contributing runoff to the eroded site. In most cases this area is a field surrounded by boundaries such as grass banks or hedges;
(3) Length of slope (in metres) contributing runoff;
(4) Type of erosion: valley bottom or valley side (a classification used by Evans and Cook, 1986);
(5) Maximum angle in catchment (measured in degrees);
(6) Relief of catchment: the difference (in metres) in height between highest and lowest point;
(7) Morphology: presence of convexity or crestal area (cf. Evans, 1980);
(8) Direction of drilling: down or across direction of maximum slope;
(9) Crop type; for example, winter cereal;
(10) Soil type; for example, chalky;
(11) Whether rolled after drilling;
(12) Measured soil loss ($m^3\ ha^{-1}$).

Some of the attributes listed above cannot, at present, be precisely defined for every site of erosion. Two in particular require further work. The rainfall index is difficult to apply accurately without precise information of the date of erosion which is often lacking. Soil type is also imprecisely characterized by the terms chalky, stony and sandy. Most South Downs soils are rendzinas with silt contents of 60–80 per cent but local variability is insufficiently known. Imprecision in site description will inevitably lead to some lessening of the predictive capacity of the system.

For the purpose of analysis, annual soil losses at sites where erosion was recorded were placed into one of seven classes:

Class	Amount ($m^3\ ha^{-1}$)	Class	Amount ($m^3\ ha^{-1}$)
a	0.00–0.99	e	10.00–19.99
b	1.00–1.99	f	20.00–49.99
c	2.00–4.99	g	>50.00
d	5.00–9.99		

As part of the testing procedure described below, these seven classes were amalgamated to form three, chosen to conform to low, moderate and high rates of erosion appropriate to the British context.

$0.0–1.99 \, m^3 \, ha^{-1}$ 'low'
$2.0–9.99 \, m^3 \, ha^{-1}$ 'moderate'
$> 10.00 \, m^3 \, ha^{-1}$ 'high'

It is suggested that low rates as defined above would be of little concern either in an on-farm or off-farm context, whereas high rates would certainly be a cause for concern and would require ameliorative action.

The frequency of recurrence of an annual rate of erosion for a specific site is not predicted. However, given the continuance of current land use and management practices and an assumption about future rainfall, this could be investigated.

The expert system approach described here aims to predict the annual rate of erosion for a specific site on the South Downs under definable conditions. Other predictive methods attempt to provide a long-term mean rate of erosion (e.g. USLE) or the effect of a rainfall event of a given magnitude (e.g. CREAMS).

EXPERT SYSTEMS: LEARNING FROM EXAMPLES

Expert systems, or Intelligent Knowledge Based Systems, stem from developments in the field of artificial intelligence and promise considerable potential for solving complex real-world problems (Fisher et al., 1988). Precise definitions of an expert system vary (Jackson, 1986; Hayes-Roth et al., 1983; Naylor, 1983), but they are based on logical reasoning and construct rules from a given knowledge base expressed in an IF–THEN form. Whereas traditional analysis has relied heavily upon statistical techniques for data analysis, expert systems adopt a somewhat different approach. Statistical techniques and mathematical modelling are obviously useful in database analysis but they focus almost exclusively on numeric, rather than symbolic, information. It has been suggested that these techniques capture only a small proportion of the actual knowledge hidden in a database (Parsaye, 1987). These approaches are not the only tools available for database analysis. Expert systems, which seek to uncover logical relationships within a database expressed in the form of rules rather than in the form of mathematical equations, provide alternative approaches to data analysis. Thus in the same way that people learn from past experiences, use empirical information from a range of sources and adopt a trial-and-error approach, so expert systems attempt to emulate this heuristic procedure. Many examples of these heuristic classification systems, in which an expert makes decisions and selects from a number of choices based on available criteria, exist. Some of the most well-known applications have been concerned with medical and engineering diagnosis (Buchanan and Shortliffe, 1984).

Two important provisos need to be met for the knowledge-based approach to be effective. First, expert systems operate most effectively within a local domain, as in this instance of soil-erosion prediction. Second, there is an important assumption that some underlying pattern or structure actually exists in the data and that it is non-random. If the knowledge base exhibited considerable randomness then, as with any other form of analysis, the system has little chance of extracting a reliable and effective rule base.

The expert system shell, or environment, used in this analysis was Expert Ease, which was developed at Edinburgh University and launched commercially in 1983. The shell provides the structure and the wherewithal to construct an operational expert system without recourse to time-consuming and complex computer programming. The system is of sufficiently generalized character that it can be implemented in many differing subject areas, as, for example, in finance, science, administration, engineering, medicine and even chess (Simons, 1985, p. 166). Other induction systems, such as Crystal, are now available which possess greater flexibility than that offered by Expert Ease; for example, for interfacing with commonly available database and spreadsheet packages (cf. Wallsgrove, 1988).

At the centre of an expert system is the inference engine which drives the system and acts as the interface between the user and the rule tree. Expert Ease constructs, or induces, from a number of examples a series of rules which are displayed in the form of a rule, or decision, tree. Once generated, the rule tree may be used for predictive purposes. The 334 examples detailed above comprise environmental information, the 'attributes', and their associated 'values', and erosion rates which form the 'outcome'. Knowledge representation and what the expert knows about the particular domain are here systematically codified in the form of the examples, and it is these that determine the knowledge base of the system. The expert system divides the examples in such a way as to construct relationships, or rules, between the attributes and outcomes. The system is based on Quinlan's ID3 or Interactive Dichotomizer 3 which sorts the examples by using the attribute which discriminates best between the examples. From these examples the production rules, sometimes called 'condition–action' rules or 'situation–action' rules, are generated. These rules formalize the empirical associations or patterns which exist within the data presented to the system. The rules consist of the left-hand side, or the 'premise', and the right-hand side, being the 'conclusion'. The system is forward chaining, or data driven, in that it moves from conditions which are known to be true towards conclusions which the facts allow us to establish (Jackson, 1986).

EVALUATING THE PREDICTIVE CAPABILITY OF THE EXPERT SYSTEM

During the development stage of the expert system *ad hoc* testing was undertaken. This involved generating a rule tree from all the 334 soil erosion examples and comparing the predicted results against expected values estimated by an expert in soil-erosion studies. The initial results were sufficiently promising to warrant further development of the system and a detailed evaluation of the system's predictive capabilities.

The testing procedure employed involved the extraction of a random sample of erosion sites, or examples, from the knowledge base which were subsequently tested against a rule tree generated from the remaining examples. To gauge the accuracy of the system predicted values for each of the sample sites were obtained by querying the expert system and compared with the actual recorded soil loss as measured by field survey. Such an approach has an inherent weakness in that the extraction of a sample from the data set reduces the overall number of examples from which the subsequent rule tree is generated. The optimum size of the knowledge base at this stage is unknown and the effect on system accuracy which might arise as a result of reducing the knowledge base by even a few examples could not be estimated. The expectation was that the smaller the number of examples extracted from the data set for testing, the more accurate the rule tree generated from the remaining examples would be. For this reason, two sample sizes of 5 per cent and 1.5 per cent, amounting to sample sizes of 17 and 5, respectively, were determined and tested. A number of trials were run in which a random sample was extracted from the data set and subsequently tested against a newly induced rule tree generated from the remaining examples. One important practical consideration was that each new rule tree took approximately 45 min of computing time on an IBM XT to be created.

Table 26.1 contains the results of some 15 tests carried out using a sample size of 1.5 per cent. Using a sevenfold soil-loss classification the system predicted, on average, soil loss to the exact category class in 43 per cent of all instances. Given the fine soil-loss classification employed, this is by no means a poor result. In many instances the prediction fell within an adjacent class. In view of the sevenfold soil-loss classification employed, it was felt that prediction to within one class could be considered satisfactory. At this more realistic level, the system attained 76 per cent accuracy with 57 predictions out of 75 being to within one class of the actual soil-loss class. Furthermore, of the 15 tests undertaken, three quarters achieved at least 80 per cent accuracy. Predictions to within two classes of soil loss accounted for nearly 91 per cent of all occurrences. Even at this level of accuracy the predicted results are still acceptable, for they continue to reflect, if not precise levels of soil loss, at least a good indication of the magnitude. When a less detailed classification is employed, involving three classes of low-, medium- and high-magnitude events, with thresholds at

Table 26.1 Sample of 5 (1.5 per cent): predicted to within specified classes (cumulative score, maximum = 5)

Test	Exact	1 class	2 classes	3 classes	4 classes	5 classes
1	2	4			5	
2	1	4		5		
3	2	4	5			
4	2	5				
5	3	4	5			
6	2	4			5	
7	2	4				5
8	2	3	4	5		
9	1	3	5			
10	4	5				
11	3		4		5	
12	1	4	5			
13	3	4	5			
14	2	3	5			
15	2	3	4		5	
Cum. %	43	76	91	93	98	100

Table 26.2 Sample of 17 (5 per cent): prediction to within specified classes (cumulative score, maximum = 17)

Test	Exact	1 class	2 classes	3 classes	4 classes	5 classes
1	6	12	14	17		
2	6	11	17			
3	5	11	16		17	
4	6	9	12	17		
5	4	10	16	17		
6	6	11	14		15	17
7	6	13	16			17
8	7	11	16	17		
9	6	13	14	17		
10	3	11	15	16		17
Cum. %	32	72	88	97	98	100

$2 \text{ m}^3 \text{ ha}^{-1}$ and $10 \text{ m}^3 \text{ ha}^{-1}$, almost two thirds of all predictions fall within the correct soil-loss category. Only a very small number are more than one class removed from the correct soil-loss category, and here the system has clearly failed by a considerable margin. These outliers then become the subject of scrutiny, as discussed below, to determine why the rules generated produce a prediction so at odds with the actual soil loss.

The results of testing a larger sample of 17 (5 per cent) (Table 26.2) supports the earlier concern that a lower rate of predictive accuracy might result as the knowledge base is depleted by the removal of a larger number of examples for testing purposes. It is thus possible to gauge the relative effect that the number of examples contained in the knowledge base has upon system accuracy. This is a crude estimation, because the impact is not uniform across the range of soil erosion events. This aspect is discussed more fully below. Overall, system accuracy lags behind that of the smaller 1.5 per cent sample throughout the six levels of accuracy. However, while the greatest decrease in accuracy was in predicting soil loss to the exact class, falling from 43 per cent to 32 per cent, the loss of accuracy to within one class or greater was much less and broadly comparable to the 1.5 per cent sample results. What this indicates is that a relatively small change in the size of the knowledge base at this stage can have a significant effect on the levels of accuracy obtained. Necessarily, as more examples are added to the system so the level of accuracy can be expected to increase.

To pursue this point further, it is important to note that the loss of accuracy, or the reciprocal increase in system accuracy arising from the incorporation of more examples into the knowledge base, is not uniform across the whole range of low to high soil-loss events. This is indicated by a breakdown (Table 26.3) of the expert system results for the 1.5 per cent sample (Table 26.1) across the range of soil-loss events. A closer examination of predicted against actual soil loss indicates that for low-magnitude soil loss events the expert system predicted to within one class with a commendable accuracy of 88 per cent. At the other extreme of high-magnitude events, despite low sample numbers, prediction to within one class attained an accuracy of 75 per cent. However, the predictive accuracy of the system for erosion events of medium magnitude fell to below 45 per cent. Noticeably, this occurs where the range actually straddles the largest spread of erosion rates between $2\,m^3\,ha^{-1}$ and $20\,m^3\,ha^{-1}$. To some extent, this discrepancy is a reflection of the number of examples in the knowledge base covering the range of erosion rates (Table 26.4). Thus some 55 per cent of the knowledge base consisted of erosion events of low magnitude, and this has undoubtedly improved the capacity of the rule tree to accurately predict erosion at this end of the spectrum, approaching as it does almost a 90 per cent success rate. However, this does not adequately explain the equally high rates of accuracy

Table 26.3 Classification of results according to class

Soil loss class	Exact	1 class	2 classes	3 classes	4 classes	5 classes
a	25	8	2		4	
b	2	9				
c	2	1	2			
d	1	4	6	2		
e	1	1	1			
f	1	2				1
g						
Total	32	25	11	2	4	1

Table 26.4 Profile of soil-erosion events in the knowledge base

Soil loss class	Number	Percentage
a	127	38.02
b	56	16.77
c	58	17.36
d	32	9.58
e	24	7.19
f	19	5.69
g	12	3.60
Total	334	100.00

attained for high-magnitude erosion events, where the number of examples is actually lower than that for the mid-range events. What this suggests is that something more than the sheer number of erosion examples has contributed to the variation in predictive accuracy.

It is suggested that less variability exists in erosion events of low and high magnitude than is the case for medium-magnitude erosion events. The less variability which exists in the examples from which the expert system seeks patterns would suggest, *ceteris paribus*, greater accuracy in the resultant rule tree. In contrast, greater variability within the examples will lead to more variability in the rule tree and a consequent loss in accuracy. One way of overcoming this is to include a greater number of examples within the knowledge base in this range in order to improve the accuracy of the rule tree. Here an expert system approach provides a clear advantage in that it focuses attention on selective areas of the erosion process where further data gathering can be targeted to extend and strengthen the knowledge base where it is weakest. By focusing on improving selective areas of the knowledge base, scarce resources can be brought to bear where the greatest returns can be obtained and the predictive capability of the rule tree improved. If, in due course, system prediction of erosion events of a certain magnitude achieves a consistent and acceptable level of accuracy, then attention and resources can be shifted to more productive areas of data collection.

CONCLUSION

The approach adopted here has been to derive a series of rules via logical reasoning from a data set of recorded soil-erosion events. What emerges is that from the current knowledge base of some 334 recorded soil-erosion events in East Sussex, the expert system achieves a relatively high degree of accuracy in predicting soil loss. This knowledge-based approach is novel, and contrasts with the more traditional approaches employed in modelling soil-erosion prediction.

Some similarities do, of course, exist between the two approaches; for example, the expert is integrally involved in determining those factors considered to be pertinent to soil-erosion prediction. Furthermore, both approaches involve the collection of field data with which to develop and validate the models or the rules which have been generated. To this end, the authors are pursuing these two alternative approaches and are currently involved in a project validating the EPIC system in the UK (Favis-Mortlock, 1989). However, in one major respect the approaches differ markedly. The expert system actively seeks to incorporate the real world information directly into the knowledge base in the form of examples in order to develop rules which reflect the latent empirical relationships within the data. This contrasts with models that attempt to synthesize such relationships into mathematical equations and to replicate real-world scenarios with varying degrees of success.

While the use of class boundaries leads to some imprecision concerning the accuracy of the system, the above results do suggest that optimism concerning the application of expert system approaches to soil-loss prediction is warranted. The results are at least comparable with those derived from existing models outlined earlier. In the case of low- and high-magnitude erosion events the system already attains a creditable 88 per cent and 75 per cent accuracy. What is perhaps most promising is that as the knowledge base increases, as further examples are added, so the rule tree will also increase in accuracy. The task is to develop and enhance the knowledge base, particularly in the less than satisfactory mid-magnitude range, confident that data collection is not being pursued for its own sake. Each additional soil erosion event is indeed an addition to our knowledge base. If nothing else, such a rule-based approach to soil-loss prediction does lend credence to the old adage that 'there is more than one way to skin a cat'.

REFERENCES

Boardman, J. (1986). The context of soil erosion, *SEESOIL*, **3**, 2–13.
Boardman, J. (1988). Public policy and soil erosion in Britain. In Hooke, J. M. (ed.), *Geomorphology in Environmental Management*, John Wiley, Chichester, pp. 33–50.
Boardman, J. (1989). Severe erosion on agricultural land in East Sussex, UK, October 1987, *Soil Technology*, **1**, 333–48.
Boardman, J. (1990). Soil erosion on the South Downs: a review. In Boardman, J., Foster, I. D. L. and Dearing, J. A. (eds), *Soil Erosion on Agricultural Land*, John Wiley, Chichester, pp. 87–105.
Boardman, J. and Robinson, D. A. (1985). Soil erosion, climatic vagary and agricultural change on the Downs around Lewes and Brighton, Autumn 1982, *Applied Geography*, **5**, 243–58.
Buchanan, B. G. and Shortliffe, E. H. (eds) (1984). *Rule-based Expert Systems: the MYCIN Experiments of the Stanford Heuristic Programming Project*, Addison-Wesley, London.

Burwell, R. E. and Kramer, L. A. (1983). Long-term annual runoff and soil loss from conventional and conservation tillage of corn, *Journal Soil and Water Conservation*, **38**, (3), 315–19.

Crosson, P. (1984). New perspectives on soil conservation policy, *Journal of Soil and Water Conservation*, **39**, (4), 222–5.

Evans, R. (1980). Characteristics of water-eroded fields in lowland England. In De Boodt, M. and Gabriels, D. (eds), *Assessment of Erosion*, John Wiley, Chichester, pp. 77–87.

Evans, R. and Cook, S. (1986). Soil erosion in Britain, *SEESOIL*, **3**, 28–58.

Favis-Mortlock, D. (1989). Using EPIC to model erosion on the British South Downs, Poster paper, *British Geomorphological Research Group Workshop*, 'Soil Erosion on Agricultural Land', Coventry.

Fisher, P. F., Mackaness, W. A., Peacegood, G. and Wilkinson, G. G. (1988). Artificial intelligence and expert systems in geodata processing, *Progress in Physical Geography*, **12**, 3, 371–88.

Foster, G. R. (1990). Process-based modelling of soil erosion by water on agricultural land. In Boardman, J., Foster, I. D. L. and Dearing, J. A. (eds), *Soil Erosion on Agricultural Land*, John Wiley, Chichester, pp. 429–445.

Graham, I. and Jones, P. L. (1988). *Expert Systems: Knowledge, Uncertainty and Decision*, Chapman and Hall, London.

Hayes-Roth, F., Waterman, D. A. and Lenat, D. B. (eds) (1983). *Building Expert Systems*, Addison-Wesley, London.

Jackson, P. (1986). *Introduction to Expert Systems*, Addison-Wesley, Wokingham.

Morgan, R. P. C., Morgan, D. D. V. and Finney, H. J. (1984). A predictive model for the assessment of soil erosion risk, *Journal Agricultural Engineering Research*, **30**, 245–53.

Naylor, C. (1983). *Build your own Expert System*, Sigma Press, Wilmslow.

Parsaye, K. (1987). Machine learning: the next step, *Computerworld*, 23 November, 16.

Robinson, D. A., and Blackman, J. D. (1990). Some costs and consequences of soil erosion and flooding around Brighton and Hove, Autumn 1987. In Boardman, J., Foster, I. D. L. and Dearing, J. A. (eds), *Soil Erosion on Agricultural Land*, John Wiley, Chichester, pp. 369–382.

Simons, G. L. (1985). *Expert Systems and Micros*, National Computing Centre, Manchester.

Stammers, R. L. and Boardman, J. (1984). Soil erosion and flooding on downland areas, *The Surveyor*, **164**, 8–11.

Wall, G. J. and Dickinson, W. T. (1978). Economic impacts of erosion. In *Soil Erosion in Ontario, Notes on Agriculture*, **14**, 10–12.

Wallsgrove, R. (1988). Crystal, *Personal Computer World*, November, 172.

Williams, J. R. (1985). The physical components of the EPIC model. In El-Swaify, S. A., Moldenhaner, W. C. and Lo, A. (eds), *Soil Erosion and Conservation*, Soil Conservation Society of America, Ankeny, Iowa, pp. 272–84.

Wischmeier, W. H. and Smith, D. D. (1978). Predicting rainfall erosion losses—a guide to conservation planning, *US Department of Agriculture, Agriculture Handbook*, No. 537.

Modelling

27 Model Building for Predicting and Managing Soil Erosion and Transport

W. T. DICKINSON, G. J. WALL and R. P. RUDRA
Guelph, Canada

INTRODUCTION

There are now many models which have been developed for the description and prediction of soil erosion and sediment yield. These have taken into account erosion processes on a microscale (Palmer, 1965; Mutchler and Hansen, 1970; Ghadhiri and Payne, 1981; Guy et al., 1987), soil loss from sloping plots (Young and Wiersma, 1973; Walker et al., 1978; Wischmeier and Smith, 1978; Watson and Laflen, 1986; Rudra et al., 1985), soil detachment and transport from fields (Knisel, 1980), fluvial transport (Alonso et al., 1981; Gilley et al., 1985; Julien and Simons, 1985), and erosion and sediment yields from watershed areas (Beasley et al., 1980; Cook et al., 1985; Young et al., 1987). They have afforded us opportunities to improve databases, ascertain temporal trends, examine spatial variations, identify critical processes and explore the possible impacts of remedial measures and the relative effectiveness of implementation strategies for erosion and sediment control. Yet our understanding of soil detachment and transport processes remains sadly incomplete, and our ability to make predictions about soil erosion is fraught with difficulties associated with model parameter estimation, input data requirements and model validation (Rose et al., 1988).

Although erosion models are available for a range of spatial scales, as noted above, scale has posed a particular and significant problem in research investigations. As noted by Klemes (1983), different sets of physical laws dominate in plots, hillslopes and small basins, with relationships applicable to the larger units of area tending to express averages or integrals of relationships suitable for small areas. Likewise, timescaling can be a problem, with parameters developed for empirical expressions relating to daily data not necessarily being representative or obviously related to those appropriate for an hourly or other timescale. We have had difficulty not only scaling our results up or down but also taking cognizance of the differences in our models which may be attributable to differences in scale (Ketcheson *et al.*, 1973).

This chapter presents a practical approach to soil erosion being taken at the University of Guelph, Canada, in which experiments are being conducted at various scales. There is an attempt to explicitly acknowledge differences among controlling processes and results obtained at one scale are providing input to or provoking questions to be resolved by model studies at other scales. Examples of recent projects being conducted at various scales are presented, along with illustrations of cross-fertilization among the projects.

A MULTISCALE APPROACH

The approach can be represented by the simple diagram shown in Figure 27.1. Microscale studies are being conducted in a soil erosion flume in the laboratory, and on 1 m × 1 m field plots, with the aid of a rainfall simulator. The laboratory facilities have provided opportunities to explore the mechanics and transport capacity of interrill flow, infiltration and overland flow on sloping and layered soils, aggregate stability and potential seasonal differences in infiltration, overland flow and suspended sediment concentrations and sediment yields. The microscale field plots have been used to study infiltration rates, overland flow, suspended sediment concentrations and yields as functions of rainfall intensity, season, soil and tillage practices. Field plots similar to those of Wischmeier have provided data for the calibration and modification of a field-scale model; and watershed databases for a number of small catchments exhibiting wide ranges of physiography and agricultural land use have provided a resource for the development, modification, calibration, verification and application of a number of watershed models. Knowledge obtained about processes at the smaller scales is used to modify models employed at the larger ones; and problems and critical processes identified at the larger scales have prompted detailed studies at the smaller ones.

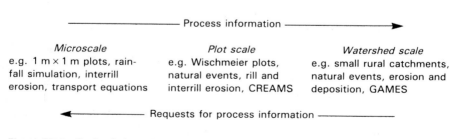

Figure 27.1 Scales being explored for erosion research, with information flows

MICROSCALE STUDIES

Rainfall Simulation

In order to expedite microscale erosion studies in the laboratory or field it is necessary to select or develop a rainfall simulator device. Two generations of simulators have been developed, calibrated and validated at Guelph: the GRS I, a fixed laboratory model (Pall *et al.*, 1983); and the GRS II, a quite portable laboratory and field model (Tossel, 1987; Tossel *et al.*, 1987). The more recent model basically employs a continuous-spray, wide-angle, low to medium flowrate, full-jet nozzle, reproducing a storm intensity from 17 to 200 mm h^{-1}.

Calibration involved the development of relationships between simulated intensity and spatial uniformity, on the one hand, and nozzle size, nozzle water pressure and height of nozzle above the study surface, on the other. Validation of the simulator and of the precipitation produced has focused on the development of methodology for validating such devices as well as application of selected methods to the GRS II. A photographic technique involving darkfield illuminated macrophotography (Beals *et al.*, 1983) and a particle measuring systems (PMS) laser probe (Knollenberg, 1970) have been used to evaluate the drop size and the drop velocity distributions of the simulated rainfall, along with the associated variables of momentum and kinetic energy. The results provide excellent input to a consideration of standards appropriate for rainfall simulators to be used for erosion research, and reveal that the GRS II generates precipitation exhibiting characteristics very similar to those of natural rainfall. The device, with its calibration and validation, has provided a vital tool and invaluable information for the development of microscale erosion models.

Hydraulics and Transport Capacity of Interrill Flow

Rainfall-induced interrill erosion is currently not well understood, primarily due to measurement difficulties and the complexity of the processes involved. At Guelph, flow measurements and the sediment transport at the capacity rate in interrill flow have been investigated for uniform flow and rainfall-induced, gradually varied flow in a 0.69 m wide by 1.50 m long flume. The influence of rainfall on the flow and transport capacity has also been examined (Guy *et al.*, 1987).

In uniform flow runs representative of shallow overland flows in southern Ontario, discharges of 0.5, 3.0 and 10.0 l min^{-1} were examined at flume slopes of 2, 9 and 20 per cent. Flow-velocity measurements indicated that velocities matched the laminar profile equation; and, therefore, discharge and flume slope were the only independent hydraulic variables, both being required for transport capacity analyses. Reynolds numbers varied from 11.0 to 240.9; Froude numbers ranged from 0.343 to 4.985; maximum flow depth was 1.543 mm; and boundary shear stress varied from 0.109 to 1.387 Pa.

For the rainfall-generated runs, simulated rainfall intensities of 45, 140 and 180 mm h^{-1} (again representative of Ontario conditions) were examined at slopes of 2, 9 and 20 per cent. The flow velocities once more matched the values required by the laminar profile equation, indicating that small Reynolds number flows even under intense rainfall may retain the appearance of laminar flow. The durations of drop impact were observed to be small, helping to explain this result; yet a comparison of the uniform and rainfall-induced flow data suggested that the simulated rainfall retarded the flow velocities, as noted by others (Li, 1972). It was also noted that raindrop angle of incidence, directed upslope over the upper flume section and downslope over the lower section due to spray pattern, respectively reduced and enhanced flow velocities in the upper and lower sections.

Both discharge and slope were found to influence transport capacity in uniform flow runs, according to the relationship

$$q_{sf} = a.q^b.So^c$$

where q_{sf} = transport capacity,
 q = discharge per unit width,
 So = bedslope, and
 a,b,c = coefficients.

The specification of both q and So, or any other hydraulic parameter with either q or So, was shown to constitute both necessary and sufficient conditions to fully account for flow hydraulics. The coefficients in the discharge and slope relationships appeared to be a function of transport capacity, and this behaviour was attributed to changing flow competence.

Transport capacities of the rainfall-generated flows were 85 per cent attributable to raindrop enhancement and only 15 per cent attributable to the surface flow. The rainfall enhancement of transport capacity was strongly related to both rainfall intensity and flume slope, and could be expressed as

$$q_{sr} = d.I^e.So^f$$

where q_{sr} = transport capacity enhanced by rainfall,
 I = rainfall intensity, and
 d,e,f = coefficients.

Although rainfall momentum and kinetic energy fluxes provided slightly better results than intensity, simulator nozzles which allow larger variations in drop velocity are required to discern such differences.

Because of the dominant role of rainfall, simple predictive transport capacity equations based on rainfall properties alone, provided about equally good predictive ability to relationships involving flow and rainfall. However, it is believed that rigorous evaluations of transport capacity must account for the variation of the flow contribution with distance down the slope, the probable variation of the rainfall contribution with depth, the problem of reduced competence in small uniform flows and low rainfall intensities, and the superiority of rainfall momentum and kinetic energy fluxes in accounting for the rainfall contribution.

The similarity of the relationships developed from the Guelph experiments and those reported previously for interrill and combined rill/interrill sediment delivery suggests that reported transport rates may have frequently represented transport near the capacity rate, and indicates the need to carefully distinguish between erosion subprocesses. Since transport rate is the integral of detachment rate over the distance, knowledge of detachment rate is required to predict interrill delivery for values of flow distance less than the critical value for attainment of transport capacity. Knowledge of transport capacity and detachment rate is required to find this critical distance. For greater values of flow distance, interrill delivery is at the capacity rate. For the successful modelling of interrill delivery, it is therefore necessary to separate and understand both detachment and transport and the factors controlling their behaviour.

Infiltration

The description and modelling of overland flow for erosion research and prediction must take into account the infiltration of water; and in many climates the infiltration process, and hence the erosion process, is dependent on the season of the year. In most parts of Canada, including southern Ontario, infiltration is affected by the freezing and thawing of soils. More than 75 per cent of the annual flow in Ontario streams occurs during late winter and early spring, when soils are experiencing freeze–thaw cycles; and it is during this same period that an even greater percentage of the annual sediment load is yielded by the basins. The extent to which runoff potential during this period depends on the characteristics of and depth to frost layers has been a topic of recent research.

The Green and Ampt equation was modified to predict infiltration and runoff from layered soils, and laboratory experiments were conducted in a soil-erosion flume under simulated rainfall (Rudra *et al.*, 1986b). A thin galvanzied steel sheet having a prescribed number and diameter of holes was used to simulate a soil layer of lower permeability than the 15 cm depth of soil in the flume. The experimental results indicated that runoff response was markedly affected by the relatively impermeable (simulated frost) layer in medium- and light-textured soils; and the Green and Ampt equation was shown to predict the infiltration well not only for the layered situation established in the laboratory flume but also a field situation in the spring of the year involving a frost layer extending to 50 cm and thawing at the surface. Results to date have indicated that infiltration, and hence runoff volume and associated erosion, are sensitive to changes in the saturated hydraulic conductivity of the frozen layer and its thickness, while time to ponding is more sensitive to the thaw penetration depth.

Seasonal K

It has been suggested that the susceptibility of a soil to erode i.e. its erodibility, can change seasonally, as a function of temperature (Mutchler and Carter, 1983) and as a function of freeze–thaw conditions (Dickinson and Wall, 1976; Dickinson et al., 1982; Kirby and Mehuys, 1987), possibly as a result of reduced aggregation (Bryan, 1971). A multi-faceted study was conducted to explore the potential seasonal variation in soil erodibility for selected soils (Wall et al., 1988) and to investigate soil-erodibility indices and their seasonal variation (Coote et al., 1988). Laboratory flume/rainfall simulation studies, field plot data, and a K-prediction equation were used to assess the potential magnitude of the seasonal variability in southern Ontario. In these phases of the study, soil erodibility was found to vary seasonally, the highest erodibility occurring for the thaw and simulated thaw conditions associated with the winter–spring period, the magnitude of seasonal differences being a function of soil texture. In the soil-erodibility index portion of the study, vane shear strength and water-stable aggregates >0.5 mm increased substantially as four soils, ranging from loamy sand to clay, dried and warmed following spring thaw. For winter thaw conditions, the index 1/shear strength averaged seventeen times greater than the same index for summer conditions. Spring values of the index averaged twice those of summer. The results clearly indicated that at least some Ontario soils are much more susceptible to erosion under thaw and spring conditions than later during the growing season, and that soil water content and soil warming affect the re-establishment of resistance to erosion in soils rendered erodible by freezing, thawing and saturation.

PLOT SCALE

Modified CREAMS

A modified version of the CREAMS model (Knisel, 1980) was calibrated with research plot data on a loam soil (Rudra et al., 1985). Test data were available for 10 plots, 44.2 m in length × 6.4 m in width with uniform slopes of between 7 and 9 per cent, which had been monitored for the last 32 years on the University of Guelph campus. The hydrology component of CREAMS was tested with the event runoff data from non-frozen periods (16–30 events per plot), and the erosion component was evaluated with the associated measured soil loss amounts. Modifications to the model included the introduction of seasonal variations in soil hydraulic conductivity and in soil erodibility, and an adjustment to the empirical runoff erosivity factor used in the model's rill-erosion equation.

The modification regarding hydraulic conductivity was made to take into account daily variations in water viscosity as a result of changes in water temperature, influencing the infiltration component of the model. Changes in the soil-erodibility parameter in CREAMS were introduced by means of a procedure for estimating seasonal erodibility (Dickinson et al., 1982). For this procedure, the model user must define the number of seasons selected per year (i.e. three or four), and assign to each a soil-erodibility scaling factor and a Julian date. An adjustment was also found to be necessary for the coefficient in the empirical equation for the erosivity factor. Calibration of the model indicated that an optimum value of this coefficient for the Guelph plot data was two to four times the value in the basic model established by Foster et al. (1977). Inclusion of this adjustment infers that overland flow plays a more significant role than originally assumed.

With the modifications and adjustment noted above, the hydrology submodel of CREAMS was judged to calibrate reasonably well for the Guelph data, and the erosion model was erratic but within acceptable limits (Rudra et al., 1985).

WATERSHED SCALE

GAMES

GAMES—the Guelph model for evaluating the effects of Agricultural Management systems on Erosion and Sedimentation (Cook et al., 1985; Rudra et al., 1986a)—was developed as a screening tool for watershed management, the purpose being to predict soil loss by water erosion and the subsequent delivery of sediment from field to stream and downstream in agricultural areas. The fundamental building blocks of the model involve estimation of (1) soil erosion, by a modification of the Universal Soil Loss Equation (Wischmeier and Smith, 1978); and (2) the percentage of the soil loss delivered to downstream fields and the main stream by a field-scale delivery ratio function which incorporates surface roughness, land slope, hydraulic condition and the length of the overland flow path (Dickinson et al., 1986). GAMES accommodates field-size land cells of variable shape and size; and the seasonal time frame allows for changes in rainfall characteristics, soil erodibility and land-management conditions which affect soil detachment and transport. The output generated for each field for each season within a watershed includes potential soil erosion, erosion rate, delivery ratio to the adjacent field, delivery ratio to the stream, sediment yield and sediment yield rate. Application of the model to a small agricultural watershed indicates the existence and location of potential soil erosion problem areas and prime sediment sources, along with the watershed and management characteristics associated with such zones (Dickinson et al., 1987).

Although GAMES (like most, if not all, other distributed deterministic watershed models) has not been quantitatively validated on a field-by-field basis, qualitative field observations have confirmed that areas estimated to be characterized by high soil loss and/or high sediment yield do exhibit large soil losses and/or large sediment yields; and areas predicted to have minimal soil loss and/or sediment yield exhibit these traits. With a capacity for 'screening' erosion problem and sediment source areas, based on seasonal climatic, soil and management variables, the model has proven very useful for:

(1) Identifying 'hot spots' in selected watersheds (Rudra et al., 1986a);
(2) Evaluating the possible effectiveness of alternative soil- and crop-management practices in those watersheds and watersheds of like conditions (Rousseau et al., 1987);
(3) Developing a framework for soil erosion and sediment problem identification (Dickinson et al., 1987);
(4) Providing insights into the spatial variability of soil erosion and sediment yield characteristics in the region of southern Ontario (Dickinson et al., 1984); and
(5) Providing a tool for determining whether soil erosion and sediment yield 'hot spots' can be discriminated with simpler data inputs (Dickinson and Rudra, 1985).

GAMES has also been used as the basis for GAMESP, a phosphorus transport model (Rousseau et al., 1988), a tool for evaluating best management practices to control phosphorus non-point source pollution (Rousseau et al., 1987).

The spatial distributions of soil loss and sediment yield have been predicted by GAMES for rolling upland and relatively level lowland watersheds in southern Ontario (Dickinson and Pall, 1982; Dickinson et al., 1987). These model results have aided in the identification of sediment sources in such upland areas, and the subsequent evaluation of the effect of implementing erosion-control measures in these source areas. Perhaps more importantly, the results have revealed the extent to which serious erosion and high sediment yield are localized phenomena in many upland watersheds, 85 per cent of the annual watershed sediment load often emanating from < 10 per cent of the watershed area; whereas in relatively flat lowland basins, erosion and sediment yield rates are spatially much more uniform. This discovery has had major implications for both the selection of appropriate remedial measures and the choice of strategy for implementing erosion and sediment controls. The targeting of controls is being shown to be not only cost-effective in upland areas but also essential for achieving sediment and associated pollution-control objectives.

Model Comparison

A study has just commenced to explore similarities, differences, strengths and weaknesses of a number of computer models now being used for investigating soil erosion and sediment problems in small rural watersheds. To date, the models being studied include AGNPS (Young *et al.*, 1987), ANSWERS (Beasley *et al.*, 1980) and GAMES. Input data files have been developed for the models for a selected upland basin in southern Ontario; the models are being calibrated on the basis of subsets of the data; and the sensitivity of model outputs to variations in input data and parameter values is being explored. The study will also compare model outputs for a range of selected inputs which include conditions representative of existing and proposed land-management techniques. It is anticipated that results from this study will clarify not only model similarities and differences but also directions for the most efficient collection of field data required for model calibration and for model development.

Sediment Load Data

Watershed suspended sediment load data have been analysed for Ontario streams with an eye primarily on the temporal occurrence and distribution of loads (Dickinson and Wall, 1976). It has become very clear that daily loads follow a distinctive seasonal pattern, the bulk of the annual suspended load being transported during the spring period, in concert with the seasonal occurrence of high water flows. This seasonal characteristic, in conjunction with the fact that daily loads exhibit a highly skewed frequency distribution, results in a very large percentage of the annual load being transported downstream in a very small percentage of the time (in most cases, 80–90 per cent of the annual load is delivered in less than 40 days each year). Therefore, the movement of suspended sediment in Ontario streams is highly event oriented, and reliable estimates of suspended loads (and of associated pollutant loads) are contingent upon the application of a sampling scheme in time and a computational method that ensures the obtaining of good samples and good load estimates during the brief periods when most of the load is delivered (Dickinson, 1981).

It has been ascertained that infrequent sediment loads (with return periods of 10 years or greater) account for up to 30 per cent of the total loads transported by Ontario streams (Conservation Management Systems, 1986). Annual peak events contribute a slightly larger, significant portion of the total load. Therefore, the reliable estimation of sediment loads requires careful consideration of all significant events.

CROSS-SCALE FERTILIZATION

Have results from model studies conducted at one scale proven to be very useful as input to studies at other scales? In other words, has the multiscale modelling approach really worked? The following commentary illustrates the flow of ideas and results which has occurred to date at Guelph, and is offered in the hope that it might stimulate more such cross-fertilization within, with and among other institutions.

The erosion and sediment work in the 1970s involved the exploration of stream-suspended sediment data and summer soil-loss data from the Wischmeier-type hillslope plot sites on the Guelph campus. The plot data confirmed that summer and autumn erosion did occur on Guelph loam for a variety of crops, and that estimates of potential soil loss made on the basis of the Universal Soil Loss Equation corresponded well with the measured quantities. However, it was also evident from the sediment data from nearby agricultural watersheds that a relatively small percentage of the annual suspended load was transported through the basins during the period of the year when soil loss had been monitored on the plots. These observations, including the apparent discrepancy regarding the timing of field erosion and sediment transport, led to the postulation of a conceptual model for watershed erosion and sediment yield, presented in Table 27.1.

This model included the hypotheses that, at least for southern Ontario agricultural conditions, spring erosion is expected to be widespread, probably with low to moderate soil-loss rates. The corresponding surface drainage system is widespread and efficient, with a resulting delivery of a relatively high percentage of eroded material. Summer erosion, on the other hand, is expected to be quite localized; and since the surface drainage system has become largely dry and is poorly developed at this time of the year, the percentage of suspended material delivered through basins and downstream is extemely small.

This conceptual framework served as the basis for the development of GAMES in which important seasonal differences in erosion and transport could be accommodated yet the extensive databases normally associated with

Table 27.1 A conceptual model of soil erosion and sediment transport

Season	Processes	
	Erosion	Sediment delivery
Spring	Spatially widespread, at low to moderate rates of soil loss	Efficient and effective micro-drainage delivery system, with high delivery percentages
Summer	Localized erosion, at up to high rates of soil loss	Extremely poor delivery system, with very low delivery percentages

deterministic event and continuous models could be avoided. The obvious lack of plot and field-scale experimental data on spring soil loss associated with snowmelt, rain on snow and post-snowmelt thawing conditions prompted detailed laboratory and field plot studies, including the consideration of possible seasonal changes in soil erodibility and infiltration rates.

The need for detailed studies prompted the development of rainfall simulators; and the apparent lack of standard calibration techniques for simulators led to the development of equipment and procedures for describing simulated rainfall characteristics, including momentum and kinetic energy properties. Preliminary results from the laboratory soil-erosion flume and field microplots supported the hypothesis that soil erodibility could vary significantly from season to season, as could infiltration and the corresponding runoff. These variations were incorporated into CREAMS, GAMES and ANSWERS, for the purpose of exploring the associated sensitivity of the model outputs and other model parameter values. These model studies confirmed the importance of taking seasonal variations in soil characteristics into account, leading to further microscale studies of soil erodibility, soil disaggregation and infiltration, and to the corresponding modifications to model components. It was during this 'give and take' exchange between the model and microscale work that the need for clarification of the hydraulics of rills and interrills emerged, prompting the project on interrill transport capacity. Results from this research are premature to incorporate into existing models; however, preliminary alterations to transport algorithms are being explored.

Application of distributed models such as GAMES and ANSWERS to a number of watersheds exhibiting a wide range of physiography revealed that spatial variability as well as temporal variability required careful consideration for predicting and managing soil erosion and sediment transport. These results have prompted microscale field studies of spatial variations in soil erodibility, infiltration, surface runoff and soil loss as a function of time of year; and CREAMS is being used as the basic field-scale model to explore the consequences of lumping spatial units exhibiting considerable variability in erosion and transport factors.

CONCLUSIONS

The desirability of using observations and models from one scale to direct and motivate work at other scales is now quite clear, even if the manner in which results may be transferred is not readily apparent and the direct transfer of results is not always possible. An integrated, multiscale approach to the development of models for predicting and managing soil erosion and transport is certainly stimulating and exciting, but also cost-effective and probably essential for advancement of our knowledge and technical capability. However, it should be stated that integration of results from various scales, i.e. 'breaking the scale barrier', can be achieved only with successful integration of people from various disciplines, i.e. 'breaking the discipline barrier'. Research at the various scales is distributed among scientists and journals associated with a variety of academic disciplines, e.g. microscale work has often been conducted by physical geographers, field plot work by soil scientists, watershed modelling by agricultural engineers and stream-sediment monitoring by hydraulic engineers. Further, scientists among the group have striven to test hypotheses and draw conclusions while technologists have sought best possible solutions given available information. Our model building will be successful only if we meet the challenge of integrating both more of the model results obtained at the various scales and more of the modellers themselves, whatever their discipline or orientation.

REFERENCES

Alonso, C. V., Niebling W. H. and Foster G. R. (1981). Estimating sediment transport capacity in watershed modeling, *Trans. ASAE*, **24**(5), 1211–20, 1226.

Beals, D., Dickinson, W. T., Helsdon D. S. and Pall, R. (1983). An approach for characterizing falling water droplets, *Can. Agr. Eng.*, **25**(2), 189–92.

Beasley, D. B., Huggins, L. F. and Monke, E. J. (1980). ANSWERS: a model for watershed planning. *Trans. ASAE*, **23**, 938–44.

Bryan, R. B. (1971). The influence of frost action on soil-aggregate stability, *Trans. Inst. Br. Geogr.*, **54**, 71–88.

Conservation Management Systems (1986). Sediment issues and data needs in Ontario, *Inland Waters Directorate Tech. Rep.* IWD-OR-WRB-SS-86-1, Environment Canada, Ottawa, Ontario.

Cook, D. J., Dickinson, W. T. and Rudra, R. P. (1985). *GAMES: Guelph Model for Evaluating the Effects of Agricultural Management Systems on Erosion and Sedimentation*, User's Manual, School of Engineering, University of Guelph, Guelph, Ontario.

Coote, D. R., Malcolm-McGovern, C. A., Wall, G. J., Dickinson, W. T. and Rudra, R. P. (1988). Seasonal variation of erodibility indices based on shear strength and aggregate stability in some Ontario soils, *Can. Jour. Soil Sci.*, **68**, 405–16.

Dickinson, W. T. (1981). Accuracy and precision of suspended sediment loads. In *Erosion and Sediment Transport Measurement*, IASH-AISH Publ. No. 133, 195–202.

Dickinson, T. and Pall, R. (1982). *Indentification and Control of Soil Erosion and Fluvial Sedimentation in Agricultural Areas of the Canadian Great Lakes Basin*, Final Report to Dep. Supp. and Serv., Contract No. 23SU.01525-1-0433, Ottawa, Ontario.

Dickinson, W. T. and Wall, G. J. (1976). Temporal pattern of erosion and fluvial sedimentation in the Great Lakes Basin, *Geoscience Canada*, 3(3), 158–63.

Dickinson, W. T., Pall, R. and Wall, G. J. (1982). Seasonal variations in soil erodibility. *ASAE Paper No.* 82-2573, Am. Soc. Agr. Eng., St Joseph, Michigan.

Dickinson, W. T., Pall, R. and Wall G. J. (1984). GAMES—a method for identifying sources and amounts of soil erosion and fluvial sediment. *In Water Quality Evolution within the Hydrological Cycle of Watersheds*, Proc. Can. Hyd. Symp., NRC Assoc. Comm. on Hydrology, Ottawa, Ontario.

Dickinson, W. T. and Rudra, R. P. (1985). Discrimination of soil erosion and fluvial sediment areas, *Can. Jour. Earth Sci.*, **22**, 1112–17.

Dickinson, W. T., Rudra, R. P. and Clark, D. J. (1986). A delivery ratio approach for seasonal transport of sediments. In *Drainage Basin Sediment Delivery*, IASH Publ. No. 159, 237-51.

Dickinson, W. T., Rudra, R. P. and Wall, G. J. (1987). Identification of soil erosion and fluvial sediment processes, *Hydrological Processes*, 1, 111-24.

Foster, G. R., Meyer, L. D. and Onstad, C. A. (1977). A runoff erosivity factor and variable slope length exponents for soil loss estimates, *Trans. ASAE*, 20(4), 683-7.

Ghadhiri, H. and Payne, D. (1981). Raindrop impact stress, *Jour. Soil Sci.*, 32, 41-9.

Gilley, J. E., Woolhiser, D. A. and McWhorter, D. B. (1985). Interrill soil erosion— part I: Development of model equations, *Trans. ASAE*, 28, 147-53, 159.

Guy, B. T., Dickinson, W. T. and Rudra, R. P. (1987). The roles of rainfall and runoff in the transport capacity of interrill flow, *Trans. ASAE*, 30(5), 1378-86.

Julien, P. Y. and Simons, D. B. (1985). Sediment transport capacity of overland flow, *Trans. ASAE*, 28, 755-62.

Ketcheson, J. W., Dickinson, W. T. and Chisholm, P. S. (1973). Potential contributions of sediment from agricultural land. *Proc. 9th Can. Hyd. Symp.*, NRC Assoc. Comm. on Hyd., Ottawa, Ontario.

Kirby, P. C. and Mehuys, G. R. (1987). Seasonal variation of soil erodibilities in southwestern Quebec, *Jour. Soil Water Cons.*, 42, 211-15.

Klemes, V. (1983). Conceptualization and scale in hydrology, *Jour. Hydrology*, 65, 1-23.

Knisel, W. G. (1980). CREAMS: a field-scale model for chemicals, runoff and erosion from agricultural management systems, *USDA Cons. Res. Rep.* No. 26.

Knollenberg, R. G. (1970). The optical array: an alternative to scattering or extinction for airborne particle size determination, *Jour. App. Met.*, 9, 86-103.

Li, R. M. (1972). *Sheet Flow under Simulated Rainfall*. MSc thesis, Colorado State University, Fort Collins, Colorado.

Mutchler, C. R. and Carter C. E. (1983). Soil erodibility variation during the year. *Trans. ASAE*, 26, 1102-4, 1108.

Mutchler, C. K. and Hansen, L. M. (1970). Splash of a waterdrop at terminal velocity, *Science*, 169, 1311-12.

Pall, R., Dickinson, W. T., Beals, D. and McGirr, R. (1983). Development and calibration of a rainfall simulator, *Can. Agr. Eng.*, 25(2), 181-7.

Palmer, R. S. (1965). Waterdrop impact forces, *Trans. ASAE*, 8(1), 69-70, 72.

Rose, C. W., Dickinson, W. T., Jorgenson, S. E. and Ghadhiri, H. (1988). Agricultural non-point source runoff and sediment yield water quality models: modeller's perspective, *International Symposium on Water Quality Modeling*, Logan, Utah.

Rousseau, A., Dickinson, W. T. and Rudra, R. P. (1987). Evaluation of best management practices to control phosphorus nonpoint source pollution, *Can. Agr. Eng.*, 29(2), 163-8.

Rousseau, A., Dickinson, W. T., Rudra, R. P. and Wall, G. J. (1988). A phosphorus transport model for small agricultural watersheds, *Can. Agr. Eng.*, 30, 213-20.

Rudra, R. P., Dickinson, W. T. and Wall, G. J. (1985). Application of the CREAMS model in southern Ontario conditions, *Trans. ASAE*, 28(4), 1233-40.

Rudra, R. P., Dickinson, W. T. and Wall, G. J. (1986a). GAMES—a screening model of soil erosion and fluvial sedimentation on agricultural watersheds, *Can. Wat. Res. Jour.*, 11(4), 58-71.

Rudra, R. P., Dickinson, W. T., Wall, G. J. and Tan, K. A. (1986b). Runoff response to frost layering, *Trans. ASAE*, 29(3), 735-40.

Tossel, R. W. (1987). *A Comparison of Three Methods Used to Characterize Simulated Rainfall Properties*, MSc thesis, University of Guelph, Guelph, Ontario.

Tossel, R. W., Dickinson, W. T., Rudra, R. P. and Wall, G. J. (1987). A portable rainfall simulator, *Can. Agr. Eng.*, 29, 155-62.

Walker, P. H., Kinnell, P. I. A. and Green, P. (1978). Transport of a noncohesive sandy mixture in rainfall and runoff experiments, *Soil Sci. Soc. Am. Jour.*, **42**, 793–801.

Wall, G. J., Dickinson, W. T., Rudra, R. P. and Coote, D. R. (1988). Seasonal soil erodibility variation in southwestern Ontario, *Can. Jour. Soil Sci.*, **68**, 417–24.

Watson, D. A. and Laflen, J. M. (1986). Soil strength, slope, and rainfall intensity effects on interrill erosion, *Trans. ASAE*, **29**(1), 98–102.

Wischmeier, W. H. and Smith D. D. (1978). Predicting rainfall erosion losses—a guide to conservation planning, *USDA Agr. Handbook*, No. 537, Washington, DC.

Young, R. A., Onstad, C. A., Bush, D. B. and Anderson, W. P. (1987). AGNPS, agricultural non-point source pollution model, *USDA, ARS Cons. Res. Rep.* No. 35, Washington, DC.

Young, R. A. and Wiersma, J. L. (1973). The role of rainfall impact in soil detachment and transport, *Wat. Res. Res.*, **9**, 1629–39.

28 Process-based Modelling of Soil Erosion by Water on Agricultural Land

G. R. FOSTER
University of Minnesota

INTRODUCTION

For more than four decades, soil-erosion prediction has been a powerful tool used by soil conservationists. Its most frequent use has been to guide farmers' choice of conservation practices that best fit their needs and interests in specific fields (Hayes, 1977). In the last decade, erosion prediction has also been used to develop regional and national inventories and assessments of erosion and its impact (Heimlich and Bills, 1986). With the passage of legislation requiring that US farmers prevent excessive erosion to participate in certain federal agricultural programmes, erosion prediction will be used as a tool to implement policy (Ritterbusch, 1988). It will help to identify problems, appropriate control practices and the degree that erosion has been controlled. The importance of erosion prediction as a tool will expand as concern grows for soil erosion on agricultural land.

Two main types of technology, empirically based and process-based, are available for predicting soil erosion. Although empirically based technology has been widely used in the past, process-based technology is emerging and is likely to become the erosion-prediction technology of choice. Therefore, this chapter mainly deals with process-based erosion-prediction technology, but empirically based prediction technology is also briefly discussed because it has been widely used in the past and it will continue to be in the near future. This chapter discusses only soil erosion by water and not erosion by wind.

EMPIRICALLY BASED TECHNOLOGY

Empirically based technology for predicting soil erosion by water is represented by methods such as the Universal Soil Loss Equation (USLE) (Wischmeier and

Smith, 1978) and the Soil Loss Estimator for Southern Africa (SLEMA) (Elwell, 1981). The USLE is given by:

$$A = RKLSCP \qquad (28.1)$$

where A = soil loss, R = erosivity factor, K = soil-erodibility factor, L = slope length factor, S = slope steepness factor, C = cover-management factor and P = supporting practices factor. The USLE estimates sheet and rill erosion using values for indices that represent the four major factors affecting erosion: R—climatic erosivity, K—soil erodibility, L and S—topography, and C and P—land use.

The USLE was derived from a large database of more than 10 000 plot-years of data. These field plots were 'long' (22 m for most of the plots), but other plot lengths included 11, 44, 81 and 189 m, and covered a range of soils, steepness, crops, management practices and climates over the eastern United States. Most of the data were produced by natural rainfall, but some were obtained by simulated rainfall. Given its large database, the USLE is very well validated (Foster *et al.*, 1981).

Since the plots producing the USLE data were of uniform slope, the USLE is restricted in the degree that it applies to non-uniform slopes. However, it can be applied to irregular slopes where erosion is occurring, but not to those portions of slope, such as the toe of concave slopes, where deposition occurs (Foster and Wischmeier, 1974).

A major limitation of the USLE is that it does not explicitly represent hydrologic and erosion processes. For example, if an adjustment is made in the USLE to account for an effect of runoff, every USLE factor, except perhaps R, must be modified.

Furthermore, the USLE's equation structure is extremely limiting. The equation does not represent (and cannot be easily modified to represent) fully the form observed in experimental data for the effect of cover and steepness on erosion and deposition in furrows (Foster and Lane, 1987a). Therefore, no major improvements in erosion prediction technology are likely to come from the USLE or similar empirically based technology. Major improvements are much more likely to originate from erosion-prediction technology based on fundamental hydrologic and erosion processes.

PROCESS-BASED EROSION-PREDICTION TECHNOLOGY

Process-based technology computes erosion using mathematical representations of fundamental hydrologic and erosion processes. Fundamental erosion processes are detachment by raindrop impact, detachment by flow, transport by raindrop impact, transport by flow and deposition by flow. Detachment processes remove soil particles from the soil mass producing sediment while transport processes move sediment from its point of origin. Detachment adds sediment to the sediment load being transported while deposition removes sediment from the sediment load and adds it to the soil mass.

Scale is an issue, even for erosion-prediction models that apply to scales as small as landscape profiles and fields. One approach to dealing with scale is to define the source areas of interrill, rill, ephemeral gully, impoundment and unit source watershed areas. Use of fundamental processes to develop and apply erosion models requires a conceptual framework for the model and a physical study area that isolates processes in a hierarchical relationship having physical meaning.

Plots as long as 22 m (standard USLE plot length) generally give no specific information about the processes that determine the amount of sediment measured at the lower end of the plot. The amount of sediment reaching the lower end of a long plot can be controlled by detachment by raindrop impact, detachment by flow, transport capacity of the runoff or a combination of these processes, but the measured data generally do not indicate the controlling processes. In contrast the source area concept isolates processes so that they can be studied individually.

Two main concepts apply to every source area. The first is conservation of mass, expressed by (Bennett, 1974):

$$\partial G/\partial x + \partial(cy)/\partial t = D_i + D_f \qquad (28.2)$$

where G = sediment load, x = distance, c = sediment concentration in the runoff (mass of sediment/unit volume of water), t = time, D_i = lateral inflow of sediment and D_f = detachment or deposition by flow. Equation (28.2) is for unsteady flow, typical for natural rainfall. However, for some modelling purposes, experiments under steady rainfall and interpretation of erosion processes the steady state continuity equation is useful (Foster and Meyer, 1975):

$$dG/dx = D_i + D_f \qquad (28.3)$$

Equation (28.3) mathematically states that sediment load increases or decreases along the slope, dG/dx, at the rate that detachment or deposition, $D_i + D_f$, adds to or takes from the sediment load. The additional term, $\partial(cy)/\partial t$, in equation (28.2) represents dynamic storage of sediment in the runoff.

The other main concept concerns the relationship between sediment load and the transport capacity of the flow. The Meyer–Wischmeier (1969) version of this concept states that if the sediment load reaching an interval on the slope plus the sediment detached within the interval is greater than the transport capacity of the transport agents on the interval, the sediment load actually transported equals the transport capacity of the transport agents. Otherwise, the sediment load equals the amount of sediment produced by detachment within the interval plus the amount of sediment arriving from upslope.

The Foster–Meyer (1975) version of the concept assumes a relationship between detachment by flow and sediment load in the flow, and is given by:

$$D_f = D_c (1 - G/T_c) \qquad (28.4)$$

where D_c = detachment capacity of the flow and T_c = transport capacity of the flow. For deposition, the equation is (Renard and Foster, 1983):

$$D_f = (\beta V_f / \sigma x)(T_c - G) \qquad (28.5)$$

where β = a turbulence factor related to the tendency of rainfall to keep sediment in suspension, V_f = fall velocity of the sediment and σ = excess rainfall rate (rainfall intensity–infiltration rate). The Foster–Meyer version better describes observed erosion rates under idealized conditions (Foster and Meyer, 1972, 1975) and gives more continuous forms for computed landscape profiles evolved by erosion than does the Meyer–Wischmeier version. Also, equation (28.5) is very powerful for computing the selectivity of deposition, because sediment eroded on typical farm fields has a wide range of fall velocities (Foster *et al.*, 1985).

Interrill Areas

The soil surface of most agricultural fields is non-uniform producing areas of flow concentrations. As an erosion event continues, certain flow concentrations can become more distinct. These areas of flow concentration, typically about 50–200 mm wide, are called rills and the areas between the rills are called interrills (Foster and Meyer, 1975). Interrill areas are defined such that all detachment that occurs on these areas is by raindrop impact. The row side sideslope of the ridge–furrow configuration typical of many management practices for row crops is an ideal example of an interrill area (Meyer and Harmon, 1985). Interrill areas can also be easily identified on a heavily rilled slope. However, in pastures and fields having close-growing vegetation, interrill areas may not be obvious. Nevertheless, the concept is critical, because it isolates processes and separates two distinct flow regimens: thin, broad, sheet flow on the interrill areas and channellized flow on the rill areas. Even though both interrill and rill areas occur where hydrologists treat overland flow as broad sheet flow, the rill–interrill concept requires envisaging these two types of flow.

By definition, detachment on interrill areas is by raindrop impact. A typical expression for this process is (Foster, 1982; Foster and Lane, 1987a):

$$P_i = i^2 K_i C_i C_g C_s \qquad (28.6)$$

where P_i = detachment rate on interrill areas, i = rainfall intensity, K_i = erodibility factor for detachment by raindrop impact, C_i = factor for the effect of canopy on detachment by raindrop impact, C_g = factor for the effect of groundcover on detachment by raindrop impact and C_s = factor for the effect of soil biomass and soil disturbance on detachment by raindrop impact. Equation (28.6) is empirical, which illustrates that every model intended for field applications contains empirical parameters whose values must be determined by experiment. Also, equation (28.6) would be more process-based if the erosivity term i^2 was replaced with a term based on kinetic energy of raindrop impact (Al-Durrah and Bradford, 1982). However, kinetic energy could be replaced with an even more process-based term derived from the integration of the spatial and temporal forces produced by an impacting raindrop (Huang et al., 1983; Nearing et al., 1987). Like beauty, terms considered to be process-based vary among beholders.

The value for P_i calculated by equation (28.6) is the sediment delivered from the interrill area if transport capacity on the interrill area exceeds the amount of sediment produced by raindrop impact. On very steep interrill slopes, transport capacity is great as Figure 28.1 illustrates, and sediment delivered from the interrill area, D_i, equals the sediment produced by detachment, P_i. This result is given directly from equation (28.2), where detachment by flow, D_f, is zero. On flatter slopes sediment delivery is limited by sediment transport capacity, so that sediment delivered from the interrill area is less than the amount of sediment detached by raindrop impact (Foster and Meyer, 1975).

When transport capacity limits sediment delivery, equation (28.5) must be included in the solution to equation (28.2), which results in an equation similar to (Renard and Foster, 1983):

$$D_i = (\phi dT_c/dx + P_i)/(1 + \phi) \tag{28.7}$$

where $\phi = \beta V_f/\sigma$.

As equation (28.7) shows, sediment delivery D_i, from low interrill slopes is a function of interrill detachment, P_i; the change in the transport capacity along the slope, dT_c/dx; fall velocity, V_f; and excess rainfall rate, σ. For large and dense particles that have large fall velocities, D_i nearly equals

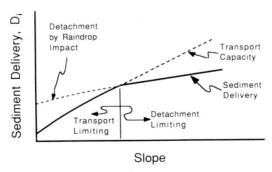

Figure 28.1 Relation of fundamental erosion processes on interrill areas

the change of transport capacity along the interrill area, dT_c/dx. The product of dT_c/dx and length, λ_i, of the interrill slope equals transport capacity at the end of the interrill slope when dT_c/dx is constant along the interrill area. The product of A_i and λ_i equals the sediment load G_i from interrill areas. Therefore, for coarse particles, sediment load from interrill areas is almost equal to transport capacity on the interrill areas. For fine particles where V_f approaches zero, sediment delivery from interrill areas nearly equals the amount of fine sediment produced by detachment, as equation (28.7) shows. The combined result for non-uniform sediment is computed selectivity where coarse particles remain on the interrill area and fine particles leave the area.

An important need is an equation for transport capacity for interrill flow impacted by raindrops. Without raindrop impact, the transport capacity of thin, interrill flow is very low, but raindrop impact greatly increases the transport capacity of interrill flow (Moss *et al.*, 1979). The data of Moss *et al.* show some of the important factors affecting transport capacity on interrill areas. Scientists at the University of Guelph have research in progress to develop a transport capacity for interrill flow (Dickinson, 1989). Hairsine and Rose (1989) have also proposed a theory that deals with dynamic and steady state detachment, deposition, entrainment and transport processes on interrill areas.

Rill Areas

Rills typically are numerous, small flow concentrations that tend to be parallel across the slope except where land form converges or diverges. Whereas runoff and sediment tend to move laterally on interrill areas toward the rills, runoff and sediment move directly downslope in rills unless the rills are forced around the slope by tillage marks. When the rilling pattern is random, flow varies greatly among the rills, but when tillage marks completely define the rills, such as in a ridge–furrow system, discharge is similar for all rills.

Several detachment processes occur in rills. These processes include headcutting, undercutting of sidewalls with subsequent slumping and erosion along the wetted perimeter (Meyer *et al.*, 1975). Research on these processes is underway at the University of Kentucky (Barfield, 1988) and the University of Leuven (Poesen, 1989), but widely accepted equations for these processes have not emerged in erosion modelling.

Typically, detachment capacity in rills is given by (Foster, 1982):

$$D_c = K_r(\tau_s - \tau_c) \tag{28.8}$$

where K_r = an erodibility factor for detachment by flow, τ_s = shear stress of flow acting on the soil and τ_c = critical shear stress for the soil. Foster and Lane (1983) used equation (28.8) and an equation for the distribution of shear stress around a channel to compute the eroded geometry of a rill and erosion rate for steady flow for the time before and after a rill reaches a non-erodible layer. A non-erodible layer causes rills to widen and erosion rate to decrease.

The erodibility terms K_r and τ_c are functions of both static and temporal soil properties (Foster *et al.*, 1982b). In particular, critical shear stress τ_c is related to static soil properties such as texture, but for many soils, their condition at the time of the erosion event is more important than static properties. Some soils have a very low critical shear stress immediately following tillage and after thawing (Foster *et al.*, 1982b). Over time, the soil consolidates and critical shear stress increases significantly. However, the relationship surely varies among soils. Understanding this effect and developing relationships to describe it is an important research need.

Erosion processes in rills depend on the runoff and sediment delivered to them from the interrill areas, as Figure 28.2 illustrates. The erosion processes illustrated in this figure resulted from a classic ridge–furrow study by Meyer and Harmon (1985), where they varied grade along the furrow and separately measured interrill erosion.

At low furrow grades, interrill erosion on the row sideslopes delivered more sediment to the furrow than the flow could transport. The result is deposition, which can be described by (Renard and Foster, 1983):

$$A = (\phi dT_c/dx + D_i)/(1 + \phi) \qquad (28.9)$$

where A = soil loss at the end of the furrow expressed as sediment yield/area. Equation (28.9) is similar to equation (28.7) for interrill areas where transport capacity limits sediment delivery. A reasonable approximation for the change in transport capacity with distance, dT_c/dx, is that it is constant. With this

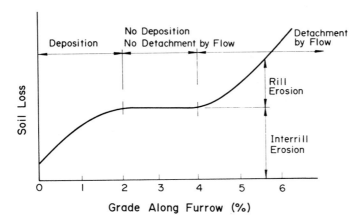

Figure 28.2 Relation of interrill and rill erosion on a ridge–furrow system (after Meyer and Harmon, 1985)

assumption, equation (28.9) suggests that soil loss per unit area is independent of slope length when deposition occurs in furrows, a result supported by Meyer and Harmon's (1985) data.

As furrow grade increased in Meyer and Harmon's study, transport capacity of the flow in the furrow rose to where flow transported all the sediment reaching it from the interrill areas. However, at the intermediate grades, the shear stress of the flow was less than the critical shear stress of the soil and thus no detachment by flow occurred. At steep grades, the shear stress of the flow exceeded the critical shear stress of the soil and detachment by flow (rill erosion) took place.

Rill erosion does not occur at the detachment capacity rate given by equation (28.8) but at the reduced rate indicated by equation (28.4). The presence of τ_c in equation (28.8) prevents an analytical, closed-form solution to the equations. However, a closed-form equation for a uniform slope where $\tau_c = 0$ and $D_c = K_r \tau_s^{3/2}$ is given by (Foster and Meyer, 1975):

$$A = dT_c/dx - (dT_c/dx - D_i) [1 - \exp(-\alpha)]/\alpha \quad (28.10)$$

where $\alpha = \lambda D_{co}/T_{co}$, λ = slope length, D_{co} = detachment capacity by flow at the end of slope and T_{co} = transport capacity of flow at the end of the slope. The variable α is a measure of the capacity of detachment by flow to fill transport capacity. Though not explicitly shown in equation (28.10), the amount of sediment produced by detachment by flow is significantly reduced by the presence of sediment from interrill erosion and by that arriving from upslope.

Groundcover and depressional storage can greatly reduce soil erosion. Its effect can be considered by assuming that the total shear stress of the flow is distributed between the cover, the depressional roughness and the soil (Foster and Meyer, 1975; Foster et al., 1982a). The part acting on the soil is assumed to be responsible for detaching and transporting sediment. The shear stress acting on the soil can be computed from:

$$\tau_s = \gamma V_{cr}^2 f_s/8g \quad (28.11)$$

where γ = weight density of the runoff, V_{cr} = flow velocity when cover and form roughness is present, f_s = friction factor related to grain roughness of soil and g = acceleration of gravity. The concept is similar to grain and form roughness used to analyse channel hydraulics (Graf, 1971). Total friction factor, the sum of friction factors related to cover, form roughness such as depressional areas and grain roughness, is given by:

$$f_t = f_c + f_r + f_s \quad (28.12)$$

where f_t = total friction factor, f_c = part of friction factor related to cover, f_r = part of friction factor related to form roughness on the soil surface and

f_s = part of friction factor related to grain roughness of the soil. The relation of velocity to the friction factor f_t is given by (Foster, 1982):

$$f_t = 8gsq/V_{cr}^3 \qquad (28.13)$$

where s = land slope and q = discharge per unit width. The relationship of shear stress acting on the soil with cover and form roughness to the shear stress of flow over bare, smooth soil is given by (Foster, 1982):

$$\tau_s/\tau_b = (V_{cr}/V_b)^2 \qquad (28.14)$$

where τ_b = shear stress for bare, smooth soil and V_b = velocity of flow over bare, smooth soil. This approach is very powerful for dealing with the effect of surface cover and surface roughness on detachment and transport capacity by flow.

Although several equations are used to compute transport capacity of flow in rills, one of the best seems to be that of Yalin (Alonso et al., 1981). This gives reasonable results for a broad range of sediment sizes and densities without calibration. Before an equation is chosen for use in an erosion model its behaviour for the hydraulic and sediment characteristics typical of agricultural fields should be checked. Some common sediment transport equations give very poor results for these conditions (Alonso et al., 1981). Development of an equation for sediment transport capacity for these conditions is an important research need.

Ephemeral Gully Areas

Many landscapes are shaped such that runoff collects in a few major natural waterways, called ephemeral gullies, before leaving farm fields (Foster and Lane, 1983). Also, constructed channels such as terrace channels behave as ephemeral gullies. Most of the runoff and sediment produced in a field leave it in these channels.

These channels are tilled across annually unless they are grassed waterways. Tillage moves soil from adjacent areas and fills voided areas where erosion has occurred in the channel. This leaves many soils highly erodible, but over time the soil consolidates and becomes less so (Foster et al., 1982a,b). The untilled soil immediately beneath the surface-tilled layer often acts as a non-erodible layer restricting the depth of erosion. Ephemeral gullies frequently have large depth-to-width ratios in the order of 30 to 1.

Ephemeral gullies can contribute significant amounts of sediment—in erosive climates as much as that produced by rill and interrill erosion (Foster and Lane, 1983). In other climates where rainfall intensity is low, the proportion of runoff to rainfall is high and soils are erodible, ephemeral gully erosion can contribute most of the sediment produced in a field.

The grade along some ephemeral gullies is concave, causing much deposition within the field. Transport capacity near the outlets of ephemeral gullies with concave profiles is a major factor determining sediment yield from a field.

Ephemeral gully erosion processes are like those in rills except for the relative importance of the processes. Therefore, equations used to describe rill erosion can also be used for ephemeral gully erosion (Foster and Lane, 1983). Sediment delivery from the rills is the lateral input (D_i in equation (28.3)) to the ephemeral gullies, while detachment or deposition in the ephemeral gully is given by D_f.

Impoundments

Both natural and constructed impoundments can occur in fields. Constructed impoundments include farm ponds and tile outlet terraces designed to impound runoff and to discharge it through underground tile. The primary processes in impoundments are sediment transport and large rates of deposition. These processes may be modelled using settling tank and batch reactor mixing theory (Barfield *et al.*, 1981).

Unit Source Watershed

A unit source watershed can be defined as a watershed having a first-order ephemeral gully or similar concentrated flow channel. The response of the next-order watershed is analysed by application of the continuity equation where the input from the upstream watershed is treated as a point input rather than as input from rills uniformly distributed along an ephemeral gully.

HYDROLOGY

Modelling fundamental hydrologic processes is necessary in process-based erosion modelling because hydrologic inputs are needed to drive the erosion equations such as those for detachment by raindrop impact and by flow. The two principal hydrologic variables needed in the erosion equations are rainfall intensity and runoff rate. Infiltration must be considered to convert rainfall intensity to runoff rate, and computation of infiltration requires considering the hydrologic processes affecting antecedent soil moisture. As with erosion, runoff modelling is moving from empirically based methods such as the SCS curve number method (USDA-SCS, 1972) to more process-based methods like the Green–Ampt infiltration equation (Skaggs and Khaleel, 1982).

After values for hydrologic variables are computed, they must be converted to hydraulic variables. For example, runoff rate must be converted to shear stress that applies to the particular source area. Values for hydraulic variables for interrill areas can be computed assuming broad sheet flow but channel geometries should be used to compute values for hydraulic variables for flow in rills and ephemeral gullies. Careful consideration of hydraulics can greatly minimize distortion in parameter values between ideal laboratory studies and the field (Lane *et al.*, 1975).

NON-UNIFORM LANDSCAPES

Most landscapes are spatially non-uniform in shape and soil properties and sometimes land use. The effect of spatial non-uniformity is considered by solving the governing equations with variables for steepness, soil and other properties written as functions of distance along the slope. Figure 28.3 illustrates how slope shape causes important erosion processes to vary along a complex slope.

The complexity of the governing equations often requires numerical solutions, especially when unsteady rainfall and runoff is considered. Although closed-form solutions for steady and unsteady conditions can be derived from special cases (Singh, 1983), analysing three-dimensional landscapes usually requires numerical approaches. These are often built around one of two concepts, a grid approach, illustrated in Figure 28.4, like that used by the ANSWERS model (Beasley *et al.*, 1980), or a equipotential-stream tube approach (Figure 28.5) (Moore *et al.*, 1988). A major advantage of the stream tube approach is that

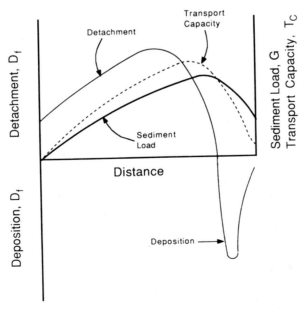

Figure 28.3 Variation in erosion processes along a complex slope

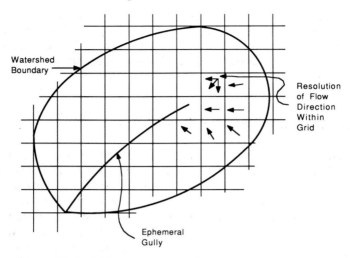

Figure 28.4 Grid representation of a unit source watershed

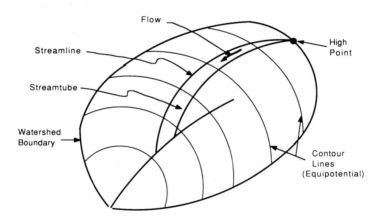

Figure 28.5 Streamtube representation of a unit source watershed

flow paths on the slope are one-dimensional, which greatly simplifies solution of the governing equations. Both the grid and stream tube approaches rely heavily on contour maps. In the future, digital terrain models will provide the core of erosion models for representing dimensional landscapes (Moore *et al.*, 1988).

CRITICAL ISSUES

Models are developed for a variety of purposes. Some like the USLE are used by action agencies in inventories, assessments and conservation planning while others are employed as research tools to study processes and the behaviour of hydrologic and erosion systems. The science in research models becomes the basis of applied models developed a decade or two later.

Critical issues related to development of research models include a better understanding of the basic processes, creative schemes for dealing with spatial non-uniformity and scale (including ways of representing detailed processes where needed) and much better relationships for describing soil erodibility and how it is affected by climate, vegetation and management. Some of the critical issues related to applied models concern availability of an array of parameter values so that the model can be applied to a broad range of conditions, reasonable effort to acquire input values, computationally efficient computer programs implementing the model, and easy-to-use models and computer programs.

EMERGENCE OF PROCESS-BASED TECHNOLOGY

Process-based erosion-prediction technology is not new. Horton's (1938) and Ellison's (1947) work in the 1940s provided many of the basic ideas for process-based erosion modelling. However, these technologies did not emerge until the 1970s, after mainframe computers became readily available and interest in erosion was stimulated by concern for surface-water quality. CREAMS (Knisel et al., 1980), ARM (Donigian and Davis, 1978) and ANSWERS (Beasley et al., 1980) are typical models from this era. More recently, Morgan et al. (1986) and others in Europe have progressed in developing process-based erosion models.

A major shortcoming of many of the existing process-based models is that some, like CREAMS, are partly based on the USLE and thus have some of the same limitations as does the USLE. Other models, such as ANSWERS, are not easily used or convenient as replacements for the USLE. In addition, all current models using equations based on fundamental erosion processes include parameters that require an extensive database of values needed to apply the models to the full range of field conditions encountered by conservationists. The required database does not exist for any current process-based model. Therefore, if the USLE is to be replaced as a conservation tool for field applications a major research effort is needed to formulate a process-based model in a convenient, easy-to-use form and to develop the necessary parameter values.

Such a major effort, the USDA-Water Erosion Prediction Project (WEPP), is currently underway to develop new-generation, process-based technology for use by action agencies (Foster and Lane, 1987b). This, led by the USDA-Agricultural Research Service (ARS), involves several co-operators, including government agencies and universities. The first version of the WEPP model is expected in August 1989. After field testing and further development, a version for use by local conservationists is expected in 1992.

This project, led by Dr L. J. Lane, USDA-Agricultural Research Service, Aridland Watershed Management Research Unit, 2000 East Allen Road, Tucson, AZ 85719, is a team effort between scientists and users. The project also involves a large experimental programme on cropland, rangeland, and disturbed forest lands at about 50 locations in the United States and at some foreign locations. If fully successful, WEPP will be a major milestone in erosion research and will shift much of erosion research from the USLE empirically based approach to a fundamental, process-based one. In many applications the USLE will be replaced by the WEPP model.

USLE REVISION AND UPDATE

The USLE is currently being revised and updated in a USDA-ARS project led by Dr K. G. Renard, USDA-Agricultural Research Service, Aridland Watershed Management Research Unit, 2000 East Allen Road, Tucson, AZ 85719, in cooperation with the USDA-Soil Conservation Services, other agencies, and universities. The changes in the USLE factors include a greatly improved erosivity map for the western United States, a review of worldwide data on soil erodibility, a slope length factor related to the ratio of rill to interrill erosion, a slope steepness factor that is more linear than the current factor, use of the subfactor method to compute the cover-management factor and revision of contour factor values. This revision and update of the USLE may be the last major effort that the USDA-ARS invests in the USLE.

SUMMARY

Erosion prediction is a powerful tool that has been widely used to deal with sheet and rill erosion problems on agricultural land. The Universal Soil Loss Equation (USLE) is by far the most widely used method for estimating soil erosion by water. The USLE, derived from a large mass of data, is empirically based and does not explicitly represent erosion processes. As a result, the method has severe limitations.

Erosion-prediction technology based on fundamental hydrologic and erosion processes overcomes many of the limitations of the USLE. The major fundamental erosion processes are detachment by raindrop impact, detachment by flow, transport by raindrop impact, transport by flow and deposition by flow. Erosion prediction by this technology is accomplished using mathematical equations that describe these processes and their interrelationships.

Most problems of practical consequence for erosion on agricultural lands are field scale, but the equations representing the processes are on a much smaller scale. The concept of linking source areas hierarchically is one way to deal with scale. Important source areas are interrill areas, rills, ephemeral gullies, impoundments and unit source watersheds.

A major international team project (WEPP) is underway to develop new-generation, process-based technology for estimating soil erosion. The technology is being developed for use by action agencies on practical, field problems. This project, led by the USDA-Agricultural Research Service, is expected to produce its field product by the early 1990s. If expectations are realized, the WEPP model will replace the USLE as the principal method for estimating soil erosion by water on agricultural lands.

REFERENCES

Al-Durrah, M. M. and Bradford, J. M. (1982). The mechanism of raindrop splash on soil surfaces, *Soil Sci. Soc. Amer. J.*, **46**, 1086-90.

Alonso, C. V., Neibling, W. H. and Foster, G. R. (1981). Estimating sediment transport capacity in watershed modeling, *Trans. of the Amer. Soc. of Agric. Engr.*, **24**, 1211-20, 1226.

Barfield, B. J. (1988). Personal communication, University of Kentucky, Lexington, Kentucky.

Barfield, B. J., Warner, R. C. and Haan, C. T. (1981). *Applied Hydrology and Sedimentology for Disturbed Areas*, Oklahoma Technical Press, Stillwater, Oklahoma.

Beasley, D. B., Monke, E. J. and Huggins, L. F. (1980), ANSWERS: A model for watershed planning, *Trans. of the Amer. Soc. of Agric. Engr.*, **23**, 939-44.

Bennett, J. P. (1974). Concepts of mathematical modeling of sediment yield, *Water Resources Research*, **10**, 484-92.

Dickinson, T. (1989). Personal communication, University of Guelph, Guelph, Ontario.

Donigian, A. S. and Davis, H. H. (1978). *User's Manual for Agricultural Management (ARM) Model*, EPA-600/3-76-083, US Environmental Protection Agency, Washington, DC.

Ellison, W. D. (1947). Soil erosion studies, *Agric. Engr.*, 145-6, 197-201, 245-8, 297-300, 349-51, 402-5, 442-4.

Elwell, H. A. (1981). A soil loss estimation technique for Southern Africa. In Morgan, R. P. C. (ed.), *Soil Conservation: Problems and Prospects*, John Wiley, Chichester, pp. 281-92.

Foster, G. R. (1982). Modeling the erosion process. In Haan, C. T., Johnson, H. P. and Brakensiek, D. L. (eds), *Hydrologic Modeling of Small Watersheds*, Amer. Soc. of Agric. Engr., St Joseph, Michigan, pp. 297-382.

Foster, G. R., Johnson, C. B. and Moldenhauer, W. C. (1982a). Hydraulics of failure of unanchored cornstalk mulches for erosion control, *Trans. of the Amer. Soc. of Agric. Engr.*, **25**, 940-47.

Foster, G. R. and Lane, L. J. (1983). Erosion by concentrated flow in farm fields. In *Proc. of the D. B. Simons Sympos. on Erosion and Sedimentation*, Colorado State University, Fort Collins, Colorado, pp. 9.65-9.82.

Foster, G. R. and Lane, L. J. (1987a). Beyond the USLE: Advancements in soil erosion prediction. In Boersma, L. L. *et al.* (eds), *Future Developments in Soil Science Research*, Soil Sci. Soc. of Amer., Madison, Wisconsin, pp. 315-26.

Foster, G. R. and Lane, L. J. (1987b). *USDA-Water Erosion Prediction Project (WEPP): User Requirements*, USDA-Agricultural Research Service, National Soil Erosion Research Laboratory, West Lafayette, Indiana.

Foster, G. R., Lane, L. J., Osterkamp, W. R. and Hunt, D. W. (1982b). Effect of discharge on rill erosion, *Paper No. 82-2572*, Amer. Soc. of Agric. Engr., St Joseph. Michigan.

Foster, G. R. and Meyer, L. D. (1972). A closed-form soil erosion equation for upland areas. In Shen, H. W. (ed.), *Sedimentation (Einstein)*, Colorado State University, Fort Collins, Colorado, Chapter 12.

Foster, G. R. and Meyer, L. D., (1975). Mathematical simulation of upland erosion by fundamental erosion mechanics. In *Present and Prospective Technology for Predicting Sediment Yields and Sources*, ARS-S-40, USDA-Sci. and Educ. Admin., Washington, DC, pp. 190–204.

Foster, G. R., Simanton, J. R., Renard, K. G., Lane, L. J. and Osborn, H. B. (1981). Discussion of application of the universal soil loss equation to rangelands on a per storm basis, *J. of Range Management*, **34**, 161–5.

Foster, G. R. and Wischmeier, W. H. (1974). Evaluating irregular slopes for soil loss prediction, *Trans. of the Amer. Soc. of Agric. Engr.*, **17**, 305–7.

Foster, G. R., Young, R. A. and Niebling, W. H. (1985). Sediment composition for nonpoint source pollution analyses, *Trans. of the Amer. Soc. of Agric. Engr.*, **28**, 133–9, 146.

Graf, W. H. (1971). *Hydraulics of Sediment Transport*, McGraw-Hill, New York.

Hairsine, P. and Rose, C. (1989). Personal communication, Griffith University, Brisbane, Queensland, Australia.

Hayes, W. A. (1977). Estimating water erosion in the field. In Foster, G. R. (ed.), *Soil Erosion: Prediction and Control*, Soil and Water Conserv. Soc. of America, Ankeny, Iowa, pp. 6–11.

Heimlich, R. E. and Bills, N. L. (1986). An improved soil erosion classification system: update, comparison, and extension. In Carlson, C. (ed.), *Soil Conservation: Assessing the National Resources Inventory*, Vol. 2, National Academy Press, Washington, DC, pp. 1–17.

Horton, R. E. (1938). The interpretation and application of runoff plot experiments with reference to soil erosion problems, *Soil Sci. Soc. Amer. Proc.*, **3**, 340–49.

Huang, C., Bradford, J. M. and Cushman, J. H., 1983. A numerical study of raindrop impact phenomena: the elastic deformation case, *Soil Sci. Soc. Amer. J.*, **47**, 855–61.

Knisel, W. G. and Foster, G. R. (1980). CREAMS: A system for evaluating best management practices. In Jestse, W. E. (ed.), *Economics, Ethics, Ecology: Roots of a Productive Conservation*, Soil Conserv. Soc. of America, Ankeny, Iowa, pp. 177–94.

Lane, L. J., Woolhiser, D. A. and Yerjevich, V. (1975). *Influence of Simplification Geometry in Simulation of Surface Runoff*, Hydrology Paper no. 81, Colorado State University, Fort Collins, Colorado.

Meyer, L. D., Foster, G. R. and Nikolov, S. (1975). Effect of flow rate and canopy on rill erosion, *Trans. of the Amer. Soc. of Agric. Engr*, **18**, 905–11.

Meyer, L. D. and Harmon, C. (1985). Sediment losses from cropland furrows on different gradients, *Trans. of the Amer. Soc. Agric. Engr*, 298, 448–53, 461.

Meyer, L. D. and Wischmeier, W. H. (1979). Mathematical simulation of the process of soil erosion by water, *Trans. of the Amer. Soc. of Agric. Engr*, **12**, 754–8, 762.

Moore, I. D., O'Loughlin, E. M. and Burch, G. J. (1988). A contour-based topographic model for hydrological and ecological applications, *Earth Surface Processes and Landforms*, **13**, 305–20.

Morgan, R. P. C., Morgan, D. D. V. and Finney, H. J. (1986). A simple model for assessing annual soil erosion on hillslopes. In Giorgini, A. and Zingales, F. (eds), *Agricultural Nonpoint Source Pollution: Model Selection and Application*, Elsevier, Oxford, pp. 147–59.

Moss, A. J., Walker, P. H. and Hutka, J. (1979) Raindrop-stimulated transportation in shallow water flows: an experimental study, *Sedimentary Geology*, **22**, 165–84.

Nearing, M. A., Bradford, J. M. and Holtz, R. D. (1987). Measurement of force vs. time relationships for waterdrop impact, *Soil Sci. Soc. Amer. J.*, **50**, 1532–6.

Poesen, J. (1989). Personal communication. University of Leuven, Leuven.

Ritterbusch, J. (1988). How goes conservation compliance planning?, *J. of Soil and Water Conserv.*, **43**, 376–8.

Renard, K. G. and Foster, G. R. (1983). Soil conservation: Principles of erosion by water. In Dregne, H. E. and Willis, W. O. (eds), *Dryland Agriculture*, Agronomy Monograph no. 23, Amer. Soc. of Agron., Madison, Wisconsin, pp. 155–76.

Singh, V. P. (1983). Analytical solutions of kinematic equations for erosion on a plane. II. Rainfall of finite duration, *Advances in Water Resources*, **6**, 88–95.

Skaggs, R. W. and Khaleel, R. (1982). Infiltration. In Haan, C. T., Johnson, H. P. and Brakensiek, D. L. (eds), *Hydrologic Modeling of Small Watersheds*, Amer. Soc. of Agric. Engr., St Joseph, Michigan, pp. 121–66.

USDA-Soil Conservation Service (SCS) (1972). *National Engineering Handbook*, Hydrology section, Chapt. 4–10, US Dept. of Agric., Washington, DC.

Wischmeier, W. H. and Smith, D. D. (1978). *Predicting Rainfall Erosion Losses*, Agric. Hbk 537, USDA-Sci. and Educ. Admin., Washington, DC.

29 Prediction of Ephemeral Gully Erosion on Cropland in the South-eastern United States

COLIN R. THORNE
Department of Geography, University of Nottingham

and

LYLE W. ZEVENBERGEN
Water Engineering and Technology, Fort Collins, Colorado, USA

INTRODUCTION

For soil conservation and management purposes, rill and interrill erosion in arable fields can be estimated quickly and simply using the Universal Soil Loss Equation (USLE). No similar approach to estimating soil loss due to ephemeral gullying is available. This is unfortunate, because the erosion due to ephemeral gullying may be comparable to rill and interrill erosion. Under these circumstances, application of the USLE alone results in soil loss estimates which are about half the true loss. This is a serious problem, because accurate erosion estimates are necessary for effective management of the land for runoff and erosion control.

This chapter presents a simple, easy-to-apply technique to estimate ephemeral gully erosion on the basis of a single field visit, consultation with the farm and practical experience. Either a hand-held calculator or a nomograph is used to aid calculations. The method is intended to predict ephemeral gully erosion only for the first year of gully development—that is, from gully initiation following seedbed preparation to its eradication about one year later by annual tillage. The method could be used to estimate ephemeral gully erosion in subsequent years, when the gully has not been eradicated at the end of the first year, but this would require modification of the equations employed to predict gully cross-sectional area.

This method is intended for use by field personnel, such as District Conservationists in the USDA Soil Conservation Service. Such persons are called upon daily to make estimates of ephemeral gully erosion. In theory, good estimates could be made using process-based computer models such as CREAMS (Knisel, 1980). In practice, limitations of data availability, computer hardware and the specialized expertise necessary to run such models successfully, preclude this. In the future, more straightforward process-based approaches such as WEPP (Water Erosion Prediction Program), currently being developed by the USDA Soil Conservation Service and Agricultural Research Service, should supersede this method. Meanwhile, it presents a simple alternative to guesswork.

GENERAL APPROACH

The basis of the method lies in the analysis of field topography. Ephemeral gullies usually form in topographic lows (swales) in the field, where streamlines of surface and subsurface runoff converge to produce concentrated overland flow. This has long been recognized by researchers and conservationists, but the problem has been to convert this qualitative recognition into a quantitative erosion-prediction technique, without introducing complex modelling of water and sediment processes. It is known that ephemeral gullies occur where there is a sufficient concentration of streampower to initiate and maintain erosion. Streampower is defined by the product of discharge, slope and unit weight of water. Its concentration is a function of the flow convergence.

In our method a pragmatic approach is adopted, aimed at producing a technique which is reasonably easy to use, does not require comprehensive data collected over extended periods, can be applied by field personnel, but maintains acceptable accuracy.

A Compound Topographic Index (CTI) is used to predict the intensity (or streampower) of concentrated surface runoff at any point in a swale. The CTI incorporates the upstream drainage area (a surrogate for discharge) and the local slope, which, taken together, represent total streampower because the unit weight of water is about constant. The degree of flow convergence (planform curvature), at each point determines the concentration of streampower on the soil surface. A critical or threshold value of CTI is necessary for the concentrated flow to initiate ephemeral gully erosion. This value varies, depending on the complex interaction of variables, including climate, soil type, cropping, management and conservation practice. At present, the critical value cannot be calculated reliably from basic principles. Instead, it is calibrated for the specific site in question, using measurements of CTI at the known locations of critical conditions—that is, around gully heads where erosion is initiated. The size of ephemeral gully likely to develop at a point in an average year is then predicted using an empirical equation expressing gully cross-sectional area as a function of local CTI value,

Ephemeral Gully Erosion on Cropland

where that value is greater than the critical CTI. The empirical equation is based on field data collected by the US Army Engineers, Waterways Experiment Station, in central Mississippi between 1984 and 1985. These data were supplied to the authors by Dr Lawson Smith. The method is illustrated in this report using data independently collected by the first author at one of Dr Smith's field sites in 1987.

The limiting assumptions are that soil properties and rainfall distributions are reasonably uniform over the site in question and that gullies are the result of overland flow erosion rather than subsurface processes. The theory of the topographic analysis used here has been described in a recent paper (Zevenbergen and Thorne, 1987) and application of topographic analysis to ephemeral gully erosion production using a computerized analysis of gridded altitude data has been outlined elsewhere (Thorne et al., 1986).

The whole basis to the approach is described in great detail in a thesis by the second author, copies of which may be obtained from either author. Interested readers are referred to that thesis for further information (Zevenbergen, 1989).

In this chapter, first, the field data needed are outlined; second, the methods of undertaking the necessary calculations are presented; and third, a worked example is given. Finally, the main points are summarized.

FIELD DATA COLLECTION

The procedure is straightforward and, with experience the necessary field measurements and calculations can be completed in less than one hour per acre of arable field. The procedure may be broken down into a series of steps as follows.

Locating Ephemeral Gully or Gullies

Ephemeral gullies usually form in swales (topographic lows) in the field, where surface runoff concentrates due to ground surface convergence. Therefore, the first step in the procedure is to walk the field and pick out the significant swales. A scaled sketch map of the field, showing swales and drainage divides between swales, is then prepared (Figure 29.1(a)). A particular swale may or may not contain an ephemeral gully at the time of the visit, depending on many factors such as the time of year in relation to cropping and tillage practices and the severity of rainfall events prior to the visit. Hence, unless the visit is made just prior to annual tillage, after an average year, it will not be possible to estimate the ephemeral gully erosion directly by simply measuring the volume of gullies present. Instead, the farmer is consulted as to the usual location of gullies and *in particular* the position of the gully heads just prior to tillage. Each gully head

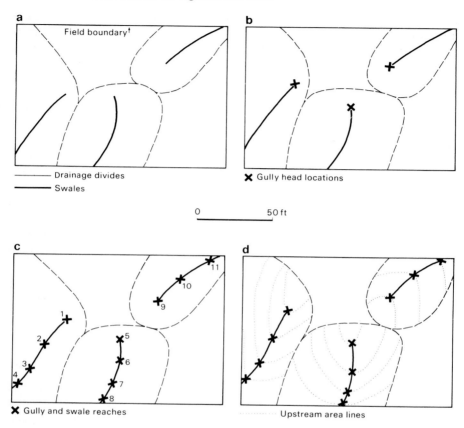

Figure 29.1 Sketch map for field data collection

location is then identified with a survey flag and these points are marked on the sketch map (Figure 29.1(b)). Gully head points are used to calibrate the critical CTI for the field. The method is quite robust in this regard, and is not very sensitive to the exact location of the gully heads. Hence, the information needs to be accurate but not especially precise.

Selecting Representative Reaches

Between two and ten additional flags are used to mark the course of any ephemeral gullies present, from their heads to the point where they leave the field. Flags should be spaced so as to divide the length into characteristic reaches, so that within a reach the gully's size and shape are relatively constant. Any swales not containing a gully are similarly flagged, the spacing in this case being

Ephemeral Gully Erosion on Cropland

determined so as to divide the swale into characteristic reaches of slope, width and depth (Figure 29.1(c)). The points are the locations for measurement of the topographic indices that go into the CTI.

Measuring Topographic Parameters

The required topographic parameters are:

(1) Stake height (H ft);
(2) Left swale width (A ft);
(3) Right swale width (B ft);
(4) Local slope (S ft/ft);
(5) Upstream area, ($AREA$ ft^2);
(6) Distance downstream from gully or swale head (L ft).

These parameters are measured in the field at each flagged point, starting at each gully or swale head and working downstream. The values are recorded in a results table. Table 29.1 is an example of a completed results table. US customary units are strongly preferred to SI units because they are much more amenable to use by non-technical and field personnel.

The measurement methods are as follows:

(1) For the measurement of stake height (H) and swale widths (A and B), insert a stake next to the flagged point, so that its top is 1 ft above the ground. (*Note*: If an ephemeral gully is present, position the stake on the bank top, not the gully bed (Figure 29.2))
(2) Measure the horizontal distance across the swale, at right angles to the downstream direction, from the top of the stake to the ground on the left and right sides of the swale (A and B in Figure 29.2). It is not necessary for A and B to be equal to each other. If either A or B is undesirably large and therefore difficult to measure, reduce the stake height to 0.5 ft. Record the values of H, A and B in the results table. The purpose of these measurements is to define the planform curvature (PLANC) of the swale,

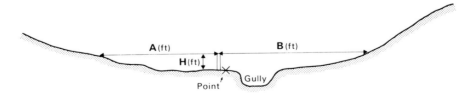

Figure 29.2 Field measurement of H, A and B

Table 29.1 Lawson Smith's Site 3, Warren County, Mississippi: results table
Date: 20 November 1987   **Start:** 12.00   **Finish:** 13:45

Swale number (length, ft)	Section number (flag)	Distance from gully head (ft)	H (ft)	A (ft)	B (ft)	Slope (ft/ft)	Upstream area (ft²)	CTI (gully head only)	X-AREA (ft²)	Volume voided (ft³)
1	1	0	0.5	14.5	15.5	0.064	1481.25	42	0.51	
	2	34	0.5	22.0	7.0	0.074	2637.50		0.67	
	3	55	0.5	20.0	9.8	0.100	3343.75		0.72	
	4	74.5	0.5	33.0	21.0	0.126	3943.75		0.58	
	5	110.5	0.5	24.0	24.0	0.049	5056.25		0.51	
2	1	0	0.5	12.0	20.0	0.050	3568.75		0.59	67
	2	20	0.5	17.0	22.0	0.090	4418.75	74	0.64	
	3	58	0.5	26.0	22.0	0.065	5793.75		0.57	
3a	1	0	0.5	26.0	28.0	0.029	1612.50		<CRIT	35
	2	34	0.5	28.0	23.0	0.068	5587.50	6.4	0.55	
	3	77	0.5	16.0	29.0	0.079	7537.5		0.67	
3b	1	0	0.5	21.0	26.0	0.041	3850.00		<CRIT	36
	2	31	0.5	13.5	25.0	0.039	5762.50	29	0.57	
	3	76	0.5	30.0	17.0	0.042	8000.00		0.57	35
								(CTIcrit = 38)	T.VOL	173 CUFT
									EROS.	8.2 TNS
									FIELD.A	1.3 ACRES
									UERO.	6.3 T/ACR

Observed rate (Lawson Smith) 7.0 T/ACR

which controls the convergence of surface runoff and hence the concentration of erosive streampower per unit bed area. The calculation of PLANC is incorporated into the nomograph, and is also presented explicitly in the section on hand calculation.

(3) The local slope down the swale is measured in the field with a hand level, survey staff and tape, or it may be estimated by eye by a person of great experience. It is expressed as a decimal (feet per foot) and noted in the results table. Slope is an important factor determining the total streampower and hence the erosivity of surface runoff.

(4) The scaled sketch map is used to determine the upstream area draining to each point. On the basis of the field reconnaissance interpretation of the drainage basin shape a line defining the drainage area for each point is sketched in Figure 29.1(d). The area enclosed by the drainage divide and area lines is measured by planimeter, by counting the squares on graph paper, or by dot counting, and is recorded, in square feet, in the results table. Upstream drainage area is an important variable determining the volume of runoff at a point, which, in turn, affects gully size.

(5) The distance downstream from gully or swale head (L) is measured by tape, by range finder or by pacing, and is noted in the results table. If the first point is located at the gully or swale head, the downstream distance is zero for that point. Subsequent points on the same gully or swale have cumulative downstream distances. This completes the measurements for that flagged point, and the same measurements are made for the next flagged point and so on, to the last point in the first swale. If the last point is at the end of the gully or swale or where the gully leaves the field, then its downstream distance corresponds to the swale length (to be noted in column one of the results table). If the last point is not at the end of the swale, measure the distance to the end and add this to the downstream distance to obtain the swale length. This completes measurements for the first swale.

The whole process is repeated for the second swale, and so on until all the swales in the field have been measured. When the last swale has been measured and the results recorded, the topographic parameters are complete and the ephemeral gully erosion can be calculated. Calculation may be undertaken in the field or later in the office, but the former is recommended to allow checking of data which appear to be in error.

CALCULATIONS

The necessary calculations may be carried out most easily using the nomograph (Figure 29.3). However, if desired, the basic equations may be used in hand calculation of the expected ephemeral gully erosion. The following section illustrates how this is done, and also gives the background to the derivation of the nomograph.

Basic Equations and Hand Calculation

The CTI incorporates the upstream area (AREA), slope (S) and planform curvature (*PLANC*) parameters and is defined by:

$$CTI = AREA.S.PLANC \tag{29.1}$$

where *PLANC* is given by:

$$PLANC = \frac{200H}{AB} \tag{29.2}$$

CTI is calculated for each point and the value noted in column nine of the results table. The critical or threshold value of CTI necessary for the initiation of ephemeral gully erosion in the field in question is calculated by averaging the gully head CTIs for that field. Mathematically:

$$CTI_{crit} = \frac{\sum_1^n [CTI_{gh}]}{n} \tag{29.3}$$

where

CTI_{crit} = critical CTI n = number of gully heads,
CTI_{gh} = gully head CTI.

Points with CTI values less than the critical value are excluded from the cross-sectional area calculations because they would not have an ephemeral gully in an average year. For swale head locations this suggests insufficient upstream area, slope or convergence to initiate erosion. For points lower down a swale, it indicates intermediate deposition of soil eroded upstream, due to flattening of slope or opening out of the swale (decreasing or zero convergence). Mathematically, this test is written

$$CTI < CTI_{crit} \quad ?$$

At this stage all CTIs which fail the test are crossed through in the results sheet and excluded from further analysis.

Cross-sectional area (*X-AREA*) for the ephemeral gully at each point with a CTI greater than the critical value is found from the experimentally derived equation:

$$X\text{-}AREA = \frac{(CTI)^{0.25}}{5} \text{ ft}^2 \qquad (29.4)$$

The results are noted in column ten of the results table.

Ephemeral gully erosion is calculated by multiplying the cross-sectional area for each point by the length of the reach which it represents. For the first point on a gully or swale:

$$\text{Volume voided}_1 = (X\text{-}AREA \text{ Pt } 1)\left(\frac{\text{Downstream distance to Pt 2}}{2}\right) \qquad (29.5a)$$

For intermediate points:

$$\text{Volume voided}_n = (X\text{-}AREA \text{ Pt } n)\left[\frac{(L_n - L_{n-1})}{2} + \frac{(L_{n+1} - L_n)}{2}\right] \qquad (29.5b)$$

For the last point (L):

$$\text{Volume voided} = (X\text{-}AREA \text{ Pt Last})\left[\frac{(L_L - L_{L-1})}{2} + (SL - L_L)\right] \qquad (29.5c)$$

Where L_n = Downstream distance to point in question,
L_{n-1} = Downstream distance to previous point,
L_{n+1} = Downstream distance to next point,
L_L = Downstream distance to last point,
SL = Total swale length.

The results are recorded in column eleven of the results sheet.

The total volume eroded (in cubic feet) is found by summing the reach volumes for the whole field. If a weight of soil eroded is required, multiply this volume figure by the soil unit weight in pounds per cubic foot and divide by 2000 to obtain tons. If the weight per acre is required, divide by the total field area in acres to obtain tons/acre.

Using the Nomograph

If no calculator is available, the nomograph shown in Figure 29.3 can be used to speed up the calculation procedure. The nomograph is entered on the left

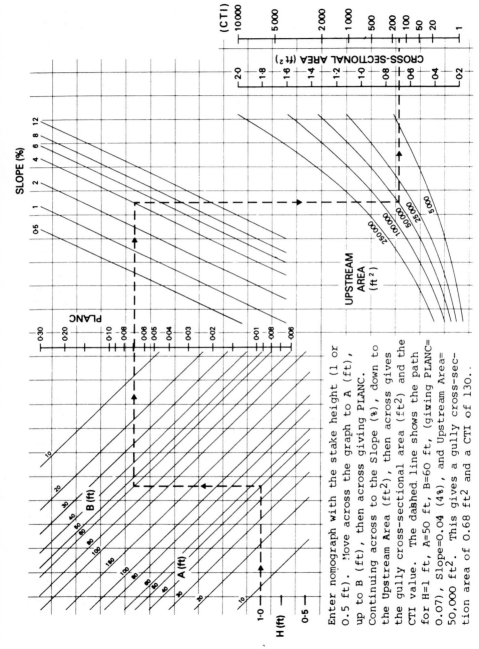

Figure 29.3 Nomograph for determination of ephemeral gully cross-section area and CTI value

scale of stake height (0.5, 0.75 or 1.0 ft). Then the *A* and *B* distances are used to find *PLANC*. Next the Local Slope (per cent) and Upstream Area are used to determine cross-sectional area and CTI (Compound Topographic Index). When the area and CTI have been determined for all points, the gully head CTI values are averaged to estimate the critical value for the field, as already discussed. Any points with CTIs less than the critical value are given zero cross-sectional area because, for the field in question, they would not be expected to have an ephemeral gully in an average year, as explained above. The ephemeral gully erosion is calculated using the method outlined in the previous section.

EXAMPLE

This example is based on an analysis of ephemeral gully erosion at one of Dr Lawson Smith's field sites in central Mississippi. This site was visited on 20 November 1987. The drainage divides, swales and observed ephemeral gullies were marked on a sketch map (Figure 29.4). For the experimental area 4 ephemeral gullies were identified. For each of the swales containing a gully, topographically similar reaches were identified and points selected to be representative of the gully head, gully end or exit from the field and intermediate reaches between. Distance downstream from the swale or gully head was measured for each point, and lines defining the upstream drainage area were sketched (Figure 29.4). The upstream area was then measured by planimeter. The slope at each point was measured in the field using a hand level and staff. The planform curvature was measured in the field, using the method previously described.

A figure of 38 was calculated for the CTI_{crit}. This is typical for the soils and climate in Mississippi. Other, different, values will be found for any particular field and crop combination and it is impossible to arrive at any universal values. However, the strength of the method is that the particular value for a field will always be representative, because it is calibrated for that field on the basis of the actual combination of streampower and convergence needed to initiate a gully head in that field.

A calculator was used to complete columns ten and eleven in Table 29.1. The predicted unit erosion rate of 6.3 tons per acre may be compared to that of 7.0 tons per acre, found by Lawson Smith based on 5 years of detailed monitoring. The computed erosion rate is about 10 per cent too low, but is well within the limits of acceptability for a simple field method such as this, and compares favourably to predictions based on much more sophisticated analyses. Further tests on sites throughout the state of Mississippi have been undertaken by the second author and by Soil Conservation Service Personnel. These support the conclusion that the method generally produces erosion estimates within about ± 10 per cent of the true value.

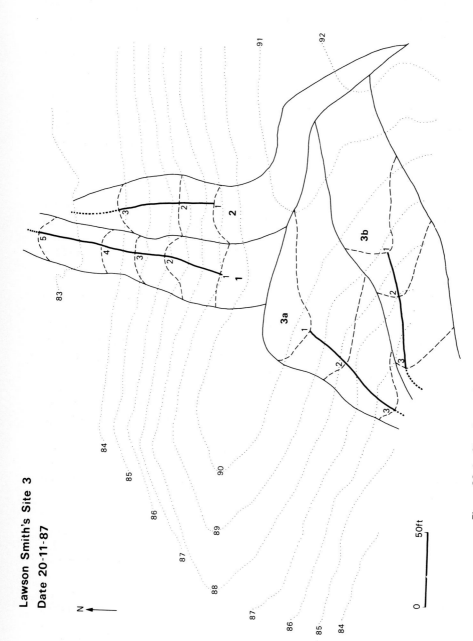

Figure 29.4 Sketch map for worked example, Warren County

The method is very robust and it is difficult to produce wildly inaccurate estimates, even when the quality of the field data is limited. This makes it a very reliable tool for use by non-specialist field workers.

SUMMARY AND CONCLUSIONS

It has been recognized that ephemeral gully erosion is not accounted for in Universal Soil Loss Equation estimates of erosion in arable fields. A simple, on-site method to estimate the additional soil erosion due to ephemeral gullies has been developed, based on the following inputs:

(1) A site visit by a field scientist;
(2) Consultation with the farmer;
(3) Professional experience and judgement.

The procedure is straightforward and can be completed in less than 1 hour per acre of field area. In summary, the steps are:

(1) Walk the field, noting the locations of swales, drainage divides and ephemeral gullies.
(2) Consult the farmer as to the usual location of gullies and particularly the location of the gully head in each swale. Flag each gully head location.
(3) Make a sketch map (to scale) of the field, showing the swales, drainage divides, ephemeral gullies and gully head locations.
(4) Select representative reaches of each swale based on local topography and the size of any gully present. Flag intermediate points in each reach. Measure the distance downstream from the gully or swale head to each flag and note this on the results table.
(5) Determine the upstream drainage area, local slope and planform convergence for each point. Area is measured on the sketch map, slope measured or estimated in the field, and planform convergence measured in the field.

If hand calculation is being undertaken:

(6) Calculate the Compound Topographic Index for each point using equation (29.1).
(7) Calibrate the CTI_{crit} by calculating the average CTI value for the gully head points. Point CTI values less than the critical value are excluded from further analysis.
(8) Calculate the gully cross-sectional area from equation (29.4), based on the CTI.

(9) Calculate the eroded volume for each reach from the cross-sectional area and the reach length using equation (29.5).
(10) Determine the total annual ephemeral gully erosion for the field by summing the volume eroded in each reach. This may be expressed in cubic feet, cubic yards, tons or tons per acre as desired.

If using the nomograph:

(6) Find the CTI and gully cross-sectional area for each point using Figure 29.3
(7) Calibrate the Critical CTI by calculating the average gully head value.
(8) Give zero cross-sectional area to all points with CTIs less than the critical value.
(9) Calculate the ephemeral gully erosion.

ACKNOWLEDGEMENTS

This procedure was developed while the first author was an Associate Professor and the second author a graduate student at Colorado State University, as part of research into ephemeral gully erosion funded by a specific co-operative agreement with the Agricultural Research Service Sedimentation Laboratory, Oxford, Mississippi. The advice, support and input of the following individuals is gratefully acknowledged: Bill Mildner and Pete Forsythe (USDA-SCS), Lawson Smith (US Army, WES) and Neil Coleman, Earl Grissinger and Joe Murphey (USDA-ARS).

REFERENCES

Knisel, W. G. (1980). CREAMS: A field scale model for chemicals runoff, and erosion from agricultural management systems, *US Department of Agriculture, Conservation Research Report* No. 26.

Thorne, C. R., Zevenbergen, L. W., Grissinger, E. H. and Murphey, J. B. (1986). Ephemeral gullies as sources of sediment, *Proceedings of the Fourth Interagency Sedimentation Conference, Las Vegas, Nevada*, **1**, 3-152–3-161.

Zevenbergen, L. W. (1989). *Modelling Erosion Using Terrain Analysis*, Unpublished PhD dissertation, University of London April 1989, 345 p.

Zevenbergen, L. W. and Thorne, C. R. (1987). Quantative analysis of land surface topography, *Earth Surface Processes and Landforms*, **12**, 47–56.

30 Assessing the Impact of Erosion on Soil Productivity Using the EPIC Model

J. R. WILLIAMS,
USDA-ARS, Temple, Texas

A. N. SHARPLEY
USDA-ARS, Durant, Oklahoma

and

D. TAYLOR
USDA-ARS, Temple, Texas

INTRODUCTION

Erosion of agricultural land remains one of the major non-point source concerns in the United States (USEPA, 1984). This increased awareness has stimulated an urgency in obtaining detailed information on the change in soil productivity as a result of erosion due to agricultural land use and management. Soil productivity is lowered through loss of storage capacity for plant-available water, loss of plant nutrients, degradation of soil structure and decreased uniformity of soil conditions within a field. Off-farm impacts of erosion include loss of lake capacity and recreational value, blockage of navigable waterways and destruction of fish-spawning beds and aquatic biota. In the United States, annualized costs of cropland erosion on soil productivity, in terms of increased agricultural chemical applications and tillage, are approximately $1.7–1.8 billion and produce off-farm erosion damage of $3.4–13.0 billion at 1983 prices (Crosson, 1986). The accurate prediction of the relationship between long-term (50 years or more) erosion and soil productivity is, therefore, needed from economic and environmental standpoints. Until this relationship is adequately developed, selecting management strategies to maximize long-term crop production will be impossible.

A model called EPIC (Erosion–Productivity Impact Calculator) was developed for use in determining the relationship between erosion and soil productivity (Williams et al., 1984a). EPIC is composed of physically based components for simulating erosion, plant growth and related processes and economic components for assessing the cost of erosion and determining optimal strategies. The model simulates the physical processes simultaneously, using a daily time step and readily available inputs. As erosion can be a slow process, EPIC is capable of simulating hundreds of years if necessary. The EPIC model has been shown to provide accurate predictions of soil nutrient cycling (Jones et al., 1984) and crop production on soils ranging in soil water capacity, plant nutrient content and fertilizer application rates, under various agricultural management systems and erosion rates (Williams et al., 1984b; Sharpley and Williams, 1990).

The relationship between erosion and soil productivity (E/P) is quantified in EPIC by calculating erosion-productivity index (EPI) values as the ratio of annual crop yields with erosion to those without. This chapter presents the E/P for three agriculturally important soils in the United States, to demonstrate its potential use in assessing the impact of long-term erosion on soil productivity.

METHODS

The EPI is calculated as the ratio of the mean crop yield for an eroded soil profile compared with that for the profile at the start of an EPIC simulation. Simulations were performed with the initial soil properties held constant for a 10-year period (Step 1), providing an estimate of the mean annual crop yield as well as frequency distribution of crop yields. Erosion was then simulated for a 50-year period to obtain an estimate of the eroded soil properties (Step 2). The Step 1 simulation was then repeated, except that the original soil profile was replaced by the eroded profile. Comparing the resulting mean annual crop yield to that obtained with the original profile gives one EPI point on the E/P. The process was repeated four times to give sufficient EPI values to adequately define E/P.

These simulations required generation of weather data, thereby providing the same weather for each of the simulations. This ensures a realistic comparison of crop yields and provides a convenient method for estimating a crop-yield frequency distribution. The growth of corn on three soils (Hagerstown silt loam, Norfolk sandy loam and Webster clay loam—Table 30.1) was simulated, with and without erosion. Fertilizer application rates of 50 kgN and 10 kgP ha^{-1} yr^{-1} and 200 kgN and 50 kgP ha^{-1} yr^{-1} were simulated. These rates represent low and recommended rates, respectively, for these soils and crops.

RESULTS AND DISCUSSION

A decrease in corn yield was simulated with erosion compared to no erosion

Table 30.1 Physical and chemical properties of the soils used

	Hagerstown	Norfolk	Webster
A horizon			
Depth (mm)	280	180	430
Clay (%)	19	2	37
Silt (%)	67	22	33
Sand (%)	14	76	30
Organic C (%)	1.4	0.6	4.8
Sat. cond. (mm hr^{-1})	3.1	5.1	6.2
B horizon			
Depth (mm)	710	1400	530
Clay (%)	35	22	30
Silt (%)	45	20	30
Sand (%)	20	58	40
Organic C (%)	0.2	0.1	1.0
Sat. cond. (mm h^{-1})	0.3	1.2	1.9
C horizon			
Depth (mm)	500+	700+	560
Clay (%)	70	35	35
Silt (%)	20	24	25
Sand (%)	10	41	40
Organic C (%)	0.4	0.1	0.2
Sat. cond. (mm h^{-1})	1.1	1.0	7.2

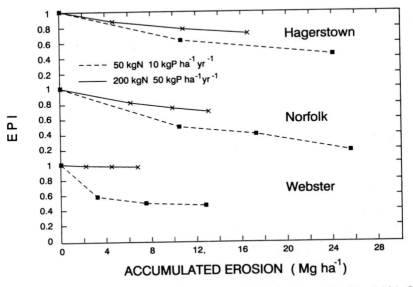

Figure 30.1 Simulated changes in relative corn yield (EPI) under low (5 kgN and 10 kgP ha^{1} yr^{-1}) and recommended (200 kgN and 50 kgP ha^{-1} yr^{-1}) fertilizer applications as a result of 200 years of erosion for three soils

for each soil, as represented by EPI (Figure 30.1). The decrease in corn yield was significantly greater (at the 5 per cent level) for the lower fertilizer application rate. With application of 50 kgN and 10 kgP ha^{-1} yr^{-1}, a 57, 83 and 55 per cent reduction in soil productivity for Hagerstown, Norfolk and Webster soils, respectively, was simulated. The greater decrease in productivity for the Norfolk soil is consistent with the shallower A horizon (Table 30.1), compared to the other soils.

Erosion of surface soil had little effect on texture, available water capacity and bulk density of the soil in the root zone (19, 10 and 5 per cent change, respectively, averaged for the three soils) during the simulation. Consequently, the decrease in corn yield was mainly a result of the decrease in organic C, total N and organic P content of the A horizon (88, 90 and 92 per cent, respectively, averaged for the three soils). With recommended rates of fertilizer application (200 kgN and 50 kgP ha^{-1} yr^{-1}), changes in the nutrient content of the A horizon were minor. Organic C, total N and organic P decreased only 29, 29 and 21 per cent, respectively, while plant available P increased 82 per cent, averaged for the three soils. As a result, recommended fertilizer application rates ameliorated the negative effect of erosion on soil productivity, with corn yield reductions of only 29, 31 and 4 per cent simulated for Hagerstown, Norfolk and Webster soils, respectively (Figure 30.1).

The EPIC model can thus provide a reasonable representation of the change in surface soil fertility and subsequent soil productivity as a result of long-term erosion. Once the erosion–productivity relationship has been developed for a soil, as described in this chapter, it can be used to evaluate other agricultural management strategies. The use of EPIC thus enables quantification of the effect of agriculture on erosion, from both water quality and soil productivity standpoints.

REFERENCES

Crosson, P. (1986). Soil conservation, *Choices*, 1: 33–8.

Jones, C. A., Sharpley, A. N. and Williams, J. R. (1984). A simplified soil and plant phosphorus model. III Testing, *Soil Science Society of America Journal*, **48**, 810–13.

Sharpley, A. N. and Williams, J. R. (eds). (1990). EPIC—Erosion–Productivity Impact Calculator: model documentation, *US Department Agricultural Technical Bulletin*, #1768 US Govt Printing Office, Washington, DC (in press).

US Environmental Protection Agency. (1984). *Report to Congress: Nonpoint source pollution in the US* Washington, DC.

Williams, J. R., Jones, C. A. and Dyke, P. T. (1984a). A modeling approach to determining the relationship between erosion and soil productivity, *Transactions of the American Society of Agricultural Engineers*, **27**, 129–44.

Williams, J. R., Putnam, J. W. and Dyke, P. T. (1984b). Assessing the effect of soil erosion on productivity with EPIC. In *Proceedings of the National Symposium on Erosion and Soil Productivity* (Chm D. K. McCool), New Orleans, LA, December 1984, Publ. by American Society Agricultural Engineers, St Joseph, Michigan, No. 8-85.

31 THEPROM—An Erosion Productivity Model

YVAN BIOT
School of Development Studies, University of East Anglia

INTRODUCTION

Knowing the importance of land degradation is a prerequisite for the planning of soil- and water-conservation programmes. A number of methods have been devised to assess the intensity of these processes, but none of them deal with the impact of the land-degradation processes on the capacity of the land to produce crops, let alone support a household—two factors which rate highly on the agenda of criteria taken into consideration by those who earn a living from the land: the farmers.

Recently a number of models have emerged in the United States which try to forecast the impact of erosion on the productivity of land, for instance: EPIC (Williams *et al.*, 1984) and the PI-method (Larson *et al.*, 1983). Elwell and Stocking (1984) introduced the Soil Life concept: a measure of the productive lifespan of a soil given specific management practices and rates of erosion. Biot (1988a,b) developed a modelling technique to forecast productivity decline using the results of routine land resource surveys and linked the assessment of land degradation to land evaluation (FAO, 1976) through the concept of the *residual suitability* (Biot, 1988c)—a measure of the time the soil will remain suitable for a given land-utilization type.

A three-step procedure has been proposed for the calculation of the residual suitability of agricultural land (Biot, 1988c):

(1) Definition and calibration of a productivity index;
(2) Design and calibration of a land-use model which contains the productivity index and the way in which this is affected by the prevailing land-degradation processes;
(3) Integration of the land-use model between the present and minimum allowable productivity levels of land—the latter as determined by socio-economic criteria.

Soil Erosion on Agricultural Land
Edited by J. Boardman, I. D. L. Foster and J. A. Dearing
©1990 John Wiley & Sons Ltd

Different degrees of sophistication of the modelling component are allowed for, depending on resource and data availability, and an example of a simple land-use model (EPROM: Erosion Productivity Model) has been used to illustrate the technique in a case study in the communal rangelands of Botswana.

This chapter introduces a more sophisticated model of the land-use system: THEPROM (Theoretical Erosion Productivity Model) which, in addition to its use as a calculator of the residual suitability of land, can also be employed to study the dynamics of the land-use system and its responses to management.

PRODUCTIVITY INDEX

Productivity can be viewed as integrating those attributes of the environment which determine the availability of water, air and nutrients to plant roots. Nutrients being replaceable (at least to some extent), it is mainly the availability of water which determines the *intrinsic* productivity of land. In this chapter, the Available Water Storage Capacity (AWSC) of the soil is proposed as an index of productivity. Effective rooting depth (a measure of quantity) and the available water capacity (AWC) of a unit volume of soil (a measure of quality) are used to calculate AWSC.

THE MODEL

The land-use system is here considered as a cascading system involving vegetation, litter, organic matter and mineral soil, similar to the model proposed by Nye and Greenland (1960). The link between soil and vegetation is made through the proposed productivity index (AWSC). The model is illustrated in Figure 31.1.

Cascading systems are governed by the laws of energy (mass) conservation and transport laws which can be expressed by differential equations. The proposed land-use model is then expressed mathematically as follows.

Litter

Changes in the amount of litter at the soil surface are governed by litter production, litter decomposition and litter erosion:

$$\frac{dL}{dt} = B_{decay} - L_{dec} - L_{eros}$$

Litter production is primarily dependent on standing biomass. Litter decomposition (humification and oxidation) is achieved by soil fauna and microfauna whose numbers are related to the amount of litter (food) present.

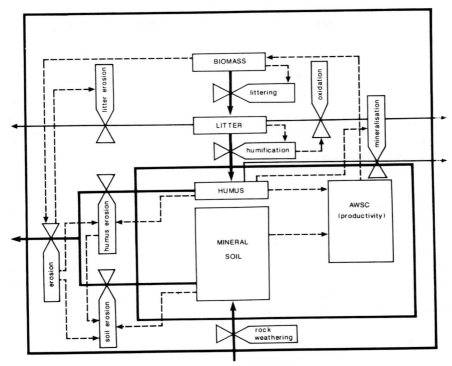

Figure 31.1 THEPROM

Accumulation of litter in the absence of erosion has been found to proceed in an exponential way (Jenny et al., 1949), characteristic for a rate of accumulation proportional to the difference between the amount of litter at any time and its equilibrium value. This behaviour can be modelled by assuming linear relationships between litter production and standing biomass and between litter decomposition and amount of litter (see, for instance, the example on pp. 82-9 in Thomas and Huggett, 1980), or:

$$\frac{dL}{dt} = k_1.B - k_2.L - L_{eros}$$

where k_1 = litter production coefficient,
k_2 = litter destruction coefficient,
L = litter,
B = biomass.

According to Young (1986), the ratio between the amount of litter used by the soil fauna and microfauna as energy and the amount of litter used in tissue building is a constant for a given environment. This means:

$$k_2 = k_3.k_2 + (1 - k_3).k_2$$

where $k_3.k_2$ = amount of litter used for tissue building
= humification coefficient, and

$(1 - k_3).k_2$ = amount of litter used as energy
= oxidation coefficient

Litter erosion has rarely been studied. It is likely, though, that the amount of litter lost through erosion is directly proportional to the rate of erosion. In this chapter we assume:

$$L_{eroded} = k_4.E$$

where k_4 = litter erosion coefficient, and
E = amount of soil erosion.

The total amount of soil eroded is a function of soil detachment and translocation which depend on the terrain configuration, the erosivity of the rain, the erodibility of the soil and the percentage cover by the vegetation. SLEMSA (Elwell and Stocking, 1982) is a model which calculates annual rates of erosion based on these variables:

$$E = 1.16.K.C.SL$$

where

$K = e^{((0.4681 + 0.7663.F).\ln(E_n) + 2.884 - 8.1209.F)}$

F = soil erodibility, based on soil type and land utilization,
E_n = rainfall erosivity, measured using rainfall intensity records or derived from annual rainfall ($J\ m^{-2}$).

$C = e^{-0.06.i}$

where

i = percentage interception by herbaceous cover and litter.

Percentage cover by the herbaceous layer is related to standing biomass (B):

$i = f(B)$

$SL = Le^{1/2}.(0.76 + 0.53.An + 0.076.An^2)/25.65$

Le = slope length (m)
An = slope gradient (per cent)

In order to simulate the strongly reduced rates of erosion at very shallow soil depth where soil erosion is determined by the capacity of the rock to produce soil rather than by the capacity of the overland flow to transport sediment (i.e. the weathering limited case of erosion of Thornes, 1987), a correction factor is added to the equation for erosion:

$$E = 1.16.K.C.SL.(1 - S_{min}/S)$$

where $(1 - S_{min}/S)$ = a factor which reduces erosion when the actual amount of soil (S) comes close to a minimum amount of soil (S_{min}), at which stage the soil surface is protected against further erosion.

and E = mineral soil formation = k_6 for $S < S_{min}$.

The final equation characterizing the changes within the litter storage component is:

$$\frac{dL}{dt} = k_1.B - k_2.L - k_4.E \qquad (31.1)$$

Humus

Organic matter is formed by humification of litter, and lost through mineralization and erosion:

$$\frac{dH}{dt} = L_{humif} - H_{mineral} - H_{erosion}$$

Accumulation (humification) and decomposition (mineralization) of humus within the soil have been the subject of numerous studies (Henin and Dupuis, 1945; Koepf, 1953; Nye and Greenland, 1960; Henin *et al.*, 1960). Young (1986) has proposed a carbon-cycle model based on the findings of Nye and Greenland (1960).

Studies of microbial activity have demonstrated that mineralization is directly proportional to the amount of organic matter in the soil ($k_5.H$). The amount of humus eroded each year can be derived from erosion figures if the original amount of organic matter in the soil and the enrichment ratios are known: amount of organic matter eroded = $E.(H/S_t).ER$, where E = total amount of soil eroded, H = total amount of organic matter in the topsoil, S_t = total amount of soil in the topsoil and ER = enrichment ratio for organic matter. Thus:

$$\frac{dH}{dt} = k_3.k_2.L - k_5.H - E.(H/S_t).ER \qquad (31.2)$$

Mineral Soil

Mineral soil is formed from the geological substratum through weathering—a process determined by climate and kind of rock. Given stable climatic and uniform geological conditions, weathering can be considered constant (k_6). Estimates of rock weathering are based on geomorphological, biochemical and geochemical studies and can be found in the literature (e.g. Owens, 1974).

The amount of mineral soil which is eroded can be found by subtracting the amount of eroded organic material from the total amount of erosion:

$$\frac{dS}{dt} = k_6 - (E - E.(H/S_t).ER) \qquad (31.3)$$

Available Water Storage Capacity

Available water-storage capacity has been found to be directly proportional to the amount of organic material and the amount of mineral soil, hence:

$$\frac{dAWSC}{dt} = k_7 \cdot \frac{dH}{dt} + k_8 \cdot \frac{dS}{dt} \qquad (31.4)$$

Biomass Production

At the present stage of development, biomass production in THEPROM is generated through a direct relationship between standing biomass at the end of the growing season and the productivity index. This relationship can be established empirically, or derived on the basis of the following reasoning. At low productivity level (low AWSC) biomass production is very low and a unit increase in productivity is not expected to cause substantial increases in annual levels of biomass production. Once over a threshold value, however, crop response to a unit rise in productivity increases rapidly. At very high productivity levels, further increases in crop production are unlikely because of other constraining agents such as light, CO_2 and plant physiology. The result is a crop response function of a logistic nature which can be expressed by:

$$B = \frac{B_{max}}{(1 + C.e^{-k_9.AWSC})} \qquad (31.5)$$

where B = standing biomass at the end of the growing season.

SIMULATION

Functions (31.1) to (31.4) represent a set of simultaneous differential equations which, together with equation (31.5), describe the model depicted in Figure 31.1

THEPROM — An Erosion Productivity Model

fully. This model can be integrated numerically between known boundary values on condition that the parameters are calibrated.

To illustrate this, the model has been calibrated for the case of a hypothetical semi-arid grass savanna in the Hard-Veld of Botswana, based on field data collected for the calibration of a simplified erosion-productivity model (Biot, 1988b).

The Case of an Eroding Landscape Protected from Cattle

The following biomass, litter, organic matter and soil data are derived from the above study.

Optimum production level

Soil depth = 120 cm and a bulk density = 1.5 g cm^{-3}, hence:
$S = 18 \cdot 10^6$ kg ha^{-1}
AWC (mineral soil) = 10% (volumetric)
AWC addition for each 1% organic matter: 5% (volumetric)
$B = 2500$ kg ha^{-1}
$L = 2000$ kg ha^{-1}
Organic matter = 0.50% over 10 cm or $H = 7500$ kg ha^{-1}

Mineral soil formation

$k_6 = 100$ kg ha^{-1} yr^{-1}

Relationship AWSC versus S and H

$$k_7 = \frac{2.5}{7500} = 0.000333 \text{ mm kg}^{-1}$$

$$k_8 = \frac{120}{1.2 \cdot 15 \cdot 10^6} = 0.00000667 \text{ mm kg}^{-1}$$

Rainfall erosivity, soil erodibility, slope angle and slope length

Consider the following typical values for a pediment landscape in the Hard Veld of Botswana (Biot, 1988b):

$F = 3$ $\qquad En = 8700$ J m^{-2}
$An = 2.5\%$ $\qquad Le = 200$ m
$S_{min} = 2.25 \cdot 10^6$ (soil depth = 15 cm, $BD = 1.5$ g cm^{-3})
$ER = 4.5$ $\qquad i = 10.7 \cdot \ln(B) - 30.4$

Litter erosion

Assume $k_4 = 0.05$.

Mineralization:

This can be calculated using the formula proposed by FAO (1979). In a semi-arid climate with $R = 450$ mm, $k_5 = 0.038$.

Oxidation of litter

Young (1986) has proposed 75–85% oxidation losses, hence $k_3 = 0.20$.

Litter decomposition

At equilibrium $\quad \dfrac{dH}{dt} = k_2.k_3.L - k_5.H - E.(H/S_t).ER = 0$, hence:

$$k_2 = \frac{k_5.H + E.(H/S_t).ER}{k_3.L}$$

At an effective rooting depth of 120 cm, $B = 2500$ kg ha^{-1}, $i = 53\%$, $E = 2231$ kg ha^{-1} yr^{-1} and $H/S_t = 0.005$, hence:

$$k_2 = \frac{0.038 \cdot 7500 + 2231 \cdot 0.005 \cdot 4.5}{0.20 \cdot 2000} = 0.8375$$

Litter production

At equilibrium $\quad \dfrac{dL}{dt} = k_1.B - k_2.L - k_4.E = 0$, hence:

$$k_1 = \frac{k_2.L + k_4.E}{B} = \frac{0.7125 \cdot 2000 + 0.05 \cdot 2231}{2500} = 0.7146$$

Biomass versus AWSC

In equation (31.5) B_{max}, C and k_9 can be derived from measured biomass and AWSC data. In our case we assume $B_{max} = 2550$ kg ha^{-1}, $B = 2500$ for AWSC = 120 and $B = 0$ for AWSC = 0. With these conditions, $C = 254$ and $k_9 = 0.787$. These parameters define the relationship between B and AWSC given in Figure 31.2.

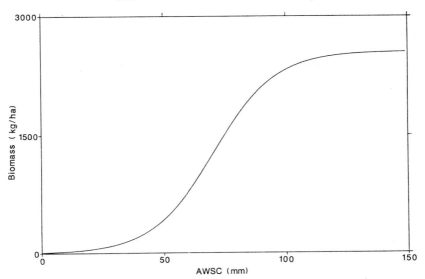

Figure 31.2 Relationship between the productivity index and biomass production

Integration

A computer program was developed which integrates the set of simultaneous differential equations defined above. The program is written in FORTRAN 77 and makes use of a NAG subroutine (NAG Group Ltd, 1987) to solve differential equations numerically based on a variable order–variable time step method using backward differentiation formulae (Hall and Watt, 1976). Plots are drawn using subroutines developed by the computing centre at the University of East Anglia based on GINO-F (Cadcentre Ltd, 1986). A simulation of biomass, litter, humus and soil from $t=0$ until equilibrium is given in Figure 31.3

The Case of an Eroding Landscape in the Presence of Cattle at 5 ha LU^{-1}

Algorithms for the grazing component within the ecosystem have been proposed by Noy-Meir (1978) and Thornes (1987). In its simplest form, grazing can be considered to take a given amount of biomass from the system depending on the number of animals and their daily intake rate.

Animal density is usually given as stocking rate (S_{rate} = area (ha) to feed one tropical Livestock Unit of 350 kg). Daily dry matter intake of 1 LU is 2 per cent of its live-weight, or 7 kg day^{-1}. For a grazing season of 10–11 months 2205 kg of grass biomass will be consumed by a livestock unit.

Equation (31.5) can now be corrected for offtake through grazing:

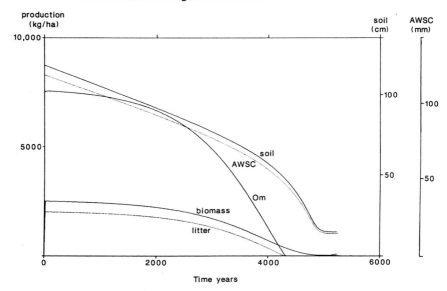

Figure 31.3 Productivity decline of a hypothetical grass savanna caused by erosion

$$B = \frac{B_{max}}{(1 + C \cdot e^{-k_9 \cdot AWSC})} - \frac{2205}{S_{rate}}$$

A simulation of biomass, litter, humus and soil from $t=0$ until equilibrium is given in Figure 31.4.

Response of the Environment to a Change in the Intensity of Use

Of particular interest to land-use planners is the study of the impact of a change in land use (intensity) on the potential of the environment to support this. Once calibrated, THEPROM can be used to study such impacts. Figures 31.5 and 31.6 illustrate the short- and long-term reaction of the environment where there is a change in land use from un-utilized grass savanna to grazed rangeland at 5 ha LU^{-1}.

DISCUSSION

The model described above remains a simplification of the real world and its responses have to be analysed with due care. Especially the erosion component, the enrichment ratios, the litter erosion and the biomass–AWSC curve are fraught with uncertainties and the proposed mathematical expressions might be

THEPROM—An Erosion Productivity Model 475

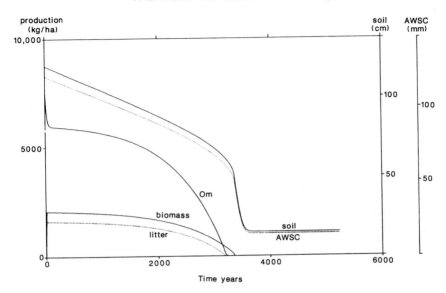

Figure 31.4 Productivity decline of a hypothetical grass savanna caused by erosion and grazing

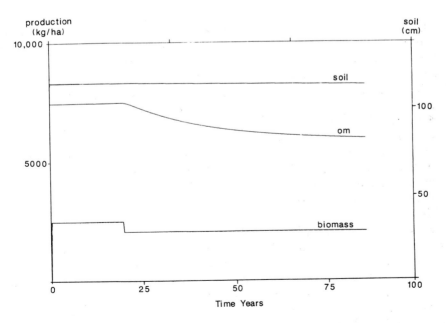

Figure 31.5 Short-term reaction of the environment to a change in land use

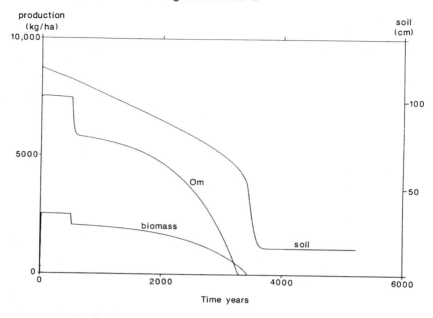

Figure 31.6 Long-term reaction of the environment to a change in land use

inappropriate. In the case of the enrichment ratio, for instance, Massey and Jackson (1952) and Stocking (1986) have indicated much more complex controls involving sediment concentration in the runoff waters, amount of erosion and original amount of carbon in the soil. The erodibility factor (F-value) in the SLEMSA model is not constant as the soil erodes and is influenced by rooting and the presence of trees. No data were available to estimate litter erosion and both the algorithm and its calibration were guessed at by comparing results from simulations with data from Botswana.

The general behaviour of the different outputs from the model, however, seem to correspond with what is observed on the field. This enables the use of the model to analyse interactions which would not necessarily become obvious through experimental work.

The way in which the model responds to erosion has been illustrated in Figure 31.3. At high productivity levels erosion is kept low by high crop-cover values. Erosion, however, increases with decreasing productivity, and biomass, organic matter, litter and the productivity index decline increasingly rapidly down to a minimum below which the system becomes resistant to further abrasion. The end stage is reached after about 4000 years.

The impact of cattle on the system has been illustrated in Figure 31.4. In this case the offtake by cattle causes a more rapid decrease in biomass, litter and organic matter and leads to a markedly faster rate of degradation, with the system 'crashing' after about 3000 years.

Both short- and long-term reactions of the environment to a change in land utilization from a un-utilized savanna to a grazed rangeland at 5 ha LU^{-1} are illustrated in Figures 31.5 and 31.6.

The difference between these two figures is only one of focus: in both figures THEPROM was run using the same system parameters and initial conditions. In Figure 31.5 the reaction of the environment (i.e. organic matter, litter, AWSC, soil and biomass) to a sudden change in land utilization (i.e. the introduction of cattle at a stocking rate of 5 ha LU^{-1}) is monitored over a period of about 50 years. In Figure 31.6 the focus is decreased to encompass a much longer time period (in the order of centuries).

It is obvious from the comparison between both responses that the results of short-term experiments monitoring productivity decline caused by erosion can lead to a mistaken interpretation of both the nature of the impact of erosion on productivity and of the long-term implications of erosion. Indeed, the negative exponential curve in Figure 31.5 is typical for an environment which is readjusting to a change in land utilization: biomass is taken away, and litter and organic matter production slow down until a new equilibrium is established. Beyond this new balance, erosion proceeds as before, but at an accelerated pace. The negative exponential curve is not caused by erosion but by the change in land utilization which involves a decrease in biomass.

CONCLUSION

The proposed theoretical erosion productivity model (THEPROM) predicts a productivity decline which is similar to the forecasts made using a simpler model (EPROM), elaborated upon earlier (Biot, 1988a,b). THEPROM requires substantially more calibration of several component flows and storages which restricts its use to cases where time and resources are less constraining than under routine survey conditions.

Apart from its employment as a land-use model to calculate the residual suitability of land (Biot, 1988c), the model can also be used to simulate the effects of different management options, and hence could serve as a first-line screener of potential interventions before field trials are embarked upon, which, in the case of rangeland systems, are costly and time consuming. The model also makes possible the analysis of the reaction of different components of the land-use system to changes in land utilization, and leads to a better understanding of the way in which the natural world operates.

Finally, the model has clearly demonstrated that the response of the natural environment to soil erosion depends on whether a change in land utilization is considered or not, and on the focus of the analysis: short or long term.

ACKNOWLEDGEMENTS

Theoretical work and the final analysis and write-up were supported by a research grant from the University of East Anglia. Field work was funded by the International Livestock Centre for Africa (ILCA—Addis Ababa) under the auspices of the Animal Production Research Unit (APRU), Botswana.

REFERENCES

Biot, Y. (1988a). Modelling Productivity Losses Caused by Erosion. In: Rimwanich, S. (ed.), *Land conservation for future generations*. Proceedings of the Vth International Soil Conservation Conference, Bangkok, Thailand: pp. 177–198.

Biot, Y. (1988b). *Forecasting Productivity Losses caused by Sheet and Rill Erosion in Semi Arid Rangelands: A Case Study from the Communal Areas of Botswana*, PhD thesis, University of East Anglia.

Biot, Y. (1988c). Calculating the Residual Suitability of Agricultural Land based on Routine Land Resource Surveys. In Bouma, and Brept, (eds.): *Land qualities in space and time*. Proceedings of the Symposium on Land Qualities in Space and Time, Wageningen: pp. 261–264.

Cadcentre Ltd (1986). *GINO—F*, Cadcentre Ltd.

Elwell, H. A. and Stocking, M. A. (1982). Developing a simple yet practical method of soil loss estimation, *Trop. Agriculture*, **59**(1), 43–8.

Elwell, H. A. and Stocking, M. A. (1984). Estimating soil-life span for conservation planning, *Tropical Agriculture*, **61**/2, 148–50.

FAO (1976). *A Framework for Land Evaluation*, Soils Bulletin No. 32, FAO, Rome.

FAO (1979). *A Provisional Methodology for Soil Degradation Assessment*, FAO, Rome.

Hall, G. and Watt, J. M. (1976). *Modern Numerical Methods for Ordinary Differential Equations*, Clarendon Press, Oxford.

Henin, S. and Dupuis, M. (1945). Essai de bilan de matière organique du sol, *Ann. Agr.*, **15**, 17–29.

Henin, S., Feodoroff, A., Gras, R. and Monnier, G. (1960). Le profil cultural: les matières organiques et leur evolution, *Soc. d'Edit. des Ing. Agr.*, 274–85.

Jenny, H., Gessel, S. P. and Bingham, F. T. (1949). Comparative study of decomposition rate of organic matter in temperate and tropical regions, *Soil Science*, **68**, 419–32.

Köpf, H. (1953). Die Temperatur/Zeit Abhängigkeit der Bodenatmung. *Z. Pflanzenärn. Düng.*, **61**, 29–48.

Larson, W. E., Pierce, F. J. and Dowdy, R. H. (1983). The threat of soil erosion to long term crop production, *Science*, **219**, 458–65.

Massey, N. F. and Jackson, M. L. (1952). Selective erosion of soil fertility constituents, *Soil Science Society of America, Proc.*, **16**, 353–6.

NAG Group Ltd (1987). *The NAG FORTRAN Library Manual—Mark 12*, The Numerical Algorithms Group Ltd.

Noy-Meir, I. (1978). Stability in simple grazing models: effects of explicit functions, *J. of Theor. Biology*, **71**, 347–80.

Nye, P. H. and Greenland, D. J. (1960). *The Soil Under Shifting Cultivation*, Commonwealth Bureau of Soils. Techn. Comm. No. 51, Harpenden.

Owens, L. B. (1974). *Rates of Weathering and Soil Formation on Granite in Rhodesia (Zimbabwe)*, Department of Agriculture, University of Rhodesia (Zimbabwe).

Stocking, M. (1986). *The Cost of Soil Erosion in Zimbabwe in Terms of the Loss of Three Major Nutrients*, Consultant's Working Paper No. 3. Soil Conservation Programme, Land and Water Development Division. AGLS, FAO, Rome.

Thomas, R. and Huggett, R. J. (1980). *Modelling in Geography: A Mathematical Approach*, Harper and Row, London.

Thornes, J. (1987). Erosion equilibria under grazing. In Bintcliff, J., Davidson, D. and Grant, E. (eds), *Environmental Archaeology*, Edinburgh University Press.

Williams, J. R., Jones, C. A. and Dyke, P. T. (1984). A modelling approach to determining the relationship between erosion and soil productivity, *Transactions of the ASAE*, **27**(1), 129–44.

Young, A. (1986). Effects of trees on soils. In Prinsley and Swift (eds), *Amelioration of Soils by Trees*, Commonwealth Science Council, 10–19.

32 Probability Distribution of Event Sediment Yields in the Northern Negev, Israel

JONATHAN B. LARONNE
Department of Geography, Ben Gurion University of the Negev, Israel

INTRODUCTION

Several methods are presently available to predict soil erosion rates on agricultural land (e.g. Kirkby and Morgan, 1980). Because the models are either insufficiently accurate or require a database too large and expensive for all but academic exercises, empiricism in the form of the USLE and its derivatives is very common (Laronne and Mosley, 1982). There is, therefore, a real need for the development of prediction tools to assess long-term erosion rates, as well as for the prediction of soil loss due to large rainfall events. This contribution focuses on the use of one such tool—stochastic analysis of laminated reservoir deposits.

STUDY AREA

Two stock pond-sized reservoirs built in 1978 were chosen to demonstrate the proposed methodology. Locally termed Limans, they are hereafter referred to as the Southern and Northern Limans. Their location in the semi-arid northern part of the Negev, central Israel, is shown in Figure 32.1.

The Limans are situated in an area with a very sharp rainfall gradient. Mean annual rainfall at the southern site is 210 mm and at the northern site 235 mm, with a coefficient of variation at both sites of 30 per cent (Katsnelson, 1979). They are close to the main highway connecting Beer Sheva to the north, but the larger, Northern Liman was constructed upstream of the highway. Therefore, large runoff volumes produced by the paved surface do not affect it. This effect is present in the smaller, southern Liman, where the drainage density of gullied

Soil Erosion on Agricultural Land
Edited by J. Boardman, I. D. L. Foster and J. A. Dearing
©1990 John Wiley & Sons Ltd

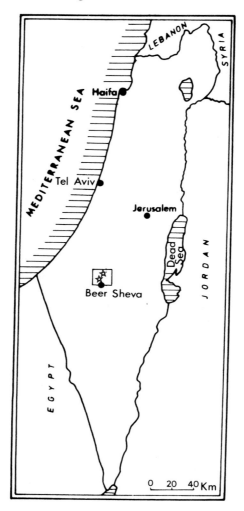

Figure 32.1 Location of the study sites (☆) in the northern Negev

channels is also large (Table 32.1). Using an additional set of aerial photographs dated June 1981, it appears that the land-use pattern had not changed appreciably with the exception of new terracettes (Table 32.1).

Land use patterns at the study sites differ (Table 32.1). The Northern basin is larger, allowing for greater sediment storage (i.e. a smaller delivery ratio), and a small but defineable area contributes neither runoff nor sediment downstream. More significantly, almost half of the basin is contour-ploughed, bringing about increased depression storage and infiltration, with a decrease

Table 32.1 Characteristics of the study sites

	Northern Liman	Southern Liman
Co-ordinates:	1301/0835	1299/0784
Elevation above msl (m):	300–360	300–360
A_L = Liman area (m^2):	7100	2800
A_D = basin area (km^2):	2.94	0.23
$AD/AL(-) : .10^6$:	414	82
Drainage density of gullies (km^{-1}):	0.7	1.4
Land use (percentage area of total drainage basin)[a]		
Agricultural, contour-ploughed (winter wheat)	47.3	0
Young pine trees	0	72.4
Loess-mantled steeper slopes	51.4	7.2
Paved roads	0	2.4
Wide dirt roads[b]	0.4	2.3
Trails[c]	0.2	0
Recent terracettes	0	15.7
Non-contributing area[d]	0.7	0

[a] Land use was derived from 1:5000 vertical aerial photographs dated September 1987.
[b] Average road width 3 m.
[c] Average trail width 1 m.
[d] Dammed by earth obstructions.

in sediment yield. The density of dirt roads is also smaller in the larger, northern basin.

Loessial soils attain a considerable thickness in the valley bottoms, where they may exceed 2 m. These are Loessial Serozems with an AbcaBb or an A(B)caC profile (Yaalon et al., 1968). The A horizon is a light yellowish loamy sand, invariably developing a pronounced crust upon drying. The B horizon is somewhat less sandy and contains limestone concretions. Deeper loess is often saline and sodic. The entire profile is limey or has limestone fragments. The loess cover thins towards the local divides, where Stony Serozems and Brown Lithosols predominate.

MAGNITUDE AND PROCESSES OF EROSION

Sediment yield from the study sites results from the magnitude of the relative effects of all processes of soil erosion and storage. The yield is calculated by multiplying the volume of exported sediment that has accumulated in a reservoir by its bulk density, divided by the length of time during which deposition took place.

The undisturbed dry bulk density was determined using an inverted sand cone and Soiltest's Volumeasure™. Density did not appear to increase with depth

Table 32.2 Sedimentation data for the Limans

	Northern Liman	Southern Liman
Average sediment thickness (m)	0.66	0.20
Liman area (m^2)	7100	2800
Sediment volume (m^3)	4686	560
Dry undisturbed bulk density (t m^{-3})	1.62	1.55
Sediment weight (t)	7590	858
Sediment yield[a] (t km^{-2} yr^{-1})	259	377

[a] During one decade, 1978-9 to 1987-8.

due to the high strength of each clayey unit and due to the overall shallow thickness of the sediments. Assuming that trap efficiencies approached 100 per cent (see the following section), the volume of sediment exported from each basin is equal to the volume of accumulated sediments, i.e. the product of the Liman area and the average sediment thickness (Table 32.2). This thickness was calculated by linear interpolation: averaging thicknesses of accumulated sediment at each pit using a procedure identical to the calculation of average rainfall depth with Thiessen polygons.

Several processes of erosion appear to be dominant in the ploughed agricultural fields as well as on the steeper hillslopes. The effect of drop impact is large but has not been measured in the study area. Its relative magnitude can be assessed from drop-impact miniature craters which occur on the soil crusts, as well as by the considerable quantity of loamy sediment that often accumulates on the leaves of grasses and on the lowermost wheat stems. Because the loess crusts it also cracks upon drying. Rilling is, therefore, common. Most rills are discontinuous but some are deepened, thereby allowing piping to form uphill and tunnelling to develop sideways. The gradual increase in pipe diameter and depth contribute considerable quantities of sediment directly into the main channel as well as to minor gullies. Based on the comparison of aerial photographs it appears that the banks of the main channel and the large gullies are not receding greatly.

THE SEDIMENTS

Sediment is deposited whenever runoff enters the reservoirs. The resultant depositional body is composed of a lower, fine sandy unit and an upper unit of sandy-silt fining upwards to silty-clay. The clay dries and desiccates, forming a resistant couplet top which is minimally reworked (Laronne, 1988). Hydraulic interpretation of such event deposits is found elsewhere (Laronne, 1987).

Texture was determined by wet sieving for the sand fraction and by X-ray diffraction (Malvern 2600/3600 Particle Sizer VA.6) for the range 1.9–188 μm.

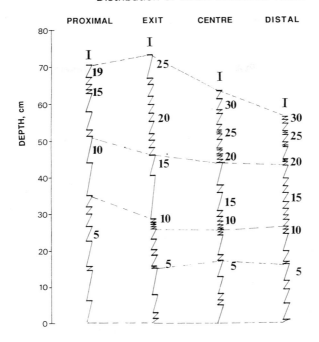

Figure 32.2 Correlation among event sedimentation units (couplets) in the Northern Liman. Each horizontal line denotes the top of an event couplet

Overall, the deposited sediments composed nearly equal amounts of sand and silt with a small clay fraction. The fraction of sand did not vary appreciably (42–61 per cent) throughout the Liman area with the exception of proximal tongue-like deposits rich in coarse sand. The fraction of clay increased from 4 per cent at the entrance to 9 per cent near the earth dam. Within each event deposit the amount of clay varied markedly: less than 0.5 per cent in the sandy bottom to 87 per cent within the couplet top.

A sequence of event deposits in the Northern Liman is depicted in Figure 32.2. The correlation between pits is based on the relative thickness of the units and on their number. It shows that some event deposits were formed at the exit but are missing at the upstream end of the pond (e.g. those numbered 10–13), which, in turn, indicates that these correspond to small events when the pond was only fractionally filled with runoff. Hence, the number of event units increases towards the lowest part of the pond, but their thickness attains a maximum slightly beneath the entrance (see 'exit' sequence). The logged sections show that distal deposits found near the dam are deficient in sandy laminae, where the entire depositional sequence is shallower and more difficult to interpret. Due to these spatial trends in sediment accumulation, the total

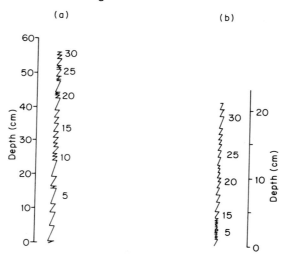

Figure 32.3 Comparison between event couplets deposited in the Northern (a) and Southern (b) Limans

number of sedimentation events in every pond has been equated with the number of event deposits at its centre.

Comparison of the event deposits at the two ponds (Figure 32.3) shows that it is inaccurate to correlate events, even though their total number is similar. That the number of identifiable events was slightly larger (7 per cent) in the Southern Liman may be ascribed to the faster response of its very small basin and smaller transmission losses in its narrow, short channel. The two sequences depicted in Figure 32.3 demonstrate that the response of the basins is different, even though they are only located 5 km apart. The effect of rainfall spottiness variability as opposed to real differences in hydrological response cannot be evaluated from this data set.

PREDICTION

The detailed description of a sequence of event deposits logically leads to the next step, namely, to relate the depositional sequence to that of runoff or rainfall events, to regress sediment volumes against peak discharges or rainfall intensities and to predict the probabilities of occurrence of given volumes of eroded sediment by transposing respective rainfall probabilities. This requires either the use of very broad, empirical relations between rainfall or runoff production and sediment yield (Hereford, 1987) or the generation of a synthetic runoff record coupled with a sediment transport function (Soares *et al.*, 1982). In both these approaches it is necessary to relate erosion to runoff.

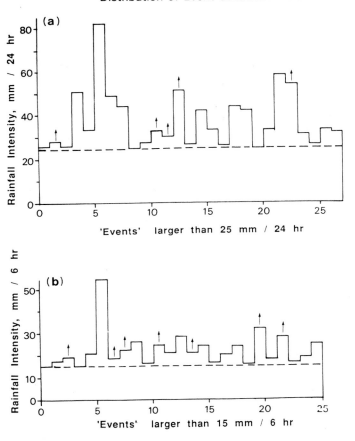

Figure 32.4 Sequences of 24-h (a) and 6-h (b) rainfall events at Lahav Station during 1978-88. Events of magnitude smaller than the indicated threshold were excluded

Rainfall Intensity and Sediment Accumulation

Lacking a runoff record, an attempt was made to relate rainfall events to given sedimentation units. Sequences of 25-30 rainfall events for a variety of intensities and durations have been derived from continuous rainfall charts from the Lahav meteorological station located 5 km north-east of the Northern Liman. It was assumed that all the intervening events of smaller magnitude did not generate runoff. As expected, the pattern of the time series of rainfall events varied with intensity and duration. For instance, Figure 32.4 depicts two sequences of rainfall events that occurred during 1978-88. Arrows indicate that a rainfall event was followed by at least one additional event of the given, minimal intensity,

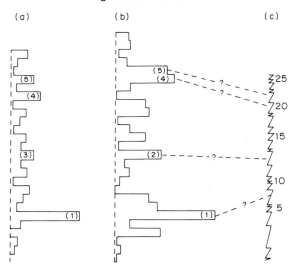

Figure 32.5 Comparison of 6-h (a) and 25-h (b) rainfall sequences (1978–88) with that of event deposits (c) at the distal end of the Northern Liman. See text for details of rainfall record dates

presumably generating larger runoff volumes and peaks. There is a general agreement between the relative location of the largest events but, self-evidently, the correlation between the sequences is poor.

The sequences of rainfall events have been compared to the erosional history of the Northern Liman. Only one sedimentary record was chosen for this comparison (Figure 32.5) because others demonstrate similar trends. The sequences are not strictly comparable because the patterns are rather dissimilar. It appears, though, that four to five large rainfall events do relate to the thicker event couplets. If this correlation is accepted, it is possible to date the deposits. Numbers 1 to 5 in the depositional sequence shown in Figure 32.5 denote respective rainfall events (29/11/1979, 10/12/1980, 8/11/1982, 2/4/1986 and 17/11/1986).

Beyond the broad similarity, it is questionable whether most of the depositional events can be accurately related to rainfall events. Only 60 per cent of the variability of runoff in similar catchments located 20 km north-west of the Northern Liman is explained by rainfall 'events', increasing to 70 per cent if the event is defined in terms of individual rainbursts (Tamir, 1972). Because sediment/runoff explanations drop to 50 per cent (Tamir, 1972), it is not surprising that attempts to relate sediment volumes to rainfall intensities (e.g. Figure 32.5) are not encouraging for prediction purposes.

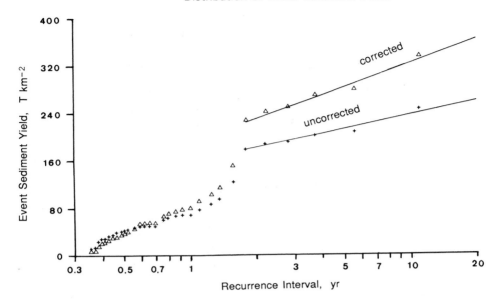

Figure 32.6 The probability distribution of event sediment yields from the Northern Liman catchment

Stochastic Analysis of Accumulated Sediment

If prediction of sediment yield cannot rely on rainfall or runoff data, or because both are lacking or representing distant areas, other approaches to prediction may be suggested. For instance, it is common knowledge that surface-water hydrologists use records of runoff volumes and peaks to predict appropriate probabilities of floods and droughts (Linsley et al., 1982). Merely because rainfall records are more available than runoff records has not detracted from the usefulness and reliability of stochastic methods in surface-water hydrology. Similarly, it is hereby suggested that the availability of event-based sediment yield records may be utilized to predict event erosion rates *without recourse to rainfall or runoff data*.

The depositional events manifested by the couplets were of sufficient magnitude to transport sediment into the reservoirs. Hence, the couplets may be regarded as a partial duration sequence of runoff events that erode and transport sediment from the catchment to the reservoirs. Similar partial duration series have been constructed for all flows above the mean annual flow or above a given flow volume or peak discharge (Haan, 1977).

The probability distribution of event sediment yields from the catchment draining to the Northern Liman is depicted in Figure 32.6. The probabilities for the partial duration series were calculated using the Weibull plotting positions

(Haan, 1979) in semilogarithmic space. Note that a probability abscissa may not be used, nor can exceedance probabilities be defined for all the values in a partial duration series that includes events occurring more often than once a year.

The uncorrected values of event sediment volumes were obtained by multiplying the thickness of each couplet at the central pit by the area of the Liman. Nevertheless, during small events sediment was deposited merely in the lower, deeper part of the reservoirs. The 12 thinnest couplets at the centre of the Northern Liman were not deposited at the entrance to the Liman, although seven among them were identified midway at the exit (Figure 32.2). The average thickness of the larger couplets across the entire reservoir area is slightly smaller than that of the correlative couplets at the Liman centre. This can also be shown by reference to Figure 32.2, where the combined thickness of the 19 thickest couplets at the centre is 51 cm, decreasing imperceptively towards the dam and increasing to 71 cm at the entrance.

The correction of event volumes consisted of a linear decrease from nil for the 13th smallest event down to 50 per cent for the smallest of all, and a more gradual linear increase from the 13th upwards to a maximum increase of 25 per cent.

Additionally, consideration must be made for the gradual increase in loss of fine sediment out of the reservoirs during large events. The trap efficiency is, indeed, dependent on the ratio of the inflow volume to the reservoir capacity, denoted I/O (Brune, 1953). For example, analyses of runoff plot experiments in relation to rainfall intensity, duration and frequency data for Beer Sheva Loessial soils have shown that the largest 10-year runoff event from plots is equal to 42 mm (Morin *et al.*, 1979). Given that the runoff/rainfall ratio decreases with increase in catchment size, the actual depth of the 10-year runoff event would be smaller, say 35 mm. For the Northern Liman drainage basin this is equivalent to an I/O of 11 when the Liman was first conducted and an I/O of 21 a decade later. From Brune's calculations (1953, his Figure 6) it appears that the appropriate trap efficiencies would be 90 and 87 per cent. For the Northern Liman catchment this would be equivalent to 9650 or 5200 m^3 one decade apart, or a runoff depth of 3.2 and 1.7 mm, respectively, over the catchment. These latter magnitudes occur several times every year. A linear trapping efficiency correction function for sediment volumes has been adopted. It increases from nil for the 12th smallest couplet to 10 per cent for the largest, irrespective of their vertical location within the depositional sequence.

An additional complexity may arise when two or more large events occur within a short period of time. Because the entire infiltration curve for the Liman substrate is often high, most reservoirs dry up completely within 5 days after their capacity has been filled. The probability of two large events recurring within such a short time interval is rather low.

The corrected sediment yield values (Figure 32.6) allow prediction of the magnitude of erosion rates for various exceedance probabilities. Observe that infrequent events carry more sediment than previously suggested. Still, predicted sediment yields from infrequent events are not too large. Based on the exceedance probabilities derived from Figure 32.6, half of the average soil loss occurs during a single event once every 18 months, and twice the average is mobilized in one event once every 3 years.

Care should be taken from unduly stretching the data, for example, to the 100-year event. Also, the correction factors for trap efficiency are not only empirically fitted, but it is questionable whether Brune's (1953) relations hold locally. Experiments are presently underway to evaluate factors affecting the trap efficiency and sediment reworking within reservoirs.

ACKNOWLEDGEMENTS

This contribution is dedicated to the memory of Ran Gerson, who suggested a decade ago that prediction of erosion may be derived from stratigraphic evidence. Surin Laovanu and Khalel Shahadah assisted in the field and Danuna Keil with data compilation. The Israel Meteorological Service kindly supplied rainfall records with the help of Ora Kadvil. This study was funded by the Jo Alon Centre. Additional financial assistance was granted by the Ben Gurion University of the Negev.

REFERENCES

Brune, G. M. (1953). Trap efficiency of reservoirs, *Am. Geophys. Union Trans.*, **34**(3), 407–17.
Haan, C. T. (1977). *Statistical Methods in Hydrology*, Iowa State University Press, Ames.
Hereford, R. (1987). Sediment-yield history of a small basin in Southern Utah, 1937–1976: Implications for land management and geomorphology, *Geology*, **15**, 954–7.
Katsnelson, J. (1979). Rains of the Negev. *In* Shmueli, A. and Gradus, Y. (eds), *Land of the Negev*, Ministry of Defence, Tel Aviv [in Hebrew], 51–73.
Kirkby, M. J. and Morgan, R. P. C. (eds) (1980). *Soil Erosion*, John Wiley, New York.
Laronne, J. B. (1987). Rhythmic couplets. *In* Berkofsky, L. and Wurtele, M. G. (eds), *Progress in Desert Research*, Rowman & Littlefield, Totowa, 229–44.
Laronne, J. B. (1988). Comment on 'Sediment-yield history of a small basin in Southern Utah, 1937–1976: Implications for land management and geomorphology', *Geology*, **16**, 956–7.
Laronne, J. B. and Mosley, M. P. (eds) (1982). *Erosion and Sediment Yield*, Benchmark Papers in Geol., Series No. 63, Hutchinson and Ross, Stroudsburg.
Linsley, R. K. Jr, Kohler, M. A. and Paulhus, J. L. H. (1982). *Hydrology for Engineers*, 3rd edition, McGraw-Hill, New York.
Morin, J., Benyamini, Y. and Dorfman, C. (1979). The rainfall–runoff relationship in the southern area of Israel, *Israel Ministry Agr. Soil Cons. Res. Unit Res. Rpt.*, R41 (in Hebrew).

Soares, E. F., Unny, T. E. and Lennox, W. C. (1982). Conjunctive use of deterministic and stochastic models for predicting sediment storage in large reservoirs. Pts, 1–3 *Jour. Hydrol.*, **59**, 49–121.

Tamir, R. (1972). Runoff and erosion in small watersheds in the South, *Israel Ministry Agr. Soil Cons. Res. Unit. Res. Rpt.* R34 (in Hebrew).

Yaalon, D. J. Koyumdjisky, H. and Raz, Z. (1968). The soil association map of Israel at a scale of 1:1,000,000, *Ktavim*, **18**, 61–8 (in Hebrew).

PART 3
CONSERVATION AND POLICY

Conservation Practices

33 Improved Management of Drylands by Water Harvesting in Third World Countries

ANDERS RAPP and ANNELI HÅSTEEN-DAHLIN
University of Lund, Sweden

INTRODUCTION

Repeated disasters of drought, land degradation, crop failures and famines have affected many countries in Africa in the last twenty years. These environmental and social catastrophes have caused much concern and debate, for instance resulting in activities such as the UN Conference on Desertification (UNCOD, 1977) and the report of the World Commission on Environment and Development (1987, Brundtland Commission). The key concept of the last-mentioned report is 'sustainable development' as a strategy to solving the formidable problems facing us in the coming decades, with doubling world population, continuing poverty and environmental decline.

The slow progress in erosion control and soil conservation in the Third World has led to critical debates and recommendations of a shift of emphasis from engineering to biological approaches in soil conservation (cf. Blaikie and Brookfield, 1987; Hudson, 1988; Lal, 1988; Stocking *et al.*, 1988). It has also been claimed that quantification of soil loss by water erosion is a too-narrow approach as a basis for cost/benefit evaluations of soil conservation. We agree with this view and argue in this chapter for integration of soil and water conservation for improved land husbandry. A sustainable land use in the drylands of Africa has to counteract four types of resource losses common in wasteful land use of today: losses of *soil*, *water*, plant *nutrients* and water *storage capacity*. Low-cost methods of water harvesting in drylands are fundamental in this context, as indicated in our chapter with examples of cases in Tanzania, Zimbabwe, Burkina Faso, and also cases of ancient water harvesting in Israel and Tunisia.

Soil Erosion on Agricultural Land
Edited by J. Boardman, I. D. L. Foster and J. A. Dearing
©1990 John Wiley & Sons Ltd

SUBSISTENCE FARMING IN AFRICAN DRYLANDS

Since the late 1960s the senior author has supervised research and environmental assessments on problems of soil erosion and sedimentation in Tanzania (Rapp et al., 1972; Rapp, 1976; Sundborg and Rapp, 1986). These studies were performed with a team of collaborators in reference field areas of highlands such as the Uluguru mountains and the Usambara mountains, and also in semi-arid drylands of low relief, for instance near Dodoma in central Tanzania, at Arusha in the north and near Mwanza (Sukumaland) in the north-west of the country. The last mentioned three areas were in semi-arid zones of $c.400-800$ mm mean annual precipitation.

Our Sukumaland report (Rapp, 1976) is based on field studies in 1975, when the entire rural population had been moved into new village sites from their earlier scattered homesteads. We made the following summary and recommendations: 'Traditional land use has mainly been one of shifting cultivation and moving settlements. People have exploited the land until soil exhaustion and then moved to new areas, a practice which could be called exploit-and-move.'

The rapidly growing population in much of dryland Africa, and the new villages mean a new situation. The new villages have purposefully been located near roads, water points and service institutions. Each village has a fixed land area belonging to it, and cannot be abandoned for a new site if the land resources should become exhausted.

The new villages must therefore be based on the concept of stay-and-maintain land fertility, instead of the traditional exploit-and-move. This basic principle, however, must allow rational adaptations of village size and out-migration of population surplus, when necessary.

Thus the concept of viable villages presupposes a planned, intensive land use, for long-sustained yields. Soil and water conservation, including manuring, controlled grazing, controlled use of firewood and other resources, is necessary for every viable village. The other alternative, an unplanned, individual exploitation of nearby land resources, will lead to a more rapid exhaustion of soil, etc. near the village, due to the concentration of pressure, and hence to a collapse of the village system (Rapp, 1976).

This plea of 1976 for a change from traditional land use of exploit-and-move type to a new sustainable land use of stay-and-maintain soil fertility did not have much impact and needs to be repeated many times.

To the list of measures for improved land use suggested above we now add water harvesting and storage for domestic use and small-scale irrigation. In tropical and subtropical drylands with less than 600–700 mm of annual rainfall, water harvesting is a particularly important and appropriate technology together with other measures of conservation. In the following text we will present and discuss two methods of water harvesting and storage: runoff-collecting ridges on slopes and groundwater dams in wadi-beds.

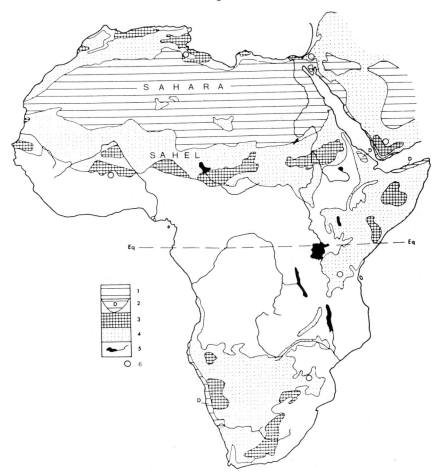

Figure 33.1 Generalized map of desertification zones in Africa, simplified from UNCOD's world map (1977). 1-2, Hyperarid zones (deserts); 3, arid to subhumid zones with a very high risk of desertification; 4, arid to subhumid zones with a moderate degree of desertification hazard; 5, major lakes and rivers; 6, location of case studies discussed in this chapter

The areas discussed are marked in Figure 33.1, a simplified map of desertification hazards in Africa according to UNCOD (1977). Three of our selected case studies are from areas in tropical Africa, in the dryland zone of 400-700 mm of annual precipitation (central Tanzania, Zimbabwe and Burkina Faso). Two other case studies are from the subtropical desert fringe with only about 80-150 mm of annual precipitation: Avdat in Israel and Matmata in Tunisia.

498   Soil Erosion on Agricultural Land

The present discussion on the cost/benefit of soil conservation in the Third World is unfortunately too focused on the counteraction of soil losses only. It cannot be stressed too much that the concept of soil conservation has to be seen in a greater perspective. It should be widened to counteract not only losses of soil but also losses of runoff water, of plant nutrients and of water-storage capacity.

CASE STUDIES IN SMALL CATCHMENTS IN TANZANIA

A close vegetation cover, either natural or cultivated, is the best protection against soil erosion. Climates with poor vegetation and strongly fluctuating rainfall are liable to high soil losses by water erosion. Semi-arid lands in tropical to temperate climates are in this category. The case studies of soil erosion and sedimentation presented below are from a tropical semi-arid area, the Dodoma district of central Tanzania, with high rates of erosion due to natural environmental factors and strong human pressure.

The importance of land use and vegetation cover in controlling soil erosion and water runoff can be demonstrated by results obtained from soil erosion tests on ground with different vegetation covers in Tanzania (Figure 33.2). The test results in the figure are representative of semi-arid pediment slopes on Precambrian bedrocks in central Tanzania. The annual precipitation is 400–600 mm, and the dry season lasts from April to November.

The upper slope profile of Figure 33.2 shows an inselberg landscape in savanna climate with traditional dryland cropping of maize, and cattle grazing on the common land of the upper pediment zones and the floodplains (mbugas, dambos) along the stream channels. The situation at the same site before and after the rainy season is a reflection of overgrazing and locally severe degradation of the upper pediment slopes. These slopes lack grass cover even after the rainy season, due to overgrazing, soil erosion, high runoff and low infiltration of water into the soil. On the lower pediments and the floodplains, on the contrary, the maize crops and the floodplain grasses have recovered. An irreversible form of land degradation that occurs downstream of the area is the filling of the man-made reservoir with sandy sediments (Figure 33.3). The expected life of the reservoir is only 30–40 years (Rapp *et al.*, 1972; Christiansson, 1981).

A general strategy to reduce soil losses and improve land use in this area would be to reduce grazing pressure, to use more ridging, mulching and tree crops on cultivated land and to encourage water harvesting.

From the colonial period until the 1950s, small-sized dams were thought to be the main answer to the water-supply problems of semi-arid Tanzania and many similar countries. Dams were built for irrigation and flood protection and to supply drinking water to people and cattle.

In the 1960s and 1970s it was found that sedimentation of such reservoirs

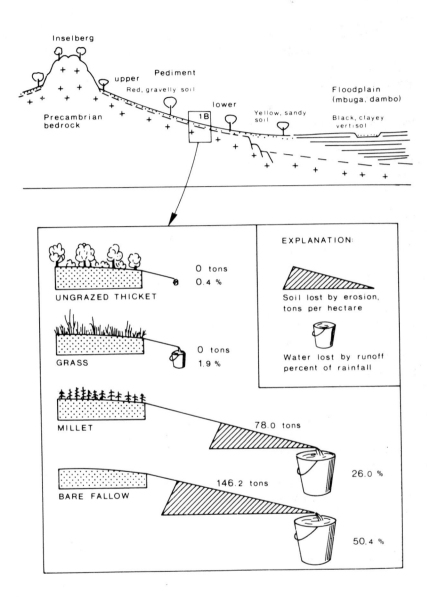

Figure 33.2 Slope profile of an inselberg plain (top) and results of soil erosion and runoff tests on land with different vegetation covers in Mpwapwa, Tanzania (bottom). Annual average of two-year recording period. The erosion plots were 50 m² of red sandy loam soil on a slope of 3.5° gradient. (Reproduced by permission of *AMBIO* from Sundborg and Rapp, 1986). Data from Staples, 1938

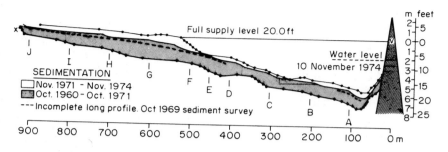

Figure 33.3 Matumbulu Dam, Tanzania. Accumulation of mainly sandy sediments in a small reservoir during the period 1960–74. (From Christiansson, 1981)

was much faster than previously thought. This drawback, in combination with large evaporation losses, reduced interest in surface-water dams and increased support for drilled groundwater wells as the solution to water-supply problems. Evaluations of the many groundwater wells and pumps provided by the Swedish International Development Agency and other agencies in the late 1960s and 1970s were then made. Sadly, wells with diesel pumps do not currently function in Tanzania's semi-arid areas because of either inadequate maintenance or lack of diesel oil. About half the wells with hand or foot pumps are out of order due to lack of maintenance or spare parts. All the simple bucket-and-rope wells, on the other hand, seem to be functioning without failures (Andersson, 1982).

One lesson we may learn from past experience of rapidly silting surface dams is to utilize the sandy riverbeds and sand fans of the temporary streams in semi-arid areas. These growing sand fans should be explored and utilized as groundwater aquifers, as has been done for a long time by local people in many regions of the world. They can be mapped and monitored using remote sensing, such as aerial photographs or satellite images with high resolution. The sandy riverbeds or sand fans are probably recharged by temporary flash floods and can keep large quantities of cool and clean water without large evaporation losses or much contamination by disease vectors.

A CASE STUDY OF WATER HARVESTING AND STORAGE IN CHABWIRA, ZIMBABWE

Chabwira is situated 200 km east of Bulawayo in Mberengwa Communal Land. It is an area on the border between Middleveld and Lowveld, with an elevation of 900 m asl. The geology consists of granites and gneisses of Precambrian age. The granitic parts are characterized by whalebacks and castle kopjes, while the gneisses give rise to a gently undulating landscape with lower relative relief.

Light-coloured sandy soils dominate and their inherent fertility is low as well as their waterholding capacity. These soils are prone to erosion (Dahlin et al., 1984). The junior author (Håsteen) is working on her PhD project on water management by development of riverbed aquifers in this area. Mean annual rainfall is about 650 mm and the rainy season starts in mid-November and ends in March.

The people are mainly Shonas and the Chabwira area of 150 km^2 has a population of about 10 000. Rainfed agriculture is the main source of living, with maize as the dominant crop. Nearly every part of the area is used either for cultivation or grazing. Cattle are an important element of the Shona culture.

Land-use Policy

The land-use policy played an important role in the history of soil and water management in former Rhodesia. Shortly after European colonization in the 1890s, 'native reserves' were established for the African population. Generally, they were located in remote areas with low rainfall and poor agriculture potential. Traditionally, the Shonas practised shifting cultivation for subsistence, but the existence of the reserves (later Tribal Trust Land) limited the land available, which meant that this form of agriculture gradually had to be abandoned. On the sandy soils, the fields became exhausted after four or five years. To sustain productivity, fertilizers, crop rotation and contour ridging had to be introduced (Bourdillon, 1982). A rapidly growing population coupled with no extension of the African areas soon resulted in soil erosion. Soil and water conservation schemes were introduced in the 1930s, but they were never totally accepted by the rural people. In these areas, now known as Communal Land (CL), the majority of the population still live and struggle with problems of land degradation and insufficient water supply. A recent soil-erosion survey of Zimbabwe (Whitlow, 1988) showed that the most extensive and severe erosion occurs within the CL, where 83 per cent of the eroded land is located. In a programme on soil conservation by Elwell et al. (1984), they state that: 'If we allow current rates of soil erosion and runoff to continue, our soil and water resources will be damaged beyond recovery within our own lifetimes.'

Small Surface-water Reservoirs

Degradation of the land results not only in substantial rates of soil loss but also in a high runoff. The rapid runoff during the start of the rainy season means low water infiltration into the soil and flash floods in the riverbeds during and shortly after the rainy season. This, together with the erratic rainfall, creates a pressing need for storage of water from the rainy to the dry seasons. A common solution is to build small surface-water dams, but they are not ideal in this environment since evaporation losses are high and the rivers carry heavy sediment

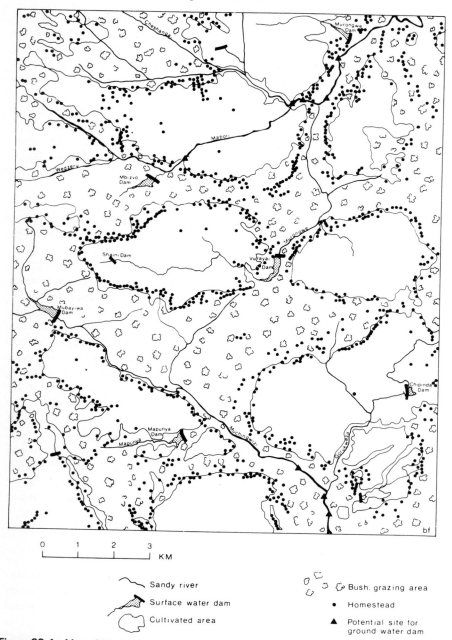

Figure 33.4 Map of land use, location of existing dams and reservoirs, and possible sites for sand storage dams (black triangles). Chabwira district, Zimbabwe. (Redrawn from Dahlin *et al.*, 1984)

loads causing siltation problems in the reservoirs. The reservoirs are also hazardous to health in several ways; malaria, bilharzia and other diseases are spread by use of surface water (Dahlin et al., 1984). An advantage is that the technique required is quite simple, and local labour and materials can be used. Small surface-water dams are still built to a large extent in the rural areas to secure water supply, since the basement aquifers are limited and are expensive to exploit and maintain (Figure 33.4).

The present water supply in Chabwira is based on various sources of varying quality (e.g. boreholes, wells, surface-water dams and riverbeds). The distance to a protected waterpoint is several kilometres, for many people, which makes them use nearby sources instead. Often they dig for water in the sandy riverbeds and scoop out the water. These open wells are, however, very sensitive to pollution. If the riverbed aquifers could be developed in a more systematical way, providing protected wells, this could form an important complement to the wells which tap the basement aquifers. In many cases the resource is sufficient for a local need, while in others an improvement by constructing underground dams and/or sand catchment dams is necessary.

Subsurface Dams

The groundwater flow is arrested in some places by rock barriers formed by dykes, faults, etc. These barriers can be considered as natural groundwater dams. In such locations the riverbeds keep water all the year round. The presence of a natural groundwater dam is often revealed by green vegetation along dried-up watercourses.

There are basically two different types of groundwater dams: subsurface dams and sand storage dams. A subsurface dam is built below ground level and arrests the flow of a natural aquifer, whereas a sand storage dam impounds water in sediments caused by the dam itself. (Hanson and Nilsson, 1986). In practice there are often combinations of the two types. Thus this technique of water harvesting can be considered as a means of conjunctive use of shallow groundwater and surface-water resources.

In cases where the aquifers are drained during the dry season due to the natural flow of groundwater a subsurface dam can secure an adequate water supply all year round. An impermeable wall is built down to bedrock or another impervious layer in order to stop the groundwater flow and store it in the river sand. Wells can be constructed either on the riverbank or in the sand deposit. The former is preferable, if there is good hydraulic contact between the riverbank and the riverbed, since wells in the riverbed risk being damaged during flooding.

Sand Storage Dams

In places where a potential aquifer is not available it is often possible to create one by building a sand storage dam. When a weir is built across a non-perennial

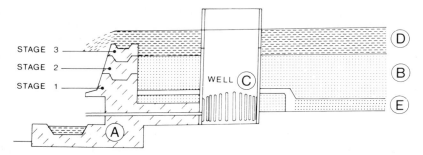

Figure 33.5 Sketch of a sand storage dam, built from local-made brick in three stages (1-3). A Wall of brick or other local material; B river sand deposit; C concrete lining with slots for water intake and with outlet pipe to cattle trough; D water; E gravelly filter. (Redrawn from IRC, 1983)

river, particles carried by the flow during the rainy season will settle and eventually fill the reservoir. A step-wise construction of the weir is often necessary to avoid the deposition of a clayey sediment and to obtain a coarse-grained aquifer. The water can be extracted from the well in the sand bed through a pipe in the dam wall, thus avoiding installation and maintenance of pumps (Hanson and Nilsson, 1986). The principle of a sand storage dam is demonstrated in Figure 33.5.

In many cases exploitation of riverbed aquifers can secure the water supply to a great extent. The advantages of using these aquifers compared to surface water dams are obvious:

(1) Evaporation losses are almost eliminated.
(2) The risk of siltation problems is reduced.
(3) The water quality is normally superior.
(4) The risk of bilharzia transmission and mosquito breeding is reduced.
(5) Infilling by vegetation is avoided.

Furthermore, the development can generally be done on a low-cost basis with large amounts of local labour and materials. Simple methods can be used for water extraction which minimizes maintenance requirements. One advantage of using the shallow groundwater compared to deeper-lying aquifers is that there is generally less risk of overexploitation of the resource, implicating ecological soundness. Figure 33.4 presents possible sites for riverbed extraction by groundwater dams in the Chabwira area. The sites have been identified by studies of aerial photographs, satellite images (SPOT), maps and field inspections. The map is compiled from Dahlin *et al.* (1984) and Nordesjö (1988).

Large parts of the Communal Lands in Zimbabwe may have considerable groundwater resources in the riverbed aquifers and could be suited for the application of groundwater dams.

Groundwater dams have been used with success in several countries (Nilsson, 1984). Dryland areas to be mentioned are, for example, in Namibia, Kenya and Ethiopia.

PROJECTS OF WATER HARVESTING IN BURKINA FASO

Burkina Faso in the West African Sahel region is one of the dryland countries where harvesting and soil conservation by low-cost methods are widely tried and applied (Mietton, 1986; Fries, 1988). Several simple methods to fight water erosion and degradation can be developed from traditional farming in those areas. They are either mechanical measures such as ridges or rows of stones for water harvesting or cultivation practices such as mixed crops, mulching and ridging.

Different types of low ridges on gentle slopes and small check dams in channels are described by Mietton (1986). He states that useful lessons for the northern regions of Burkina may be drawn from the experimentation conducted by several non-governmental organizations in the countries where up to 50 000 ha were ridged in the period 1977–85. This was done by digging or by use of small tractors. The cost was less than 1000 FF ha^{-1}.

Figure 33.6 Simple device for levelling of low ridges for water harvesting on gentle slopes in Western Sahel. (From J. Fries, 1988, after Harrison, 1987)

Fries (1988, p. 104) describes cases of low-cost technology introduced for water harvesting in northern Burkina Faso by Oxfam. After discussions with the local farmers it was concluded that the most useful method for water harvesting and soil conservation would be to build low ridges of earth and stones running parallel to the contours on gentle slopes of a gradient below 2°. In order to align the ridges parallel to the contours a simple levelling device is used. This consists of a 10 m long, transparent plastic pipe. At both ends the pipe is attached to a wooden rod, so the water level can be read at both ends of the nearly filled pipe and the direction of the horizontal line on the ground be defined. The low ridges are called 'diguettes' in French. They dam surface water and thus increase infiltration into the soil and later become zones of thicker vegetation, trapping seeds, water and suspended silt. Increased production by more than 50 per cent of fodder and tree crops is common on land improved with catchment ridges on slopes (Fries, 1988; Figure 33.6).

DESERT RUNOFF FARMING IN THE NEGEV

In the Negev Desert in today's Israel there are many ancient structures of small-scale water harvesting and irrigation, dated to the time of the Nabateans.

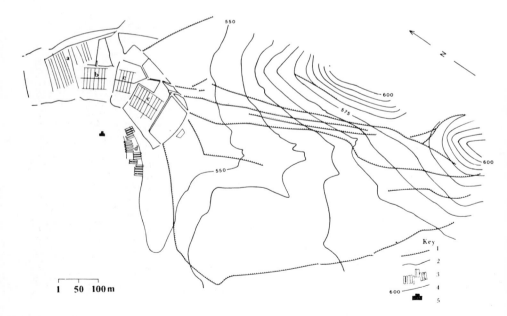

Figure 33.7 Map of the Avdat runoff experimental farm, Negev Desert. 1 Water-collecting ridge; 2 contour line; 3 irrigated plots; 4 altitude in m.asl; 5 buildings. (From Evenari et al., 1968)

Water Harvesting in Third World Countries 507

In the desert valleys of the 'Negev Highlands' they developed irrigation systems based on surface runoff water in areas with less than 100 mm of annual rainfall and built long, water-collecting ridges of stone and earth-fill, which led runoff water and suspended sediment down to dams and terraces in the valley bottoms. (Figure 33.7).

A modern, experimental farm has been established at Avdat using the old, still-functioning water-collecting ridges as a model, which were recently restored and strengthened (Shanan *et al.* 1967). The cultivated plots cover 5 ha, which are irrigated by the runoff water from an area of 375 ha of hillslopes. The hills are of Eocene limestone with a soil cover of loess on the lower slopes and alluvium in the valleys. (Figures 33.7 and 33.8).

The limestone bedrock seems to yield high amounts of surface runoff, even in rainstorms of rather low intensity. According to observations at Avdat, a surface layer of rainfall can be developed after only 2 mm of rainfall. The average surface runoff from plots without vegetation is about 20 per cent of the annual precipitation decreasing to about 5 per cent in catchments of first-order wadis (Shanan, pers. comm.).

In the restored parts of the experimental farm the ratio between catchment area : irrigated cropland is about 25 : 1, similar to the Nabatean ancient system.

Figure 33.8 Avdat experimental farm for runoff irrigation (foreground) in the Negev Desert, Israel. On the skyline are the ruins of the Roman town of Avdat. The stone dams (jessour) at the mountain foot are from Nabatean times. (Photograph: A. Rapp)

508 Soil Erosion on Agricultural Land

The crops are fodder crops, onions and tree crops such as almonds, apricots and figs. The runoff water harvesting has been successful in spite of the low and fluctuating annual precipitation. The experiments have been extended to other parts of northern Negev and also to northern Kenya (Lowenstein, pers. comm.).

The ancient Nabatean system of water harvesting was developed in the Negev from 800 BC until *c*.AD 300–600. Still older and larger systems for desert runoff farming were developed by the Sabeans at Marib in today's Yemen Arab Republic (cf. Figure 33.1). It existed from about 2000 BC and was destroyed by a dam failure about AD 600. That disaster ended the irrigation of an area covering nearly 10 000 ha (Brunner and Haefner, 1986). It is an ancient example of appropriate technology in dryland water management, which contains much useful information for reconstruction and rehabilitation today.

THE CASE OF MATMATA, SOUTH TUNISIA

The climate, relief and geology of the Matmata Hills in south Tunisia are rather similar to the Negev Highlands. Trap dams of stone with a core of mud for water harvesting are common in these areas. They are still in use for cultivation and are called 'jessour' dams in Arabic (pers. comm. by G. Novikoff). Figures 33.9 and 33.10 show their form, location and effect as a foundation for

Figure 33.9 Matmata Mountain, south Tunisia. Stone dams (jessour dams) for trapping water and sediment in desert climate. Cf. Figure 33.10. (Photograph A. Rapp)

Figure 33.10 Aerial view of series of check dams for runoff farming. Matmata mountains, south Tunisia. Cf. Figure 33.9. (Photograph A. Rapp)

sustainable agriculture producing barley, olives and dates. They have a long tradition, probably many hundreds of years, but they do not show signs of salinization, probably because of sufficient drainage through seepage of water and overflow over stable spillways in the dam construction (Figure 33.9).

It may happen that strong floods cause damage and breaks in jessour dams, so they require much work for maintenance and repair, but the fact that they are still in practice and also used as models for new constructions in Tunisia means that they are very interesting cases of appropriate, small-scale technology for study and application in other areas.

CONCLUSIONS

Selected cases from African and other drylands have been described and discussed as examples of the need for emphasis in research and management on combined soil conservation and water harvesting/storage. Drylands with 100–800 mm of annual precipitation and suitable topograpahy, bedrock and soils need applications and development of site-appropriate technology of soil conservation, runoff harvesting and water storage for sustainable, small-scale land use. There is much to learn in this respect from ancient technology of land and water management.

ACKNOWLEDGEMENTS

The field studies in Zimbabwe performed by one of us (A.H.) were sponsored by the Swedish Agency for Research Cooperation (SAREC), those in other countries by SIDA and the Bank of Sweden Tercentenary Fund. Colleagues at the Department of Physical Geography, Lund, undertook the typing, drawings (Birgitta Fogelström) and photographs (T. Nihlén). Our warm thanks to these people and institutions.

REFERENCES

Andersson, I. (1982). Wells and handpumps in Shinyanga Region, Tanzania, *Res. Paper No. 77*, BRALUP. Dar es Salaam.
Blaikie, P. and Brookfield, H. (1987). *Land Degradation and Society*, Methuen, London.
Bourdillon, M. (1982). *The Shona Peoples*, Mambo Press, Gwelo.
Brunner, U. and Haefner, H. (1986). The successful floodwater farming system of the Sabeans, Yemen Arab Republic, *Applied Geography*, **6**, 77–86.
Christiansson, C. (1981). *Soil Erosion and Sedimentation in Semiarid Tanzania*, Scandinavian Inst. of African Studies, Uppsala.
Dahlin, T., Hallberg, M. and Håsteen, C. (1984). *The Water Situation around Chabwira, Mberengwa District, Zimbabwe*, Dept of Water Engineering Geology, University of Lund.
Elwell, H., Kagoro, C. D. and Norton, A. J. (1984). *A Recommended Conservation-production system for Small Scale Farming*, First working draft, Inst. of Agric. Engineering. Harare.
Evenari, M., Shanan, L. and Tadmor, N. (1968). Runoff farming in the desert. I. Experimental layout, *Agronomy Journ.*, **60**, 29–32.
Fries, J. (1988). Västafrikas savanner—är de på väg att bli öken? *Ymer 108*, 90–109, Stockholm.
Hanson, G. and Nilsson, Å. (1986). *Groundwater Dams for Rural Water Supplies in Developing Countries*, SIDA/VIAK Consulting Engineers, Stockholm.
Harrison, P. (1987). *The Greening of Africa*. Paladin, London.
Hudson, N. (1988). Soil conservation strategies for the future, *5th Int. Soil Cons. Conf. Proc.*, Bangkok.
IRC International Reference Center for Community Water Supply and Sanitation (1983). *Small Community Supply*. John Wiley, Chichester.
Lal, R. (ed.) (1988). *Soil Erosion Research Methods*, Soil and Water Cons. Soc., Ankeny, Iowa.
Mietton, M. (1986). Méthodes et efficacité de la lutte contre l'érosion hydrique au Burkina Faso, Cah. ORSTOM, ser. *Pédologie. XXII*, **2**, 181–96, Paris.
Nilsson, Å. (1984). *Groundwater Dams for Rural Water Supply in Developing Countries*, Research report TRITA-KUT 1034, Dept. Land Improvement and Drainage, R. Inst. of Technology, Stockholm.
Nordesjö, P. (1988). *Locating Groundwater Dams Using Remote Sensing*, M-Sc thesis No 1. in *Environmental and Natural Resources Info. Systems*, R. Inst. of Technology, Dept of Photogrammetry, Stockholm.
Orev, Y. (1988). *Some Agricultural Considerations in the Planning of Runoff Farming*, Desertification Control Bulletin, 16, 13–16. UNEP, Nairobi.
Rapp, A. (1976). An assessment of soil and water conservation needs in Mwanza Region, Tanzania, Lund University Dept *Phys. Geogr. Rep. No. 31*.

Rapp, A., Berry, L. and Temple, P. (eds). (1972). Studies of soil erosion and sedimentation in Tanzania, *Geografiska Annaler*, **54A**, 3-4, 105-379. Stockholm.
Shanan, L., Evenari, M. and Tadmor, N. H. (1967). Rainfall patterns in the Central Negev desert, *Israel Explor. Journal*, **17**, 163-84.
Staples, R. R. (1938). Report on runoff and soil erosion tests at Mpwapwa. *Annu. Rep. Dep. Vet. Sci. Anim. Husb.* Tanganyika.
Sundborg, Å. and Rapp, A. (1986). Erosion and sedimentation by water: Problems and prospects, *AMBIO*, **15**, 4, 215-25, Stockholm.
UNCOD, *United Nations Conference on Desertification* (1977), UNEP, Nairobi.
WCED, World Commission on Environment and Development (1987). *Our Common Future*, Final report of the Brundtland Commission. Oxford University Press.
Whitlow, R. (1988). Soil conservation history in Zimbabwe, *Journ. Soil and Water Conservation*, **43**, 4, 299-303.

34 Gully Erosion in the Loam Belt of Belgium: Typology and Control Measures

J. POESEN and G. GOVERS
Laboratory for Experimental Geomorphology, KU Leuven
National Fund for Scientific Research, Belgium

INTRODUCTION

Soil erosion by water is a serious problem in the loess loam and sand loam belt of central Belgium. Of all erosion processes operating in this belt, gully erosion has been the least studied. This chapter examines the different gully types formed in the cultivated uplands and their controlling factors as an aid in erosion risk mapping as well as in selecting and applying gully prevention and control measures.

DEFINITION AND TYPOLOGY

A gully has been defined as

> A channel or miniature valley cut by concentrated runoff but through which water commonly flows only during and immediately after heavy rains or during the melting of snow; may be dendritic or branching or it may be linear, rather long, narrow and of uniform width. The distinction between gully and rill is one of depth. A gully is sufficiently deep that it would not be obliterated by normal tillage operations, whereas a rill is of lesser depth and would be smoothed by ordinary farm tillage (Soil Conservation Society of America, 1982).

Gullies can thus be considered as large rills. In this chapter the distinction between rills and gullies was made on the basis of a critical cross-section, i.e. 1 ft^2 ($= 929\,\text{cm}^2$) (Hauge, 1977).

Soil Erosion on Agricultural Land
Edited by J. Boardman, I. D. L. Foster and J. A. Dearing
©1990 John Wiley & Sons Ltd

In a previous study (Poesen, 1989) two main gully types were distinguished on the basis of their location in the landscape: i.e. ephemeral gullies and gullies associated with banks.

Ephemeral Gullies

Ephemeral gullies form where overland flow concentrates in the landscape, i.e. either in natural drainageways (thalwegs of zero-order catchments) or in, or along, linear landscape elements (e.g. parcel borders, field roads, plough furrows, tractor ruts, etc.). They are *continuous* temporary channels, *very often erased by cultivation*, but recurring in the same place during subsequent runoff events. Ephemeral gullies result from concentrated flow erosion. In other words, sediment detachment and removal is essentially a function of flow intensity (Foster and Lane, 1983; Thorne *et al.*, 1986; Watson *et al.* 1986).

These gullies mainly form on freshly cultivated soils when the vegetation cover is very low. They may be generated by rainstorms of rather low intensity (< 10 mm h^{-1}) provided that the catchment size and/or runoff coefficients are sufficiently large. Concentrated runoff generated during such storms starts eroding the (sealed) top layer when flow shear stress exceeds a critical value. The latter is positively related to soil shear strength (Figure 34.1; Torri *et al.*, 1987; Rauws and Govers, 1988).

Ephemeral gullies can be subdivided following their width–depth ratio (WDR). Two main types can be distinguished; gullies characterized by a WDR value of about 1, or smaller than 1, and those having a WDR value much larger than 1. The shape of an ephemeral gully cross-section is controlled by a combination of different factors; those determining *concentrated flow width* (i.e. rainfall intensity, catchment size, runoff coefficient, surface roughness and slope, morphology of concentrated flow zone), those determining *concentrated flow intensity* (e.g. flow discharge and surface slope), and those determining the *resistance of the soil material against detachment and transport* (e.g. soil moisture, soil structure and soil profile characteristics). The effect of local thalweg slope gradient and of the presence of a resistant soil sublayer (i.e. a plough pan) on the WDR of an ephemeral gully is illustrated in Figure 34.2. A high-intensity low-frequency rainstorm falling on an initially dry seedbed, for instance, is likely to cause formation of ephemeral gullies having a WDR $>>1$ (Figure 34.3). High runoff discharges, generated during these storms, will wet flat bottomed depressions over a considerable width (up to 10 m). Since the strength of the soil top layer is considerably reduced upon wetting, erodibility of this material is very high. This situation is particularly frequent in spring and early summer. Flow incision will be reduced at shallow depth if a less erodible layer is uncovered; e.g. a soil layer with a higher moisture content (Govers *et al.*, 1987), a plough pan (Figure 34.4), a Bt-horizon, a compact colluvial layer (Figure 34.5), a fragipan, a(n) (iron)sandstone layer, etc.

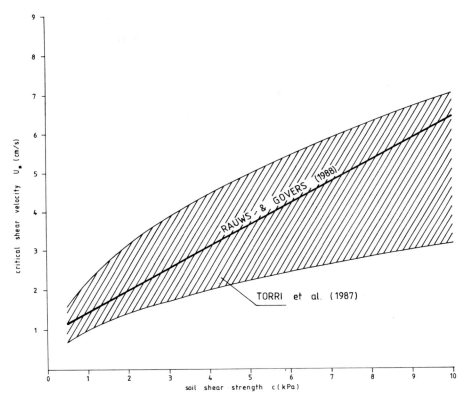

Figure 34.1 Critical flow shear velocity (u_*) for incipient rilling or gullying as a function of shear strength of soils, measured during the occurrence of runoff (wet state)

On the other hand, low-intensity high-frequency rainstorms cause runoff to flow over a limited width in the thalwegs, often causing the formation of gullies with a WDR smaller than 1. If no resistant sublayer occurs in the soil profile, flow incision can be considerable in soft sediments such as calcareous loess loam or loose sands (see Figures 34.6 and 34.7).

Once an ephemeral gully has been formed, it will be filled in with loose soil material by (deep) tillage leading to a deepening of the original thalweg and, hence, an increase of relief energy. Consequently, subsequent storms will often cause more severe gully erosion.

Ephemeral gullies with a WDR $>>1$ cause important crop damage. In addition, a high percentage of total soil lost through gullying consists of fertile topsoil material with a high organic matter and fertilizer content. However, these gullies are easily erased by normal tillage (e.g. deep ploughing). On the other hand, ephemeral gullies with a WDR = 1 or < 1 cause less crop damage

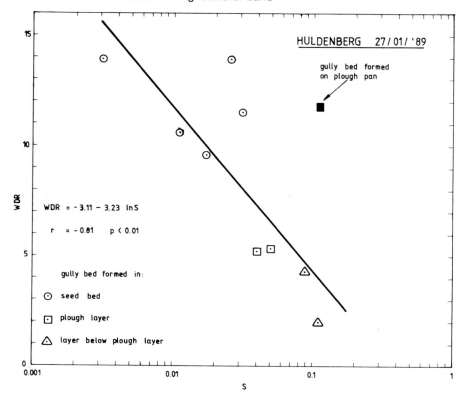

Figure 34.2 Effect of thalweg slope gradient (S) and the presence of a resistant soil sublayer (plough pan) on the width–depth ratio (WDR) of an ephemeral gully. In this case, overland flow width before gullying as well as overland flow discharge during gully erosion was approximately the same in each cross-section

and the percentage of total soil lost consisting of fertile topsoil material is less than for the other gully type. These gullies, however, are not easily erased by conventional tillage operations and often special equipment (e.g. bulldozers) is required to reshape the thalweg.

Gullies Associated with Banks

These gullies form where a wash line, a rill or an ephemeral gully crosses a sunken road bank or a lynchet (i.e. a terrace-like step, aligned parallel to the contour lines and often grown over with grass and/or brushwood). Figure 34.8 depicts such a gully.

These *discontinuous erosion features* cause considerable soil losses. Approximately 650 m^3 of soil were lost due to these gullies in a 126 ha survey

Gully Erosion in the Loam Belt   517

Figure 34.3 Shallow and wide ephemeral gully (width/depth = ± 20) formed in a sandy loam soil on a 9 per cent slope (Huldenberg, June 1986)

area. This corresponds to a mean soil loss of $5.2\,m^3\,ha^{-1}$. However, if the volume of a particular gully is divided by its catchment area, much higher figures are obtained, i.e. in the mean $44.8\,m^3\,ha^{-1}$ (Poesen, 1989).

Based on inquiries among farmers and on field observations, the genesis of these gullies can be summarized as follows. Overland flow crossing a bank will often follow existing structural voids in the banks, such as desiccation or tension cracks and/or biotic holes. In particular, the latter seem to play a very important role. As the local gradient of these pipes is often much steeper than the overall slope of the upland, severe erosion takes place in these pipes leading to their enlargement. Ultimately, the roof of the pipe collapses by itself or under the load of agricultural machinery, and a gully head is created. Next, the gully head retreats by plunge pool erosion as well as by mass movement processes acting on the gully head and walls (i.e. slumping, soil fall, creep). The processes mentioned above are also more pronounced where farmers cultivate the land very close to the rim of the bank or gully without leaving a grass strip as a buffer zone. As these discontinuous gullies *cannot be obliterated by conventional tillage operations*, they are *permanent* and, hence they seriously interfere with the cultivation of the land.

In a previous study (Poesen, 1989) it has been shown that, contrary to ephemeral gully erosion, flow intensity is not a dominating factor explaining the volume of eroded soil in a gully associated with a bank. The development

Figure 34.4 Ephemeral gully (width/depth = ± 1) formed in a loam soil on a 5 per cent slope (Grimbergen, March 1984). Note the presence of a plough pan at 25 cm depth. Length of stick equals 60 cm

of these gullies depends strongly on local site characteristics, such as the presence of biotic holes and cracks as well as the mechanical properties of the different soil layers (see below). From the preceding it follows that prediction of the location and volume of gullies associated with banks is more difficult than for ephemeral gullies.

FIELD EVALUATION OF THE SUSCEPTIBILITY OF SOIL LAYERS TO GULLY EROSION

In order to detect which soil parameter best reflects the resistance of a particular soil layer to deepening by concentrated flow erosion, ten different gully sites

Gully Erosion in the Loam Belt 519

Figure 34.5 Ephemeral gully (width/depth = ± 1) formed in a loam soil on a 9.4 per cent slope (Korbeek-Dijle, September 1988). Note the presence of a 50 cm thick compact colluvial layer. This photograph was taken at gully site No. 2 (Table 34.1: gully erodibility class = 1). Length of stick equals 45 cm

in the area between Leuven and Brussels were selected. All gullies were formed in silt–loam material. At each gully site, the gully–erodibility class was assessed by two people, using a rating procedure which took gully cross-sectional shape, gully size and gully site characteristics into account. For instance, class 1 corresponds to a very low gully erodibility and is attributed to a soil layer on which concentrated flow occurred, resulting in a very small channel cross-section, despite the relatively steep soil surface slope and the important size of the catchment (e.g. Figure 34.5). On the other hand, class 5 corresponds to a very high gully erodibility and is attributed to a soil layer on which concentrated flow occurred, resulting in a large channel cross-section, despite the small channel gradient and the small size of the catchment (e.g. Figure 34.6).

Protruding soil layers in a gully cross-section were given a lower erodibility class value than those in which cavities developed. The rating technique was chosen because it allowed a rapid evaluation of the susceptibility of a soil layer to concentrated flow erosion while taking into account all factors known to be relevant.

Next, for each gully site, a series of soil parameters were determined; i.e. percentage clay, percentage silt, percentage sand, percentage organic matter (loss on ignition method), percentage $CaCO_3$ (loss on decalcification), bulk density

Figure 34.6 Ephemeral gully (width/depth = ± 1) formed in a calcareous loess loam on a 12.6 per cent slope (Huldenberg, September 1988). Because of the presence of calcareous loess in the soil profile, the gully has become 2 m deep. Note the gully bank collapse which forces the runoff to flow against the bank, leading to its undercutting. This photograph was taken at gully site No. 1 (Table 34.1: gully erodibility class = 5)

(using 100 cm^3 cylinders), gravimetric moisture content, shear strength measured with a torvane (Soiltest CL-600, when necessary equipped with a sensitive vane adaptor CL-602), penetration resistance, measured with a pocket penetrometer (Soiltest CL-700) and angle of internal friction deduced from shear strength measurements with a portable Soil Sheargraph (Soiltest D-250). Shear strength and moisture content of the ten gully sites were determined at two different dates; on 22 September, after a relatively dry period when moisture content of the soils was well below field capacity, and on 19 December, during

Gully Erosion in the Loam Belt   521

Figure 34.7 Ephemeral gully (width/depth = ± 1) formed in a 40 cm thick sandy loam layer overlying medium sands (Brusselian sands) on a 20 per cent slope (Huldenberg, June 1986)

Figure 34.8 Gully associated with a bank caused by a rill crossing a lynchet (Huldenberg, June 1988)

a rainy period when, for most gully sites, moisture content of the soils was above field capacity. On 19 December additional shear strength and moisture content measurements were made after the soil material had been saturated by spraying distilled water onto it.

All data are shown in Table 34.1, together with the correlation coefficients between each soil property and the gully-erodibility class. Figure 34.9 depicts some scatter diagrams. From the 14 investigated soil properties, only three were significantly correlated with the gully-erodibility class at a 1 per cent significance level; i.e. moisture content measured on 22/09/1988 ($r = -0.93$), soil shear strength measured on 22/09/1988 ($r = -0.82$) and soil shear strength measured on 19/12/1988 after the soil material had been artificially saturated ($r = -0.89$).

The fact that gully-erodibility class is negatively related to shear strength after saturation of the soil layer is the most relevant of the relationships found as this condition adequately reflects the conditions under which soil detachment occurs. This finding is also in line with literature results: Kamphuis and Hall (1983), Lyle and Smerdon (1965), Partheniades (1971), van der Poel and Schwab (1985) and Zeller (1965) observed vane shear strength to be negatively related to soil erosion intensity under concentrated flow.

A significant negative relationship between gully-erodibility class and shear strength is also found in a relatively dry period. Under wet (but non-saturated) conditions no significant relationship between both parameters is observed. This is caused by the differential behaviour upon wetting of the different soil layers. The layers classified as erosion resistant show a relatively high shear strength when dry. However, water uptake appears to reduce their shear strength much more than it does for soil layers labelled as erodible (Table 34.1). On the other hand, saturation significantly reduces the shear strength of the erodible layers, while shear strength reduction of the more resistant layers is much less pronounced.

The interaction between shear strength and moisture status implies that direct shear strength measurements in the field may lead to erroneous results. Therefore, it is recommended to saturate the soil before shear strength is measured. Although this will allow a good classification of the various soil layers, it will not yield a unique value. Previous experiments clearly indicated that the reduction of shear strength upon wetting is not only a function of intrinsic soil properties but also of the initial moisture content; shear strength values obtained after saturation as well as erosion resistance are positively correlated with initial moisture content of the material (e.g. Grissinger *et al.*, 1981; Kemper *et al.*, 1985; Govers *et al.*, 1987). This phenomenon may also explain the significant negative relation found between gully-erodibility class and soil moisture content measured in a relatively dry period. Variations in gully-erodibility class between different soil layers due to soil mechanical characteristics seem to be reinforced by variations in soil moisture regime which, in turn, are strongly related to the intrinsic properties of the soil layers.

Table 34.1 Correlation between soil properties and gully-erodibility class for ten different gully sites

Gully site no.	1	2	3	4	5	6	7	8	9	10	r	p
Gully-erodibility class	5	1	3.25	2.5	4.75	1	4.75	3	1	0.75	1.00	0.00
Clay (%)	9.4	13.2	12.7	13.8	14.0	13.7	9.8	8.5	13.6	16.3	−0.61	0.06
Silt (%)	88.0	84.7	83.2	83.5	81.5	78.8	73.3	83.4	78.8	79.9	0.10	0.79
Sand (%)	2.6	2.1	4.1	2.6	4.5	7.5	16.9	8.1	7.7	3.8	0.25	0.49
Organic matter (%)	0.3	0.7	0.9	0.5	0.7	0.4	0.3	0.3	0.3	0.3	0.07	0.85
$CaCO_3$ (%)	14.8	2.2	2.2	7.9	2.8	0.9	3.0	2.6	1.0	0.7	0.56	0.09
22/09/1988												
Bulk density ($g\,cm^{-3}$) ($n=5$)	1.52	1.57	1.48	1.38	1.36	1.58	1.60	1.58	1.64	1.72	−0.53	0.11
Moisture cont. (%) ($n=5$)	13.6	19.3	18.6	16.1	14.7	20.7	13.4	16.7	20.5	19.9	−0.93	0.001
Shear strength (kPa) ($n=5$)	10.4	20.0	17.0	23.8	12.4	26.4	20.4	21.0	25.2	27.0	−0.82	0.003
Penetration resistance (kPa) ($n=5$)	257.9	254.0	292.2	211.8	213.8	349.1	215.7	392.2	239.3	358.9	−0.41	0.24
Angle of int. friction (°) ($n=5$)	47.1	36.0	42.9	34.3	36.0	44.9	44.5	44.0	43.0	47.0	0.01	0.97
19/12/1988												
Before wetting Moisture cont. (%) ($n=2$)	18.5	23.2	22.6	19.2	—	22.0	18.6	21.2	22.1	19.6	−0.61	0.08
Shear strength (kPa) ($n=5$)	11.5	11.6	12.6	30.0	—	13.2	14.3	14.8	15.5	43.0	−0.37	0.32
After wetting Moisture cont. (%) ($n=2$)	26.3	24.7	28.2	25.3	—	27.2	24.4	25.2	25.8	27.2	−0.18	0.64
Shear strength (kPa) ($n=5$)	2.5	11.0	3.9	10.9	—	9.8	4.6	6.3	10.2	14.5	−0.89	0.001

Gully-erodibility class: 1 = very low, 2 = low, 3 = moderate, 4 = high, 5 = very high;
r = Pearson correlation coefficient between each soil property and gully-erodibility class,
p = significance level,
n = number of measurements.

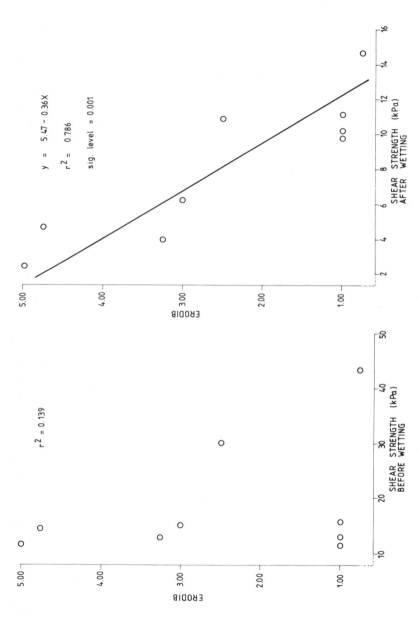

Figure 34.9 Scatter diagrams showing the relation between gully-erodibility class (ERODIB) and shear strength of the soil layers at the different gully sites before and after artificial wetting. Measurements were made on 19 December 1988. Note that, despite the fact that measurements were taken in a wet period, shear strength is in most cases greatly reduced by wetting

Gully Erosion in the Loam Belt   525

From the previous analysis it can be inferred that direct measurements of soil moisture content as well as of shear strength cannot always be used as reliable indicators of the resistance of a loamy soil layer to erosion by concentrated flow. Shear strength of artificially saturated soil material, however, adequately reflects the resistance of loamy soil layers to concentrated flow erosion. Since this parameter can be easily determined in the field, it is recommended for use in gully risk assessment. On upland areas, where overland flow concentrates, knowledge of *shear strength properties of the soil sublayers in a wetted state* as well as the *thickness of these sublayers* is of utmost importance when predicting the potential development of either gully type.

GULLY PREVENTION AND CONTROL MEASURES

Since both described gully types differ widely, prevention and control measures are also different.

Ephemeral Gullies

Prevention of ephemeral gullies can, in principle, be achieved by either preventing runoff from flowing in the depressions (e.g. by increasing the infiltrability of the soils) or by increasing the resistance of the soil top layer to concentrated flow erosion.

The first solution can be achieved in several ways; e.g. by improving the top soil structure, by mulching, etc. However, deep subsoiling, as has been proposed by Fullen (1985) on loamy sands in order to break the plough pan, is not rcommended on upland slopes as well as in the depressions. Many field observations clearly indicate that compact subsoils resist concentrated flow erosion much more than do loose, cultivated topsoils, and so they reduce the development of ephemeral gullies with a WDR = 1 or << 1.

The increase in the resistance of the soil top layer in the thalwegs to concentrated flow erosion can also be achieved in different ways; i.e. by compacting the soil top layer, by applying no-tillage, by establishing grassed waterways or erosion-resistant access roads in concentrated flow zones. Compaction of loamy topsoils in the thalweg has been successfully tested by Ouvry (1987) on gentle sloping thalwegs in northern France (Pays de Caux). The efficiency of compaction in reducing ephemeral gully development, however, will largely depend on the moisture content at both the moment of compaction as well as at the moment of rainfall and concentrated overland flow occurrence. No-tillage has also been successfully tested in central Belgium (De Ploey, 1988; Figure 34.10). Although no-tilled field plots produced as much runoff as the conventional tilled plots, and even more during some periods (Wysen, 1986), ephemeral gully development was negligible on no-tilled winter barley field plots while it was quite important (up to 20 t ha^{-1} yr^{-1}) on conventional tilled winter

Figure 34.10 Effect of no-tillage (background) and conventional tillage (foreground) on ephemeral gully development in winter barley fields (Leefdaal, November 1987)

Table 34.2 Topsoil properties of conventional tilled and no-till winter barley fields on loamy soils in central Belgium (November 1987; De Smet, 1988)

	Conventional tillage	No-tillage
Mulch-vegetation cover (%)	5–15	15–20
Gravimetric moisture content (%)	23.3 ($n=5$)	22.6 ($n=6$)
Dry bulk density (g cm^{-3})	1.35 ($n=5$)	1.56 ($n=6$)
Shear strength (kPa)	15.4 ($n=5$)	18.4 ($n=6$)

Cover is estimated using the point-grid method,
Bulk density is measured with 5 cm long cylinders,
Shear strength is measured in the 0.5 cm thick top layer with a torvane,
n equals the number of field plots.

barley plots during the same year. Similar observations have been reported by Laflen (1987) and by Spomer and Hjelmfelt (1986). The main reason for this difference lies in the mulch effect as well as in the higher bulk density and shear strength values of the topsoils on the no-tilled plots (Table 34.2). In some cases, construction of erosion resistant access roads in thalwegs conducting important runoff volumes during and after heavy rainfall events has been a suitable solution to decrease gully formation risks. Probably the most efficient solution, although not tested in central Belgium, is the establishment of grassed waterways. However, this gully prevention control measure requires a higher financial input.

Gullies Associated with Banks

Prevention of gullies associated with banks can, in principle, be achieved by a series of measures (Poesen, 1988); i.e.

(1) By preventing runoff from flowing across the bank;
(2) By eliminating lynchets and sunken road banks in the landscape (e.g. in the framework of a land-consolidation programme);
(3) By applying biological measures: i.e. by putting the land upslope of a bank under permanent grassland (increase of infiltration rates, surface roughness and shear strength of the soil) and by reinforcing the banks with deep-rooting species in order to reduce the risk of mass movements.

In many cases, only the third type of measure will be feasible.

Once a gully associated with a bank has been formed, two possible control measures can be taken;

(1) Preventing runoff from flowing through the headcut;
(2) Stabilizing the gully by structural measures and accompanying revegetation.

The first possibility can be achieved by improving the gully catchment in such a way that no runoff is produced or by diverting the surface runoff above the gully area. However, runoff elimination is difficult to achieve since one has

Figure 34.11 Structural measure to control a gully associated with a bank: i.e. an 11 m long and 2 m wide reinforced plastic sheet used as a geomembrane in order to protect the gully head (Bertem, October 1988)

to modify the response of the entire gully catchment, which is seldom the case, while runoff diversion would create similar problems elsewhere. Hence, very often, the second control measure remains as the only feasible one.

In order to test structural gully control measures, a gully was selected in Korbeek-Dijle. In order to control the gully head from further migration a plug of loose rock was installed in the gully head following guidelines given by Heede (1977) and Gray and Leiser (1982). The advantages of the loose rock plug are the low cost and its high porosity, which avoids excessive pressure. After installation, however, erosion continued on the contact between the loamy material of the gully headcut wall and the rock plug. In order to prevent further erosion and the formation of a hollow which would undermine the headcut, a plastic sheet (i.e. a reinforced polyethylene tarpaulin of 190 g m^{-2}) was installed on the bottom of the concentrated flow channel, upstream of the gully head, and extended over the rock plug (Figure 34.11). This proved to be a very effective way to conduct the concentrated overland flow safely over the gully head without eroding and undermining the gully head. When applying this simple technique to control a gully head, care should be taken to extend the plastic sheet or other geomembrane-type material sufficiently far downslope, where the erosive forces can be dissipated on any kind of structure without creating new erosion problems. Next, willow staking was applied in order to reinforce the gully walls permanently.

CONCLUSIONS

This chapter has shown that knowledge of gully typology, as well as of factors controlling gully development, is crucial for erosion risk mapping as well as for selecting and applying gully prevention and control measures. Contrary to interrill and partly also rill erosion, knowledge of *the thickness* and of *the resistance properties of soil sublayers* in areas where overland concentrates seems to be crucial for gully risk assessment. Results from this study indicate that vane shear strength of the soil sublayers, measured after artificial wetting, is an adequate indicator of their resistance to concentrated flow erosion. Finally, testing of structural measures to control a gully associated with a bank reveals the potential of geosynthetics.

REFERENCES

De Ploey, J. (1988). No-tillage experiments in the central Belgian loess belt, *Soil Technology*, 1, 181–84.

De Smet, H. (1988). *De invloed van de compactie op de erosiegevoeligheid: laboratoriumexperimenten en terreinmetingen*, MSc thesis, KU Leuven.

Foster, G. R. and Lane, L. J. (1983). Erosion by concentrated flow in farm fields. In Ruh-Ming Li, Lagasse, P. F. and Simons, Li & Associates (eds), *Proceedings of the D. B. Simons Symposium on Erosion and Sedimentation*, Colorado State University, Fort Collins, pp. 9.65–9.82.

Fullen, M. A. (1985). Compaction, hydrological processes and soil erosion on loamy sands in east Shropshire, England, *Soil and Tillage Research*, **6**, 17-29.

Govers, G., Everaert, W., Poesen, J., Rauws, G. and De Ploey, J. (1987). Susceptibilité d'un sol limoneux à l'érosion par rigoles: essais dans le grand canal de Caen, *Bulletin du Centre de Géomorphologie du CNRS*, Caen, **33**, 85-106.

Gray, D. H. and Leiser, A. T. (1982). *Biotechnical Slope Protection and Erosion Control*, Van Nostrand Reinhold, New York.

Grissinger, E. H., Little, W. C. and Murphey, J. B. (1981). Erodibility of streambank materials of low cohesion, *Transactions of the American Society of Agricultural Engineers*, **24**, 624-30.

Hauge, C. J. (1977). Soil erosion definitions, *California Geology*, **30**, 202-3.

Heede, B. H. (1977). Gully control structures, *Food and Agricultural Organisation Conservation Guide*, **1**, 181-222.

Kamphuis, J. W. and Hall, K. R. (1983). Cohesive material erosion by unidirectional current, *Journal of Hydraulic Engineering, American Society of Civil Engineers*, **109**, 49-61.

Kemper, W. D., Trout, T. J., Brown, M. J. and Rosenau, R. C. (1985) Furrow erosion and water and soil management, *Transactions of the American Society of Agricultural Engineers*, **28**, 1564-72.

Laflen, J. M. (1987). Effect of tillage systems on concentrated flow erosion. In Pla, I. (ed.), *Soil Conservation and Productivity*, 2. Universidad Central de Venezuela, Maracay, pp. 798-809.

Lyle, W. M. and Smerdon, E. T. (1965). Relation of compaction and other soil properties to erosion resistance of soils, *Transactions of the American Society of Agricultural Engineers*, **8**, 814-22.

Ouvry, J. P. (1987). *Ruissellement et Erosion des terres. Bilan des travaux campagne 86-87*, Association régionale pour l'étude et l'amélioration des sols, Boisguillaume, France.

Partheniades, E. (1971). Erosion and deposition of cohesive materials. In Shen, H. W. (ed.), *River Mechanics*, **II**, 25-1-25-91.

Poesen, J. W. A. (1989). Conditions for gully formation in the Belgian loam belt and some ways to control them, *Soil Technology Series*, **1**, 39-52.

Rauws, G. and Govers, G. (1988). Hydraulic and soil mechanical aspects of rill generation on agricultural soils, *Journal of Soil Science*, **39**, 111-24.

Soil Conservation Society of America (1982). *Resource Conservation Glossary*, Soil Conservation Society of America, Ankeny.

Spomer, R. G. and Hjelmfelt, A. T. (1986). Concentrated flow erosion on conventional and conservation tilled watersheds, *Transactions of the American Society of Agricultural Engineers*, **29**, 124-7, 134.

Thorne, C. R., Zevenbergen, L. W., Grissinger, E. H. and Murphey, J. B. (1986). Ephemeral gullies as sources of sediment, *Proceedings of the Fourth Federal Interagency Sedimentation Conference*, **1**, 3-152-3-161.

Torri, D., Sfalanga, M. and Chisci, G. (1987) Threshold conditions for incipient rilling, *Catena Supplement*, **8**, 97-105.

van der Poel, P. and Schwab, G. O. (1985). Plunge pool erosion in cohesive channels below a free overfall, *American Society of Agricultural Engineers Paper*, 85-2038.

Watson, D. A., Laflen, J. M. and Franti, T. G. (1986). Estimating ephemeral gully erosion, *American Society of Agricultural Engineers Paper*, 86-2020.

Wijsen, J. (1986). *Onderzoek naar de invloed van een mulch en van no-tillage op bodemeigenschappen, hydrologie en regenerosie*, MSc thesis, KU Leuven.

Zeller, J. (1965). Versuche der VAWE über die Erosion in Kohärenten Gerinnen, *Schweizerische Bauzeitung*, **42**, 3-8.

35 Experimental Study of Erosion and Crop Production on Bench Terraces on Sloping Land

MIROLJUB DJOROVIĆ
Forestry Faculty, Belgrade University, Yugoslavia

INTRODUCTION

The problem of the use of sloping land for crop production is very important for the area of the Central part of the Republic of Serbia. Nearly 87 per cent of the total area (55 000 km^2) is on a slope of more than 5° and 79 per cent of this area is on one of more than 10°. According to the relief classification, 80 per cent of the central part of the Serbian Republic belongs to hilly-mountainous regions.

This natural condition does not suit crop production and creates many problems due to severe water erosion. Sloping land is in constant use and will be used even more in the future, especially bearing in mind the constant growth of population. In order to protect the sloping land, the use of bench terraces for crop production is of great importance.

METHODOLOGY

The Ralja experimental station is located 40 km from Belgrade and consists of 54 experimental plots for the study of water-erosion intensity (Figure 35.1). Geological substratum is made up of Neogen lake sediment (sand and clay) of 'Sarmat' age and brown forest soil (loam) developed on this formation. The average rainfall for the period 1971–87 was 712 mm per year, ranging from 550–900 mm per year. Rainfall erosion intensities are; for 5 min duration of 0.902 mm min^{-1}; for 15 min duration of 0.502 mm min^{-1} and for 30 min duration of 0.286 mm min^{-1}. The average air temperature is 10.2° C.

Experimental plots for the determination of erosion control and crop productivity effects of bench terraces on steep slopes (14° and 20°) were set up in 1970 (Figure 35.2). The following dimensions of sample plots were adopted;

Soil Erosion on Agricultural Land
Edited by J. Boardman, I. D. L. Foster and J. A. Dearing
©1990 John Wiley & Sons Ltd

532 Soil Erosion on Agricultural Land

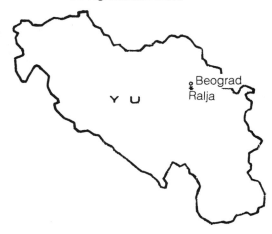

Figure 35.1 Location map of the experimental station at Ralja

Figure 35.2 Experimental plots with bench terraces, with a slope of 20 degrees at Ralja

Study of Erosion and Crop Production on Bench Terraces

length 20 and 40 m, width 2.5 and 5.0 m, with a total area of 50 and 200 m², respectively. The borders of the sample plots are constructed from aluminium sheets (0.8 mm thick) which are very durable, distribute precipitation very evenly and are easy to work with. For collection of runoff water and deposits, covered basins were used, made of metal sheets, 1.0 m³ in volume and with dimensions 2.0 × 1.0 × 0.5 m (Figure 35.3).

The terraces were constructed manually and cropped with wheat and corn in rotation. The corn (SK-1) was cultivated in the usual way for this area, fertilized with 20 t ha^{-1} of cattle manure in the autumn and with 200 kg ha^{-1} of mineral manure of N and K in the spring. The wheat (Bezostaja) was fertilized in autumn with 200 kg ha^{-1} of phosphates, 100 kg ha^{-1} of K and, in spring, 200 kg ha^{-1} of Ca and N.

Runoff, soil loss and crop yield were measured continuously as well as the deformation of the risers of terraces on a yearly basis. Some of the characteristics of the test plots are given in Table 35.1.

Runoff and soil loss were measured after each rainfall and the results are given in Table 35.2 as the mean values per year. The results given in Table 35.2 show that runoff and soil loss from terraced plots are very small and far

Figure 35.3 Runoff and sediment collector at the end of a bench terrace plot, with a slope of 14 degrees, Ralja

Table 35.1 Basic characteristics of the test plots

Sample plot no.	Slope (°)	Dimensions $b \times l$ (m)	Area (m^2)	Effective Area (m^2)	Number of terraces	Length of terraces (m)	Corn and wheat in rotation starting 1971
1	14	5×40	200	137.5	5	5.5	Corn
2	14	5×40	200	137.5	5	5.5	Wheat
3	20	2.5×20	50	25.0	4	2.5	Wheat
4	20	2.5×20	50	25.0	4	2.5	Corn

Table 35.2 Average runoff and soil loss per year from terraced and control plots for the period 1971–87, Ralja

Plot no.	Type of plot	Slope angle (°)	Runoff (m^3 ha^{-1} yr^{-1})	Soil loss (t ha^{-1} yr^{-1})	Culture
1	Terraces	14	7.5	0.14	Corn
2	Terraces	14	3.6	0.11	Wheat
3	Terraces	20	6.3	0.90	Wheat
4	Terraces	20	6.2	0.16	Corn
5	Control	14	130.0	16.15	Corn
6	Control	14	105.0	10.30	Wheat
7	Control	20	225.0	24.20	Corn
8	Control	20	210.0	17.90	Wheat

below the accepted values of erosion tolerance of 2.0 t ha^{-1} yr^{-1}, (Djorović, 1975). Terraces eliminate the slope effect almost completely, as well as the effect of the manner of soil utilization (land-use factor), the two most important erosion factors. Surface runoff and, accordingly, soil loss are observed only on the risers of bench terraces, but only during the first 2 to 3 years of observation. After that the risers were well protected by natural vegetation cover (grass). Eroded soil, mainly from the risers, is deposited on the bench (flat part of the terrace) and could not be transported further because of the inslope of 1 per cent and continuous cultivation. Deformation of the risers of terraces is shown in Figures 35.4 and 35.5 and the flat part of terraces during 17 years of continuous use decreased by 0.4 m, from 5.5 m to 5.1 m (slope 14°) and from 2.5 m to 2.1 m for a slope of 20°, which should be considered as acceptable.

These results have confirmed the great significance of bench terraces as a very successful water erosion control measure. Bench terraces may decrease erosion losses from very steep slopes from strong and medium erosion categories (strong erosion 20–50 t ha^{-1} yr^{-1}, medium erosion 10–20 t ha^{-1}, Djorović, 1978) to below the erosion tolerance of 2.0 t ha^{-1} yr^{-1}.

However, bench terraces exclude a certain area from cultivation. For slopes of 14° it is 32.0 per cent but for 20 degrees of slope it is as much as 50 per cent

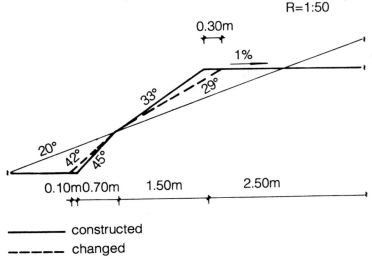

Figure 35.4 Deformation of bench terrace riser during investigation period 1971–87, on a slope of 14 degrees, Ralja

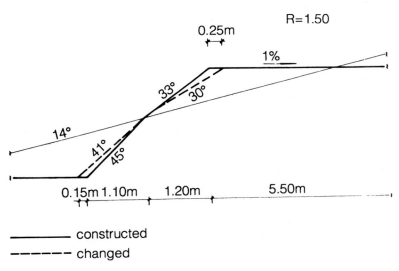

Figure 35.5 Deformation of bench terrace riser during the period 1971–87, on a slope of 20 degrees, Ralja

(Table 35.1). This decrease of cultivation area has no significant influence on total crop yield, for production on terraces almost compensates for it (Table 35.3).

The average crop yield (net ha) is obtained on the area covering only the flat part of terraces, i.e. on 68 per cent or 50 per cent of the total area of the slope

Table 35.3 Average crop production on terraced and non-terraced (control) plots for the period 1971-87, Ralja

Type of plot	Terraced plot				Non-terraced plot	
Slope (degrees)	14		20		14	20
Crop production	Gross ha	Net ha	Gross ha	Net ha	ha	ha
Wheat (kg)	2703	3920	1930	3860	2600	2200
Corn (kg)	3172	4600	2200	4400	3000	2600

at 14 and 20° respectively. Gross crop production includes the area that covers not only the flat part of terraces but also the risers, which are not in production.

Terraces were built manually and up to now, in Yugoslavia, there is no suitable machine to make terraces on slopes of more than 10°. Manual construction is very expensive and sometimes impossible to apply because of the lack of labour. If this problem could be solved in a satisfactory way, there would be no further obstacles for the wide use of terraces in hilly–mountainous regions as a very effective erosion-control measure on steep slopes with agricultural claims.

CONCLUSION

On the basis of 17 years of experimental observation of runoff, soil loss and crop production on bench terrace plots, the following conclusions are drawn:

(1) Bench terraces almost completely eliminate soil erosion and enable constant use of soil on steep slopes for crop production, which is of great importance for the hilly–mountainous region of the central part of the Serbian Republic.
(2) Though the construction of bench terraces excludes a certain area of arable land, the total crop production is quite satisfactory, especially as it is stable and permanent;
(3) Construction of bench terraces is rather expensive, especially when it is manual, which is usually the case on slopes of 10° and more. However, when the soil is in question and its protection from water erosion is required, in addition to providing stable and permanent crop production in hilly–mountainous regions, then even such high expenditure may have economical and social justifications.

REFERENCES

Djorović, M. (1975). Gubici zemljišta i vode dejstvom erozije s raznih tipova zemljišta u SR Srbiji, *Posebno izdanje I*ŠDI br, 37, Beograd.
Djorović M. (1978). Slope effect on runoff and erosion. In *Assessment of Erosion*, John Wiley, Chichester, 215–227.

36 Degradation of Dambo Soils and Peasant Agriculture in Zimbabwe

NEIL ROBERTS and ROBERT LAMBERT
Departments of Geography and Civil Engineering, Loughborough University

INTRODUCTION

Soil erosion is one of the most serious forms of environmental degradation that threatens agriculture in Africa and other parts of the Third World tropics (Stocking, 1984). Historically, acceleration in the rate of soil and nutrient loss has most often been associated with colonial and post-colonial impacts upon either largely undisturbed ecosystems or upon indigenous agro-ecosystems. The type of impacts that have been involved range from land alienation (e.g. in South Africa), through the introduction of a cash economy, to discouragement of traditional forms of land use such as shifting cultivation (Richards, 1985; Blaikie and Brookfield, 1986; Darkoh, 1987). While broad-ranging, often sweeping statements concerning tropical soil degradation abound in the literature, the empirical basis for these statements is frequently far from adequate. There is a pressing need not only for more field-based studies of soil change but also for studies that are environmentally and culturally specific. It is essential to know which combinations of land use and soil conditions are ecologically sound and which are ecologically hazardous. As this chapter will endeavour to show, there are some land resources which are fragile under certain forms of human use, but which may nonetheless be safely exploited with appropriate land-management practices. Conservation policies need to be applied sensitively and selectively in order to incorporate this environmental and cultural complexity. The case study employed to illustrate these issues relates to the degradation and conservation of dambo wetland soils in the communal areas of Zimbabwe (formerly Rhodesia).

Soil Erosion on Agricultural Land
Edited by J. Boardman, I. D. L. Foster and J. A. Dearing
©1990 John Wiley & Sons Ltd

DEFINITION AND DISTRIBUTION OF DAMBOS

Dambos are prominent geomorphic features of Africa's tropical plateau savannas, particularly in central southern Africa, when they occupy around 10 per cent of the total land surface. They are shallow, seasonally waterlogged valleys at or near the head of a drainage network (Acres *et al.*, 1985; Mackel, 1985) and are generally treeless with vegetation dominated by grasses and sedges. Under natural vegetation cover a dambo land unit is most easily identified by the often abrupt contact between dry, typically miombo woodland on the dryland and open herbaceous vegetation on the dambo itself (Malaisse *et al.*, 1972). This contrast is particularly clear on aerial photographs and satellite imagery. Dambo catchments are thought to act as hydrological reservoirs, storing water in the rainy season and releasing it for evapotranspiration on the dambo surface and for dry season streamflow (Hough, 1986). Individual dambos are much smaller than many other forms of wetland, being typically 0.1–1.0 km wide and 0.5–5.0 km long. This means that they have tended to be overlooked in development plans and soil surveys as being individually too small to be significant, despite the fact that in total area they are more important than many other land classes.

In Zimbabwe dambos are known by the names *bane* (Shona) and *vlei* (Afrikaans). It should be noted that in Zimbabwe the term *vlei* is applied to any seasonally wet area of ground, including pans and downstream alluvial valleys, as well as headwater depressions (Thompson, 1972). While forming part of a larger set of wetland habitats, dambos have certain characteristics which place them apart from environments such as alluvial valleys, and which justify their study as a separate resource (Turner, 1986; Adams and Carter, 1987; Roberts, 1988). In comparison with floodplains, dambos receive water and nutrients from upslope not from upstream. Therefore the control of land use which affects the soil and water budgets lies locally with the community that uses the dambo rather than with those that are far upstream.

Whitlow (1984) has mapped Zimbabwe's dambos using 1:80 000 scale aerial photograph cover and has shown that they are located mainly on the gently undulating highveld plateau above 1200 m (Figure 36.1(a)). Within this central watershed region, dambo density is greater towards the north where mean annual rainfall is above 800 mm. In Chiota Communal Area, where much of the present research was carried out, dambos occupy about one-third of the total land surface. Whitlow calculated that in Zimbabwe as a whole dambos cover 1.28 million ha or 3.6 per cent of that country's land area. Because of inequitable colonial land allocation in Zimbabwe, the majority of the highveld, which is the area with most dambos, is now used for large-scale commercial farming (Figure 36.1(b)). Nonetheless, the estimated 0.26 million ha of dambo land in Communal Areas (formerly Tribal Trust Lands), where small-scale peasant farming predominates, represents an important resource for Zimbabwe's peasant farmers.

Dambo Soils and Peasant Agriculture 539

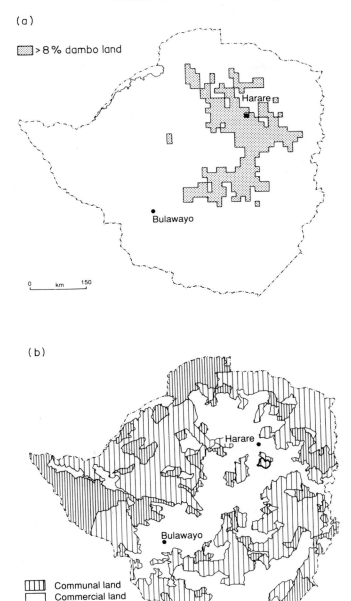

Figure 36.1 Zimbabwe. (a) Dambo distribution (derived from Whitlow, 1984); (b) land classification

BACKGROUND TO DAMBO USE AND DEGRADATION IN ZIMBABWE

Dambos are a multiple resource, fulfilling three main purposes in Zimbabwe's communal lands—water supply, livestock grazing and garden cultivation. Because they remain moist during the dry season, dambos represent a reliable, near-surface water supply for human and animal consumption and for dry-season irrigation. They are especially important in areas where alternative water sources are remote or prone to drying up. Because dambos support a fresh growth of grasses when other forms of graze and browse are in short supply, grazing of cattle and other livestock is one of the most common forms of dambo land use. The other main form of dambo land use in peasant agriculture is cultivation based on indigenous water-management techniques. In many cases this takes the form of dry-season cultivation of vegetables in fenced gardens. There is a long tradition to this form of simple irrigation in southern Africa, stretching back into the pre-colonial era. This is evidenced not only from early travellers' reports but also from abandoned field systems of ridges and hollows (Whitlow, 1983). In Zimbabwe, cultivation of maize, rice and vegetables is also carried out in the wet season without irrigation.

In their natural state dambos are well protected by dense vegetation from the effects of both sheet and gully erosion. However, many dambos in the communal lands of Zimbabwe show evidence of environmental degradation in the form of erosion and desiccation. The origins of this degradation are complex, involving both natural and cultural factors. During the first half of the twentieth century, commercial (European) farmers attempted to plough up some dambos for large-scale cereal cultivation. This led directly to serious soil erosion involving sheet wash and rilling (Whitlow, 1983), which prompted legislation—still in force today—severely restricting cultivation of *all* wetlands in Zimbabwe (Bell and Hotchkiss, 1989). Two main pieces of colonial legislation were introduced which affected dambo use: the Natural Resources (Protection) Regulation (commonly known as the Streambank Protection Regulation) and the Water Act. Only the former will be referred to here.

The Streambank Protection Regulation prohibits cultivation within 30 m of a streambank, and on 'wetland', defined as:

Land that is saturated to within 15 cm of the surface for the major part of a rainfall season of average or above-average rainfall and which may exhibit one or more of the following characteristics:

(1) The presence of mottles or rustlike stains in root channels within 15 cm of the surface;
(2) A black topsoil horizon very rich in organic matter overlying pale leached sands;

(3) A dark grey or black heavy clay showing considerable surface cracking when dry (Government of Rhodesia, 1975).

Although not restricted to dambos, this definition effectively includes all potentially cultivable dambo land. Exemptions from this regulation can, in theory, be secured by farmers. However, the procedures involved are unworkable in the Communal Areas due to the scale of garden cultivation.

This legislation was prompted by three main fears (Rattray *et al.*, 1953; Elwell and Davey, 1972; Elwell, 1983):

(1) That cropping of wetland would lower the water table and reduce dry-season streamflow;
(2) That cultivation would increase the hazard of soil erosion by both gullying and sheetflow, in turn adding to the problem of siltation (e.g. in reservoirs) downstream;
(3) That cropping would 'burn up' organic matter in sandy *vleis*, rendering the soil infertile and giving it the appearance of building sand.

However, dambo degradation can be caused by overgrazing by animal stock as well as other forms of land use such as cultivation. Indeed, casual observation suggests that overgrazing of dambos may be the main factor causing gully erosion. Excessive grazing can reduce the height and percentage ground cover of grasses. This in turn leads to more surface runoff and less infiltration during heavy-rainfall events and a more peaked streamflow regimen. Cattle trampling along pathways and on saturated ground can be important in gully initiation. Similarly, burning may have a deleterious effect on soil nutrient budgets and, if over frequent, kill off valuable plant species. For these and other reasons, Zimbabwe's existing legislation, which acts solely to prevent dambo cultivation while leaving other forms of land use uncontrolled, has been criticized in some quarters as inappropriate (Theisen, 1975; Whitlow, 1983).

In terms of land-capability classification, Zimbabwe's dambo soils are included in Class V, which includes all soils subject to permanent wetness (Ivy, 1981). Land classes I–IV are considered arable according to this classification, while VI–VIII are non-arable. The severe waterlogging of class V is normally considered by Zimbabwe's agricultural extension service, Agritex, to preclude cultivation except with special measures. This evidently bears little relation to the apparent success with which small-scale garden cultivation of dambos takes place in many communal areas.

Previous research (e.g. Rattray *et al.*, 1953; Theisen, 1975) has recognized that under an appropriate form of management dambos should represent an important agricultural resource. Many authors have been hesitant, however, to recommend action to promote their use because data on the environmental consequences have been inadequate or ambiguous. Elwell and Davey

(1972, p. 156), for example, argued that 'we have not yet proven that we can use them safely, without damage to the soil and water resources'. With definitive empirical data on the environmental consequences of dambo utilization still lacking, debates over carrying capacity and soil erosion have continued. Meanwhile, information from aerial photographs suggests that in spite of legal restrictions, the area under dambo gardens has greatly increased since Zimbabwe became independent in 1980.

This chapter addresses the question of whether existing dambo cultivation is significantly depleting the soil resource base. In order to assess the effects of cultivation on rates of soil loss, field studies on sheet and gully erosion were carried out in selected dambos, along with an analysis of associated changes in soil organic matter. The present study forms part of wider interdisciplinary investigations into the role played by dambo resources in communal farming in Zimbabwe (Bell et al., 1987).

DAMBO SOILS AND FIELD STUDY SITES

Dambos have been divided into non-calcic 'sand*vlei*' and calcic 'clay*vlei*' types, on the basis of the physico-chemical characteristics of these soils (Elwell, 1983). Of these two types the former is the more widespread in Zimbabwe, accounting for $c.89$ per cent of all dambos (Whitlow, 1985, Table 3). In addition to these two main types Whitlow recognizes 'perched' dambos with sodic soils and peaty dambos underlain by Kalahari sands, neither of which is suitable for cultivation. These variants are not considered in this chapter.

The main field sites studied lie in Chiota Communal Area, $c.60$ km southeast of Harare, and in Zimbabwe's natural region II (intensive farming; 750–1000 mm mean annual rainfall). Additional field data have been collected from two other Communal Areas, Zwimba and Gutu, but are not reported here (see Bell et al., 1987, for details). Most dambos in Chiota are of the broad headwater type, in which the lowest part of the dambo is dry except immediately after rain. The wet dambo area in which gardens are located lies further up the slope catena in an 'out-seepage zone'. Aerial photograph mapping indicates that in 1984 there were $c.2500$ ha of irrigated damo gardens in this Communal Area (Bell et al., 1987). The dambo sites studied in greatest detail were in the

Table 36.1 Areas of dambo catchments studied (ha)

	Total catchment (including dambo)	Dambo	Dambo under cultivation (1984)
Chizengeni	274	86	28.5
Chigwada	289	101	5.9
Bumburwi	303	103	6.5

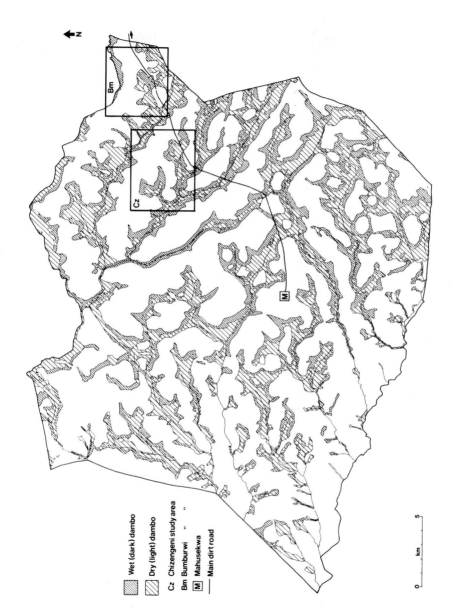

Figure 36.2 Dambos in Chiota Communal Area, showing study sites

Table 36.2 Soil profile and analysis, Bumburwi dambo

Profile: Bb4, Bumburwi, Chiota C. A. Zimbabwe.
Dambo bottom, 150 m above gully head.; slope < 1%; land use, grassland grazing

Description
    0–15 cm   :Black (10YR 2/1) organic clay, cohesive, clear transition to . . .
    15–70 cm   :Black (7.5YR 2/0) clay, sticky cohesive homogenous

Analytical data	Bb4A	Bb4B
Depth (cm)	0–15	15–70
Texture	Coarse sandy clay	clay
Coarse sand (%)	18.10	12.90
Medium sand (%)	10.80	10.40
Fine sand (%)	7.35	7.04
Coarse silt (%)	10.75	6.46
Fine silt (%)	17.00	9.00
Clay (%)	36.00	54.00
pH (H_2O)	7.00	8.20
pH ($CaCl_2$)	6.00	6.90
EC (mhos)	70.00	90.00
Loss on ignition (%)	17.80	5.85
Carbon (%)	4.85	0.83
TEB m.e. (%)	32.80	46.32
CEC m.e. (%)	38.22	61.91
Base saturation (%)	85.80	74.80
Exchangeable Ca	21.63	33.46
Exchangeable Mg	10.52	12.23
Exchangeable Na	0.56	0.58
Exchangeable K	0.09	0.05
ESP	1.50	0.90
EKP	0.20	0.10
Available P (ppm)	29.00	7.00
E/C value	107.30	114.60
S/C value	92.12	85.80

Profile description: R. Lambert (3/86)

Laboratory analyses: Research and Specialist Services, Harare

Soil textural classes:
    Sand: less than 8% clay, more than 85% sand
    Sandy clay: 35–55% clay, more than 45% sand, less than 20% fine silt
    Clay: more than 40% clay, less than 50% sand or 40% fine silt

Cations
    CEC, TEB, Na, K, Ca, Mg in milligram equivalents per cent
    N and P in parts per million
    E/C value denotes CEC per 100 gm of clay
    S/C value denotes TEB per 100 gm of clay
    Base saturation = TEB as a percentage of CEC

ESP/EKP
    = 100 (Ex.Na/K (m.e.%)—water soluble Na/K)/CEC (m.e.%)

north-eastern part of Chiota (Figure 36.2) at Chizengeni/Chigwada and Bumburwi. The latter site has rather few irrigated gardens, and its soils are of 'clay*vlei*' vertisolic type devoted on mafic (basic) rocks (Tables 36.1 and 36.2). The main land use type on Bumburwi dambo is cattle grazing. An extensive and apparently active gully network runs down the centre ('eye') of the dambo.

Chizengeni and Chigwada lie in an area of high dambo density and numerous gardens. Taken together, they have a proportion of their wet dambo land under cultivation that is close to the mean for Chiota as a whole (*c.*30 per cent). However, this gross figure disguises the fact that Chizengeni dambo is used intensively for irrigated gardens, while the adjacent Chigwada dambo had few gardens and is instead given over mostly to dry-season grazing. The gardens in Chizengeni dambo have been used over several decades, sequential time-series analysis of aerial photographs showing that the main gardens were created between 1947 and 1965. It was further known that Chizengeni village included farmers with long experience of the benefits and problems of dambo cultivation. The site therefore offered the possibility of assessing the effects of cultivation on soil and water resources over more than the short term. The soil catena sequences at Chizengeni and Chigwada are both of 'sand*vlei*' type developed on granitic rocks (Table 36.3).

SHEET EROSION

A standard method for the investigation of sheet erosion is through the use of erosion plots. In Zimbabwe, work has been done at the Institute of Agricultural Engineering on erosion plots. These plots measure $30\,m \times 10\,m$ and the total runoff is collected and analysed. The results of these plot studies show, for example, that over a 10-year period on a 4.5 per cent slope and with a mean annual rainfall of 750 mm, soil losses on a bare slope averaged $127\,t\,ha^{-1}\,yr^{-1}$, while those with a complete grass cover lost only $0.7\,t\,ha^{-1}\,yr^{-1}$. Dryland soils are believed to form at around $1\,t\,ha^{-1}\,yr^{-1}$, and acceptable target levels for soil loss are between 3 and $5\,t\,ha^{-1}\,yr^{-1}$ (Elwell, 1984). These studies were carried out on deep, reddish-brown kaolinitic clay soils, and were related primarily to large-scale mechanized agricultural practices on commercial farms. Erosion plot studies on 'sandveld' soils, where a large proportion of communal peasant farming is located, were carried out under N. Hudson between 1953 and 1963 (Barnes and Franklin, 1970). However, very few have been undertaken on 'sandveld' soils during the 1970s and 1980s and none at all have been done on dambo or other wetland soils.

Rather than base field experiments under the rigorously controlled but 'artificial' condition of an agricultural research station, it was felt most appropriate to study soil changes in dambos used by working farming communities with existing farming methods. By doing this it was possible to

Table 36.3 Soil profile and analysis, Chizengeni dambo

Profile: EP3, Chizengeni dambo, Chiota C.A., Zimbabwe
Lower dambo; slope c.2%; land use, garden.

Description
0–25 cm	: Black (10YR 2/1) organic sandy loam, very weak crumb structure, homogenous, clear sharp transition to . . .
25–55 cm	: Dark greyish brown (10YR 4/2) sand, homogenous, indistinct transition to . . .
55–90 cm	: Greyish brown (10YR 5/2) sand, homogenous.
90– cm	: Gravel

Analytical data	EP3A	EP3B	EP3C
Depth (cm)	0–25	25–55	55–90
Texture	Coarse sand	Coarse sand	Coarse sand
Coarse sand (%)	37.72	46.21	46.81
Medium sand (%)	29.19	33.57	31.82
Fine sand (%)	14.15	9.69	12.23
Coarse silt (%)	10.10	7.73	6.24
Fine silt and clay (%)	8.84	2.80	2.90
pH (H_2O)	5.80	6.30	6.30
pH ($CaCl_2$)	5.00	4.90	5.00
EC (mhos)	25.00	10.00	11.00
Loss on ignition (%)	2.50	0.80	0.60
Carbon (%)	1.57	0.03	0.07
TEB m.e. (%)	2.20	0.60	0.50
CEC m.e. (%)	3.30	1.00	2.40
Base saturation (%)	66.00	63.00	22.00
Exchangeable Ca	1.60	0.30	0.20
Exchangeable Mg	0.40	0.20	0.20
Exchangeable Na	0.10	0.12	0.14
Exchangeable K	0.10	0.02	0.02
ESP	3.00	12.00	5.80
EKP	3.00	2.00	0.80
Available P (ppm)	10.00	3.00	1.00
N (ppm) initial	6.00	4.00	3.00
N (ppm) after inc	14.00	3.00	3.00
E/C value	37.30	35.70	82.80

Profile description: R. Lambert (9/86)

integrate environmental studies of soil and water balances with socio-economic surveys of those same communities. One consequence of this decision was that the standard plot technique had to be modified in a number of ways. The small size of dambo gardens, for example, did not always allow a 30 m × 10 m plot to be conveniently located. Also dambos are waterlogged during the rains and the construction of tanks large enough to collect runoff as well as soil would have presented a major engineering challenge. Because of these constraints, it was decided to use a plot size of 10 m × 5 m and a small perforated collection tank lined with a fine-mesh cloth to collect soil. Small plots such as were used in this study cannot adequately model the effect of slope length. The length of slope effect is complicated by the fact that dambo gardens, being fenced, ridged and cultivated in a variety of ways, will greatly change the surface runoff characteristics. Grazed dambo areas, being generally without contours, often have slope lengths in excess of 200 m. Detailed modelling of slope length was not carried out, although some attempt to calibrate soil plot losses for the standard 30 m slope length was made using SLEMSA (see below).

Six plots were installed at Chizengeni, two on grazed dambo areas, two on cultivated dambo areas and two on non-dambo areas, one on a maize field and one on scrub grazing. Because plots were located on the basis of already existing land use, they could not be placed precisely side by side. However, the four dambo plots were located within 250 m of each other on slopes of similar gradient around 2 per cent. Similarly, the two erosion plots located on dryfields above Chizengeni dambo lay adjacent to each other (see Figure 36.4). Erosion plots were not set up at Bumburwi dambo. Sediment from the tanks was collected at monthly intervals during one full rainy season from November 1985 to April 1986, dried and weighed. Daily rainfall and other meteorological parameters were also recorded at Chizengeni for this study period. The results, details of which can be found in Bell *et al.* (1987, appendix 5), are summarized in Table 36.4.

The sediment yields from the erosion plots indicate that measured erosion never exceeded $1.0 \, t \, ha^{-1} \, yr^{-1}$, which is well within the 'acceptable' figures of $3.0–5.0 \, t \, ha^{-1} \, yr^{-1}$ mentioned above. The lowest rate of erosion was on a dambo plot which had the highest slope but was well covered with short grass. The highest rate of erosion on the dambo was on the maize field. On the non-dambo dryland areas the cultivated maize field gave a higher rate of erosion, as might be expected.

Table 36.4 Monthly erosion plot losses (t ha^{-1}) from Chizengeni catchment, season 1985/6

Plots	Slope (%)	Nov.	Dec.	Jan.	Feb.	Mar./Apr.	Total	SLEMSA prediction
Dambo								
1 Grazing	1.9	0.04	0.01	0.05	0.03	0.01	0.14	0.3–0.6
2 Maize	2.3	0.03	0.13	0.14	0.03	0.03	0.36	2.0–11.9
3 Weed fallow	1.5	0.07	0.04	0.04	0.02	0.01	0.18	1.0–2.5
4 Grazing	2.3	0.02	0.01	0.01	0.01	0.00	0.05	0.3–0.6
Non-dambo								
5 Scrub grazing	1.8	0.04	0.06	0.08	0.07	0.02	0.27	0.40
6 Maize	1.8	0.04	0.12	0.29	0.19[a]	0.07	0.71	8.3–11.0
Rainfall (mm)		44	220	309	135	205	922	

[a]Estimate.

An alternative approach to the measurement of sheet erosion involves the use of predictive soil-loss equations, such as the Universal Soil Loss Equation (USLE). A local model termed SLEMSA (Soil Loss Estimation for Southern Africa) has been developed using data from Zimbabwe (Elwell and Stocking, 1982; Elwell, 1984). SLEMSA was applied to the erosion plot sites at Chizengeni. This was easier for the non-dambo, 'sandveld' soils, than for the azonal dambo soils, for which erodibility factors (Fb) are not provided in SLEMSA. Two further modifications were necessary for comparison with recorded erosion plot losses; first, because rainfall as measured during 1985–6 at Chizengeni (941 mm) was above the 19-year average at the nearest meteorological station, Grasslands research station (863 mm); second, because of the small plot size used, so that the slope length factor (L) required adjustment. In fact, these two modifications effectively counteracted and cancelled out each other. Details of the SLEMSA calculations can be found in Bell et al. (1987, appendix 5), but summary results are shown in Table 36.4.

It is clear that SLEMSA gives a considerably higher value for the erosion losses on all but the non-dambo grazed areas. On the dryland maize field erosion was overpredicted by a factor of between 12 and 16 times as compared with the measured rate. We consider it unlikely that this could be accounted for by errors in the collection of samples eroded from the plots. Note that SLEMSA was designed to predict long-term mean soil loss and one year's data are not adequate for a rigorous comparison. Although dryland sheet erosion was not the focus of the present study the difference between the erosion plot results and the SLEMSA values, particularly on dryland maize where the difference is at least one order of magnitude, emphasizes that more field research on erosion under these conditions is required. Significantly, SLEMSA values for dryland scrub grazing gives good agreement with the erosion plot results.

GULLY EROSION

Gully erosion is one of the most serious threats to dambos in Zimbabwe. In this project a number of gullies were monitored on dambos in Chiota in two

Table 36.5 Gully change in selected sites

Gully	Retreat ($m\ yr^{-1}$)		Volume lost ($m^3\ yr^{-1}$)		Percentage of dambo under cultivation (1984)
	A	B	A	B	
Chizengeni	1.2	<0.4	1.0	<0.8	32.9%
Chigwada	0.2	<0.4	0.0	<0.8	5.6%
Bumburwi	5.1	6.9	20.1	203.2	6.5%

A = monitored change (1985–6).
B = Mean change on aerial photographs (1965–84).

Table 36.6 Gully retreat, Bumburwi dambo, Chiota

Period	Retreat (m yr^{-1})	Volume lost (m^3 yr^{-1})
1947–65	5.9	60.67
1965–76	9.7	216.27
1976–84	3.1	185.25
Mean (1947–84)	6.4	133.86

ways. First, the positions of gully heads at the start and finish of the 1985–6 rainy season were recorded by survey from marker points fixed in concrete. The results of this monitoring are summarized in Table 36.5. The gullies in the Chigwada dambo were inactive, notwithstanding the fact that a flood event occurred during January 1986 sufficient to wash away a masonry-built gauging weir on Chigwada dambo outlet stream! While there was some gullying in Chizengeni dambo it was quite minor in comparison with the rapid retreat of the gully head measured in the Bumburwi dambo.

The second form of measurement involved examination of gully head positions on old aerial photographs. In Chizengeni and Chigwada dambos, no significant gully retreat was visible on aerial photographs between 1947 and 1984. The most rapid retreat occurred in Bumburwi dambo (Table 36.5), and this has been investigated further using historical aerial photographs (Table 36.6 and Figure 36.3).

Figures for the rate of retreat using aerial photography and monitoring are comparable. The much higher volumetric losses estimated from the former technique is a result of monitoring being restricted only to one or two gully heads per site. As Figure 36.3 shows, there may be many more active gully heads than this and soil is also lost through side-wall collapse downstream. The 37-year mean of 134 m^3 yr^{-1} is equivalent to 13 mm lowering/100 yr over the 103 ha of the Bumburwi dambo, or 1.69 t ha^{-1} yr^{-1}.

Table 36.5 shows no correlation between soil loss through gully erosion and dambo land use. Intensely cultivated dambos such as Chizengeni have experienced rather insignificant gully erosion (equivalent to a soil loss of less than 0.015 t ha^{-1} yr^{-1} over the whole dambo). By far the most serious gully erosion problem exists in Bumburwi dambo, used largely for grazing, but the prime cause of the erosion is the heavy cracking clay in the dambo bottom, which breaks off in large blocks once gullying has been initiated.

Many areas of southern Africa experiencing severe gully (donga) erosion have been found to be associated with highly dispersive soils and sediments (Stocking, 1979). This parameter, as measured by exchangeable sodium percentage (ESP), was recorded on samples from dambo soil profiles at Chizengeni-Chigwada and Bumburwi. Surprisingly, the latter dambo has the lowest ESP values of the sites investigated, never exceeding 3 per cent (i.e. in the cohesive range). In contrast, some of the A2 and B horizons of seasonally wet sandy dambo soils at

Chizengeni have ESP values in the 'dispersive' range (>7 per cent). Although these soils are potentially vulnerable to erosion, current land-use practics under garden cultivation at this site would appear to conserve rather than threaten the soil against gully erosion. Most of these soils have been continuously cultivated for over 20 years. In short, dispersive soils need not suffer degradation given appropriate, in this case traditional, land-use practices.

ORGANIC MATTER

The erodibility of most soils is closely linked to their organic content. When soils are cultivated a drop in organic matter is to be expected as the natural root structure is broken down and exposed to greater weathering and decomposition. Dambo soils, with a high level of organic matter on which their fertility depends, would seem particularly vulnerable. If the normally high level of organic matter in dambos is greatly reduced, their stability and fertility will be seriously affected, particularly in the case of sandy dambo soils. In clay dambo soils the effect of a reduction in organic matter may be less serious.

In order to assess the effects of cultivation on dambo soils investigation of organic matter content were carried out on dambos in Chiota and elsewhere in Zimbabwe. A smaller number of samples was taken on non-dambo, dryland soils to serve as a basis for comparison. The method chosen was the loss-on-ignition expressed as a percentage of the dry soil mass (see Bell *et al.*, 1987, for details). Samples were also tested for percentage of organic carbon, so it is possible to examine the relationship between percentage loss-on-ignition and percentage organic carbon. Conventionally, organic carbon is assumed to be 58 per cent of total organic matter (Landon, 1984, p. 139). For topsoil samples from 'wet' dambo sites, proportions ranged between 36 per cent and 77 per cent, with a modal value around 50 per cent. Loss-on-ignition values for these samples therefore correlate fairly closely with total organic matter. On the other hand, in clay-rich samples from B and C horizons, organic carbon fell to 10 per cent or less of loss-on-ignition values. Topsoil samples from the dambo margin also gave loss-on-ignition values which overestimated total organic matter, although the former never exceeded 2 per cent. Because loss-on-ignition best reflects total organic matter in sandy topsoil, samples for OM% determinations were collected only from the top 10 cm of soil, with the largest number of samples from Chizengeni and Chigwada and nearby dambos at Kanjanda and Chiwanzamarara, all sites of predominantly sandy texture. The loss-on-ignition method measures only the quantity of organic matter. The type or quality of the organic matter may also be important and this will be affected by cultivation and soil husbandry practices.

Ideally, measuring the effect of cultivation on dambo soils would involve detailed long-term monitoring extending over five to ten years. In this

Dambo Soils and Peasant Agriculture 551

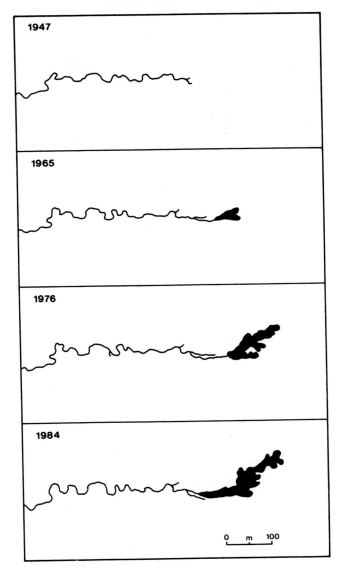

Figure 36.3 Historical gully development, Bumburwi dambo

shorter-term project it was decided to select paired sites of cultivated and uncultivated soils for sampling and analysis. These sites were chosen with the aid of local field assistants who had detailed knowledge of the local environment. In most cases the uncultivated sites were chosen outside but close to a garden

Table 36.7 Total organic matter values

Site name	Uncultivated		Cultivated	
	Number of tests	Mean OM%	Number of tests	Mean OM%
Chizengeni				
Dambo	37	5.3	46	3.8
Non-dambo	17	1.7	14	1.5
Chigwada				
Dambo	40	2.7	22	2.9
Non-dambo	14	0.9	0	—
Chiwanzamarara				
Dambo	30	3.3	32	3.0
Kanjanda				
Dambo	13	3.3	14	3.6
Bumburwi				
Dambo	12	5.8	11	4.2
Non-dambo	1	1.3	0	—
Total				
Dambo	132	3.9	125	3.5
Non-dambo	32	1.3	14	1.5

fence with the cultivated site being nearby but inside the fence. Samples were not taken where there was evidence of past cultivation on a site which was currently uncultivated. A summary of the results is given in Table 36.7.

Figure 36.4 shows the location of sites on Chizengeni dambo. Organic matter values exceeding 2 per cent are very rare outside the dambo. Within the dambo itself organic matter values vary greatly, with a value of 19.3 per cent being recorded less than 200 m from a site where the value was 4.3 per cent, both sites being uncultivated grazed dambo. Only in the upper 'tail' of the dambo are values of less than 2 per cent consistently found.

As Table 36.7 illustrates, mean organic matter values are consistently slightly lower for cultivated than for uncultivated dambo soils (3.9 per cent versus 3.5 per cent). On the other hand, only rarely does the organic content of cultivated dambo soils decline below 2.5 per cent. Furthermore, more detailed analysis of paired sites from Chiota and elsewhere in Zimbabwe (Figure 36.5) shows no significant difference between cultivated and uncultivated soils in samples whose organic content was below 5 per cent. In other words, the apparently higher mean values for uncultivated soils are a product of a decline in organic content in soils with over 5 per cent total organic matter. A fall in already relatively organic-rich soils is clearly less deleterious to soil fertility and stability than one in soils with lower organic matter. Having declined from, say, 7 per cent to 4 per cent organic matter with initial cultivation, dambo soils subsequently seem not to decline further, but stabilize at still-adequate levels.

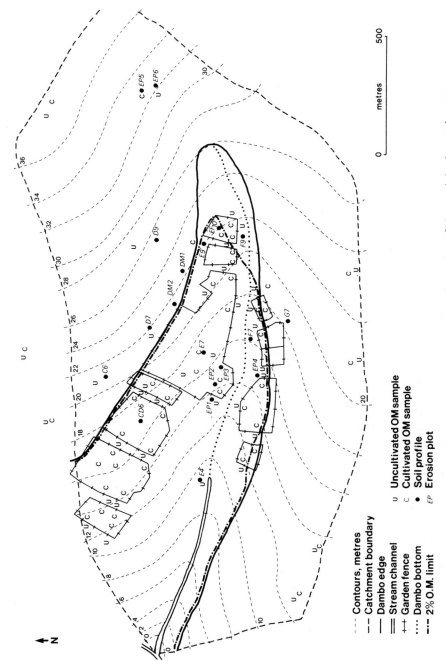

Figure 36.4 Location of erosion plots and other soil sampling points, Chizengeni dambo catchment

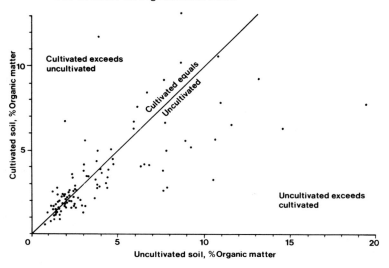

Figure 36.5 Correlation of soil organic matter at paired dambo sites

DISCUSSION

Close control of land-use practices in stream source areas, such as dambos, is considered to be a key element in tropical watershed management (Pereira, 1973). There is little doubt that under 'natural' vegetation of tall grasses on dambo land and miombo woodland on the interfluves, soil losses through sheet and rill erosion would be minimized. In Zimbabwe, this land-use state is approached on some 'European' commercial farms, where only limited patches of woodland have been cleared and dambos are subject to light grazing pressure. On communal farmland where the African population was concentrated via enforced land alienation the natural vegetation has been progressively cleared during the course of the twentieth century and land-use patterns have changed irrevocably. In the catchments studied here, namely Chizengeni-Chigwada and Bumburwi in Chiota Communal Area, historical aerial photographs show that woodland cover on non-dambo land had been reduced to under 10 per cent by 1947, and is currently around 5 per cent. While reafforestation may restore some of this woodland it is unlikely to involve the major part of the interfluves.

In order to assess the effects of dambo garden cultivation on soil conditions it is appropriate to consider existing forms of land use as well as the situation prior to cultivation. In the Communal Areas of Zimbabwe the main alternative form of dambo use to cultivation is now livestock grazing. This means that the dambo is covered with short grass for most of the year and in dry years often very poorly covered due to grazing pressure. Thus in Communal Areas the soil conditions under dambo grazing are as important a model for comparison as those under undisturbed conditions. In this chapter soil degradation through sheet and gully erosion and *in situ* loss of organic matter have been compared for grazed and cultivated dambo land in order to establish the environmental impact of each.

Erosion plot results indicate that soil loss through sheet erosion is not a significant problem on low-slope dambo catchments. While sheet erosion on the cultivated areas of the dambo was higher than on the grazed areas, which were quite well covered with short thick grass, soil losses were still below $0.5 \text{ t ha}^{-1} \text{ yr}^{-1}$, and therefore well within acceptable limits. Although garden cultivation exposes the soil to rainsplash impact, crops are usually well established before the rains begin, and these may provide better cover than overgrazed dambo areas. In addition, dambo gardens are bounded by ditches and fences, in some cases being completely surrounded by elaborate earthbanks covered in sisal. Surface runoff inside the gardens will experience frequent interruptions in flow, reducing the velocity and discharge and hence the erosive power. The main area that is subject to sheet wash is the dambo margin, and soil removal here is a function of slope length and agricultural practice on the *non-dambo* part of catchments. The greater lengths of slope which exist on some interfluves can cause a greater volume and velocity of water to runoff, in turn leading to greater potential for erosion. Although most dry fields on interfluves are subject to mechanical conservation measures in the form of contour ridges set at 30 m intervals, current environmental legislation in Zimbabwe does not constrain land-use activities on the interfluves in the way that it does on dambo land.

The major erosion problem for Zimbabwe's dambos relates to gullying. Gully erosion was investigated by historical analysis of aerial photography and by the use of surveyed erosion pegs. The results show that gully erosion could not be correlated with the intensity of cultivation; rather, garden cultivation appears to be an agent of soil conservation. For example, gullies have not developed even on dispersive (high ESP) dambo soils at Chizengeni. On this intensively cultivated dambo gully erosion was under 1 m^3 per year while at Bumburwi, which is used mainly for grazing, gully erosion rates were over 100 m^3 per year. Our data suggest that serious gully erosion is most often associated with factors other than cultivation, such as soil type and a loss of groundcover caused by overgrazing. Cattle climbing in and out of stream channels exacerbate the latter effect. It is also important to recognize that gully development is affected by land-use patterns over the entire catchment and that the effects of garden cultivation on dambos cannot be separated from land use on the catchment as a whole. Woodland, for example, would have intercepted much of the rainfall which now runs off to reach the dambo. If there are a large number of protected gardens, runoff from these interfluve areas will be concentrated along alleys between the gardens, so increasing the gully erosion hazard.

The organic matter of dambo soils is vital, especially for those that are sandy, and must be maintained. It is this organic matter which gives the soil both its stability against erosion and its fertility for agriculture. This research considered the effects of cultivation on the amount of organic matter in cultivated plots compared with that in adjacent grazed areas. Some drop in organic content is to be expected with any soil that is cultivated as the roots and vegetation of the natural vegetation are broken down. Results show that there is a drop in organic matter related to cultivation of dambo soils but that this drop is most marked where organic content was originally high (i.e. above 10 per cent). When the organic content is below 5 per cent there is no significant difference between organic matter values on cultivated and uncultivated sites. Cultivation consequently has less effect on the medium and lower OM% values, and the organic content of these cultivated soils stabilizes at acceptable levels. Organic matter is also maintained through cultural practices such as the transfer of organic material, in the form of cattle manure, from the grazing areas to the gardens.

Given the stresses imposed through enforced population relocation and concentration, the peasant farming communities studied appear to be relatively efficient and conservation-conscious 'land managers'. Some of the most intensively cultivated dambos in Zimbabwe are found in Chiota Communal Area, which has a long history of garden cultivation. Good soil husbandry practices such as ploughing across the slope and applying cow manure, anthill and compost are long-established here. A mix of cultivated gardens and limited livestock grazing permits utilization of dambos as an agricultural resource within environmentally acceptable limits, at least in Zimbabwe's highveld. The fact that gardens occupy only one-fifth of the dambo area that could safely be brought under cultivation in Zimbabwe's Communal Areas suggests that there is considerable potential for expanding this type of land use here and elsewhere in Africa.

ACKNOWLEDGEMENTS

This research was carried out under ODA grant project R3869. The authors are pleased to thank the other members of the Dambo Research Unit, namely Morag Bell, Richard Faulkner, Patricia Hotchkiss and Alan Windram. Help and advice were also received from Richard Whitlow and Henry Elwell. The Zimbabwe Department of Research and Specialist Services kindly assisted with laboratory analyses of soil samples. Field assistance was provided by Collins Chizengeni and Patricia Tambudze. The maps were drawn by Anne Tarver.

REFERENCES

Acres, B. D. et al. (1985). African dambos: their distribution, characteristics and use, *Zeitschrift fur Geomorphologie Supplementband*, **52**, 63–86.

Adams, W. and Carter, R. C. (1987). Small-scale irrigation in sub-Saharan Africa, *Progress in Physical Geography*, **11**, 1–27.

Barnes, D. L. and Franklin, M. J. (1970). Runoff and soil loss on a sandveld in Rhodesia, *Proceedings of the Grassland Society of South Africa*, **5**, 140–44.

Bell, M. et al. (1987). *The Use of Dambos in Rural Development with Reference to Zimbabwe*, Final report of ODA project R3869, Loughborough University/University of Zimbabwe.

Bell, M. and Hotchkiss, P. (1989). Political interventions in environmental resource use with reference to dambos in Zimbabwe, *Land Use Policy*, October 313–23.

Blaikie, P. and Brookfield, H. (1986). *Land Degradation and Society*, Methuen, London.

Darkoh, M. B. K. (1987). Socio-economic and institutional factors behind desertification in Southern Africa, *Area*, **19**, 25–33.

Elwell, H. A. (1983). *Notes on Conservation Aspects of Vlei Use*, Institute of Agricultural Engineering, Borrowdale, Zimbabwe, Mimeo.

Elwell, H. A. (1984). Soil loss estimation: a modelling technique. In Hadley, R. F. and Walling, D. E. (eds) *Erosion and Sediment Yield: Some Methods of Measurement and Modelling*, Geobooks, Norwich, pp. 15–36.

Elwell, H. A. and Davey, C. J. N. (1972). Vlei cropping and the soil and water resources, *Rhodesia Agricultural Journal, Technical Bulletin*, **15**, 155–68.

Elwell, H. A. and Stocking, M. A. (1982). Developing a simple yet practical method of soil-loss estimation, *Tropical Agriculture (Trinidad)*, **59**, 43–8.

Government of Rhodesia (1975). *Chapter 264, Rhodesia Government Notice No. 1190*, Harare.

Hough, J. (1986). Management alternatives for increasing dry season base flow in the miombo woodlands of Southern Africa, *Ambio*, **15**, 341–6.

Ivy, P. (1981). *A Guide to Soil Coding and Land Capability Classification for Land Use Planners*, Agritex, Harare.

Landon, J. R. (ed.) (1984) *Booker Tropical Soil Manual. A Handbook for Soil Survey and Agricultural Land Evaluation in the Tropics and Subtropics*, Longman, London.

Mackel, R. (1985). Dambos and related landforms in Africa—an example for the ecological approach to tropical geomorphology, *Zeitschrift fur Geomorphologie Supplementband*, **52**, 1–24.

Malaisse, F. et al. (1972). The miombo ecosystem: a preliminary study. In Golley, P. M. and Golley, F. B. (eds), *Symposium on Tropical Ecology*, Athens, Georgia, pp. 363–405.

Pereira, H. C. (1973). *Land Use and Water Resources*, Cambridge University Press, Cambridge.

Rattray, J. M., Cormack, R. M. M. and Staples, R. R. (1953). The vlei areas of Southern Rhodesia and their uses, *Rhodesia Agricultural Journal*, **50**, 465–83.

Richards, P. (1985). *Indigenous Agricultural Revolution*, Hutchinson, London.

Roberts, N. (1988). Dambos in development: management of a fragile ecological resource, *Journal of Biogeography*, **15**, 141–8.

Stocking, M. A. (1979). Catena of sodium-rich soil in Rhodesia, *Journal of Soil Science*, **30**, 139–46.

Stocking, M. (1984). Rates of erosion and sediment yield in the African environment. In Walling, D. E. et al. (eds), *Challenges in African Hydrology and Water Resources*, IAHS Publication No. 144, pp. 285–94.

Theisen, R. J. (1975). Development in rural communities, *Zambezia*, **4**, 93–8.

Thompson, J. G. (1972). What is a vlei? *Rhodesia Agricultural Journal, Technical Bulletin*, **15**, 153–4.

Turner, B. (1986). The importance of dambos in African agriculture, *Land Use Policy*, October 343–7.

Whitlow, R. (1983). Vlei cultivation in Zimbabwe, *Zimbabwe Agricultural Journal*, **80**, 123–35.

Whitlow, R. (1984). A survey of dambos in Zimbabwe, *Zimbabwe Agricultural Journal*, **81**, 129–38.

Whitlow, R. (1985). Dambos in Zimbabwe: a review, *Zeitschrift fur Geomorphologie Supplementband*, **52**, 63–86.

37 Erosion Control for the UK: Strategies and Short-term costs and Benefits

C. A. FROST, R. B. SPEIRS and J. McLEAN
East of Scotland School of Agriculture, Edinburgh

INTRODUCTION

Numerous studies in recent years have pointed to the increasing incidence of accelerated erosion on arable land in the United Kingdom by water (e.g. Reed, 1979; Evans, 1980; Colborne and Staines, 1985; Fullen, 1985; Speirs and Frost, 1985; Boardman, 1986). These studies suggest that a widespread problem may exist which is, locally at least, of sufficient severity to merit some action on the part of the farmer. Few, however, of these studies have addressed themselves to what such action should be or what it will cost. The problem of soil erosion, insofar as it may involve the destruction of a resource which is both vital and also non-renewable, goes beyond mere economic considerations. It is, nevertheless, economic considerations which will spring immediately to mind of any farmer faced with the prospect of having to introduce erosion-control measures. It is only with an appreciation of the economic implications of possible action that any national erosion-control programme could be implemented. While it is likely that the costs of soil erosion off the farm exceed the cost on the farm by many times, it is on-farm that any costs of erosion-control work will be borne.

This chapter aims to examine the likely costs of a variety of possible erosion-control strategies and to see the likely effectiveness of such strategies and any economic benefits which may occur, on-farm, to offset these costs. The off-farm benefits are not examined in detail, as, for the moment at least, any such benefits will not contribute to the cost of the erosion-control measures.

Soil Erosion on Agricultural Land
Edited by J. Boardman, I. D. L. Foster and J. A. Dearing
©1990 John Wiley & Sons Ltd

LAND AND FARM TYPE

In order to make realistic costing of various possible measures it is necessary to make some assumptions about the type of farm enterprise being considered. A 200 ha all-arable farm was assumed. A seven-course crop rotation of winter barley, winter oilseed rape, winter wheat, winter wheat, potatoes, winter wheat and spring barley was used as a typical continuous arable rotation. Different rotations would give somewhat different costs for some options.

POSSIBLE EROSION-CONTROL TECHNIQUES AND THEIR EVALUATION

Five possible erosion-control techniques have been chosen for evaluation in terms of cost and likely effectiveness. These range from very simple changes in cultivation to fairly major civil engineering works and are all costed both in terms of initial capital outlay and in terms of continuing cost of maintenance or loss of cropping income. These costings can be regarded as being at least in the correct order of magnitude, although there will clearly be considerable variation.

They are also assessed in terms of likely effectiveness. This assessment is much more open to argument. It is based on a combination of considerations, including reported UK experience and relevant work from elsewhere including the United States Department of Agriculture work on the Universal Soil Loss Equation (Wischmeier and Smith, 1978).

The assessment of likely on-farm benefits from erosion control work is limited to considering that crop which would otherwise have been lost by the direct immediate effects of erosion. These have been assessed at a maximum of £20 ha^{-1} yr^{-1}. This figure is probably an overestimate, as it represents 2.5 per cent of the crop's value. There is good evidence (Darwinkel, 1984) that crops have remarkable abilities of compensating for loss of plants at an early date. Therefore, to reduce yield by 2.5 per cent, considerably more than this proportion of the field will have to be washed out as rills. Even in very severe events, the area of rills is unlikely to exceed 2.5 per cent.

Long-term effects on soil productivity are not considered. This is not because the authors do not believe that such effects occur but rather that, in general, they occur so slowly that from the standpoint of the individual farmer, their effects are negligible.

This is a contentious statement which requires some further justification. Work on the long-term effects of soil erosion in the UK is very limited. One study (Frost and Speirs, 1984) concluded that on the site studied, rates of soil loss in the order of 25 t ha^{-1} yr^{-1} could be tolerated for in excess of two hundred years before the land would suffer significant yield loss as a result of soil erosion. This eventual downgrading of the land would be due to increasing droughtiness as the available depth of suitable material began to be reduced to the point where it affected crop-rooting depth. The catastrophic rapid yield reductions associated with soil erosion in some other countries will not, in general, occur in UK systems of agriculture as they are caused by a reduction in the chemical fertility of the soil. No significant reduction in chemical fertility occurs with the UK systems of farming as this is maintained by large regular additions of fertilizer. The time taken for such intensively managed land to suffer significant yield loss due to erosion is largely associated with the available depth of suitable soil-forming material. Even in the presence of rates of soil loss as high as 25 t ha^{-1} yr^{-1}, the soil is only decreasing in depth at a rate of about 30 mm per decade. As the average farmer probably only manages his land for three or at most, four decades, he is unlikely to be affected by the rates of soil loss generally observed in the UK. He is thus unlikely to regard soil erosion as an immediate threat to the productivity of his land, and for this reason, prevention of long-term decline in productivity has not been considered here as a benefit to the farmer. However, the authors realize that from a national standpoint, this is without doubt the most important on-farm benefit. It all depends on one's perspective. As it is the farmer who is currently likely to have to pay for any erosion-control measures, it is his perspective which is considered here.

EROSION-CONTROL TECHNIQUES

The five erosion-control techniques being considered are outlined below. Table 37.1 summarizes the authors' estimate of costs and likelihood of effective erosion control being achieved. Much of the information used in the costings in Table 37.1 was drawn from *The Farm Management Handbook* (1988). It was treated as a farm management exercise and all likely factors were considered, including such diverse items as building requirements, machinery depreciation, labour and cash flow problems, etc. The detailed calculations are not included here.

Contour Cultivations

By this, what is meant is carrying out all operations along the contour rather than straight up and down the slope, as is currently the common method. Some authors have, perhaps rather naively, regarded this practice as the cause of soil erosion and contour cultivations as a cure. This is unlikely to be the case for a variety of reasons.

Wischmeier and Smith (1978) state that such contouring is most effective for slopes from 3 to 8 per cent (2–5°). Beyond this narrow range the soil loss ratio (that is, the ratio of soil loss with contouring to soil loss without contouring)

Table 37.1 Assessment of erosion-control techniques

Method	Likelihood of effective erosion control	Initial capital cost	Continuous annual cost	Likely[a] on-farm benefit
No action	Nil	Nil	£20 ha^{-1}[b]	Nil
Contour cultivations	Low	£0-25 ha^{-1}[c]	Nil	£5 ha^{-1}
Change to coarser seedbeds	Low	Nil	£30 ha^{-1}[d]	£5 ha^{-1}
Change to all spring-sown crops	Low to moderate	Nil	£60 ha^{-1}	£10 ha^{-1}
Introducing grass to rotation				
25% grass	Moderate	£370 ha^{-1}[e]	£15/ha + £55 ha^{-1}[f]	£10 ha^{-1}
50% grass	High	£575 ha^{-1}[e]	£0/ha + £85 ha^{-1}[f]	£20 ha^{-1}
100% grass	Very high	£1400 ha^{-1}[e]	£35/ha + £210 ha^{-1}[f]	£20 ha^{-1}
Diversion terraces	High	£1000 ha^{-1}	£28 ha^{-1}[g] + £150[f]	£20 ha^{-1}

[a] Direct effects of reducing loss of seedlings, etc.
[b] Direct yield loss by removal of seedlings, etc. Rarely exceeds 2.5 per cent of potential yield.
[c] Cost of pneumatic fertilizer spreader (averaged over 200 ha).
[d] Increased costs of weed and slug control.
[e] Capital cost of stock plus buildings, improving fences, etc.
[f] Interest on borrowed capital.
[g] Loss of cropping land on terrace faces and in grassed waterways.

approaches 1.0. They also state that fields should be generally free from depressions other than grassed waterways. This is rarely the case in the UK and the land smoothing advocated by USDA is an additional, expensive civil engineering operation.

Changes in Cultivation Techniques

It has been considered (Speirs and Frost, 1985) that current cultivation techniques contribute to the observed increase in soil erosion. In particular, the production of firm, level seedbeds with a very fine tilth produced by powered harrows and rollers has been regarded as a factor predisposing soils to surface capping and subsequent soil loss. There is no evidence that such fine seedbeds are necessary for the satisfactory establishment of cereal crops. They do have certain other agronomic advantages that would be sacrificed were coarser seedbeds to be adopted. Firstly, they are necessary for the satisfactory functioning of pre-emergence herbicides. They also have value in reducing the risks to growing crops from slugs. Both problems can be tackled in other ways (post-emergence herbicides and slug pellets) at an additional cost which is shown in Table 37.1. The likelihood of effective erosion control by this technique has been assessed in Table 37.1 as low. It is considered that on some relatively stable soils, leaving a coarser seedbed may be sufficient to allow adequate rates of infiltration of rainwater to avoid the risk of soil erosion. On other, less stable soils, however, the coarse seedbed is rapidly broken down under raindrop action and the effective protection from soil loss is short lived.

Another approach, that of direct drilling cereals, has been suggested by Robinson and Boardman (1988). This may well work in some situations. No attempt has been made to cost this approach here, as the yield penalty (or gain) of direct drilling appears to be very soil dependent and would vary widely from site to site. However, it should be pointed out that those fine sandy loam soils which appear to be most prone to erosion are also among those where significant yield loss is frequently recorded due to soil compaction.

Change from Autumn-sown to Spring-sown Crops

Many authors (Boardman and Robinson, 1985; Speirs and Frost, 1985; Arden-Clarke and Hodges, 1987) have related the increased incidence of soil erosion to increases in the proportion of autumn-sown crops. During the late 1970s and early 1980s there was a strong shift from spring barley to winter barley, winter wheat and oilseed rape in many of the areas currently suffering soil loss. The reasons why this may have contributed to increased soil loss are twofold.

First, autumn-sown crops develop much more slowly than spring-sown ones and thus leave the soil exposed and at risk for longer periods. Soils are not adequately covered by crops such as winter wheat throughout the months of

October through to March. On the other hand, soils left in stubble or left rough ploughed very rarely erode. Spring-sown crops, once drilled, develop complete groundcover far more rapidly.

Secondly, climatic statistics show that throughout the UK heavy rainfall is more likely to occur during the risk period October/November than during the period March/April. This, in turn, contributes to the greater risk of soil loss under autumn-sown crops.

Erosion can and does occur, however, in spring barley crops if there are rainfalls at the wrong time in spring. Potato crops also frequently show soil erosion. For this reason, the likelihood of effective erosion control for spring-sown cropping is assessed here as low to moderate. It will almost certainly reduce the incidence of erosion but will not prevent it entirely.

Farmers cannot change to this option at no cost. While the inputs to autumn-sown cereals are generally higher than spring-sown, the yields are also likely to be appreciably higher. The cost of reverting to entirely spring-sown crops will depend on many factors, but is here estimated at £60 ha^{-1} yr^{-1}.

Introducing Grass to the Rotation

The surest way of effectively preventing soil erosion is to sow the land to grass. Water erosion of established grassland in UK conditions is virtually unknown. At the extreme, erosion could be entirely prevented by putting a farm into permanent pasture.

The cost of doing this for an all-arable farmer is extremely high. Grass has to be grazed or fed in other ways to stock which must be bought. The capital cost of stocking an arable farm from scratch is likely to be in the order of £1250 per hectare. In addition, the farmer may be faced with fencing costs and provision of water supply and winter housing. He also may not possess the necessary skills of stockmanship which will have to be acquired. Once established as a stock farm, the annual income may well be broadly comparable to the original arable farm (excluding interest charges on capital).

At present, some scope exists for farmers to reduce the high capital requirement of putting land to grass, while still maintaining some income. This is by accepting the government's offers of 'set-aside'. This would involve sowing the land to grass but not stocking it. Such an option is currently available for the next five years.

Very few farmers could contemplate the very large capital demands of a complete transition to grassland farming. A more likely response to soil erosion along these lines is to introduce a certain proportion of grass into the arable rotation. This will have two effects. First, if the rotation is, for example, half grass, at any one time half the farm is protected absolutely from soil erosion. Second, the grass has the effect of increasing both the soil organic matter level and the soil's structural stability, both of which will tend to protect the soil from erosion while it is in the arable part of the rotation. The degree of this protection will depend largely on the relative time each field spends in grass and in arable cropping. When grass accounts for less than 25 per cent of any rotation such protection is likely to be negligible. When, however, it accounts for 50 per cent or more of the rotation the protection is likely to be good.

Table 37.1 considers three variations: rotations containing 25 per cent grass, 50 per cent grass and 100 per cent grass. The likelihood of effective erosion control is assessed as moderate, high and very high, respectively, but at a rapidly increasing initial capital cost.

Diversion Terraces

The final option considered here involves the construction of physical erosion-control structures to take runoff water from fields and carry it safely away to the nearest stream. The system of diversion terraces which has proved to be most acceptable to mechanized farmers is that described as parallel, variable-grade diversion terraces. These consist of parallel ditch and bank structures constructed across fields at spacings calculated to meet the erosion-control need. The major factor governing spacing is land gradient, with spacings of 40–60 m being typical. Once the approximate spacing is calculated it is generally adjusted to a suitable multiple of implement width to lessen the problems caused to the farmer by the terraces. The terraces are constructed not on a contour but with a slight fall not exceeding 1 per cent. Some cut-and-fill work on land grades is often necessary to achieve this.

Such terraces have been widely used in North America, and are, for example, being actively promoted in the Atlantic Provinces of Canada. However, they are expensive to build and result in some loss of cropping land to the farmer. While in many cases the farmer may be able to crop the terraces themselves with some difficulty, the terraces must eventually discharge their water into permanently grassed watercourses which are clearly not in crop. With costs of up to or exceeding £1000 per hectare being typical, such work is only done with government assistance. In Eastern Canada, for example, the state pays for two-thirds of the costs (Canada, 1986). It is highly unlikely that UK farmers would adopt any such measures without a substantial input from the state.

COMBINATIONS OF TECHNIQUES

It is likely that were erosion-control techniques to be introduced, that a combination of some of the above techniques would be applied, rather than single techniques alone. For example, contour cultivations and rougher seedbeds might be introduced together. In general, the costs would simply be the sum of the cost of individual techniques.

A particular example of combined techniques is strip cropping. This consists of alternating strips of crops with a high-risk factor such as, in a UK context, winter wheat, with strips of grass. In effect, it is a combination of the above methods. Because of fencing constraints, the grass could only be used for conservation cutting, that is, for hay or silage. The minimum cost of establishing such a system would be equivalent to that of putting between 25 per cent and 50 per cent of the farm into grass as this proportion of grass would be needed. Additional cost is almost inevitable, however, as the stock which will consume the silage or hay would have to be housed since a zero grazing system would be almost unavoidable.

DISCUSSION

For a farmer facing problems of soil erosion, various choices are possible. The first is to ignore the problem, to accept the crop damage which is generally slight in any case, and to leave the problem of soil degradation for future generations to solve.

The second choice is to adopt some low-cost alteration in cultivation methods in the hope of reducing the scale of the problem. On sites only marginally prone to soil loss, this approach may be enough. There are, however, many sites where such an approach is unlikely to be satisfactory. Further research, probably in the form of standard erosion plots, is needed to differentiate those sites where minimal changes will work from those where they will not.

The third choice is to adopt some major change in cropping regimen with significant increases in spring-sown crops or grass. Such a course of action will have significant costs either in loss of annual revenue or in initial capital costs. While such a course of action will probably significantly reduce the erosion loss on most sites, it will not, in the short or medium term be of economic benefit to the farmer.

The fourth choice is to continue with present cropping regimens but to install expensive erosion-control structures to deal with runoff water.

The third or fourth choices are those likely to be successful on most sites. They are, however, unlikely to be adopted by any UK farmers unless either the government compels farmers to reduce soil loss or pays the costs of them doing so. While farmers may be as concerned as anyone about the long-term effects of soil loss, if the necessary changes to control that loss will result in the farm business going bankrupt, no farmer will adopt those changes.

It has been the experience of soil-conservation workers worldwide that farmers do not adopt adequate conservation methods until it is too late unless they are either paid or compelled to do so. As it is society in general rather than the individual farmers in particular who are the main beneficiaries of good soil conservation, it is arguable that society in general (that is, the state) should pay for such measures.

CONCLUSIONS

On many sites in the UK the cost of effective soil-conservation methods will greatly exceed their immediate benefits to the landowner. It is therefore unlikely that any such methods will be adopted without a significant input, in the form of either cash or compulsion, from the government. Some marginal sites may, however, benefit from much cheaper approaches which might well be adopted. More research is needed to identify those marginal sites.

REFERENCES

Arden-Clarke, C. and Hodges, D. (1987). Soil erosion: the answer lies in organic farming, *New Scientist*, **12**, 42–3.

Boardman, J. and Robinson, D. A. (1985). Soil erosion, climatic vagary and agricultural change on the Downs around Lewes and Brighton, Autumn 1982, *Applied Geography*, **2**, 243–58.

Boardman, J. (1986). Soil erosion in south-east England, *London Environmental Bulletin*, **3**, 5–7.

Canada (1986). Canada/New Brunswick Agri-Food Development Agreement, Land resource development initiative implementation guidelines, Appendex F.

Colborne, G. J. N. and Staines, S. I. (1985). Soil erosion in south Somerset, *Journal of Agricultural Science, Cambridge*, **104**, 107–12.

Darwinkel, A. (1984). Yield responses of winter wheat to plant removal and to wheelings, *Netherlands Journal of Agricultural Science*, **32**, 293–300.

Evans, R. (1980). Characteristics of water-eroded fields in lowland England, In De Boodt, M. and Gabriels, D. (eds), *Assessment of Erosion*, Wiley, Chichester, 1980, 77–87.

Farm Management Handbook (1988). Scottish Agricultural Colleges, Edinburgh.

Frost, C. A. and Speirs, R. B. (1984). Water erosion of soils in south-east Scotland—a case study, *Research and Development in Agriculture*, **1**, 145–52.

Fullen, M. A. (1985). Erosion of arable soils in Britain, *International Journal of Environmental Studies*, **26**, 55–69.

Reed, A. H. (1979). Accelerated soil erosion of arable soils in the U.K. by rainfall and runoff, *Outlook on Agriculture*, **10**, 41–8.

Robinson, D. A. and Boardman, J. (1988). Cultivation practices, sowing season and soil erosion on the South Downs, England: a preliminary study, *Journal of Agricultural Science, Cambridge*, **110**, 169–77.

Speirs, R. B. and Frost, C. A. (1985). The increasing incidence of accelerated soil water erosion on arable land in the east of Scotland, *Research and Development in Agriculture*, **2**, 161–7.

Wischmeier, W. H. and Smith, D. D. (1978). *Predicting rainfall erosion losses—A guide to conservation planning*, USDA, Handbook No. 537, Washington, DC.

38 Soil Water Management and the Control of Erosion on Agricultural Land

A. C. ARMSTRONG, D. B. DAVIES and D. A. CASTLE
ADAS, Cambridge

BACKGROUND

The extent of erosion within agricultural areas in the UK has gradually become evident over the last two decades. The MAFF/Soil Survey aerial photographic survey of agricultural land conducted between 1982 and 1986 (Evans and Cook, 1986; Evans and Skinner, 1987; Evans, 1988; Evans et al., 1988) has indicated that in susceptible areas about 5 per cent of agricultural land is eroding. In some areas this may increase to nearer 15 per cent in some years. However, because of the rotations, not all susceptible land erodes in each year. This observation, coupled with studies of more obviously eroding areas (e.g. Boardman and Hazelden, 1986) and reports of spectacular erosion events (Evans and Nortcliff, 1978; Boardman, 1986) have done much to increase the awareness of the extent of the problem.

The occurrence of erosion has not led to widespread adoption of erosion-control practices on farms in susceptible soils. Nevertheless, control measures have been advocated (Boardman, 1984; MAFF, 1984) in localized areas where erosion is known to present particular problems. For example, MAFF have issued guidelines for the reduction of erosion within the South Downs Environmentally Sensitive Area (ESA).

CONSEQUENCES OF EROSION

The consequences of erosion can be separated into two major areas of impact, within the farm and in the wider context. Although the effects of erosion on agriculture within the UK are generally small, and even spectacular events (e.g. Evans and Nortcliff, 1978) affect a relatively small proportion of the fields

Soil Erosion on Agricultural Land
Edited by J. Boardman, I. D. L. Foster and J. A. Dearing
©1990 John Wiley & Sons Ltd

within which they occur, many farmers consider erosion to be at least a moderately severe problem (Skinner, 1986; Evans and Skinner, 1987). It is theoretically possible to relate crop yields to soil depth, and thence to indicate that a significant shallowing of soils may reduce crop yields. In addition, variation in soil fertility as a consequence of local patterns of erosion and deposition of soils make management difficult, and reduces the efficiency of agricultural inputs. The many changes in British agriculture since the war (Davies, 1984) have increased crop yields to almost double their pre-war levels. Farm efficiencies have increased to a degree to mask any small losses that might result from localized soil erosion. Soil erosion is consequently not seen by the farming community as a major threat to their livelihoods.

In contrast, the off-farm costs of erosion can be high. Where spectacular erosion events lead to localized flooding and deposition of sediment, the costs are easily documented (e.g. Boardman, 1986). Probably more significant, however, is the release of sediment into receiving watercourses, with attendant costs: riverbeds may aggrade, with a potential increase in flood risk; sedimentation behind impounding structures will reduce their life, or increase maintenance costs; suspended solids may have to be removed from water if it is used for potable supply, imposing costs on the water supply industry; and sediment in water has deleterious effects on aquatic life. Additionally, suspended sediment carries with it other chemicals, which may provide inputs into nutrient-poor lakes which then suffer algal blooms, or contain bonded agrochemicals which may cause pollution in the affected aquatic environment. Lastly, erosion, particularly gullying, has an immediate visual impact of the landscape, which reduces the visual amenity of the rural environment.

PRINCIPLES OF EROSION CONTROL

As erosion is not a universal problem, the first stage must be the identification of potential risk areas. This can be both at national (strategic planning) scale and on a within-farm basis.

National identification of erosion risk areas requires the specification of the controlling factors. These include topography (notably land slope), soil type and rainfall intensity. The survey by Evans *et al.* (1988) has identified soil, land use and slope factors, and the data presented by Morgan (1980) have indicated general patterns of rainfall erosivity in UK. Consideration of soil properties has enabled Evans (pers. comm.) to produce erosion risk maps, in which, however, rainfall is subsumed as part of the environmental conditions which comprise the soil association. Such maps can assess risk only, and actual rates of erosion are controlled by individual storm rainfall intensities.

Within a single farm, examination of landscape units permits the identification of potential problem areas, where soil types and topography are likely to be important. More significant, however, is the combination of field layout and cultivation method which can lead to concentration of surface runoff. Local man-made impermeable areas, (e.g. farm buildings, roads and tracks) can generate surface runoff in sufficient volumes which lead to erosion problems if allowed to run downslope without adequate control.

Erosion prevention thus centres round the need to control surface runoff. This can be achieved both by the adoption of agricultural practices that help to reduce the generation of runoff and the prevention of that runoff from accumulating to erosive volumes. Equally, adaptation of farming systems that maintain a vegetation cover, particularly at critical times of the year, will do much to prevent surface runoff from eroding the soil over which it flows.

EROSION-CONTROL PRACTICES

Wherever possible, the adoption of agricultural practices which reduce erosion risk should not prejudice viable farming systems. Initial attempts at erosion control should therefore adjust existing practices within the farming system. Only if these fail to provide adequate control will it then be necessary to contemplate structural changes on the farm enterprise.

First and foremost comes the positive control of surface runoff, both in terms of its generation and its conveyance once generated. Simple measures include the maintenance of all ditches and culverts to prevent overflow. Observations on clay soils have suggested that underdrainage can reduce surface runoff volumes significantly. Consequently, drainage can be undertaken as a technique for the control of erosion by surface runoff in some areas.

Cross-slope interceptors also provide a positive approach, either by encouraging infiltration or by diverting surface runoff to a controlled structure (ditch or pipe). Such interceptors, whether grass strips, ditches or hedges, offer considerable potential for breaking up flows before they become erosive (e.g. Evans and Nortcliff, 1978; Boardman, 1984). In effect they reduce the slope length by breaking it into a series of shorter sections. Cross-slope breaks can also be extended to include structures such as hedges and ditches. Reintroducing hedgerows may restore rural amenity, but at the expense of making fields smaller and therefore less efficient for modern machinery.

Cultivation practices can also be altered to reduce the erosion risk. Considerations of safety and mechanical efficiency have led to the almost universal practice of performing field operations up and down the slope, thereby minimizing the risk of overturning farm machinery (Hunter, 1981). However, in some cases, it is possible to implement cross-slope management.

Some current practices have a particular potential for erosion, notably the use of tramlines aligned up and down the slope, which both provide ready routes for surface runoff to be concentrated and, by virtue of the compacted soil within them, a locus for the further generation of runoff. One way of reducing the erosion potential within tramlines is to cultivate along the tramline with a single tine, thereby both reducing the compaction (and so increasing infiltration) and roughening the surface which will reduce the velocity of the flows. However, if the soil in the tramline slakes and then receives heavy rainfall, erosion may be exacerbated, so that adoption of the technique of cultivating tramlines needs careful consideration before being widely adopted.

The minimizing of soil compaction and the encouragement of infiltration throughout the field can be achieved by avoidance of overcultivation and overcompaction, and the incorporation of crop residues (notably straw). Compacted zones and plough pans prevent the downward movement of water and thus encourage the generation of surface runoff. Equally, the creation of increasingly fine seedbeds has been blamed for increasing the erosion risk by reducing the infiltration capacity (Speirs and Frost, 1985).

The strength of the soil surface can be controlled by using the stabilizing characteristics of vegetation. A good vegetation cover will bind the soil, increase the surface roughness and reduce the kinetic energy of the incident rainfall. A growing crop will also increase the rate of soil water loss from the surface through transpiration, which will again help to reduce the incidence of soil saturation and hence reduce runoff generation.

Such vegetation cover may need only be local, being restricted to particularly vulnerable areas. Grassing valley floors which are known to be particularly prone to erosion (Evans and Cook, 1986) fall within this category. Equally, the vegetation cover may need only be temporary; the practice of using cover crops falls within this category (e.g. Auzet, 1987).

The need for protection by a growing crop may lead to a change in the cropping system. As a last resort, land may be returned to permanent grassland or even woodland to minimize erosion. Such changes, however, may require structural alterations in the economy of the farm. More moderate changes include the adoption of cropping systems that maintain crop cover at critical times. Robinson and Boardman (1988) have suggested that direct drilling of autumn cereals through unburnt stubble may offer a way of obtaining high yields with low erosion risk, but the agricultural difficulties of such a technique prevent its widespread adoption. Nevertheless, the greater use of autumn cereals has appeared to increase the erosion risk where fine-rolled seedbeds and tramlines provide a ready route for water movement, and late sowing can leave the soil vulnerable for long periods. The early sowing of autumn crops and delay in rolling until the spring is indicated for soils at risk of erosion.

In general, the choice of cropping patterns can be indicated by the following rough (subjective) league table of erosion risk:

Least risk
   Forestry/woodland
   Permanent pasture

Spring cereals
Autumn cereals
Short-term grass ley
Sugar beet/potatoes/horticultural crops
Greatest risk

CONCLUSION

The major impact of soil erosion in the UK is to be found outside the farm. In order to minimize erosion levels, farm practices need to be adapted while maintaining farm incomes. Positive soil water management coupled with careful choice of cultivation and land-management techniques should reduce erosion risk to acceptable levels in most circumstances. However, if these are not sufficient, then farm structural change (involving, for example, a change in land use) is indicated.

REFERENCES

Auzet, A. V. (1987). *L'erosion des sols par l'eau dans les regions de grande culture: aspects agronomiques*, Centre d'etudes et de recherches eco-geographiques (CEREG), Strasbourg.

Boardman, J. (1984). Erosion on the South Downs, *Soil and Water*, **12**, 19–21.

Boardman, J. (1986). The context of soil erosion, *SEESOIL*, **3**, 2–13.

Boardman, J. and Hazelden, J. (1986). Examples of erosion on brickearth soils in East Kent, *Soil Use and Management*, **2**, 105–8.

Davies, D. B. (1984). Trends in mechanisation in the lowlands. In Jenkins, D. (ed.), *Agriculture and the Environment*, Institute of Terrestrial Ecology (NERC), Cambridge, pp. 44–8.

Evans, R. (1988). *Water Erosion in England and Wales*, Internal report, Soil Survey and Land Use Centre, Silsoe.

Evans, R., Bullock, P. and Davies, B. D. (1988). Monitoring soil erosion in England and Wales. In Morgan, R. P. C. and Rickson, R. J. (eds), *Agriculture: Erosion Assessment and Modelling*, Commission of the European Communities EUR 10860, Luxembourg, pp. 73–91.

Evans, R. and Cook, S. (1986). Soil erosion in Britain, *SEESOIL*, **3**, 28–58.

Evans, R. and Nortcliff, S. (1978). Soil erosion in north Norfolk, *Journal of Agricultural Science, Cambridge*, **90**, 185–92.

Evans, R. and Skinner, R. J. (1987). A survey of water erosion, *Soil and Water*, **15**, 28–31.

Hunter, A. G. M. (1981) Tractor safety on slopes, *The Agricultural Engineer*, **36**, 95–8.

MAFF (1984). *Soil Erosion by Water*, Leaflet 890, Ministry of Agriculture, Fisheries and Food, Alnwick.

Morgan, R. P. C. (1980). Soil erosion and conservation in Britain, *Progress in Physical Geography*, **4**, 24–47.

Robinson, D. A. and Boardman, J. (1988). Cultivation practice, sowing season and soil erosion on the South Downs, England: a preliminary study, *Journal of Agricultural Science, Cambridge*, **110**, 169–77.

Skinner, R. J. (1986). A survey of water erosion in England and Wales, *SEESOIL*, **3**, 60–61.

Speirs, R. B. and Frost, C. A. (1985). The increasing incidence of accelerated soil water erosion on arable land in the east of Scotland, *Research and Development in Agriculture*, **2**, 161–7.

39 A Field Study to Assess the Benefits of Land Husbandry in Malawi

M. B. AMPHLETT
Hydraulics Research, Wallingford

INTRODUCTION

The economy of Malawi, a small landlocked country in southern Africa, is very dependent on agriculture. This forms the livelihood for over 85 per cent of the population with agricultural products accounting for 90 per cent of export earnings. Like many developing countries, increasing population pressures on the land, often coupled with poor cultivation methods as practised by subsistence farmers, has led to accelerated soil erosion which poses a serious threat to long-term agricultural productivity. To combat the problem of soil erosion and to promote good land-use planning and environmental conservation a Land Husbandry Branch was set up within the Ministry of Agriculture in 1968/9. It has a very comprehensive manual for land-use planning and physical conservation (Shaxson *et al.*, 1977) and a Land Husbandry training Centre. Despite these commendable institutional arrangements, very little practical soil conservation is to be seen in the field, particularly at the subsistence farming level.

The reasons for this are both social and political, underlying which is a lack of quantitative data which can be used to demonstrate the benefits of conservation measures the Land Husbandry Branch are trying to promote. Much of the information upon which conservation strategy is based in Malawi is derived from the results of studies carried out in other countries such as Zimbabwe, South Africa and the United States. Adopted conservation measures need to be verified and, where necessary, modified for local agro-climatic conditions. The potential benefits also need to be quantified if land users and policy makers are to be convinced of the desirability for erosion control, particularly when it must compete with other necessary development policies for scarce technical, financial and other resources.

Soil Erosion on Agricultural Land
Edited by J. Boardman, I. D. L. Foster and J. A. Dearing
©1990 John Wiley & Sons Ltd

As a first step in providing such data, a field project was initiated in 1981 by Hydraulics Research (HR) in collaboration with the Land Husbandry Branch of the Ministry of Agriculture. The formulated objectives of the research project were to assess the relative benefits of different levels of land use and management practice in reducing runoff and soil loss (Amphlett, 1980). Interim results of the study were published in a paper in 1984 (Amphlett, 1984) and a final technical report published in 1986 (Amphlett, 1986). This chapter summarizes the principal results of the field study and a statistical analysis of the data which has subsequently been carried out.

PROJECT DESCRIPTION

The land-management practices proposed for study were uncontrolled arable agriculture, plantation cover and physical conservation with (and physical conservation without) land-use planning. In consideration of the scale of study required to quantitatively assess the four land-use practices, plots, which have been the most commonly applied erosion research tool and which still dominate much of the work in developing countries, were discounted as too small to physically represent their effects (Amphlett, 1984). The research approach adopted was thus one of a comparative catchment study.

The catchment or drainage basin as a fundamental unit for studies in fluvial geomorphology has become more prominent in the last two decades (Burt and Walling, 1984). These can provide valuable information on the landform controls of runoff and erosion which cannot be replicated at the plot scale. While representative and paired basins have been extensively used in the United States and Europe as a means of studying hydrological processes and for contrasting land-use changes (Rodda, 1976), very few quantitative catchment studies have been conducted to date in Third World countries. Two well-documented catchment studies in East Africa are that by Rapp and co-workers in Tanzania (Rapp *et al.*, 1973) and the East African Agricultural and Forestry Research Organization experiments in Kenya, Tanzania and Uganda (Blackie *et al.*, 1979). Representative and paired catchments of varying size were used in these studies to obtain information on erosion and sedimentation rates and to evaluate the hydrological effects of land-use changes in forested areas.

Compared to plots, catchment-scale experiments are expensive, and to yield worthwhile quantitative results need to be studied over a reasonable period of time to account for climatic and other seasonal variations that can occur. It is also a relatively young science, and so appropriate measurement techniques are still evolving. These are just some of the reasons why catchment-scale studies have not become more prominent in developing countries. If, as in the Malawi study, the objectives are to assess the relative benefits of different land uses then comparative catchments do provide a quick method for evaluation. If more detailed quantitative results are required then a longer period of study is necessary. Such a comparative catchment approach does, however, require that the catchments are as similar as possible in all respects other than land use. For the Malawi study, special attention was thus given to the selection of the catchments which do resemble one another very closely in area, soil type, slope, altitude and mean annual rainfall. The catchments, their areas and the land-use systems represented by them are as follows:

Bvumbwe (7.8 ha) : a full land-use plan as set out in accordance with the Malawi Government's land husbandry manual.
Mindawo (5.3 ha) : under intensive uncontrolled arable farming for subsistence crops.
Mindawo II (6.7 ha): with complete physical conservation works but with no planned system of land use.
Mphezo (17.2 ha) : under eucalyptus plantation.

These catchments are located at Bvumbwe, on the rift valley escarpment of southern Malawi (Figure 39.1). They lie within a radius of 3 km at a mean altitude of 1 150 m. The general landform of the study area is one of dissected plateau with moderate slopes ranging from 5 per cent to 25 per cent. There are two distinct seasons in Malawi, a dry season which extends from May to October and a wet one from November to April. Mean annual rainfall for the area is of the order of 1 000 mm. The soils are medium-altitude ferruginious-ferrallitic integrades of clay, sandy clay, sandy-clay-loam and sandy-loam texture and are classified as suitable for intensive arable cultivation.

LAND USES

The Bvumbwe and Mindawo catchments represented two extremes in land use; a full conservation technique, including both biological and physical control measures, and traditional uncontrolled agriculture, respectively. To ensure complete implementation of the land husbandry plan, the Bvumbwe catchment was selected within a government agricultural research station; the catchment being managed by a 'unit' farmer under the supervision of research staff. In contrast, Mindawo is held by private smallholders whose farming practices were not adjusted in any way for the study. The Mindawo II catchment represented an intermediate level of management strategy with physical conservation works but which, like Mindawo, was otherwise traditionally cultivated. This was included in the study to enable the benefits due to physical and biological conservation to be assessed. The fourth catchment, Mphezo, which is under eucalyptus plantation, was intended to give some indication of a base level of erosion. While eucalyptus trees themselves do not provide very dense foliage

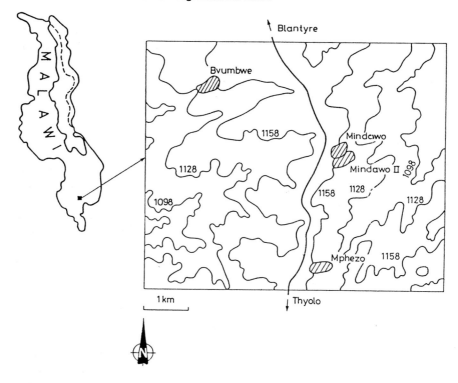

Figure 39.1 Location of Malawi research catchments

cover, the under-canopy vegetation and ground litter of the selected catchment was very good throughout the year. This was the best alternative to a natural vegetative cover that could be found in the immediate area.

The physical conservation works implemented on the Bvumbwe and Mindawo II catchments consisted of graded contour bunds to control and channel excess runoff water to grassed waterways, together with an aligned ridge and furrow cultivation system. For the Bvumbwe catchment, biological conservation was achieved through suitable tillage practices and with a balance between annual and longer-term crops grown in a rotational sequence with fallow grass areas. On the Mindawo and Mindawo II catchments, which have no land-use plan and are intensively cultivated, no crop rotation can be expected from one season to the next with no areas left fallow. While crop ridging is the traditional farming practice, the ridges, particularly on the Mindawo catchment where there were no bunds to define the contour, were often not aligned properly, thus increasing the erosion risk.

On all three cultivated catchments, the principal crop was maize with secondary crops of beans, sweet potatoes and groundnuts. In the earlier part of the rainy season these were intercropped with the maize but became more prominent later in the season when the maize had been harvested. On the Bvumbwe catchment other cash crops such as tomatoes, pineapples and cucumbers were also grown.

DATA COLLECTION

Measurements of rainfall, runoff, soil and nutrient loss, soil moisture and crop cover formed the basis of the field work; this being carried out under a daily monitoring programme during the wet season. Rainfall was measured by two tilting-syphon autographic rain gauges, supplemented by one standard rain gauge on each catchment. At the catchment outlets, H-flume structures were employed for the measurement of runoff; water level being recorded by means of a conventional float-operated gauge. For the measurement of sediment and nutrient losses at the catchment outlets, the need was seen for an automatic sampling system because of the ephemeral nature of the flow and also because of the large volumes of water and sediment leaving the catchments during runoff events which precluded the use of storage devices. To meet this requirement, an automatic pump sediment sampler was developed by HR (Amphlett, 1986). The sampler is triggered on water level and can take up to 25 one-litre bottles during a given sampling period. It is also programmable so that any desired frequency of sampling can be selected to match the runoff characteristics of the catchment. Once triggered by a certain threshold of flow, sampling continues until either the full supply of sample bottles are used up or the flow again drops below the set threshold level. If another storm event occurs before a site visit and not all the sample bottles have been used, then sampling begins again from the start of the programme. Because of the vertical variations in sediment concentration that occur in a naturally flowing stream, for single-point sampling it is essential to ensure that the sample taken is representative of the mean concentration. To do this, one-metre drop-stilling basins were constructed below the flume outlets to provide a turbulent zone of fully mixed water and sediment from which the samples could be abstracted (Figure 39.2). An event marker on the water-level recorder, linked to the sampler, provided a record of when each sample had been taken relative to the flow rate. Sample bottles collected in the field were transferred to a laboratory for analysis of sediment concentration and for concentrations of total nitrogen, phosphate and potassium.

For the measurement of soil moisture, four arrays of tensiometers were installed within each catchment. Each array covered soil depths of 15, 30, 60, 90, 120 and 150 cm below ground level.

580 Soil Erosion on Agricultural Land

Figure 39.2 H-flume and pump sampler

To obtain some estimate of percentage crop or vegetative cover throughout the wet season, a ground-based photographic technique was used; a 35 mm camera being elevated to a fixed height of 4.3 m above ground level by means of a tripod. With the exception of Mphezo, on which cover was assumed to be 100 per cent, on each of the catchments the camera was set up at selected locations and a series of photographs taken at approximately weekly intervals.

The Bvumbwe and Mindawo catchments were fully instrumented for the 1981/1982 rainy season with the Mindawo II and Mphezo catchments being included in the monitoring programme for the 1982/3 season.

RESULTS

Data collection was carried out for four consecutive rainy seasons, 1981–5. The results obtained in terms of total annual rainfall, runoff and soil loss are summarized in Table 39.1.

Annual rainfall totals experienced during the monitoring period varied not only between seasons but also between catchments, reflecting the highly localized nature and small areal dimensions of storm events. The results show that annual runoff and soil loss totals do not correlate with rainfall for a particular catchment. For example, in 1983/1984 all four catchments exhibit much lower losses of water and sediment when compared to the 1981/2 and 1982/3 seasons,

Table 39.1 Seasonal rainfall, runoff and soil loss totals

Catchment	Season	Rainfall (mm)	Runoff (mm)	Soil loss (t ha^{-1})
Bvumbwe (full land plan)	1981/2	957	29.7	0.12
	1982/3	822	19.0	0.21
	1983/4	836	6.7	0.03
	1984/5	1334	62.2	0.13
Mindawo (traditional cultivation)	1981/2	890	81.0	10.06
	1982/3	910	156.7	13.70
	1983/4	865	42.0	4.44
	1984/5	1191	154.5	14.32
Mindawo II (physical conservation)	1981/2	—	—	—
	1982/3	975	54.8	2.31
	1983/4	914	17.4	1.18
	1984/5	1207	85.6	5.11
Mphezo (plantation)	1981/2	—	—	—
	1982/3	951	7.7	0.09
	1983/4	923	4.2	0.03
	1984/5	1137	8.4	0.06

even though rainfall totals are similar. This is because both runoff and sediment yield are very much dependent on the characteristics of rainfall events such as intensity, duration and seasonal distribution. In 1983/4 a greater proportion of the rainfall occurred in the latter half of the season, a time when vegetative cover on the catchments was higher. This may have been a dominant factor in the reduced water and soil losses experienced that year.

Despite these climatic variations, both between seasons and between catchments, a very consistent pattern does emerge which clearly illustrates the benefits of the different land uses in operation on the catchments in retaining water and sediment. In terms of annual runoff volume, in all seasons the unmanaged Mindawo catchment exhibited the highest. Compared to this catchment, the physically conserved Mindawo II catchment shows an average runoff reduction of 56 per cent, the full land-use plan on Bvumbwe a 74 per cent reduction and the Mphezo plantation catchment a 93 per cent reduction. While the benefits of physical conservation measures on Mindawo II are quite significant in retaining runoff water, the additional benefits gained when physical measures are supplemented by biological conservation are clearly demonstrated on the Bvumbwe catchment. Cumulative runoff records for the four catchments in 1983/1984 and 1984/1985, the seasons with the highest and lowest runoff volumes, are shown in Figure 39.3 which graphically illustrates this consistent pattern of runoff.

As may be expected, a pattern similar to that of runoff also emerges for soil loss, although the benefits are far more substantial. Compared to the Bvumbwe catchment, soil losses on the unmanaged Mindawo catchment are of the order

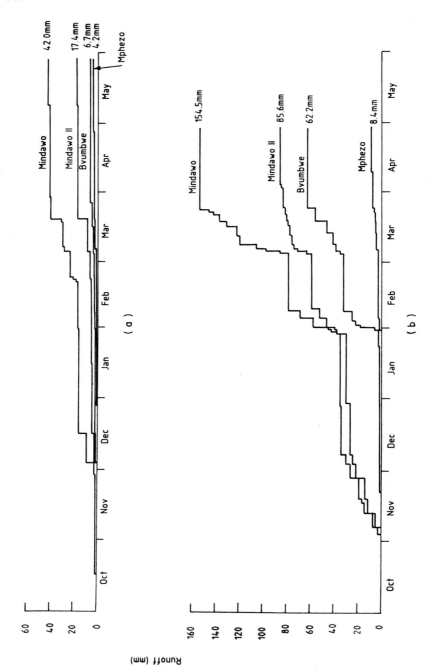

Figure 39.3 Cumulative runoff records for (a) 1983/4 and (b) 1984/5 seasons

of a hundredfold higher in each season. Physical conservation measures on the Mindawo II catchment do reduce these losses, but not to the extent seen on the Bvumbwe catchment, further emphasizing the additional benefits of biological conservation measures. It is significant also that soil losses from the Bvumbwe catchment are only marginally higher than for the well-covered Mphezo catchment. This result shows that, under a good land-use plan, erosion rates are comparable with those approaching natural vegetation cover.

The relative differences in runoff and soil loss between the various land-use practices are comparable with the results obtained from other studies in Africa. In Tanzania (Rapp et al., 1973) a maize plot was found to lose around thirty times more soil and 43 per cent more runoff water than a grass plot. There was also a marked difference between maize grown with closely spaced trash bunds and that grown without any conservation, the latter losing 70 per cent more water. In a comparative catchment study at Mbeya in Tanzania (Blackie et al., 1979) sediment yield from a 20.2 ha cultivated catchment was estimated as $9 \text{ t ha}^{-1} \text{ yr}^{-1}$ compared to practically nil from a 16.2 ha forested catchment. The measured erosion rates on the Mindawo and Mphezo catchments in Malawi, which most closely correspond with these land-use practices, present a similar picture.

The relative benefits of the different land-use practices were further substantiated through crop cover measurements and, to a lesser degree, by soil-moisture storage conditions. In Figure 39.4 mean monthly crop cover percentages are shown for three seasons, 1981–4. Percentage cover on the Bvumbwe

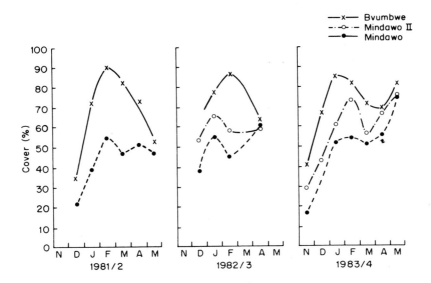

Figure 39.4 Mean monthly crop cover

catchment was consistently higher throughout each season, reaching a peak of around 90 per cent in mid-season compared to a mean of about 70 per cent on Mindawo II and 55 per cent on Mindawo. The results characteristically showed a drop in cover following the initial peak, increasing again towards the end of the season. This later increase in cover corresponded with the planting and maturing of secondary crops following harvesting of the maize.

In terms of soil-moisture storage the upper soil profile, as expected, exhibited very large fluctuations in storage capacity, corresponding to the wetting and drying sequences of rainfall and evaporation. No clear distinction between the catchments could be made for the shallower soil depths. It was only at the deeper monitored soil profile depths, where short-term climatic variations and their effect on water storage were damped out, that the benefits of the different land-use practices began to emerge. In general, soil-moisture conditions were marginally better on the Bvumbwe catchment compared to the other catchments, although the results were not conclusive and did not reflect the land-use practices as clearly as other measured parameters.

The results obtained from the analysis of runoff water for the principal nutrients nitrogen (N), phosphate (P) and potassium (K) are given in Table 39.2. The nutrient losses, represented as ratios relative to that on the Bvumbwe catchment, are shown in Figure 39.5. The results are not as consistent as found for runoff and soil loss and exhibit much more variability between seasons and between catchments. Nevertheless, the general pattern of the land-use practices can again be seen. The variability in the results may be attributable to differences in the amount of fertilizer applied to the catchments in any given season. For instance, the P loss from the Bvumbwe catchment during 1984/5 was

Table 39.2 Seasonal nutrient losses

Catchment	Season	Nutrient losses (kg ha^{-1})		
		N	P	K
Bvumbwe	1981/2	0.093	0.068	0.621
	1982/3	0.157	0.046	0.030
	1983/4	0.032	0.010	0.012
	1984/5	0.222	0.673	0.182
Mindawo	1981/2	0.373	0.088	1.626
	1982/3	0.531	0.235	0.208
	1983/4	0.056	0.050	0.059
	1984/5	0.304	0.160	0.282
Mindawo II	1982/3	0.297	0.090	0.041
	1983/4	0.031	0.020	0.021
	1984/5	0.185	0.073	0.070
Mphezo	1982/3	0.034	0.021	0.006
	1983/4	0.019	0.011	0.003
	1984/5	0.042	0.028	0.005

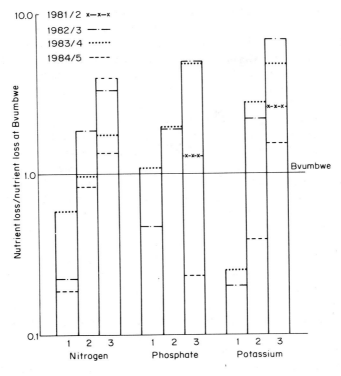

Figure 39.5 Ratio of nutrient losses to that in the Bvumbwe catchment. 1, Mphezo; 2, Mindawo II; 3, Mindawo

exceptionally high compared to other seasons, as was the K loss on Mindawo in 1981/2. A full and complete record of what fertilizers had been applied was, however, not obtained, so the reasons for these high nutrient levels cannot be ascertained. The laboratory analysis of samples was only undertaken for those nutrients carried in solution in the runoff water. Nutrient loss can also occur through absorption to soil particles and can be many orders of magnitude greater in the eroded soil than the runoff water (Kang et al., 1981). Thus, while showing the benefits of the different land-use practices on the catchments, the nutrient loss results do not give a true reflection of the likely losses in soil fertility due to erosion.

MULTIPLE REGRESSION ANALYSIS

A detailed statistical analysis of the data was also undertaken to illustrate the practical application of multiple regression and to try and obtain a better understanding of which input variables were important in determining runoff and soil loss from the Malawi catchments (White *et al.*, 1988). Empirical (databased) models, such as regression models, often have to be used when it is not possible to find a suitable mechanistic model. Multiple regression is also a very useful technique when the database is limited, as was the case for the Malawi study. Because of the limited database, the individual storm had to be taken as the unit of analysis.

Total storm runoff and soil loss were taken as the response variables and various parameters relating to rainfall, soil moisture, vegetation, topography, soil characteristics and farming practices as the explanatory variables in predicting the response variable. In all, 15 explanatory variables were considered together with their variation across the four catchments. The models were developed and parameters estimated using the data from the 1983/4 season; the 1984/5 season being used for validation tests.

In predicting runoff for 1983/4 a six-parameter model with explanatory variables of storm date, rainfall total, storm length, time since last storm and maximum 30-min rainfall intensity gave the best results. However, in validating the model it was found to underestimate runoff for 1984/5, indicating that a year effect was present. The distribution of rainfall, both throughout the year and relative to other storms, was suggested as the cause of discrepancy.

In predicting the response variables of soil loss the main problem was the very sparse database. An analysis was, however, attempted using the two Mindawo catchments which had more sediment measurements. This showed the only important parameters to be time since last storm and some measure of rainfall intensity. However, if runoff was included as a predictive variable then this negated the effect of time since last storm. The best-fit models were those using 5-min rainfall intensity for the Mindawo catchment and 30-min intensity for Mindawo II.

The simple regression models developed, while not taking account of all the information that was available, gave results that were encouraging. Further development of prediction methods, based on statistical analysis, is being pursued using data obtained from a similar study conducted by Hydraulics Research in the Philippines.

CONCLUSIONS

The results obtained from the project clearly showed the relative benefits of the different land management practices on the four catchments in controlling water, sediment and nutrient losses. They were also very consistent over the four-year study period. While quantitatively erosion rates varied from season to season, principally due to climatic variations, in comparative terms the pattern between catchments remained the same. The reduction in soil loss between traditionally farmed and fully managed land use was of the order of 100 to 1. For physical conservation measures alone, with no planned system of land use, the ratio was 10 to 1. This illustrated the additional benefits that can be achieved when physical conservation is supplemented by biological conservation. Furthermore, erosion rates for the fully managed catchment were minimal and comparable with that under forest cover. That such benefits could be demonstrated from a relatively short monitoring programme justified the comparative catchment approach adopted for the study.

The methodology does, however, rely on the selected catchments being as similar as possible in all respects other than for the principal variant being investigated, in this case land use. The catchments being more or less contiguous reduces the likely variability of climate, soils, topography, etc. However, true similarity is obviously impossible to achieve and there will undoubtedly be differences in measured parameters not directly attributable to the land-use practices in operation on the catchments. For this reason, emphasis has been placed more on a comparative rather than a quantitative assessment. The significant differences in runoff and soil loss observed between the catchments, expressed as ratios, would be difficult to explain other than by the different land uses. In this respect, the project has given the Ministry of Agriculture in Malawi valuable evidence of the benefits of good land-husbandry measures.

The project, being site-specific, means that care must be exercised in extrapolating the results to other areas and in using them for general land-use planning. Modelling is one means by which the results can be interpreted and extended. A statistical analysis, using multiple regression techniques, was thus undertaken to try to obtain a better understanding of which variables were important in determining runoff and soil loss from the catchments. Analysing the data on a storm-by-storm basis, variations in runoff and soil loss were explained by simple regression models and differences between catchments were found to be statistically significant. However, a year effect was encountered in applying the models for different seasons. This highlighted the problems involved in trying to develop a model with a limited database which will predict runoff and soil loss from a particular catchment in a particular year. To develop a predictive model in which confidence could be placed, a much more extensive database than that available from the Malawi study would be required. The analysis also showed that, for predictive purposes, some important variables were not adequately monitored due to the practical difficulties of so doing. The analysis has, however, given an indication of the relative importance of variables which would benefit from further investigation in future studies and which might ultimately help in the development of more reliable predictive techniques.

ACKNOWLEDGEMENTS

The collaboration and participation in the project by the Ministry of Agriculture,

Government of Malawi is gratefully acknowledged. The research project was undertaken within the Overseas Development Unit of Hydraulics Research headed by Dr K. Sanmuganathan. The work formed part of the Unit's research programme into erosion control which is funded by the Overseas Development Administration of the Foreign and Commonwealth Office, UK.

REFERENCES

Amphlett, M. B. (1980). *Runoff/sediment Monitoring Study on Malawian Catchments*, Research Proposal OD 19/4, Hydraulics Research, Wallingford, UK.

Amphlett, M. B. (1984). Measurements of soil loss from experimental basins in Malawi. In Walling, D. E., Foster, S. S. D. and Wurzel, P. (eds), *Challenges in African Hydrology and Water Resources*, IAHS Publication No. 144, 351–62.

Amphlett, M. B. (1986). *Soil Erosion Research Project, Bvumbwe, Malawi: Summary Report*, Report OD 78, Hydraulics Research, Wallingford, UK.

Blackie, J. R., Edwards, K. A. and Clarke, R. T. (1979). Hydrological research in East Africa, *East African Agricultural and Forestry Journal*, **43**, Special issue.

Burt, T. P. and Walling, D. E. (1984). Catchment experiments in fluvial geomorphology: a review of objectives and methodology. In Burt, T. P. and Walling, D. E. (eds), *Catchment Experiments in Fluvial Geomorphology*, Geo Books, Norwich, pp. 3–18.

Kang, B. T. and Lal R. (1981). Nutrient losses in water runoff from agricultural catchments. In Lal, R. and Russell, E. W. (eds), *Tropical Agricultural Hydrology*, John Wiley, Chichester, pp. 153–61.

Rapp, A., Berry, L. and Temple, P. (eds) (1973). *Studies of Soil Erosion and Sedimentation in Tanzania*, Research Monograph No. 1, Bureau of Resources Assessment and Land Use Planning, University of Dar es Salaam.

Rodda, J. C. (1976). Basin studies. In Rodda, J. C. (ed.), *Facets of Hydrology*, John Wiley, Chichester, pp. 257–97.

Shaxson, T. F. Hunter, N. D., Jackson, T. R. and Alder, J. R. (1977). *A Land Husbandry Manual—Techniques of Land Use Planning and Physical Conservation*, Land Husbandry Branch, Ministry of Agriculture and Natural Resources, Malawi.

White, S. M., Coe, R. and Nice, S. (1988). *Multiple Regression in Runoff and Erosion Studies: a Case Study from Malawi*, Report OD/TN 37, Hydraulics Research, Wallingford, UK.

Politics and Policy

40 Issues on Soil Erosion in Europe: the Need for a Soil Conservation Policy
=====

R. P. C. MORGAN and R. J. RICKSON
Silsoe College, Bedford

INTRODUCTION

The last decade has seen an increasing awareness by European governments and the Commission of the European Communities of the problem of soil erosion. Evidence presented by European scientists at two workshops on soil erosion on agricultural land (Prendergast, 1983; Chisci and Morgan, 1986) shows that erosion rates in the Mediterranean countries and on the sandy and loamy soils of northern Europe can reach 10–100 t ha^{-1} annually whereas the maximum acceptable annual rate for conservation of the soil as a productive resource is only around 1 t ha^{-1}. Attention is also drawn to the lower crop yields obtained on eroded land and to the unwanted downstream effects of erosion such as pollution and sedimentation. Awareness, however, does not necessarily mean that anything is being done to control the problem. The question therefore arises of whether a soil conservation policy is required and, if so, what form it should take.

This chapter reviews the problem of soil erosion in Europe, emphasizing the areas affected and the major controlling factors, as a basis for accepting the need for a soil conservation policy. A framework is proposed for implementing such a policy and the implications for future research and action by the scientific community are discussed. Particular emphasis is given to the measures that might be used to control erosion.

THE EUROPEAN PROBLEM

Based on the research of European scientists engaged in monitoring and measuring erosion, three broad areas at risk may be identified. These are the Mediterranean lands, the lowlands of northern and central Europe and the European uplands.

Soil Erosion on Agricultural Land
Edited by J. Boardman, I. D. L. Foster and J. A. Dearing
© 1990 John Wiley & Sons Ltd

Mediterranean Lands

Accelerated erosion has been recognized as a problem in the Mediterranean belt since the earliest times. About 80 per cent of the land is potentially erodible at unacceptable rates and about 40 per cent of the area of Spain and Greece has had its productivity seriously reduced (Yassoglou, 1987). Italy (Chisci, 1982) and Portugal are also badly affected.

The causes of erosion in the Mediterranean belt are still not clearly understood. Vita Finzi (1969) considers the erosion as largely climatically induced. This allows for the view that man has played a positive role whereby the traditional soil- and water-management systems have served to control rather than enhance erosion (Shaw, 1981). Policies to reduce the number of people working on the land and to increase both farm size and the level of mechanization have resulted in the abandonment of many of these traditional systems. In particular, terraces have not been maintained and the mixed cropping of trees, cereals and grass has given way to crop specialization and monoculture. The result has been an increase in erosion by both surface runoff and shallow mass movements, particularly on the clayey and silty-clay soils derived from late Tertiary materials (Chisci, 1986). The severity of these changes, combined with the effects of forest fires (Sanroque, 1987) and, in some cases, soils prone to pipe erosion (Lopez Bermudez and Torcal Sainz, 1986), is sufficiently great for erosion to be considered a major factor alongside salinization contributing to the desertification of southern Spain (Rubio, 1987a).

Northern and Central Lowlands

Erosion in northern and central Europe affects mainly the sandy, loamy, loessial and chalky soils devoted to continuous arable production (Richter, 1980; Leser, 1980; Rohrer, 1985; De Ploey, 1986; Morgan, 1986). There is at least circumstantial evidence to support the view that the problem has become worse with the increase in land area under winter-sown cereals and sugar beet (Boardman, 1988). Higher levels of mechanization, greater use of tramlines and chemical rather than organic fertilizers, together with consolidation of land into larger fields, may all have played a role in increasing the area affected by erosion. However, the problem should not be thought of as entirely recent in origin. There is historical evidence for severe erosion on arable land in France in the eighteenth century (Vogt, 1957) and in Germany in the early fourteenth century (Bork, 1988). Although the latter has been interpreted as climatically induced, it also coincided with a period of declining rural population and increasing poverty following the Black Death. Thus the interplay of natural and man-induced factors in causing erosion is not entirely clear.

Soil erosion is also important on clay soils with a similar recent agricultural history, where winter freezing of the subsoil is followed by spring snowmelt.

Such erosion is being increasingly recognized as a problem on the arable lands of Norway (Skøien, 1988), Sweden (Alström and Bergman, 1988) and Finland (Mannsikkaniemi, 1982).

Uplands

Erosion affects the upland pastures of northern, western and Alpine Europe. Severe wind erosion on rangeland in Iceland has been a problem since the time of the country's settlement in the period AD 875–930 (Fridriksson, 1972). Periods of overgrazing have induced water erosion at various times in Wales (Thomas, 1965) and the southern Cévennes in France (Cosandey et al., 1987) while parts of upland England are showing increasing evidence of susceptibility to erosion (Evans, 1977). Afforestation can also increase erosion in upland areas (Newson, 1988). Again, the exact role of land-use change is not always clear, and Metailie (1987) argues that the erosion in the Pyrenees in the mid-nineteenth century was more the result of exceptionally heavy rainfall acting on land that had been made more sensitive to erosion through deforestation and overgrazing.

Review

Although the broad areas with greatest erosion risk or known occurrence of erosion have been identified, the extent of erosion at any one time or in any one year is not known for any country in Europe. Without this information it is difficult to evaluate the problem. Enough is known, however, to indicate that the areal extent, frequency and rates of erosion, at present and in the past, are sufficient to cause concern.

The evidence presented by European researchers shows that erosion is not a simple problem of pressure on land resources and misuse of land. Indeed, there is much evidence to suggest that, at least in the uplands and the Mediterranean, man's management of the land has actually reduced erosion in areas of potentially high erosion risk. Rural depopulation rather than population pressure may be a major factor in increasing erosion rates. This may operate either directly because fewer people are available to maintain terraces and ditches or indirectly because policies aimed at keeping people in rural areas are encouraging unwise use of the land such as overgrazing of the uplands. Misuse or poor management of the land may not be enough to explain the severity of erosion; it seems likely that these conditions need to coincide with sufficiently strong climatic events to trigger erosion on land which has become less resistant. This implies that erosion will only be understood if the research of geomorphologists into the mechanics of the processes is integrated with studies of the social and economic factors which control the way in which the soil is managed and thereby condition the soil state.

EFFECTS OF EROSION

Few studies have been carried out in Europe on the effect of erosion on the productivity of the land. Although long-term decreases in productivity might be expected, they have undoubtedly been more than offset by the higher crop yields associated with modern agricultural practices. The question arises as to how long this masking effect can be expected to continue. For example, studies by Becher et al. (1985) show that yields of silage maize, which average 76 t ha^{-1} of fresh weight biomass on loessial soils in south Bayern, West Germany, can be reduced by 3 t ha^{-1} after ten years of an average annual soil loss of 30 t ha^{-1}. The decline in yield is irreversible, since it is largely a result of a reduction in the water-holding capacity of the soil. Calculations based on the Stocking and Pain model (1983) indicate that sandy loams shallower than 30 cm and clay loams shallower than 25 cm may become uneconomic to farm within 50–75 years at current erosion rates (Morgan, 1987). Even though such reductions in productivity are arguably not of great economic significance in today's context of overproduction of most agricultural products, they are important to the individual farmers whose future incomes depend on the land and to the maintenance of the soil resource for future generations.

The off-site effects, resulting from sedimentation downstream or downwind, are even more significant economically. Sedimentation reduces the capacity of rivers and drainage ditches, enhancing the risk of flooding, and shortens the design life of reservoirs. There is a cost to the community for clearance of roads and drains blocked by sediment and for removal of sediment from houses. Sediment is a pollutant in its own right and through the chemicals adsorbed to it. Much recent interest in soil erosion is generated by concern about levels of nitrogen and phosphorus in water bodies (Hasholt, 1988).

The economic consequences of erosion, both on- and off-site, have not been evaluated nationally for any country in Europe but they must be considerable. Reductions in erosion could save money currently spent on maintenance and repair of infrastructure due to damage by erosion events. The need for action on erosion control is being increasingly recognized by the inhabitants of downstream areas who are affected directly by the flooding and sedimentation, as events in recent years near Valencia (Rubio, 1987b) and in the South Downs of England (Robinson and Williams, 1988) testify.

POLICY DEFINITION AND FORMULATION

There is sufficient qualitative evidence on the extent of erosion and its economic consequences to support the view that some form of soil erosion control policy in Europe is necessary. Without data on the size of the affected area at any one time or on the true economic costs it is difficult to put a level of priority on the policy, at least nationally. Where people are being regularly affected by the consequences of erosion, however, a high level of local priority must be given. This priority must be recognized by including within government policies on the environment the objectives of (1) preventing unwanted downstream pollution and sedimentation arising from soil erosion on agricultural land and (2) preserving the soil resource for future generations. This is in keeping with the view expressed in the Green Paper of the Commission of the European Communities (1985) that the role of agriculture is increasingly perceived to include conservation of the rural environment.

Since agriculture occupies 61 per cent of the land in the countries of the European Community and most erosion seems to be associated with agricultural use, it follows that any policy for erosion control must be set within an agricultural context. It should therefore take account of the heterogeneity of European agriculture in terms of incomes, agrarian structures and physical environment. This can be best achieved by setting the framework for a policy at the European or national government level but deciding on the appropriate erosion-control measures and how to implement them at the regional or local level. Support for this approach can be found in the two most successful examples of policy making and implementation on erosion control within Europe to date, namely those operating in Denmark and Iceland to control wind erosion.

The foundation and subsequent activities of the State Soil Conservation Service of Iceland (Landgraedsla Rikisins) provide a good example of how a policy on soil conservation can emerge. Erosion became particularly bad between 1860 and 1890, when drifting sand seriously affected the livelihood of farmers, especially in the southern and north-eastern parts of the country where land was being abandoned and the land surface lowered at rates in excess of 2 mm yr^{-1}. Some farmers attempted their own control measures by building stone walls or wooden fences, but these were generally ineffective. Under pressure from the Agricultural Society of Iceland (Bunadarfelag Islands), the Icelandic Parliament (Althing) made available a small grant to bring in Danish specialists to investigate the problem. In 1895, the Althing passed the Act for Resolution on Sand Erosion and Reclamation which contained authority for District Commissions to take action on the problem. Since no incentives to do so were provided, however, this was of little value (Runolfsson, 1978, 1987).

Pressure from farmers for action continued, and in 1907 the Act on Forestry and Prevention of Erosion of Land was passed, effectively marking the foundation of the State Soil Conservation Service. Erosion control was still largely ineffective, however, because the available technology of barbed-wire fencing to keep sheep off the land and allow the vegetation to regenerate naturally was inadequate. Following further severe erosion in the late 1940s, a major reclamation experiment was carried out near Gunnarsholt, using technology based on assisting the ecological succession of plant communities, supported by aerial seeding and fertilizer application (Arnalds *et al.*, 1987).

The successful reclamation of 1300 ha of land provided the basis for all subsequent work on land reclamation by the Service. The Soil Conservation Act of 1965 extended the remit of the Service beyond sand reclamation to include protection of vegetation and soil conservation in general. It also provided incentives to farmers to reduce sheep numbers in accordance with the grazing capacity of the land.

Several factors emerge from this review. First, a high level of political pressure is necessary in order to bring about a policy on soil conservation. Second, that policy will not be effective unless the technology also exists for controlling erosion in the specific physical and socio-economic environment. Third, farmers are unlikely to take up soil conservation unless there is a financial incentive to do so and some demonstration exists that the conservation measures actually work. The value of the Gunnarsholt experiment in that respect should not be underestimated. Fourth, erosion control should not be seen as an end in itself but should form part of an overall protection programme for soil and vegetation resources.

Soil protection programmes exist in a number of European countries (Barth and L'Hermite, 1987) but soil erosion is specifically mentioned only in the programme for the Federal Republic of Germany (Kloster, 1987). A great deal of work needs to be done to get erosion incorporated formally into the programmes of other countries. It is encouraging, however, that, over the last decade, European governments have increasingly recognized the importance of erosion and most now make some money available for research. Even greater awareness has been shown by the European Commission, which not only provides money for research (Table 40.1) but, in 1985, set up within the Directorate General for Agriculture (DG VI) a Steering Group of European scientists to advise on the initiatives that should be taken to achieve better co-ordination of European research on soil erosion and conservation. As a result, the number of workshops on erosion organized by the CEC has increased (Table 40.2). Unfortunately, with the present reorganization of the programme committees in DGVI, the Steering Group is likely to be abandoned. The foundation of the European Society for Soil Conservation, however, provides a new body from which the scientific community can exert pressure on the policy makers. Similar roles may also be played by the recently founded European chapters of the World Association of Soil and Water Conservation and the International Erosion Control Association.

RESEARCH NEEDS

The development of a sound soil conservation policy must be based on knowledge of the areas affected by erosion, the nature of the erosion processes, the factors controlling them and the technical, social and economic suitability

Table 40.1 Research projects related to soil erosion in Europe funded by the Commission of the European Communities

DIRECTORATE GENERAL VI

Soil Conservation in hilly areas of the Apennines, Italy

Project carried out near Cesena, Italy, by the Istituto Sperimentale per lo Studio e la Difesa del Suolo.

DIRECTORATE GENERAL XI

Soil erosion risk in the southern countries of the European Community.

Project under the CORINE programme aimed at producing a map of soil erosion risk. Lead contractor is Aquater Spa, Pesaro, Italy.

DIRECTORATE GENERAL XII

Development of a soil erosion model for Europe.

Project under the framework for soil protection. Lead contractor is Silsoe College, Silsoe, UK, with funding also to the Laboratorium voor Experimentele Geomorfologie, Leuven, Belgium and the Instituto de Agroquimica y Tecnologia de Alimentos, Valencia, Spain.

Soil erosion in response to forest fire.

Joint project involving Universidade de Aveiro, Portugal, University College Swansea, CNR Istituto per la Chimica del Terreno Pisa, Italy, and Universidad de Santiago de Compostela, Spain.

Use of remotely sensed data from satellites for studies of erosion and sedimentation.

Joint project involving Ibersat, Spain, and Bureau de Recherche Géologiques et Minières, France.

Table 40.2 Workshops on soil erosion organized and funded by the Commission of European Communities

Soil erosion and conservation: assessment of the problems and the state of the art in CEC countries. Held in Florence, 19–21 October 1982. Abridged proceedings published (Prendergast, 1983).

Land degradation due to hydrological phenomena in hilly areas: impact of change of land use and management. Held in Cesena, 9–11 October 1985. Proceedings published (Chisci and Morgan, 1986).

Erosion assessment and modelling. Held in Brussels, 2–3 December 1986. Proceedings published (Morgan and Rickson, 1988).

Strategies to combat desertification. Held in Valencia, 7–9 July 1987. Proceedings in preparation (Rubio and Rickson).

Practices to control soil erosion by water and wind. Held in Freising, 24–26 May 1988. Proceedings published (Schwertmann, Auerswald and Rickson, 1989).

of various control measures in the different European environments. It is up to the scientific community to provide this base for decision making.

A recent comprehensive review of research on soil erosion and conservation carried out in the European Community (Rickson, 1988) lists work by 282 scientists in 101 institutions. This shows that most work is on the processes of water erosion and on the relationships between erosion, land use and vegetation. There is also considerable research on soil conservation, erosion survey and mapping, soil erodibility and erosion modelling. The basis therefore exists for answering questions on the extent of erosion in Europe even though the work still needs to be done. The current state of the art on erosion assessment and modelling was reviewed at the CEC workshop in Brussels, December 1986 (Morgan and Rickson, 1988). In contrast, there is virtually no work on the social and economic factors which contribute to erosion and influence the take-up of conservation measures. Hence, questions on the economic consequences of erosion and its relationship to current agricultural policies cannot be answered.

The importance of social and economic factors was emphasized at the Brussels Workshop by Napier (1988) using work in the United States as examples. Wherever possible, soil conservation needs to be based on utilizing or modifying protection measures that exist in the present agrarian system and on understanding the reasons why such measures are breaking down. Thus, terraces cannot be expected to provide a basis for soil conservation in England, where they have not been part of the agricultural system since Roman times, but they may still be acceptable in the Mediterranean lands if less labour-intensive and cheaper ways of maintaining them can be found.

Research is also needed into the political implications of soil conservation. Farmers might be expected to bear the cost of conserving the soil if erosion is reducing their incomes but, as already seen, this is rarely the case. Conservation measures are only adopted where, as with shelterbelts and in-field shelter systems for wind-erosion control, they also increase yields in the short term. Measures such as terracing which may increase yields or prevent yield decline in the long term are not attractive. There is consistently a dilemma between recognizing a long-term economic need for soil conservation to protect the soil resource and its incompatibility with the short-term need of the farmer to maximize living standards. Since soil erosion often results from a complex interplay of climatic conditions acting on a landscape where resistance to erosion has been lowered by man's activities, it may be difficult legally to support a case that the farmer is solely responsible. For this reason, the principle that the polluter pays for off-site damage cannot be easily sustained. Some form of community support for soil conservation therefore seems inevitable if the downstream consequences of erosion are to be controlled and the soil resource preserved for the future. The various options to bring this about should be investigated.

CONSERVATION MEASURES

As already indicated, soil conservation has formed a part of traditional agricultural practice in Mediterranean Europe for many centuries. Landi (1989) has reviewed the many methods based on terraces and ditches developed to control runoff and soil erosion in Italy. Virtually no research is taking place on structural methods of erosion control today. With the breakdown of these structures, greater attention is being paid to agronomic and tillage practices with the advantage that these measures may have validity throughout Europe. Particular emphasis is being given to crop-residue management and minimum tillage with work being carried out in Belgium (De Ploey, 1989), West Germany (Buchner, 1988; Goeck and Geisler, 1989) and Spain (Giraldez et al., 1989). Results at present are still very limited, but generally show that the measures are very effective in controlling erosion on-site. There is some concern, however, that they may not give adequate control of runoff unless the tractors are fitted with large terra-tyres, as in the work of Kainz (1989), to reduce compaction of the soil. Without proper control of runoff, the risk of erosion downslope is increased.

The importance of water management for control of water erosion must be stressed. On sandy and sandy loam soils, the problem is one of crusting of the surface soil which reduces infiltration and enhances runoff generation. Runoff represents a loss of water to the crop on what becomes essentially a drought-prone soil. Attempts to overcome the drought by sprinkler irrigation often serve only to increase the crusting. From fundamental studies of the effect of cropping systems on the crusting process and the hydrological behaviour of the soils (Monnier and Boiffin, 1986), tillage and cropping practices to control erosion are being proposed and tested (Papy and Boiffin, 1989). Erosion control on clay soils needs to be directed at both water erosion and mass movement. The most effective technique appears to be drainage, as proposed for the clay soils on the hills of the central Apennines in Italy (Zanchi, 1989) and in Ullensaker, Norway (Øygarden, 1988).

Wind erosion can be effectively controlled on arable land by the use of shelterbelts and strip-cropping systems. There is considerable European experience in these techniques. Shelterbelt planting was established in western Denmark in the middle of the last century as the sandy heathlands were cleared of woodland for agriculture. A replanting programme has been in progress for the last fifteen years. With about 900 km of belt being replaced annually, the programme is expected to last until the end of the century (Als, 1989). The modern belts consist of three rows of plants, each row being a mixture of fast-growing nurse trees, such as grey alder, shade-tolerant bushes and long-living tall trees such as maple, elm, rowan and oak (Olesen, 1979). These plants are chosen to give an ecological succession, with the rapidly establishing species making the belt effective after about four years and the slower-growing trees giving the belt its design life of 80–100 years. In-field shelter systems of live barley strips and planted straw strips are used in England on lowland peat soils to control wind erosion under horticulture (Morgan, 1989).

There is less work on erosion control on the upland pastures in Europe, but experience in Iceland supports the importance of an agronomic approach based on understanding the role played by livestock within the ecosystem (Jeffers, 1986; Arnalds, 1987) and adjusting stocking levels to give sustained production from the land. Its success depends, however, on being able to keep stock off the area being reclaimed. This is not always feasible because of the high cost of fencing and its incompatability with open access to the land. Also, removal of stock from an area creates pressure on grazing land elsewhere.

Much of the recent work on soil conservation techniques has taken place on research stations and experimental farms, but only the measures for wind erosion control and rangeland restoration are widely used in practice. In some cases this is because insufficient data are available as yet to know how well particular techniques work. However, in most instances, measures have not been taken up because they result in an increase in costs to the farmer without compensating benefits. As already indicated, shelter can benefit a farmer through increased yield. Also, in Denmark and Iceland, the programmes of wind erosion control are supported all or in part by the government. Until research into soil conservation links the technology of erosion control to the social and economic aspects of the agrarian system and, where necessary, modifies the technology to make it more attractive, the relatively low uptake of conservation measures in Europe is likely to continue.

CONCLUSIONS

Soil erosion is a very active field of research in Europe at present. Sufficient data exist to show that erosion is serious enough to warrant a policy for its control. This policy is best set within a framework for soil protection at a European or national government level but implemented locally. Before a policy can be implemented successfully, more research will be required on the compatability of soil conservation measures with local agrarian systems. Various options for financing erosion control need to be investigated.

REFERENCES

Als, C. (1989). How to succeed in planting 900 km of shelterbelts a year in a small country like Denmark. In Schwertmann, U., Rickson, R. J. and Auerswald, K. (eds), *Soil Erosion Protection Measures in Europe*. Soil Technology Series 1, pp. 25–27.

Alström, K. and Bergman, A. (1988). Sediment and nutrient losses by water erosion from arable land in south Sweden: a problem with nonpoint pollution? *Vatten*, **44**, 193–204.

Arnalds, A. (1987). Ecosystem disturbance in Iceland, *Arctic and Alpine Research*, **19**, 508-13.
Arnalds, O., Aradottir, A. L. and Thorsteinsson, I. (1987). The nature and restoration of denuded areas in Iceland, *Arctic and Alpine Research*, **19**, 518-25.
Barth, H. and L'Hermite, P. (eds) (1987). *Scientific Basis for Soil Protection in the European Community*, Elsevier Applied Science, London.
Becher, H. H., Schwertmann, U. and Sturmer, H. (1985). Crop yield reduction due to reduced plant available water caused by erosion. In El-Swaify, S. A., Moldenhauer, W. C. and Lo, A. (eds), *Soil Erosion and Conservation*, Soil Conservation Society of America, Ankeny, Iowa, pp. 365-73.
Boardman, J. (1988). Public policy and soil erosion in Britain. In Hooke, J. M. (ed.), *Geomorphology in Environmental Planning*, John Wiley, Chichester, pp. 33-50.
Bork, H. R. (1988). Bodenerosion und Umwelt: Verlauf, Ursachen und Folgen der mittelalterlichen und neuzeitlichen Bodenerosion. Bodenerosionsprozesse. Modelle und Simulationen, *Landschaftsgenese und Landschaftsökologie*, **13**, Technische Universität Braunschweig.
Buchner, W. (1988). Cultivation techniques for reducing soil erosion in the Rhineland. In Morgan, R. P. C. and Rickson, R. J. (eds), *Erosion Assessment and Modelling*, Commission of the European Communities Report No. EUR 10860 EN, pp. 283-97.
Chisci, G. (1982). Physical soil degradation due to hydrological phenomena in relation to change in agricultural systems in Italy. In Boels, D., Davies, D. B. and Johnston, A. E. (eds), *Soil Degradation*, Balkema, Rotterdam, pp. 95-103.
Chisci, G. (1986). Influence of change in land use and management on the acceleration of land degradation phenomena in Apennines hilly areas. In Chisci, G. and Morgan, R. P. C. (eds), *Soil erosion in the European Community: impact of changing agriculture*, Balkema, Rotterdam, pp. 3-16.
Chisci, G. and Morgan, R. P. C. (eds) (1986). *Soil Erosion in the European Community: impact of changing agriculture*, Balkema, Rotterdam.
Commission of the European Communities (1985). *Perspectives for the Common Agricultural Policy*, Green Paper of the Commission No. 33.
Cosandey, C., Billard, A. and Muxart, T. (1987). Present day evolution of gullies formed in historical times in the Montagne du Lingas, southern Cévennes, France. In Gardiner, V. (ed.), *International Geomorphology 1986 Part II*, John Wiley, Chichester, pp. 523-31.
De Ploey, J. (1986). Soil erosion and possible conservation measures in loess loamy areas. In Chisci, G. and Morgan, R. P. C. (eds), *Soil Erosion in the European Community: impact of changing agriculture*, Balkema, Rotterdam, pp. 157-63.
De Ploey, J. (1989). Erosional systems and perspectives for erosion control in European loess areas, In Schwertmann, U., Rickson, R. J. and Auerswald, K. (eds). *Soil Erosion Protection Measures in Europe*. Soil Technology Series 1, pp. 93-102.
Evans, R. (1977). Overgrazing and soil erosion on hill pastures with particular reference to the Peak District, *Journal of the British Grassland Society*, **32**, 65-76.
Fridriksson, S. (1972). Grass and grass utilization in Iceland, *Ecology*, **53**, 785-96.
Giraldez, J. V., Laguna, A. and Gonzales, P. (1989). Soil conservation under minimum tillage techniques in Mediterranean dry farming. In Schwertmann, U., Rickson, R. J. and Auerswald, K. (eds). *Soil Erosion Protection Measures in Europe*. Soil Technology Series 1, pp. 139-147.
Goeck, J. and Geisler, G. (1989). Erosion control in maize fields in Schleswig-Holstein (FRG), In Schwertmann, U., Rickson, R. J. and Auerswald, K. (eds), *Soil Erosion Protection Measures in Europe*. Soil Technology Series 1, pp. 83-92.
Hasholt, B. (1988). Methods for measuring soil erosion applied in the Danish NPO-Project, *Proceedings of the International Symposium on Water Erosion*, 19-24 September, Varna, Sofia, Bulgaria, pp. 83-8.

Jeffers, J. N. R. (1986). The role of ecosystem theory in upland land use management. In Chisci, G. and Morgan, R. P. C. (eds), *Soil Erosion in the European Community: impact of changing agriculture*, Balkema, Rotterdam, pp. 51-65.

Kainz, M. (1989). Runoff erosion and sugar beet yields in conventional and mulched cultivation. Results of the 1988 experiment. In Schwertmann, U., Rickson, R. J. and Auerswald, K. (eds), *Soil Erosion Protection Measures in Europe*. Soil Technology Series 1, pp. 103-114.

Kloster, G. (1987). Research needs in soil protection: the Federal Ministry of Research and Technology's Soil Research Programme. In Barth, H. and L'Hermite, P. (eds), *Scientific Basis for Soil Protection in the European Community*, Elsevier Applied Science, London, pp. 569-79.

Landi, R. (1989). Revision of land management systems in Italian hilly area. In Schwertmann, U., Rickson, R. J. and Auerswald, K. (eds), *Soil Erosion Protection Measures in Europe*. Soil Technology Series 1, pp. 175-188.

Leser, H. (1980). Soil erosion measurements on arable land in north-west Switzerland, *Geography in Switzerland*, Bern, pp. 9-14.

Lopez Bermudez, F. and Torcal Sainz, L. (1986). Procesos de erosion en tunel (piping) en cuencas sedimentarias de Murcia (España). Estudio preliminar mediante difraccion de rayos X y microscopio electronico de barrido, *Papeles de Geografia Fisica*, **11**, 7-20.

Mansikkaniemi, H. (1982). Soil erosion in areas of intensive cultivation in southwestern Finland, *Fennia*, **160**, 225-76.

Metailie, J. P. (1987). The degradation of the Pyrenees in the nineteenth century: an erosion crisis? In Gardiner, V. (ed.), *International Geomorphology 1986 Part II*, John Wiley, Chichester, pp. 533-44.

Monnier, G. and Boiffin, J. (1986). Effect of the agricultural use of soils on water erosion: the case of cropping systems in western Europe. In Chisci, G. and Morgan, R. P. C. (eds), *Soil Erosion in the European Community: impact of changing agriculture*, Balkema, Rotterdam, pp. 17-32.

Morgan, R. P. C. (1986). Soil degradation and soil erosion in the loamy belt of northern Europe. In Chisci, G. and Morgan, R. P. C. (eds), *Soil Erosion in the European Community: impact of changing agriculture*, Balkema, Rotterdam, pp. 165-72.

Morgan, R. P. C. (1987). Sensitivity of European soils to ultimate physical degradation. In Barth, H. and L'Hermite, P. (eds), *Scientific Basis for Soil Protection in the European Community*, Elsevier Applied Science, London, pp. 147-57.

Morgan, R. P. C. (1989). Design of in-field shelter systems for wind erosion control. In Schwertmann, U., Rickson, R. J. and Auerswald, K. (eds), *Soil Erosion Protection Measures in Europe*. Soil Technology Series 1, pp. 15-23.

Morgan, R. P. C. and Rickson, R. J. (eds) (1988). *Erosion Assessment and Modelling*, Commission of European Communities Report No. EUR 10860 EN.

Napier, T. L. (1988). Socio-economic factors influencing the adoption of soil erosion control practices in the United States. In Morgan, R. P. C. and Rickson, R. J. (eds), *Erosion Assessment and Modelling*, Commission of the European Communities Report No. EUR 10860 EN, pp. 299-327.

Newson, M. D. (1988). Upland land use and land management: policy and research aspects of the effects on water. In Hooke, J. M. (ed.), *Geomorphology in Environmental Planning*, John Wiley, Chichester, pp. 19-32.

Olesen, F. (1979). *Collective Shelterbelt Planting*, Hedeselskabet, Viborg.

Øygarden, L. (1988). Praktisk utprøving av erosjonshindrende tiltak i Ullensaker 1985-1988, *Informasjon fra Statens Fagtjeneste for Landbruket*, **13**, 55-60.

Papy, F. and Boiffin, J. (1989). The use of farming systems for the control of runoff and erosion. In Schwertmann, U., Rickson, R. J. and Auerswald, K. (eds), *Soil Erosion Protection Measures in Europe*. Soil Technology Series 1, pp. 29-38.

Prendergast, A. G. (ed) (1983). *Soil Erosion*, Commission of the European Communities Report No. EUR 8427 EN.
Richter, G. (1980). On the soil erosion problem in the temperate humid area of central Europe, *GeoJournal*, **4**, 279-87.
Rickson, R. J. (1988). Introduction to the EC Network. In Rickson, R. J. and Morgan, R. P. C. (eds) *The EC Network: a catalogue of institution research workers and research projects concerned with soil erosion and soil conservation in the European Community*, Commission of the European Communities Report No. EUR 11388 EN, pp. 3-11.
Robinson, D. and Williams, R. (1988). Making waves in Downland Britain, *The Geographical Magazine*, **60**(10), 40-45.
Rohrer, J. (1985). Quantitative Bestimmung der Bodenerosion unter Berücksichtigung des Zusammenhanges Erosion-Nährstoff-Abfluss im Oberen Langete-Einzugsgebiet, *Physiographica*, **6**.
Rubio, J. L. (1987a). Desertificacion en la Comunidad Valenciana: antecedentes historicos y situacion actual de erosion, *Revista Valenciana d'Estudis Autonomics*, **7**, 231-58.
Rubio, J. L. (1987b). La desertificacion del territorio valenciano. In *El Medio Ambiente en la Comunidad Valenciana*, Generalitat Valenciana, Valencia, pp. 188-93.
Runolffson, S. (1978). Soil conservation in Iceland. In Holdgate, M. W. and Woodman, M. J. (eds), *The Breakdown and Restoration of Ecosystems*, Plenum, New York, pp. 231-8.
Runolffson, S. (1987). Land reclamation in Iceland, *Arctic and Alpine Research*, **19**, 514-17.
Sanroque Muñoz, P. (1987). La erosion del suelo. In *El Medio Ambiente en la Comunidad Valenciana*, Generalitat Valenciana, Valencia, pp. 184-7.
Shaw, B. D. (1981). Climate, environment and history: the case of Roman North Africa. In Wigley, T. M. L., Ingram, M. J. and Farmer, G. (eds), *Climate and History: studies in past climates and their impact on man*, Cambridge University Press, Cambridge, pp. 379-403.
Skøien, S. (1988). Omfang og utbredelse av jorderosjon i Norge. Spørreundersøkelse blant landbrukskontorene, *Informasjon fra Statens Fagtjeneste for Landbruket*, **13**, 7-23.
Stocking, M. and Pain, A. (1983). Soil life and the minimum soil depth for productive yields: developing a new concept, *University of East Anglia, School of Development Studies Discussion Paper* No. 150.
Thomas, T. M. (1965). Sheet erosion induced by sheep in the Pumlumon (Plynlimon) area, mid-Wales. In *Rates of Erosion and Weathering in the British Isles*, Symposium of the British Geomorphological Research Group, pp. 11-14.
Vita Finzi, C. (1969). *The Mediterranean Valleys*, Cambridge University Press, Cambridge.
Vogt, J. (1957). La dégradation des terroirs lorrains au milieu du XVIIIe siècle, *Bulletin de la Section de Géographie, Comité des Travaux Historiques et Scientifiques (Actes du Congrès National des Sociétés Savantes, Bordeaux, 1957)*, 111-16.
Yassoglou, N. J. (1987). The production potential of soils: Part II—sensitivity of the soil systems in southern Europe to degrading influxes. In Barth, H. and L'Hermite, P. (eds), *Scientific Basis for Soil Protection in the European Community*, Elsevier Applied Science, London, pp. 87-122.
Zanchi, C. (1989). Drainage as a soil conservation and stabilizing practice on hilly slopes. In Schwertmann, U., Rickson, R. J. and Auerswald, K. (eds), *Soil Erosion Protection Measures in Europe*. Soil Technology Series 1, pp. 73-82.

41 Soil-conservation Policy and Practice for Croplands in Hungary

ÁDAM KERTÉSZ, DÉNES LÓCZY
Geographical Research Institute, Hungarian Academy of Sciences

and

ISTVÁN OLÁH
Komárom County Plant Protection and Soil Conservation Service, Hungary

INTRODUCTION

In Hungary 70 per cent of the total area is agricultural land, and the processes of water and wind erosion affect about 35 per cent of this area. Erosion hazard on hill landscapes is increased by the fact that a considerable part of the surface is mantled by loess, loess-like deposits, sands, cultivated peat and other unconsolidated sediments.

The recognition of the importance of fertile land in national wealth initially brought about a central organization of soil conservation in Hungary in 1957. Soil-conservation planning at national, regional and farm levels ensued. Legislation on land was unified in 1987 (Land Codex Act).

As illustrated in this chapter by the example of Komárom county, soil-conservation measures are closely related to soil amelioration and environmental protection projects. Complex soil conservation embraces biological (crop rotation, sodding and shelterbelts), agronomical (cultivation techniques), technical (ridges, terraces and other artefacts) and hydro-amelioration (drainage and irrigation) interventions.

The geographer's contributions to soil-conservation planning are field and laboratory experiments, mapping of erosion hazard and estimating erosion rates under various environmental conditions.

SOIL AMELIORATION AND SOIL CONSERVATION

In Hungary erosion-control measures are usually implemented in a broader framework as one of the many tasks of soil amelioration. In K. Géczy's definition (in Szabó, 1977) *soil amelioration* means any influence on soils aimed at either increasing fertility in the long term or eliminating or reducing adverse natural effects, including erosion, on cultivation. Amelioration covers all those activities (agrotechnological, biological, chemical and technical) which are necessary to preserve the present state and fertility of the soil without causing adverse off-site effects.

Soil amelioration comprises investments and services. *Complex soil amelioration investments* embrace the preservation and improvement of soil quality and fertility (Várallyay, 1986). The complex nature of the interventions is emphasized since, although single amelioration measures may be efficient, the combination of influences generally produces a better effect. The techniques employed may also serve several purposes. Soil-amelioration services, providing expertise on fertilization, soil utilization and the application of chemicals, are the day-to-day tasks of regional soil-conservation units.

In connection with amelioration, two trends may be emphasized. In Hungary, agricultural land is gradually diminishing, while, partly as a consequence, *land utilization tends to become ever more intensive*. Every user of land is compelled to cultivate it continuously and within the land-use class registered in the land inventory. In regions with better-than-average environmental potential for agriculture this does not lead to any difficulties, but farming units which cultivate marginal lands are encouraged to undertake expensive conservation measures and other investments, sometimes beyond their resources.

Demanding investments over long periods of recovery, amelioration is state subsidized as an activity in the interest of society. (During the 5th Five-Year Plan, between 1976 and 1980, of the 7 billion Ft investments, 4 billion Ft was state subsidy.) The planning of soil conservation also takes place within the framework of the five-year plans and the system of subsidization changes from one planning period to another.

The scope of soil amelioration in Hungary has undergone substantial changes over the last decades. At first, the primary emphasis lay in reducing the influence of factors restricting productivity, while today the main task is to achieve a greater security of production through minimizing the fluctuation of yields. The present concept of soil amelioration covers various interventions, from technical ones to land reclamation.

Previously, soil amelioration was focused on erosion control, drainage, irrigation, chemical soil improvement and other local tasks. Recently there is an increasing trend to *concentrate* the resources in critical areas. The importance of this *watershed amelioration* is growing, as reflected in the distribution of subsidies: in the 5th Five-Year Plan, only farm and other local amelioration projects existed, in the 6th Five-Year Plan (1981-6) 68 per cent of subsidies went to complex watershed amelioration schemes, and in the present (7th) planning period this figure may reach 90 per cent. The increase in yields after amelioration was 30 per cent on average in the 6th Plan, and the level of security was also higher (Varga, 1985).

SOIL-CONSERVATION POLICY IN HUNGARY

Soil erosion affects 2 297 000 ha of Hungary's land, 25 per cent of the country's total area (Stefanovits, 1977). Agricultural land makes up 70 per cent and about 35 per cent of it has a water or wind-erosion hazard. As a result of the post-war reorganization of agriculture, average field size has been multiplied and mechanization has also created a new situation for the preservation of soil fertility.

Its importance having been recognized, soil conservation became part of the state agricultural policy (Stefanovits, 1977; Várallyay-Dezsény, 1979). In 1957 the National Soil Conservation Council (later renamed Amelioration Council) was formed to co-ordinate conservation activities and to identify priorities in this field.

Soil-conservation planning began at three levels:

(1) Plans for the whole country or for selected regions or watersheds were made at 1:500 000 to 1:100 000 scales;
(2) Smaller regions and partial watersheds were surveyed and recommendations were made at 1:50 000 to 1:25 000 scales;
(3) For individual farms, scales of 1:25 000 to 1:10 000 were preferred.

The national authority of soil conservation is the Ministry of Agriculture and Food. Its soil-conservation tasks include the elaboration of a long-term policy for the preservation of soil quality; the identification of the technical conditions of soil conservation and its investments; the protection of land against conversion to non-agricultural uses (for instance, in town planning); and the prescription of conditions and methods for sewage, sludge and non-toxic waste disposal on certain tracts of land.

In each county a plant-protection and soil-conservation service is in charge of supervising regional activities. These ensure that conservation requirements should be observed during soil amelioration, that conservation facilities are properly maintained and plantations are optimally located.

The protection of agricultural land had been legislated for a long time by the 1961/VI Act. It was the partly bureaucratic constraints on landownership and partly wastefulness of land use in terms of the conversion of prime land for other purposes and the resulting steady reduction of cropland that necessitated comprehensive legislation on land.

The 1987/I Act (popularly called the Land Codex) includes provisions on the soil-conserving cultivation of land, according to the physical endowments and actual land use. Cultivated crops have to reduce soil loss and, if necessary, special measures also have to be applied. Cultivation must promote the balance of nutrients and preserve soil fertility. For any construction requiring land, tracts of poorer-quality land must be used. In general, land is now gradually acquiring a commodity character in Hungary, and this process is assisted by the new land-evaluation system, decreed in 1979 and now being introduced. With the exception of soil-conservation and irrigation projects, any other-purpose construction can only acquire land under permission and after paying a set contribution.

The paragraphs on landownership also have direct implications for soil conservation. In order to establish a more rational cultivation, farming units (state and co-operative farms) are free to exchange tracts of land between each other and to form fields better adjusted to the physical conditions. If no agreement is reached between the parties involved, a farm is allowed to apply for action at the land inventory office.

SOIL-CONSERVATION TASKS IN HUNGARY

Water Erosion

Its relief and drainage conditions make Hungary rather severely affected by water-erosion processes. In the mountain and hill regions surplus runoff, loss of soil, nutrients and fertilizers and the accumulation of washed-down material present problems over an area of 1 300 000 ha.

Situated in the middle of the Carpathian basin, the territory of Hungary receives water from the encircling mountain ranges. Parallel with flood waves on rivers, groundwater levels rise in the lowlands, causing damage by excess water over 330 000 ha of agricultural land. About 100 million tonnes of soil and 1 500 000 tonnes of humus are removed every year in runoff. (Stefanovits, 1977).

The above data, however, are national averages, and significant local differences are observed. Considering just some major factors controlling water erosion, a map of areas with potential water erosion hazard in Hungary has been constructed (Figure 41.1). The following indicators and factors were included:

(1) Since erosion is most effective on agricultural hill-slopes, heavily dissected hill regions built up of unconsolidated deposits and major elevated loess surfaces were delimited.
(2) As the amount of precipitation is a measure of erosion hazard, the areas with more than 600 mm annual precipitation were also identified.

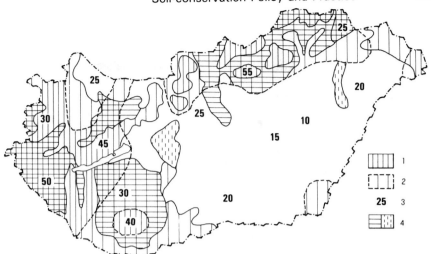

Figure 41.1 Areas with potential hazard of water erosion in Hungary identified by climatic criteria. 1, areas with annual average precipitation above 600 mm; 2, areas with more than 40 days with thunderstorm during the summer half-year; 3, annual average of maximum depth of snow cover (cm); 4, dissected hill regions of loess deposits or loess plateaus

(3) To portray rainfall intensity was rather difficult. In Hungary there are no marked differences in the regional distribution of maximum daily rainfall: values about 100 mm are possible anywhere in the country. Therefore the map shows those areas with more than 40 thunderstorm days during the summer half-year (Péczely, 1979).

(4) Field measurements (presented below) also underline the role of snowmelt in generating runoff and soil loss. Therefore, figures of annually recurring maximum snow depth are also included on the map.

The above indicators serve to give an overview of the areas which are potentially affected by water erosion. It is characteristic of Hungary that almost two-thirds of the area are covered by easily erodible rocks and erosion hazard is aggravated by climatic conditions over most of these surfaces. Figure 41.1 may be compared with Figure 41.2, where the areas actually affected by water erosion of various degree are represented. Forests are indicated as surfaces not prone to sheet wash.

Erosion control is a central task of farming on hillslopes. The techniques applied are usually grouped as either agrotechnological, biological or technical.

The most common practices of mechanical soil conservation are ridging (ridges are obliterated when ploughing on less than 12 per cent slopes and maintained on 12-17 per cent slopes) and terracing of 17-25 per cent slopes. Both types of erosion control are usually supplemented by grass waterways. Where runoff surges are particularly destructive, waterways are built of more durable materials.

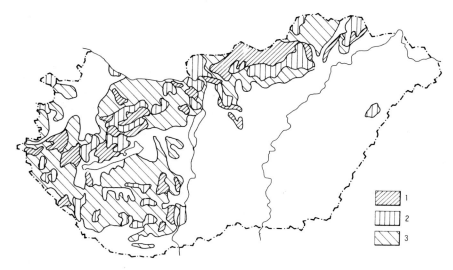

Figure 41.2 Areas with erosion by water in Hungary (after Stefanovits, 1977). 1, areas affected by strong erosion; 2, areas affected by slight to medium erosion; 3, major forests

Crop-rotation planning and agronomic techniques (e.g. subsoiling and ribbed rolling) are employed to promote the infiltration of rainwater into the soil.

Wind Erosion

Soil erosion by wind affects 16 per cent of Hungary's surface. Damage is primarily caused on sand soils, where crop yields may be reduced by up to 50 per cent. Improperly cultivated peat soils with decomposed, powdery surfaces also have low resistance to wind erosion.

To identify areas with potential wind erosion hazard, a map of critical climatic thresholds has been constructed. Over a map of sand and peat soils (Figure 41.3), the following features were superimposed:

(1) The 350 mm isohyet for the summer half-year;
(2) The boundary of the area with more than 75 'summer days' (equal to a daily maximum temperature above 25°C) to show summer drought;
(3) Annual average wind velocity (at 14.00 h) with the direction of the prevailing wind (more frequent than 25 per cent) for local meteorological stations.

The map of deflation areas (Stefanovits, 1977; Figure 41.4) confirms that wind erosion is the most active where the listed conditions are fulfilled. There is a strong seasonality in deflation with peaks in early spring and in summer. Improper farming practices may lead to a powdering of the soil surface or compaction, and ultimately to deflation.

Soil-conservation Policy and Practice   611

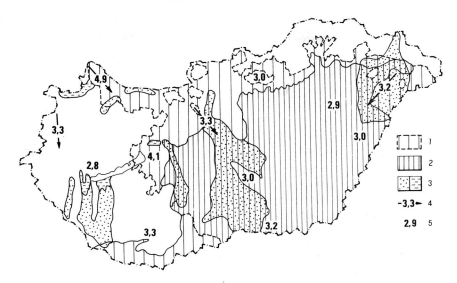

Figure 41.3 Areas with potential wind erosion hazard in Hungary identified by climatic criteria. 1, Areas with precipitation above 350 mm during the growing season of row crops (April to September); 2, Areas with more than 75 'summer days' (peak temperature above 25°C) in a year; 3, sand or cultivated peat soils; 4, annual average wind velocities (m s^{-1}) and direction of prevailing wind (if its frequency exceeds 25 per cent); 5, extremely high wind velocities

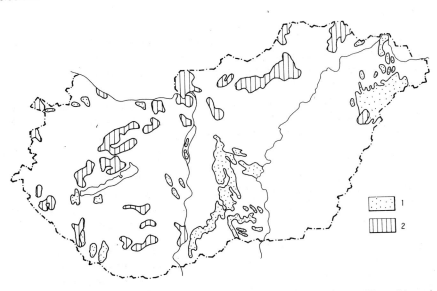

Figure 41.4 Deflated areas in Hungary (after Kertész, 1976). 1, Areas affected by wind erosion and accumulation; 2, major forests

In Hungary the major factor of wind erosion is the low cohesion of a dry soil surface. The obvious preventive measure is to ensure a proper *vegetation cover*, which reduces turbulent air motion on the surface. Rye sowing, mulching or green manuring are most often applied (Stefanovits, 1977). Inorganic materials are also suitable for sealing the soil surface (e.g. clay, bentonite injection, resins or plastic foils; see Szabó, 1977).

In order to allow mechanization after the collectivization of Hungarian agriculture, large arable fields were formed. Today where wind velocities are high and droughts are frequent, small (maximum 25 ha) plots separated by shelterbelts are recommended. The shelterbelts of trees of rapid growth (e.g. poplars and acacia) are preferred. The optimum effect, in accordance with Bagnold's formula (Kirkby and Morgan, 1980) is observed at a distance interval between three and ten times that of tree height (Tompa, 1988). Shelterbelts were introduced in Hungary following Soviet examples. They are now considered necessary on those soils with poor water retention and which are liable to drought.

THE ROLE OF PHYSICAL GEOGRAPHY IN SOIL-CONSERVATION PLANNING

Over the last decades applied geography has gained greater importance (cf. Pécsi, 1975). The rise in *experimental geomorphology* (Kertész, 1984) strengthened the links between physical geography and soil conservation. Among other targets, geomorphological experiments are directed to study hillslope evolution on agricultural surfaces, focusing on the processes of soil erosion (Pinczés, 1982; Kerényi, 1985). The investigations in test areas are always accompanied by recommendations for soil conservation.

In this chapter some soil-conservation problems of Komárom county are cited as an example and, for this reason, a study of physical geographical processes is also taken from there.

In the outskirts of the village Pilismarót a team of the Geographical Research Institute mapped soil properties. The degree of soil erosion was also revealed and conservation proposals were supplied. The severe erosion observed there motivated the initiation of continuous erosion measurements at this site.

The main objective of the measurements was to monitor soil-erosion processes to arrive at an environmentally favourable land use. The utilization of rain and meltwater, and of fertilizers was studied.

One large and five small plots were instrumented. For the large plot, runoff and sediment yield were measured and eroded soil aggregates were collected and classified by sieving. The plots were located on moderately eroded lessivé brown forest soils, at a transition to its heavily eroded variety on barren ground. Forest soil slope deposits accumulated at the footslope. During the measurement

Table 41.1 Runoff and sediment measurements, Pilismarót test area, 1982–5

Plot	23.6.1982	23.7.1982	28.5.1984		2.6.1984		9.6.1984		7–11.8.1984		9.1984		10.1984	20.5.1985	
	r	r	r	s	r	s	r	s	r	s	r	s	r	r	s
L	9.1	4.5	2.8	650	3.3	1050	6.8	2180	7.5		9.7		4.5	11.3	311
L_1				166		9		7800				47			597
L_2				208		56		7100		497		46			1232
L_3				283		78		1760		194					48
1						6									804
2			0.6	8.2	0.4	2									
3			0.3	31	0.8										274
4															530
5															

r = surface runoff (l).
s = sediment load (g).
L_1 = upper sieve.
L_2 = middle sieve.
L_3 = lower sieve.
L = large plot (688.5 m^2).
1–5 = small plots (1 = 40 m^2; 2 = 9 m^2; 3 = 4.5 m^2; 4 = 6 m^2; 5 = 19.5 m^2).
The amount of sediment (s) given in the line of L means sediment deposited in the sludge-tank.

Table 41.2 Results of rainfall simulation experiments; runoff in mm h^{-1} (after Góczán et al., 1982)

Slope (%)	Rainfall intensity (mm h^{-1})				
	5	10	20	30	40
5	0.1	0.3	1.2	2.6	4.5
12	0.1	0.9	6.4	11.1	14.0
17	0.3	2.2	10.6	17.1	22.2
25	1.3	5.4	17.6	22.8	32.7
30	1.3	5.6	17.6	24.6	35.5

period (1982–5) the instruments operated from spring to autumn without disturbing farm practices. (Table 41.1).

The measured runoff figures were compared to rainfall simulation data (Table 41.2). The latter produced about a tenfold runoff, explained by different soil moisture conditions and soil types, but principally by effective vegetation cover (wheat).

When evaluating small plot measurements, the importance of the factors LS and K was confirmed. The variable amounts of sediment retained do not show any clear trend.

Laboratory analyses of removed sediment indicate a high proportion of Ca ions. Differences in fertilization can also be detected in the runoff samples. The particle size composition of the sediment is correlated with soil and parent rock texture.

Although the most destructive rainfalls are of short duration and high intensity, the measurements also point to the role of prolonged, low-intensity rainfalls. The need for soil conservation is evident from the results.

The soil-conservation plan identifies and locates the rotation of crops providing good surface coverage to reduce overland flow. In order to increase infiltration, subsoil shattering and the addition of organic matter (peat-earth) are recommended.

SOIL CONSERVATION IN KOMÁROM COUNTY

In spite of its small area (2249 km^2), this administrative unit in North Transdanubia has a varied environment (Figure 41.5). As erosion hazard extends over 76 per cent of the county's surface, the area is suitable for demonstrating soil-conservation practices in Hungary.

In the terraced plain of the Danube, areas of blown-sand of former dunes, the cover sand of gravel terraces and the exposed sandy loess parent material of severely eroded chernozems (altogether, 38 per cent of the area) are all vulnerable to wind erosion. Water erosion is considerable on hillslopes with

Soil-conservation Policy and Practice 615

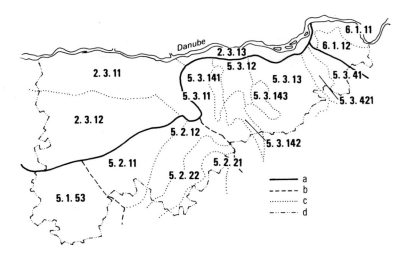

Figure 41.5 Physical geographical divisions for the area of Komárom county (Pécsi-Somogyi, 1980). (a) Boundary of macroregions; (b) boundary of mesoregions; (c) boundary of microregions. 2.3, Komárom-Esztergom plain; 2.3.11, Györ-Tata terraced plain; 2.3.12, Igmánd-Kisbér basin; 2.3.13, Danube valley; 5.1.53, Bakony foothills; 5.2, Vértes Mountains; 5.2.11, Bársonyos hills; 5.2.12, Általér valley; 5.2.21, Vértes plateau; 5.2.22, Vértes slopes; 5.3, Danube Bend Mountains; 5.3.11, W-Gerecse Mountains; 5.3.12, Central Gerecse Mountains; 5.3.13, E-Gerecse Mountains; 5.3.141, Tardos basin; 5.3.142, Héreg-Tarján basin; 5.3.143, Bajna basin; 5.3.41, Pilis horst; 5.4.421, Dorog basin; 6.1.11, Danube Bend; 6.1.12 Visegrád Mountains

Table 41.3 Land-use classes in Komárom county, 1983

Land-use class	Area (ha)	(%)	Percentage of Hungary's area
Arable	101 887	45.27	2.18
Meadow	3 650	1.62	1.16
Pasture	17 264	7.67	1.79
Vineyard	3 902	1.73	2.49
Garden	9 171	4.08	2.71
Orchard	1 280	0.57	1.11
Agricultural land	137 154	60.94	2.09
Reedbeds	253	0.11	0.64
Forest	58 847	26.15	3.60
Fish-ponds	639	0.29	2.46
Cultivated land	196 893	87.49	2.38
Uncultivated	28 159	12.51	2.72
Total	225 052	100.00	2.42

Table 41.4 Land converted into non-agricultural use (1967–80)

Years	Area (ha)
1967–71	599.0
1971–5	396.0
1975–9	1603.0[a]
1979–80	249.0
Total	2847.0

[a] Motorway construction.

brown forest soils (48 per cent of area). Large tracts of basins and valleys are prone to waterlogging.

Regarding land-use classes (Table 41.3), forests have a higher-than-average share in national comparison. However, derelict lands are also considerable (12.51 per cent), explained by the mining and industrial nature of the economy and by the extensive built-up areas. When converting agricultural land for non-agricultural uses, the principles contained in legislation, as far as the preservation of prime land is concerned, have often been neglected (Table 41.4). The Slovak-Hungarian Barrage Scheme, led to the loss of 1637 ha of cultivated land; in total there has been the dereliction of 5896 ha (4.49 per cent).

At the same time it is a positive development that uncultivated agricultural land was reduced to one-tenth over a decade.

Soil-conservation interventions in Komárom county began almost parallel with the organization of collective farms in the 1950s, but routine conservation planning dates back to the 1960s. Soil genetic maps at 1:10 000–1:25 000 scales are now available for 95 per cent of the land cultivated by large farms.

The most successful state and co-operation farms of Hungary are found in Komárom county. Rapidly spreading industrial production systems (primarily of maize) require soil-amelioration measures. Most of the farms can afford expensive projects.

In order to ensure harmony between the mining, industrial, residential, agricultural and recreational functions of the country, a long-term environmental protection and nature conservation plan was prepared in 1977.

Even before then (in 1974) the watershed of the Általér river was declared a COMECON Model Area of Environmental Protection (Figure 41.6). The selection of the area was motivated by the considerations that population density is the second highest here (250 people per square kilometre) and industrial incomes are at an outstanding level. The environment is characterized by intensive agriculture as well as by brown coal mining, quarrying, electricity generation and the aluminium industry (Molnár and Tevan-Bartalis, 1981). Industrial and residential quarters make up 20 per cent of the 521 km² model

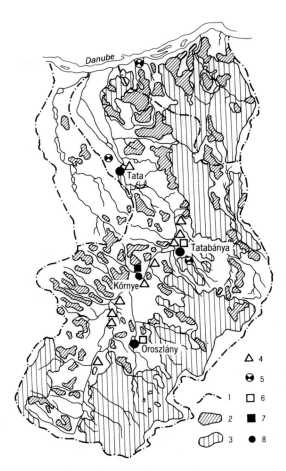

Figure 41.6 The COMECON Environmental Protection Model Area of the Általér catchment, Komárom county, Hungary (after Tata Water Management Association, 1980). 1, Boundary of catchment; 2, agricultural land with erosion hazard; 3, forest; 4, hydrological observations; 5, regional drinking-water supply system; 6, industrial source of pollution; 7, agricultural source of pollution; 8, settlement

area and this also increases runoff. Thirty-two per cent of the area is forested, partly on reclaimed land.

OFF-SITE EFFECTS OF LAKE SILTATION AND EUTROPHICATION

Accelerated soil erosion has become a major hazard recently; between 1970 and 1980, 1 million m^3 of silt accumulated in the Tata Great Lake which, along

Table 41.5 The impact of complex soil amelioration on yields in a co-operative farm, Komárom county

Main crops	Total area (ha)	6333	2214	1230
	Ameliorated area (ha)	1858	738	410
	Year of implementation	1978/80	1978	1979
	Year of evaluation	1983	1983	1983
		Specific surplus yield in percentage of the figure before amelioration		
Winter wheat		15	19	24
Maize		22	16	34
Oil flax		43	4	92
Lucerne		9	9	—

with the washed-in nutrients, increased the rate of eutrophication. Fisheries were stopped and now the lake only serves recreational purposes. Consequently, the need for soil conservation emerged on the mostly agricultural western half of the catchment. The activities completed were the rearrangement of plot boundaries, the implementation of drainage measures (channelization of watercourses and building belt ditches and pipe systems) and mechanical soil-conservation measures (terracing). In spite of the decreasing amount of fertilizers used, crop yields have risen considerably following the three stages of amelioration (Table 41.5).

REFERENCES

Chisci, G. and Zanchi, C. (1981). The influence of different tillage systems and different crops on soil loss on hilly-clayey soils, In Morgan, R. P. C. (ed.), *Soil Conservation*, John Wiley, Chichester, pp. 211–17.

Góczán, L. and Kertész, Á. (1988). Some results of soil erosion monitoring at a large-scale farming experimental station in Hungary, *Catena Suppl.*, **12**, pp. 175–84.

Götz, G. and Pápai-Szalay, G. (1966). Zivatartevékenység a nyári félévben Magyarországon (Thunderstorm activity during the summer half-year over Hungary), *Időjárás*, **70**, 106–16.

Hudson, N. (1971). *Soil Conservation*, Cornell University Press, Ithaca, New York.

Kerényi, A. (1985). Surface evolution and soil erosion as reflected by measured data, In *Environmental and Dynamic Geomorphology*, Akadémiai Kiadó, Budapest, Studies in geography in Hungary 17, pp. 79–84.

Kertész, Á. (1976). Magyarország deflációveszélyes és potenciálisan deflációveszélyes területei (Areas with potential wind erosion hazard in Hungary (with English summary)), *Földrajzi Értesítő*, **25**, 101–13.

Kertész, Á. (1985). Subject and methodology of experimental geomorphology, In Pécsi, M. (ed.), *Environmental and Dynamic Geomorphology*, Akadémiai Kiadó, Budapest, 21–9.

Kirkby, M. J. and Morgan, R. P. C. (eds) (1980). *Soil Erosion*, John Wiley, Chichester.

Molnár, K. and Tevan-Bartalis, É. (1981). A vizminöség-védelem területi teendöi a Tatai Környezetvédelmi Modellterületen (Water conservation tasks in the Tata Model Area of Environmental Protection: with English summary), *Földrajzi Közlemények*, **29**, 205-17.

Pécsi, M. (1975). *Geomorphology* Geological Institute, Budapest.

Pécsi, M. and Somogyi, S. (1980). *Magyarország tájbeosztás-térképe (Physical geographical divisions of Hungary)*, 1:500 000, Geogr. Res. Inst. Hung. Acad. Sci., Budapest

Péczely, G. Y. (1979). *Éghajlattan (Climatology)*, Tankönyvkiadó, Budapest.

Pinczés, Z. (1982). Variations in runoff and erosion under various methods of protection. In *Recent Developments in the Explanation and Prediction of Erosion and Sediment Yield. Proceedings of the Exeter symposium*, July 1982. IAHS Publications 137, 49-57.

Stefanovits, P. (1977). *Talajvédelem—környezetvédelem (Soil conservation—environmental protection)*, Mezögazdasági Kiadó, Budapest.

Szabó, J. (ed.) (1977). *A melioráció kézikönyve (Soil amelioration handbook)*, Mezögazdasági Kiadó, Budapest.

Tompa, K. (1988). *Erdészeti melioráció—a komplex talajvédelem része (Forest amelioration—a task of complex soil conservation)*, Természet Világa, Budapest, 119, 482-7.

Várallyay, G. (1986). *Soil Conservation Researches in Hungary*, Round Table Meeting on Soil Conservation Technologies 16-20, VI, 1986, USDA SCS- MÉM NAK, Budapest, 1986, 5-8.

Várallyay, G. and Dezsény, Z. (1979). Hydrophysical studies for the characterization and prognosis of soil erosion processes in Hungary. In *The Hydrology of Areas of Low Precipitation*, Proc. of the Canberra Symp., December 1979, IASH-AISH Publ. 128, 471-7.

Varga, J. (1985). *A termöföld minöségvédelmének helyzete és a VII. ötéves tervi föbb feladatok (Soil conservation and the tasks in the 7th Five-Year Plan)*, Melioráció, Öntözés és Tápanyaggazdálkodás, 1, pp. 3-11.

42 Identification and Reclamation of Erosion-affected Lands in the Emilia–Romagna Region, Italy

F. GUERRIERI and G. VIANELLO
University of Bologna, Italy

INTRODUCTION

The concept that land is a limited, non-renewable resource is a relatively recent one. Italy needs to assess the extent of degradation of its land and to identify the most appropriate conservation measures.

The present investigation, carried out within the larger reclamation project of the Emilia–Romagna region, is a feasibility study to outline a soil-conservation programme in central Italy. Although the degradation of soils is a common feature on both flat and hilly areas, only sloping lands are considered in this chapter and the evaluation of erosion-induced soil loss has been given particular consideration. Though not generally considered in such terms, soil is indeed an essential support to human life, not only in relation to food supply but also as a part of the whole ecosystem. Unlike air and water, for which anti-pollution provisions are being made in Italy, the use of land is not yet subject to any agreed standard. Soil loss is a problem of great importance, especially when soil resources remain unchanged while the demand for land increases at a very rapid rate through population growth, technological progress and industrial development (see Figure 42.1).

LAND-DEGRADATION APPRAISAL IN EMILIA–ROMAGNA

Policy Guidelines

The factors which promote land degradation are generally well known but the assessment of the problem involves both national interests and governmental

Table 42.1 Reclamation Project for Hilly Areas in Emilia-Romagna

Land-utilization type	Physical characteristics of soils					Project action		Data distribution per contour bands according to the reclamation project				
	Slope		Soil erodibility		Slope stability		Cultivation operationally sound	Type of action	Labour power (workers required)				Total
	<20%	>20%	High	Low	High	Low			100/400 m asl	400/800 m asl	800/1200 m asl	>1200 m asl	
Orchard	a	a	a	a	a	a	a		766	11	—	—	777
Crop-land with trees							Areas above 400 m asl	Conversion to orchards (altitude <400 m)					
	a	a	a		a				2 438	603	17	11	3 069
Arable land and arable land with trees	a			a	a		a		2 853	5 275	2 137	88	10 353
		a		a	a			Conversion to grassland	2 261	2 550	470	30	5 311
	a			a		a		Drainage improvement and conversion to grass-land	78	32	10	—	120
		a		a		a		Hydro-geological reclamation and coppice plantation	20	22	4	—	—
	a	a	a			a		Afforestation	14	37	5	—	50
Pasture, grassland	a	a	a	a	a	a	Areas below 1200 m asl	Conversion to forest (areas above 1200 m asl)	1 059	455	536	232	2 282
Chestnut-tree forest	a	a	a	a	a	a		Forest revitalization through coppicing	—	84	63	—	147
Coppice	a	a	a	a	a	a	Areas below 1200 m asl	Conversion to forest (areas above 1200 m asl)	100	608	393	164	1 265
Forest	a	a	a	a	a	a	a		3	20	34	20	77
Uncultivated areas	a	a	a	a	a			Partial reafforestation of riverbanks	—	—	—	—	—
Urban areas barren land									—	—	—	—	—
								Totals	9 592	9 697	3 669	545	23 451

[a] Potentially viable land use given morphological class

Table 42.1 (continued)

Data distribution per contour bands according to the reclamation project										Costs of implementation (millions of lira/ha)
Cattle capacity (number of animals)					Forest and agricultural areas included in project (ha)					
100/400 m asl	400/800 m asl	800/1200 m asl	>1200 m asl	Total	100/400 m asl	400/800 m asl	800/1200 m asl	>1200 m asl	Total	
—	—	—	—	—	8 000	100	—	—	8 100	—
—	13 300	500	200	14 000	38 300	9 700	300	200	48 000	—
88 000	132 600	49 900	1 850	272 450	75 900	96 500	28 100	1 700	202 200	—
86 700	74 200	11 300	650	172 860	55 900	42 600	6 400	600	105 500	0.7
750	700	100	—	1 750	2 700	2 700	400	—	5 800	1.5
—	—	—	—	—	8 200	4 800	900	—	13 900	2.5
—	—	—	—	—	5 400	7 500	1 000	—	13 900	2.2
11 750	15 200	15 200	3 800	44 950	29 100	31 500	25 300	16 300	102 200	2.2 (conversion)
—	—	—	—	—	500	16 800	12 700	—	30 000	0.7
—	—	—	—	—	35 600	127 500	78 700	32 900	274 700	0.8 (conversion)
—	—	—	—	—	1 400	4 200	6 700	4 100	16 400	—
—	—	—	—	—	45 300	23 300	5 300	800	74 700	2.5
—	—	—	—	—	54 600	61 300	21 500	5 600	143 000	—
187 200	236 000	77 000	6 506	506 700	360 900	428 500	187 300	62 200	1 038 400	

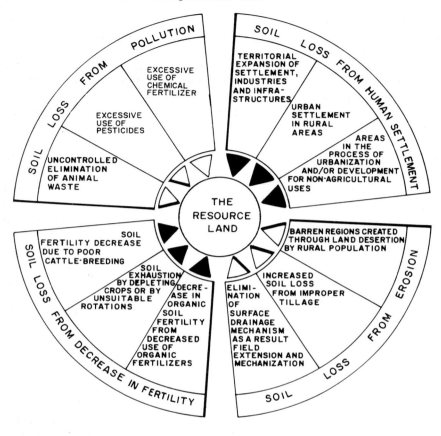

Figure 42.1 Causes resulting in the loss of land as a resource

institutions. Identification of the causes, extent and intensity of the phenomenon should be considered as a basis for the establishment of standards, control and preventive measures, and for specific problem-oriented research programmes.

When the need for conservation, although obvious, is not recognized, local authorities should be forced to provide an estimate of the consequences of inaction.

Land Degradation due to Erosion

This chapter analyses the importance of erosion in the process of land degradation. Land use and population density, as well as geomorphological and pedological data, have been used to determine the erosion hazard of the area and the most suitable conservation measures. The poster presented at the

Workshop on Soil Conservation, Policies and Practices (Coventry, UK) presented the utilized data in the form of maps; this chapter, on the contrary, only reports the final table.

The project for the management of hilly areas has been designed, the proposals of which are shown in Table 42.1. The projected action may lead to many complications and liabilities, including legal action for compensation at national and international level. Project effectiveness obviously relates strongly to the governmental policy.

Project suggestions also take into account financial aspects. An economic evaluation of the cost of implementation of proposals, once operative, will be the next step.

Each land-utilization type, in a given morphological class, has been analysed in relation to its effect on soil erosion. Although erosion criteria seem to be rather subjective, the costs of implementation appear to be economically valid in the long run. It must be realized that there are currently no definite economic standards at the national level.

Operationally sound cultivations will be retained while types of land utilization involving a high-degree of soil erosion will have to be converted to more suitable uses. The required manpower and the cost of implementation are reported, as well as the cattle capacity and the number of hectares covered by the project. All data are given per contour bands.

CONCLUSION

Generally, the cost of implementing land-degradation control is the primary inhibitor to the application of conservation recommendations. Several project proposals, although scientifically and technically sound, are hardly applicable; this is due mainly to the lack of financial incentive involved. The reclamation project presented in this chapter should be supported by an educational programme, as any regulation taken without public acceptance will fail.

43 The Evolution of US Soil-conservation Policy: From Voluntary Adoption to Coercion

TED L. NAPIER
Ohio State University, USA

Soil erosion has been perceived to be a problem in the United States for at least six decades (Halcrow *et al.*, 1982; Lovejoy and Napier, 1986; Napier, 1987a: Napier *et al.*, 1988). Severe wind erosion during the 1930s created an environmental awareness among members of society which generated considerable political support for government action to reduce erosion damage (Camboni and Napier, 1986; Camboni *et al.*, 1989; Napier, 1988a,b; Rasmussen, 1982).

All segments of society were adversely affected by soil erosion during the Dust Bowl years. Farmers lost future productivity of land resources, national supplies of food and fibre were threatened, deposition of soil particles in urban and rural areas created problems for non-farm populations and loss of farms exacerbated the economic effects of the depression.

Environmental concern was translated into social action and extensive economic and human resources were mobilized to address erosion problems (Napier, 1988b; Rasmussen, 1982). Soil erosion control agencies were created, research was funded to develop technological methods of reducing erosion, programmes were implemented to educate land operators about appropriate tillage systems, economic resources were appropriated to assist landowners in financing adoption of soil-conservation practices, and national policies were enacted to direct the efforts of government conservation programmes (Napier, 1987a).

Humid weather conditions combined with successful implementation of soil-conservation programmes significantly reduced the incidence of dust storms in a relatively short period of time. One of the most important reasons for the initial success of soil-conservation efforts during the Dust Bowl era was willingness on the part of land operators to adopt soil erosion control practices. Farmers were motivated to adopt soil-conservation practices because they were aware of the degradation of soil resources resulting from the use of inappropriate tillage systems. Erosion control practices were demonstrated to be very useful for reducing soil loss and landowners recognized that they would derive substantial benefits from adoption.

Another factor which facilitated rapid adoption of soil-conservation practices during this time period was the introduction of subsidies by the federal government. Cost sharing made it possible for many land operators to implement soil-conservation programmes which would not have been economically feasible to finance from private sources (Napier, 1987a).

The initial successes of the education-subsidy approach convinced many people within the newly formed soil-conservation agencies that these implementation methods should be institutionalized. Subsequently, policies and operating procedures were put into place which were designed to perpetuate the education-subsidy approach.

INSTITUTIONALIZATION OF SOIL-CONSERVATION PROGRAMMES AND POLICIES

While the institutionalization of conservation programmes and policies introduced during the Dust Bowl period legitimized long-term commitments to soil conservation, it also reduced priority given to soil-conservation problems. Many segments of society perceived that soil erosion was being adequately supported. Public concern was redirected towards other important issues, such as industrial growth, poverty, health and safety issues, urban growth and development, unemployment, deviance and a host of other social problems which affect developed societies. Soil erosion became one of many claimants on limited human and economic resources. While soil erosion was perceived by many members of society to be important, other issues were also defined as being worth financial support.

Soil conservation programmes in the United States changed little in terms of philosophy and implementation procedures for many years following institutionalization in the mid 1930s. Primary emphasis was placed on voluntary participation with no institutional means of encouraging landowners to continue conservation practices once highly erodible land was removed from diversion programmes.

Farmers frequently entered land in government set-aside programmes and complied with contractual obligations for the duration of the agreements, which were often for one year. Once the contracts were completed, landowners could reintroduce erosive tillage systems without penalty. The failure of federal and state conservation agencies to force land operators to continue use of erosion-control practices on previously enrolled land resulted in the loss of extensive government conservation investments.

Another factor which impeded the success of soil-conservation efforts in the United States was the attempt to achieve multiple objectives with soil-conservation programmes. Historically, soil-conservation efforts have flourished when it has been politically expedient to implement programmes to control domestic food and fibre production. Agricultural land has been periodically diverted from production to reduce supplies of agricultural products for the purpose of stabilizing commodity prices. The conservation of soil resources and the reduction of off-site environmental damages have traditionally been secondary goals to production control. The Soil Bank programme which was authorized in 1956 is a classic example.

While there were many environmental benefits derived from the Soil Bank, the political motivation for implementation was control of food and fibre production and *not* environmental concern. Land operators were primarily motivated to enroll land in the Soil Bank by their desire to secure rents from the government. Conservation of soil resources was a secondary goal which did not cost participants anything to achieve. It is interesting to note, however, that the Soil Bank programme was widely supported among the electorate primarily on the basis of its conservation impacts.

Farmers were permitted to select land to be included in the Soil Bank. This practice reduced impacts on production control because land operators frequently enrolled marginal land which was often the least productive. From an environmental perspective, however, this outcome was very desirable because the marginal land enrolled in the programme was often highly erodible.

An important feature of the Soil Bank programme was the duration of the rental contracts. Land could be enrolled in the Soil Bank for a maximum of ten years (Rasmussen, 1982) but could be entered for a minimum of three (Harmon, 1987). Many of the previous programmes diverted land from production for much shorter periods of time, such as a single crop year.

A major weakness of the Soil Bank was that no provision was made in the legislation to keep highly erodible land out of production when participation in the programme was terminated. Subsequently, many of the environmental benefits gained from retiring the land were lost. The Soil Bank programme was terminated in 1958 but land remained enrolled until the late 1960s.

The experience with the Soil Bank programme is not unique in the United States. The history of soil conservation in the United States is replete with examples of environmental gains being rapidly negated by farmers who engage in behaviour designed to maximize short-run profits. Farmers have repeatedly received direct government rents to divert land and indirect benefits from the stabilization of product prices without assuming any responsibility for maintenance of conservation practices once the diversion programmes are terminated.

The inadequacies of the existing conservation system were widely recognized but seldom challenged until the 1970s. A number of stimuli for change were in operation during the mid-1970s, but one of the most important forces for change was agitation by environmental scientists. A number of academicians began to question the return to investment in soil-conservation programmes. It was observed that decades of such programmes had eliminated massive dust storms but had not reduced soil erosion to acceptable levels (Halcrow *et al.*, 1982). Soil-conservation policies and implementation strategies began to be more closely scrutinized.

THE EMERGENCE OF NEW CONSERVATION PROGRAMMES AND POLICIES

Targeting

One of the first issues to be evaluated by critics of the traditional soil-conservation system was *targeting* of limited economic and human resources on highly erodible land. Since the inception of soil-conservation programmes in the United States, economic and human resources have been made available to any landowner who sought assistance within budget constraints (Kleckner, 1988). Since soil-conservation resources have traditionally been small and priority placed on production control, a significant proportion of highly erodible land which should have been included in conservation efforts was not enrolled. Policies which placed the highest emphasis on production control encouraged the diversion of productive land where possible. Frequently, the most highly productive land was not the most erodible.

Natural resource professionals external to the existing conservation system suggested policy changes which placed greater emphasis on reduction of soil erosion. These efforts were rewarded by the introduction of 'targeting' by the Soil Conservation Service (SCS) in the early 1980s. State SCS staff identified counties with severe erosion problems and allocated economic and human resources to reduce soil degradation. The concept was later applied to micro-targeted areas which were smaller geographic units than counties. The rationale for targeting was to allocate limited resources to areas which had the greatest need for erosion control.

Resistance to the concept of targeting was initially high because many landowners who were not in the designated areas argued that they should have equal access to conservation resources. The conflict was resolved by allocating 'new resources' to the targeted areas and not reallocating existing ones. Over time, however, reallocation of resources did occur.

Targeting was demonstrated to be a more efficient method of reducing soil loss with limited economic resources than previous approaches. The success of targeting generated enthusiasm for additional changes in the conservation system. The next issues examined were on-site and off-site damages.

On-site Versus Off-site Costs

Strategies developed by conservation agencies to convince Dust Bowl farmers to adopt soil-conservation practices were used for decades without examining their relevance to contemporary society. Soil conservation was repeatedly 'sold' on the basis of on-site damage. Education programmes were developed to inform land operators that they should adopt soil-conservation practices to protect long-term productivity of land resources. It was argued that land operators would lose future productivity if they did not conserve soil resources.

The response of farmers to the education approach was very positive during the Dust Bowl era because soil resources were being rapidly degraded. Landowners could see the loss of topsoil due to wind erosion and were aware that productivity was threatened. They were highly motivated to change farming practices to save soil resources. Recommended soil-conservation practices were quickly adopted at the farm level because land operators realized that corrective action was necessary to save future productive capacities.

The early successes using the education approach resulted in the institutionalization of this method of bringing about adoption of soil-conservation efforts. 'Historical inertia' helped to perpetuate the approach and little thought was given to the appropriateness of the strategy for modern agricultural situations.

Observation of land operator behaviours over the years, however, revealed that educational approaches tended to produce only marginal success. Most landowners did not adopt soil-conservation practices even after being exposed to information about on-site damage. Observers of this phenomenon began to question the validity of the education strategy for bringing about change.

Research conducted during the early 1980s challenged the assumptions concerning the knowledge base of farmers relative to soil erosion. Several studies revealed that land operators were aware of the causes of soil erosion and were aware of the techniques to solve erosion problems (Napier and Forster, 1982). Many farmers who had erosion problems knew that soil erosion was affecting productivity of land resources that they operated (Lovejoy and Napier, 1986; Napier and Forster, 1982; Napier and Lovejoy, 1988). Research demonstrated that most land in the United States was not being adversely affected by soil erosion. It was also revealed that farmers were aware that productivity was not being significantly reduced by soil erosion. In sum, behavioural studies basically demonstrated that information and technical assistance were relatively inconsequential in affecting adoption of soil-conservation practices at the farm level.

Research focused on the loss of future productivity due to soil erosion added to the mounting evidence that existing soil-conservation policies and programmes were inadequate to address soil degradation in the United States. Also, future productivity of cultivated land was not being threatened by soil erosion

(Crosson, 1984; Crosson and Stout, 1983; Larson *et al.*, 1983; USDA, 1981). Projections of the magnitude of losses of future agricultural productivity due to soil erosion clearly demonstrated that large commitments of human and economic resources could not be justified on the basis of on-site damage. Three independent analyses of erosion data collected by the Soil Conservation Service (National Resources Inventory) resulted in very similar results of productivity losses due to soil erosion. The Resources Conservation Act analysis indicated that continuance of 1977 erosion rates would reduce productivity by 8 per cent in 50 years. Researchers at the University of Minnesota using a different model estimated productivity losses over the next 100 years to be 3–5 per cent in the 'important parts' of the Cornbelt. Scientists at Resources for the Future used a different model to predict production losses between 1950 and 1980 using the 1977 erosion rates and estimated that soybean and corn losses would be approximately 2–3 per cent below what they would have been otherwise (Crosson, 1984). It should be noted that no assumptions were made about increases in productivity due to biotechnology. Including such estimates in the calculations would have reduced the expected production losses considerably.

These findings strongly suggested that farmers were not ignoring soil-erosion problems but knew that productivity was *not* being destroyed as a function of using conventional tillage systems on most farmland. Implementation strategies designed to convince farmers to adopt soil-conservation practices on the basis of on-site damage were shown to be questionable.

While research suggested that on-site damage was not as significant as previously thought, off-site costs were demonstrated to be high. Research conducted by Conservation Foundation scientists demonstrated that off-site damage exceeded $6 billion dollars per year in the early 1980s (Crosson, 1984; Sampson, 1985). While the actual cost of soil erosion from agricultural sources is probably much higher than this estimate, it is obvious that the off-site damages caused by erosion of agricultural land constitute a serious socio-environmental problem for society.

Several off-site costs have been identified. Some of the most important off-site damages are as follows: contamination of water supplies, loss of wildlife habitat, loss of water-based recreation opportunities, interruption of highway and water-transportation systems, and the degradation of the aesthetic quality of land and water resources (Halcrow *et al.*, 1982; Napier, *et al.*, 1983; Lovejoy and Napier, 1986).

This summary of existing research focused on the costs of soil erosion strongly suggests that off-site damage is the primary problem created by soil erosion from agricultural sources. Unfortunately, landowners are seldom motivated to adopt soil-conservation practices by off-site damage because they can treat such costs as externalities of production. It is extremely difficult to identify specific agricultural polluters and even more difficult to demonstrate that a certain land operator is responsible for specific damage resulting from soil erosion

(Napier, 1989). Thus, agricultural polluters are seldom penalized for contributing to environmental degradation.

Profitability of Conservation Practices

Another factor which tended to negate the effectiveness of the information-technical assistance strategy was the economic return to farmers who adopted soil-conservation practices. Evidence began to mount during the 1980s that adoption of many soil-conservation practices without extensive economic subsidies was *not* profitable in the short run and frequently *not* profitable in the long run (Ervin and Washburn, 1981; Buttel and Swanson, 1986; Miller, 1982; Mueller *et al.*, 1985; Swanson *et al.*, 1986). Implementation strategies designed to convince farmers to adopt soil-conservation practices on the basis of profitability were shown to have little empirical support.

The studies which examined the profitability of soil-conservation practices at the farm level were useful in explaining why the education approach is only marginally effective in modern society. The basic assumption of an education-technical assistance model is that adoption will produce profits. If profits are not forthcoming, then there is no incentive to change existing modes of behaviour. Most farmers are aware that adoption of most soil-conservation practices will cost them time, effort and money with little return on investments. Subsequently, they are often resistant to adoption unless economic incentives are used as inducements.

Voluntary Adoption

Research in the late 1970s and early 1980s acted as a catalyst for change in the programmes and policies of the existing soil-conservation system. Even the policy of voluntary participation in soil-conservation efforts was challenged. Some scholars suggested that more coercive approaches would be required to bring about adoption of soil-conservation practices (Napier and Forster, 1982; Batie, 1986; Swanson *et al.*, 1986). It was argued that six decades of voluntary conservation programmes had not reduced soil erosion to socially acceptable levels and that another decade of traditional governmental effort would probably produce only marginal improvement in the present situation.

A socially acceptable level of soil erosion refers to the rate of soil loss which is defined as being acceptable to society. The debate continues relative to what rate of soil loss is acceptable. Some people advance the position that T should be used as the criterion, while other segments of society argue for higher levels. It is evident that the present level of soil loss is not acceptable to people living in the United States, since there is considerable concern for the existing rates of soil loss from agricultural land.

Institutional Constraints

Environmental scientists have recently recognized that institutional constraints can affect resolution of soil-erosion problems in the United States (Lovejoy and Napier, 1988; Napier, 1989; Swanson *et al.*, 1986). Government farm policies which encourage maximization of agricultural productivity can exacerbate soil-erosion problems because farming systems that tend to be the most productive also tend to be the most highly erosive. Farm production policies which contribute to the maximization of agricultural productivity are often counterproductive to the control of soil erosion. Subsequently, environmentally concerned people have been pleading for policy consistency so that environmental policies and farm production policies become compatible.

Coalition for Change

As criticisms of the traditional soil conservation system increased during the past decade, political coalitions began to form and to agitate for change. Considerable political pressure was applied to Congress to reduce soil erosion from agricultural sources. The process of change culminated in the passage of the Conservation Title of the Food Security Act of 1985.

THE CONSERVATION TITLE OF 1985

The Conservation Title of the Food Security Act of 1985 began a new era of soil-conservation policy in the United States. For the first time in United States history, federal conservation policy contains elements of *coercion*. While the penalties for degrading soil resources are not sufficient to accomplish the goal of reducing soil erosion to socially acceptable levels, the fact that elements of coercion are included in the legislation is a significant shift in national conservation policy (Harmon, 1987; Napier, 1987a; Stoddard, 1987).

The Conservation Title contains several provisions which will affect soil-conservation policies and programmes in the United States for the next 10 years and perhaps for several decades. The legislation brought into existence Sodbuster, Swampbuster, the Conservation Reserve Program and Conservation Compliance.

Sodbuster Provisions

The Sodbuster provisions of the Conservation Title were designed to establish conditions under which highly erodible land could be brought into crop production (Heimlich, 1985; Myers, 1988). The legislation denies land operators access to farm programme benefits if they bring highly erodible land into crop production without implementing an approved soil-conservation plan. Land cultivated at least once between 1981 and 1985 was declared exempt from Sodbuster.

Land operators who violate Sodbuster provisions cannot participate in price-support programmes, government-sponsored crop insurance, Farmers Home Administration loans, Commodity Credit Corporation storage loans, farm storage loans, Conservation Reserve Program payments and other United States Department of Agriculture programmes which involve commodity payments. Farmers are denied access to these government farm programmes if they own any land which is in violation of the Sodbuster provisions.

Swampbuster Provisions

The Swampbuster provisions of the Conservation Title were designed to prevent conversion of wetlands to crop production (Heimlich and Langner, 1986; Myers, 1988). Land operators cannot convert wetlands to the production of agricultural products without losing access to United States Department of Agriculture (USDA) farm benefits. Farmers who initiated wetland conversion prior to the introduction of the legislation have been permitted to continue wetland manipulations if petitions for exemption were filed before 17 September 1988.

The only legitimate reasons for conversion of wetlands at present are 'third party' and 'minimum effect' actions. Examples of third party action are government projects, such as highway construction, which result in the draining of a wetland. Wetlands drained by third party action can be used to produce agricultural products by the landowner without penalty assuming that Sodbuster provisions are satisfied. Minimum effect provisions permit landowners to convert wetlands to farm production under very specified conditions. Landowners are required to specify the type of conversion requested and an environmental assessment must be conducted. A conservation plan for the wetland is developed and the state conservationist in consultation with the United States Fish and Wildlife Service must approve the conversion. It is highly unlikely that many proposals requesting extensive conversion of wetlands will be approved under this exemption.

Conservation Reserve Provisions

The Conservation Reserve Program (CRP) was designed to retire highly erodible land from agricultural production. Agents of the federal government are authorized to enter into contractual agreements with landowners for retiring highly erodible land for a ten-year period. Landowners who enter land in the CRP receive a mutually agreed rent for the duration of the contract on a yearly basis. Enrolled land cannot be used for the production of food and fibre during the contract period. Groundcover must be introduced on enrolled land and maintained. The federal government will cost-share the introduction of permanent groundcover but will not contribute financially to maintenance of the cover (Bartlett, 1987).

Unlike previous conservation programmes, violation of CRP contracts can result in substantial penalties. Landowners who violate agreements during the ten-year contract period are subject to loss of federal farm benefits. They may also be required to repay rents received from the CRP with interest.

Landowners who have cultivated highly erodible land at least once between 1981 and 1985 are eligible to submit bids on a per-acre basis to have the land enrolled in the CRP. They indicate what level of rent is required for them to enroll highly erodible land in the programme and a decision is made to accept or reject the bid.

The implementation strategy used by the federal government placed primary emphasis on minimizing the economic cost of the programme, which was probably not the best approach (Reichelderfer and Boggess, 1988). Upper limits placed on the level of rents the government would pay (caps) for land initially entered in the programme prevented the greatest proportion of highly erodible land in the Corn Belt from being enrolled.

Much of the early enrollment in the CRP was in the arid regions of the West. This posed a problem, because some of the most highly erodible farmland in the United States is very productive. Highly productive land was *not* removed during the initial implementation phases of the CRP because the rents were too low. As the caps increased, more highly erodible land in the productive areas of the Midwest was enrolled.

While the CRP has removed several million acres of cultivated farmland from production, the cost of doing so has been very high. The average rental cost per acre of land enrolled in the CRP in the United States to date has been approximately $48 (Bartlett, 1987). Approximately 23 million acres have been enrolled, which means that landowners presently enrolled in the programme will receive about $1.1 billion per year for the next 10 years (Napier, 1988a; USDA, 1988).

The cost of the next 20 million acres authorized for inclusion in the programme will be extremely high because the bid price to motivate farmers to enroll additional acreage will be much higher. For example, the CRP cap for cultivated land in the major agricultural region of Ohio presently is $85 per acre. The cap for land in the hilly region of southeastern Ohio is $65 per acre. The caps have risen sharply after each bidding period throughout the United States. It is certain that the next round of CRP bidding will be very expensive for the United States taxpayer.

Assessments of the environmental impacts of the CRP has produced mixed results. While the environmental benefits of the programme probably would have been significantly enhanced via alternative implementation strategies (Reichelderfer and Boggess, 1988; Harmon, 1987), significant reduction of soil loss has been observed. In Ohio, approximately 148 000 acres have been enrolled in the CRP with an average soil savings of approximately 15 tons per acre. Much of the CRP enrolled land in Ohio is now eroding at less than 1 ton per acre. Greater environmental benefits will undoubtedly be achieved when additional acres of cornbelt farmland are enrolled but the economic cost will be substantially greater.

Conservation Compliance Provisions

Conservation Compliance has the potential to be the most significant component of the Conservation Title in terms of long-term reduction of soil loss in the United States. Unfortunately, recent action taken by the Chief of the Soil Conservation Service may render the legislation ineffective.

The Conservation Compliance provision basically states that operators of highly erodible land must develop a conservation farm plan by 1 January 1990 which will reduce soil loss to meet the specifications of the Field Office Technical Guide for the local area. The legislation requires that the conservation plan be fully implemented by 1 January 1995. Land operators who do not comply with the provisions of the Conservation Compliance Program will lose all (USDA) farm benefits until an approved farm plan has been implemented.

While the original legislation outlined general parameters of soil loss which were defined as being socially acceptable, provision was made for administrators of the Soil Conservation Service to make modifications in the criteria to facilitate implementation of the programme at the farm level. It was recognized that certain circumstances would require exemptions. Recently, the Chief of the SCS exercised his prerogative and notified state conservationists that the level of acceptable soil loss should be changed to make it 'reasonable' and easier for farmers to comply with Conservation Compliance provisions.

The Conservation Compliance criteria were modified so extensively that a large proportion of land operators throughout the United States will not be required to significantly change existing farming practices. In Ohio, practically all farmers will be able to meet the new requirements (Alternative Conservation Systems) by eliminating deep ploughing. Unfortunately, the changes in the criteria make it possible for some farmers in Ohio to comply with Conservation Compliance, while their land contributes more than 20 tons per acre per year to erosion. The average soil loss from cultivated land in the United States is about 8 tons per acre (Elfring, 1983; Lee, 1984).

While the substitution of the Alternative Conservation Systems (ACS) approach for the original Conservation Compliance was politically expedient and popular with most farmers, it significantly reduced the effectiveness of this potentially useful conservation legislation. It is also likely to alienate conservation groups which were instrumental in bringing the Conservation Title into being (Napier, 1987b).

The legality of the ACS approach has been challenged and a decision is pending. It is possible that the ACS will be declared invalid and the more stringent criteria originally outlined in the legislation will be applied. Such a situation will create many problems for the SCS because numerous farm plans have been developed which satisfy the ACS criteria. A negative decision concerning the legality of the ACS will require reformulation of many farm plans and will probably alienate a large portion of land operators whose plans have been approved using the ACS approach.

LIMITATIONS OF THE CONSERVATION TITLE

While the Conservation Title is a significant improvement over voluntary approaches to the resolution of soil loss problems in the United States, there are several inherent problems with the legislation which reduce its effectiveness.

A very important limitation to the effectiveness of the Conservation Title in solving soil erosion problems is that a significant number of landowners in the United States do not participate in government farm programmes. The threat of losing government benefits cannot facilitate participation in soil-conservation programmes if land operators do not receive farm programme benefits.

It is estimated that 50–55 per cent of all land operators in the United States participate in one or more government farm programmes. About 85 per cent of all feed and food grain acreage in the United States is operated by farm programme participants (Henderson, 1988). These data strongly indicate that loss of government benefits due to unwillingness to conform to Conservation Title requirements is of no consequence to many farmers. This is especially true for land operators with a small acreage in production.

Another factor which may limit the effectiveness of Conservation Title programmes in the future is the possibility that farm programmes will be phased out of existence. If they are terminated, most of the penalties for violating Sodbuster, Swampbuster, the Conservation Reserve Program and Conservation Compliance will no longer exist.

Attitudes of farmers can affect implementation of the Conservation Title. Agriculturalists in the United States have traditionally been given special consideration on most issues affecting operation of their businesses. This orientation may impede implementation of the Conservation Title. Farmers have been provided with significant economic incentives to participate in soil-conservation programmes for decades with no expectations for continuance of conservation behaviours once participation is terminated. They have been encouraged to maximize production and have been rewarded for their efforts via federal farm programmes. Land operators have been free to export agricultural pollutants with immunity which has produced a belief that they should not be responsible for off-site damage. Other members of society have traditionally assumed most of the costs of agricultural production exported from fields via soil erosion. Land operators believe these traditions should be maintained and resist efforts, such as the Conservation Title, to introduce change.

Lastly, orientations of existing soil conservation agencies towards coercion may be a major barrier to effective implementation of the Conservation Title. Coercion has been an alien concept to agency personnel commissioned to implement federal conservation programmes. It is highly unlikely that existing agencies will be able to effectively implement coercive elements of the Conservation Title. Evidence to date suggests that existing government agencies are too responsive to political pressures to be successful in implementing coercive conservation policies. Farmers are aware of the tradition to protect them from government constraints and expect exemption from regulation. Action recently taken by the Chief of the Soil Conservation Service to change Conservation Compliance criteria suggests that the expectations of the farmers have again been realized. Regulatory approaches to problem resolution requires application of sanctions to bring about compliance, and it is becoming evident that existing soil-conservation agencies are not willing to apply necessary penalties to farmers.

IMPLICATIONS OF THE UNITED STATES SOIL-CONSERVATION EXPERIENCES FOR OTHER SOCIETIES

The soil-conservation experiences of the United States have significant implications for other societies which are attempting to reduce the incidence of soil erosion. While it must be noted that direct transfer of programmes and policies is impossible without consideration of the unique situations in the receiving society, knowledge gained in United States can be very useful in the development of programmes and policies to address similar problems in other societies.

Necessary Conditions for Effective Soil-conservation Programmes

Soil-conservation experiences in the United States strongly suggest that at least 12 conditions must be satisfied to some extent before such programmes can be effectively implemented. The necessary but not sufficient conditions are as follows:

(1) The development of a political constituency which supports action to reduce the social, economic and environmental costs of soil erosion.
(2) The allocation of extensive human and economic resources on a long-term basis by national governments to finance soil-conservation programmes.

(3) The creation of government agencies commissioned to address soil-erosion problems with sufficient autonomy to be immune from short-term political influences.
(4) The development of well-trained professionals to staff soil-conservation agencies.
(5) The development of an informed farm population which is aware of the causes and remedies of soil erosion.
(6) Development of a stewardship orientation among land operators to protect soil and water resources.
(7) National policies which place high priority on the protection of soil and water resources.
(8) The creation of national development, agricultural and soil-conservation policies and programmes which are consistent and complementary.
(9) The creation of national environmental policies which are consistent and complementary.
(10) The development of physical and social scientists who are committed to the generation of scientific information which will contribute to the creation, implementation and continual modification of soil and water conservation policies and programmes.
(11) The creation of an interdisciplinary professional society committed to the maintenance of environmental integrity of soil and water resources.
(12) The emergence of political leadership which will be willing to implement policies and programmes which some segments of the agricultural population will find oppressive.

CONCLUSIONS

The United States experiences with conservation programmes clearly indicate that soil conservation is *very* expensive. Individual land operators are seldom motivated to adopt soil-conservation practices without being extensively subsidized. The subsidies must be provided by the federal goverment, since state and local jurisdictions do not have adequate financial resources to address the problem. Societies wishing to reduce soil erosion must be prepared to allocate necessary financial and human resources to the problem.

Vast economic resources are required to finance permanent governmental agencies to implement national soil-conservation goals. These resources must be appropriated from federal sources. It is unrealistic to believe that effective soil-conservation programmes can be introduced in any society without permanent service infrastructure to implement them. The conservation agencies must be staffed by well-trained professionals committed to soil-conservation goals.

Policy inconsistency has been shown to be problematic in the United States. Agricultural production policies often conflict with those of conservation.

Policies which serve to maximize short-term food and fibre production can exacerbate degradation of soil and water resources. Environmental policies can also be inconsistent. Soil-conservation programmes implemented without consideration of other societal policies will encounter problems.

While implementation of soil erosion control programmes necessitates enlightened land operators, it also requires political leadership which recognizes its responsibility to protect vital natural resources. Evidence to date suggests that some form of coercion is necessary to bring about adoption of soil-conservation practices at the farm level. Many career staff persons employed by conservation agencies in the United States oppose any form of coercion because they feel that existing methodologies for bringing about adoption of soil-conservation practices are adequate. However, the evidence is mounting that the education-technical assistance approach has many limitations. It is also becoming more evident that use of economic subsidies to motivate farmers to set aside highly erodible land and to adopt soil-conservation practices is extremely expensive. The development and maintenance of infrastructures to verify compliance has become cumbersome and costly.

It is clear that society has basically two options which are as follows: (1) create a situation where farmers are willing to assume the costs of halting soil erosion; or (2) institutionalize policies that will force farmers to stop polluting. Existing research strongly suggests that farmers are not willing to assume the costs of adopting soil erosion control measures and that the education-technical assistance approach will not resolve the problem. The economic costs of removing highly erodible land from agricultural production are becoming extremely high and may soon exceed the costs of soil erosion (both on-site and off-site costs). The coercive alternative is not desirable but will probably be the only feasible option in the future.

As more non-farm people become aware of the 'hidden costs' of food and fibre production in the United States, it is almost inevitable that political pressure will be used to force land operators to internalize the pollution costs of agricultural production. There is still time for the agricultural sector to participate in the decision-making process, but continued resistance on the part of farmers to compromise will probably result in compulsory action by society. Legislative action to implement strong controls on agricultural pollution is not popular with farmers and will undoubtedly result in some land operators being displaced from farming.

Political leadership must be prepared to defend coercive policies designed to protect soil and water resources. Evidence to date suggests that agricultural interests will resist environmental legislation because most methods of reducing soil erosion usually generate costs for landowners. Given the political influence of agriculturalists in most societies of the world, this will be the most difficult problem associated with reducing agriculturally induced soil erosion. It is highly likely that resolution of soil erosion in most societies of the world will occur in the socio-political arena and environmental interests must be prepared to affect the political process. This can be accomplished by generation of relevant research and the translation of the knowledge into usable environmental policy recommendations for decision makers.

REFERENCES

Bartlett, E. T. (1987). Social and economic impacts of the Conservation Reserve Program. In Mitchell, J. E. (ed.), *Impacts of the Conservation Reserve Program in the Great Plains*, USDA Forest Service General Technical Report RM-158, Denver, Colorado, pp. 52-4.

Batie, S. S. (1986). Why soil erosion: a social science perspective. In Lovejoy, S. B. and Napier, T. L. (eds), *Conserving Soil: Insights From Socioeconomic Research*, Soil Conservation Society of America Press, Ankeny, Iowa, pp. 3-14.

Buttel, F. H. and Swanson, L. E. (1986). Soil and water conservation: a farm structural and public policy context. In Lovejoy, S. B. and Napier, T. L. (eds), *Conserving Soil: Insights From Socioeconomic Research*, Soil Conservation Society of America Press, Ankeny, Iowa, pp. 26-39.

Camboni, S. M. and Napier, T. L. (1986). Five decades of soil erosion: problems and potentials, Paper presented at the national Rural Sociological Society meetings, Salt Lake City, Utah.

Camboni, S. M., Napier, T. L. and Lovejoy, S. B. (1989). Factors affecting knowledge of and participation in the Conservation Reserve Program in a micro-targeted area of Ohio. In Napier, T. L. (ed.), *Implementing the Conservation Title of the Food Security Act of 1985*, Soil Conservation Society of America Press, Ankeny, Iowa (forthcoming).

Crosson, P. (1984). New perspectives on soil conservation policy, *Journal of Soil and Water Conservation*, **39**, 222-5.

Crosson, P. and Stout, A. T. (1983). *Productivity Effects of Cropland Erosion in the United States*, Resources for the Future, Washington, DC.

Elfring, C. (1983). Land productivity and agricultural technology, *Journal of Soil and Water Conservation*, **38**, 7-9.

Ervin, D. E. and Washburn, R. (1981). Profitability of soil conservation practices in Missouri, *Journal of Soil and Water Conservation*, **36**, 107-11.

Halcrow, H. G., Heady, E. O. and Cotner, M. L. (eds). (1982). *Soil Conservation Policies, Institutions and Incentives*, Soil Conservation Society of America Press, Ankeny, Iowa.

Harmon, K. W. (1987). History and economics of Farm Bill legislation and the impacts on wildlife management and policies. In Mitchell, J. E. (ed.), *Impacts of the Conservation Reserve Program in the Great Plains*, USDA Forest Service General Technical Report RM-158, Denver, Colorado, pp. 105-8.

Heimlich, R. E. (1985). Soil erosion on new cropland: a Sodbusting perspective, *Journal of Soil and Water Conservation*, **40**, 322-6.

Heimlich, R. E. and Langner, L. L. (1986). Swampbusting in perspective, *Journal of Soil and Water Conservation*, **41**, 219-24.

Henderson, D. (1988). Personal communication. Department of Agricultural Economics and Rural Sociology, The Ohio State University, Columbus, Ohio.

Kleckner, D. (1988). Conservation programs: mandatory or voluntary?, *Journal of Soil and Water Conservation*, **43**, 358.

Larson, W. E., Pierce, F. J. and Dowdy, R. H. (1983). The threat of soil erosion to long-term crop production, *Science*, **219**, 458–65.

Lee, L. (1984). Land use and soil loss: a 1982 update, *Journal of Soil and Water Conservation*, **41**, 304–8.

Lovejoy, S. B. and Napier, T. L. (eds) (1986). *Conserving Soil: Insights From Socioeconomic Research*, Soil Conservation Society of America Press, Ankeny, Iowa.

Miller, W. L. (1982). The farm business perspective and soil conservation. In Halcrow, H. G., Heady, E. O. and Cotner, M. L. (eds), *Soil Conservation Policies, Institutions and Incentives*, Soil Conservation Society of America Press, Ankeny, Iowa, pp. 151–62.

Mueller, D. H., Klemme, R. M. and Daniel, T. C. (1985). Short- and long-term cost comparisons of conventional and conservation tillage systems in corn production, *Journal of Soil and Water Conservation*, **40**, 466–70.

Myers, P. C. (1988). Conservation at the Crossroads, *Journal of Soil and Water Conservation*, **43**, 10–13.

Napier, T. L. (1987a). Farmers and soil erosion: a question of motivation, *Forum for Applied Research and Public Policy*, **2**, 85–94.

Napier, T. L. (1987b). Anticipated changes in rural communities due to stress in agriculture: implications for conservation programs. In Mitchell, J. E. (ed.), *Impacts of the Conservation Reserve Program in the Great Plains*, USDA Forest Service General Technical Report RM-158, Denver, Colorado, 84–90.

Napier, T. L. (1988a). Implementation of soil conservation practices: past efforts and future prospects, *Topics in Applied Resource Management in the Tropics*, **1**.

Napier, T. L. (1988b). Willingness of land operators to participate in government-sponsored soil erosion control programs, *Journal of Rural Studies*, **4**, 339–47.

Napier, T. L. (1989). Farmer adoption of soil conservation practices: lessons for groundwater protection, Paper written for the Office of Technology Assessment of the United States Congress, Washington, DC.

Napier, T. L. and Forster, D. L. (1982). Farmer attitudes and behavior associated with soil erosion control. In Halcrow, H. G., Heady, E. O. and Cotner, M. L. (eds), *Soil Conservation Policies, Institutions and Incentives*, Soil Conservation Society of America Press, Ankeny, Iowa, pp. 137–50.

Napier, T. L. and Lovejoy, S. B. (1988). Factors influencing adoption of soil conservation practices in Ohio: a typological analysis, Economic and Sociology Occasional Paper Number 1498, The Ohio State University, Columbus, Ohio.

Napier, T. L., Scott, D. F., Easter, K. W. and Supalla, R. (eds) (1983). *Water Resources Research: Problems and Potentials for Agriculture and Rural Communities*, Soil Conservation Society of America Press, Ankeny, Iowa.

Napier, T. L., Thraen, C. S. and McClaskie, S. L. (1988). Adoption of soil conservation practices by farmers in erosion-prone areas of Ohio: the application of logit modeling, *Society and Natural Resources*, **1**, 109–29.

Rasmussen, W. D. (1982). History of soil conservation, institutions and incentives. In Halcrow, H. G., Heady, E. O. and Cotner, M. L. (eds), *Soil Conservation Policies, Institutions and Incentives*, Soil Conservation Society of America Press, Ankeny, Iowa, pp. 3–18.

Reichelderfer, K. and Boggess, W. G. (1988). Government decision making and program performance: the case of the Conservation Reserve Program, *American Journal of Agricultural Economics*, **70**, 1–11.

Sampson, R. N. (1985). Government and conservation: structuring an improved public role. In *Soil Conservation: What Should Be the Role of Government?*, Indiana Cooperative Extension Service, Purdue University, West Lafayette, Indiana, pp. 1–6.

Stoddard, G. M. (1987). Implementing the conservation title of the Food Security Act: the unfinished agenda, *Journal of Soil and Water Conservation*, **42**, 93-4.

Swanson, L. E., Camboni, S. M. and Napier, T. L. (1986). Barriers to the adoption of soil conservation practices on farms. In Lovejoy, S. B. and Napier, T. L. (eds), *Conserving Soil: Insights From Socioeconomic Research*, Soil Conservation Society of America Press, Ankeny, Iowa, pp. 108-20.

United States Department of Agriculture (1981). *Soil, Water and Related Resources in the United States: Analysis of Resource Trends, 1980 RCA Appraisal, Part II*, Washington, DC.

United States Department of Agriculture (1988). Conservation Reserve Program, News release, Washington, DC.

44 Problems of Land Reclamation: Pothwar Loess Plateau, Pakistan

MARTIN J. HAIGH
Oxford Polytechnic

INTRODUCTION

The Pothwar Plateau in Pakistan's Punjab Province is mantled with a thick layer of loess which is susceptible to erosion. Today, gullies occupy 60 per cent of the total land surface and only a third of this area of potentially productive soils is employed for agriculture (Anwar *et al.*, 1986). Erosion also threatens agricultural production on the existing 'Barani' (i.e. rainfed) agricultural lands which make up about 25 per cent of Pakistan's cultivated area of 20.43 million ha but rather more of the land area of the Province of Punjab (Chaudhry and Shafiq, 1985). Barani lands lose a massive amount of topsoil to water erosion (Punjab Barani Commission, in Anwar *et al.*, 1986). Official sources mention the figure of 1 billion tonnes of soil each year, equivalent to two-thirds of the capacity of the giant Tarbela Dam (Kazi, 1984, in Chaudhry and Shafiq, 1985).

Pakistan is active in the struggle against soil and gully erosion. Gully plugging has been carried out across 14 525 ha, and reclamation treatments applied to 132 354 ha of eroded land, in the period 1970–84 (Chaudhry, 1985). This chapter examines some of the difficulties facing soil-conservation workers seeking to reclaim agricultural land in the Pothwar ravines. These difficulties result from a mixture of technical, socio-economic, logistic and cultural processes. In many respects, they are typical of the problems faced by land-reclamation teams operating in similar environments around the world.

The notes which follow are based on a series of interviews with reclamation project leaders, practitioners, researchers of the Soil Survey of Pakistan, and academics from Barani College, Punjab University, and the University of Agriculture, Faisalabad. These interviews were recorded in December 1987. This study is the third in a series on the reclamation of gullied lands in South Asia. Haigh (1989) compares the Punjab Soil Conservation Directorate approach to

---
Soil Erosion on Agricultural Land
Edited by J. Boardman, I. D. L. Foster and J. A. Dearing
©1990 John Wiley & Sons Ltd

the reclamation of gullied lands with the experience gained from two demonstration projects organized by the Pakistan Forest Research Council and Pakistan Agricultural Research Council. Haigh (1984) is a review of Indian research on gully erosion, management and reclamation.

ENVIRONMENTAL CONTEXT

The Pothwar Plateau (Figure 44.1) is a tract of 1.82 million ha which lies to the north of the Salt Range and between the Jhelum and Indus Rivers. Its average annual rainfall ranges from 1500 mm yr^{-1} in the south-east, where monsoonal rains dominate, to below 375 mm yr^{-1} in the south-west. Approximately a third of the area receives more than 750 mm yr^{-1} while around one eighth receives less than 375 mm yr^{-1} (Anwar et al., 1986). The most severe storms occur in the period June–September and peak intensities may exceed 25 mm h^{-1}. Approximately 50 per cent of rainfall may normally be converted to runoff (Anwar et al., 1986). The area is underlain by Tertiary clays, conglomerates and sandstones of the Siwalik formation which outcrop locally

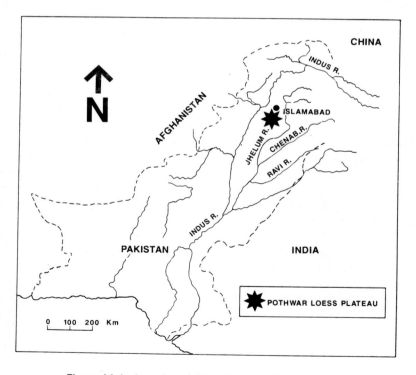

Figure 44.1 Location of Pothwar Loess Plateau, Pakistan

as ridges. These rocks, however, are mantled by a layer of Pleistocene loess, reworked loessic sediments and river gravels which can run to tens of metres in depth. Heusch (1988) suggests that the Pothwar Plateau has the highest regional erosion rate in lowland Pakistan, estimated as $37.6 \, t \, ha^{-1} \, yr^{-1}$ or $5-6 \, mm \, yr^{-1}$.

THE POTHWAR RAVINES

The origins of the Pothwar ravines are unclear (Figure 44.2). As in India, there are two schools of thought (Haigh, 1984). Conservationists and environmentalists tend to blame the problem on a long history of poor land management, the degradation of the original Acacia scrub forest, and overgrazing (Heusch, 1988). Geologists and geomorphologists argue that the ultimate cause is tectonic uplift of the Himalaya which has encouraged incision along the major river channels and their tributaries (cf. Khan, 1987).

The least-disturbed areas of the Pothwar Tablelands are mantled with soils of the Guliana Series. These fine-silty, mixed, hyperthermic Udic Haplostalfs contain a calcified horizon at depths of 1.4–1.9 m (Ahmad *et al.*, 1986). On the steeper-sloping margins of the tablelands, where these ancient soils have suffered erosion, there are less mature soils, like the Missa Series. The Missa soils are Guliana soils which have had their upper layers reworked by erosion.

Figure 44.2 Pothwar Ravine Landscape (NB: Landcruiser for scale top-right)

A new cycle of soil formation has begun in the reworked material. The result is a coarse-silty, mixed, hyperthermic Typic Ustochrept with a calcic layer at depths which decline from 0.2 to 0.01 m towards the gully rim (Ahmad et al., 1986). Here, close to the gully rim, the hard layer is at the surface. This subsoil provides the 'caprock' for the steep gully walls which usually begin with a near-vertical drop of 3–5 m. The presence of this layer coupled with the capacity of the loess to hold a steep slope and its inherent erodibility combine to permit the development of the ravineland topography.

The gullied land itself is characterized by soils of the Rajer Series. These soils are formed on reworked loess and comprise the third and final member of the Pothwar ravineland soil catena. The Rajer soils are coarse-silty, mixed calcareous, hyperthermic, Typic Ustorthents which contain no horizons beyond an ochric epipedon, except where they have been reworked by ploughing (Ahmad et al., 1986).

SOIL EROSION

The rate of soil loss from undisturbed areas can be low, as is often the case with ancient gullied lands, (cf. Wise et al., 1982). High sediment losses are prevented by the presence of vegetation and a soil crust. When this crust is disturbed by trampling, vegetation removal or agriculture, then the rate of soil loss climbs steeply. The K (soil erodibility) index value for the Guliana and Missa soil is 0.55 but that for the immature Rajer entisol is 0.64. The organic content of the topsoils of the Guliana and Missa soils exceeds 0.48 per cent while that in the Rajer is below 0.20 per cent. The context is a semi-arid climate with a rainfall erosivity index (R) of 150–200 (cf. Ram Babu et al., 1978).

Runoff increases as the surface layers of the soil are stripped away and the calcified subsoil is exposed. This runoff makes its greatest impact on the more erosive Rajer entisols of the gully margins and floor. Workers at the Soil Survey of Pakistan point out that this situation can be to the benefit of local agriculturalists downslope of the stripped areas. In this semi-arid area there is much to be said for land which collects water in droughty years. Soil erosion on high ground may help harvest water for the droughty rainfed fields. In Israel, Yair (1983) recommends stripping soil and vegetation covers from unproductive hillsides as part of a scientific approach to water harvesting for agriculture in semi-desert environments.

EXPANSION OF FIELDS

An increasing proportion of the Pothwar agricultural production comes from farming the entisols of the gully floors. For centuries, farmers working such

Problems of Land Reclamation

fields have attempted to expand them by running their ploughs hard against the ravine walls. In places where those gully walls were ploughable, up-and-down slope ploughing has been undertaken to bring water and soils down to the levelled fields below. These practices have slowly steepened the remaining gully walls and progressively expanded the area under agricultural production.

Today, the most important service provided by the Soil Conservation Directorate is the loan of bulldozers to farmers to help them level their fields and so improve water conservation. However, most outsiders agree that the farmers' enthusiasm for this programme has less to do with field levelling for better water management than with the opportunity to effect a significant increase in field area. The bulldozers are far more efficient than the old ploughs for pushing back gully walls. Fresh yellow loess cliffs are a much more common sight adjacent to agricultural fields than in the gullied wastelands (Figure 44.3). Here, most surfaces are covered with an organic crust and slopes are often less than vertical.

SOIL PIPES

The Pothwar loess is vulnerable to soil pipe erosion, especially when it has been subjected to earth-moving. Soil pipes are a common feature of many new

Figure 44.3 Punjab Soil Conservation Directorate projects involve field levelling and the provision of water-outlet drop structures

Figure 44.4 Pipe and gully erosion on new agricultural terrace front planted with Eucalypts (near Attock)

embankments and terrace fronts (Figure 44.4), but such pipes also develop from the heart of newly reclaimed fields, especially where there is standing water during the monsoon. The problem occurs, in part, because the density of the relocated loess is lower than the undisturbed soil. However, efficient agriculture demands fields which are relatively large and which can be worked economically. Larger fields mean higher terraces and a greater hydraulic gradient in the repacked loess (cf. Haigh and Rydout, 1987).

The Missa Experimental Station of the Pakistan Forest Research Institute (Haigh, 1989; Subhan, 1986) includes a paired catchment study which involves a heavily overgrazed and sparsely vegetated control catchment. The experimental catchment has been closed to allow regeneration of the vegetation cover and it has been planted with selected tree and shrub species (Figure 44.5). Preliminary results from this site indicate that the reclaimed catchment produced 39 per cent less runoff and 66 per cent less sediment than the control (Subhan, 1986; Haigh, 1989).

The reliability of data produced for environmental management in developing countries is frequently challenged (e.g. Biswas, 1987). As in other areas much national and regionally compiled data are guestimates. However, in Pakistan, the data produced by individual projects tend to be of high quality. In the case of the Pothwar scientific sites, many of the scientists have been trained in the western United States and operate their experiments with guidance and financial

Figure 44.5 Pakistan Forest Research Institute's ravineland reclamation demonstration project at Missa involves closure to grazing, the construction of checkdams and natural regeneration supplemented by planting

support from USAID personnel. Soil test results are often the work of USDA laboratories. The Missa station is instrumented with autographic hydrological equipment and has a permanently manned standard climatological station. The instrumentation is supported by sophisticated integrated stream gauging/ sediment trap structures which are an advance on most used in university-level research in Europe.

Pipe erosion is absent from the slopes which are protected by vegetation but quite common in the control catchment area. Beamish (1969) notes that, within the control catchment, piping is most prevalent in abandoned fields where the surface of the topsoil has fewer fines. Elsewhere, the surface is protected by a surface crust/colloidal layer or, on undisturbed surfaces, by a (cryptogamic) crust. Several local algae have been shown to be important in soil binding and erosion control (Anjum *et al.*, 1982; Faridi *et al.*, 1980). The rate of infiltration experienced in crusted areas is much lower so there is less infiltration and less soil pipe development. Beamish (1969) suggests that the colloidal layer may be created by a combination of raindrop impact, tillage producing a fine seedbed, and trampling and manuring by draft animals. When these practices cease, the surface colloids tend to be removed by surface wash.

Leaders of the Punjab Soil Conservation Directorate consider soil piping to be the most severe technical problem facing their land-reclamation engineering. The best solution offered to them so far involves land treatment by soil conditioners. Calcium hydroxide or calcium sulphate stabilizers have been recommended. However, this is not seen as a viable solution.

Current practice emphasizes land management to reduce the opportunities for soil pipe formation. The Soil Conservation Directorate try to impress upon local farmers the need for vigilance. The first step towards this is direct involvement of the farmers in the reclamation process. Consultation conducted during these activities stress the farmers' responsibilities towards their own land. Material advice includes the recommendation that they should not keep their land fallow, but maintain and strengthen field embankments, and be especially vigilant during the rains. Animal burrows which might trigger major pipe erosion should be plugged immediately. During years of heavy rainfall, the bunds around the fields will need constant attention and repair. Some farmers take this advice very seriously. However, others, mainly those who have an alternative source of income, are less efficient. Land remains a source of prestige in Pakistan, as elsewhere, and so may be retained long after it has ceased to be the main source of family income.

SOIL FERTILITY

Land reclamation tends to require earth-moving. Gully reclamation involves the creation of new terraces in the floor of the gully channels and sometimes also in the sidewalls (Anwar *et al.*, 1986). Typically, terrace construction involves the destruction of pre-existing soil structures and a loss of both soil humus and soil fertility (Peev, 1988). Scientists of the Pakistan Agricultural Research Council have demonstrated that economically valid high yields can be obtained through fertilizer input. However, practitioners of the Punjab Soil Conservation Directorate argue that fertility is not so easily recovered and requires large inputs of organic manure. Therefore their advice is to preserve as much of the original soil cover as possible.

FODDER AND FUELWOOD

Khan and Akram (1987) maintain that the Punjab's present shortfall in wood products is due to a defective forest policy which hinders professionals as well as farmers wishing to conduct afforestation. The Forest Department treats afforestation as routine official work. As a result, tree planting is regarded as just another onerous official duty of the forest official and there is little incentive to provide aftercare. There is no scheme of incentives to encourage farmers and no institutional support for the private tree grower. Little work has been done to establish the economic benefits of tree growing as a commercial farm enterprise. However, while officially managed linear plantations suffer neglect, in some parts of the country, notably the Peshawar Valley, commercial timber production from linear plantations of hybrid Populus and Eucalyptus cultivated along field margins is widespread. Khan and Akram (1987) estimate that Pakistan's farmers already produce nine times as much wood as the official, heavily capitalized, Forest Department, and suggest that much more might be done to promote farmers' interest in tree planting.

The Punjab Soil Conservation Directorate describe the problems of trying to establish small woodlots on overgrazed, thinly grassed, waste ground as a part of a watershed management project in the Sohawa area. Prior to reclamation, this was very poor land with only pockets of soil. However, in 1973/4 the area was fenced and grass seed was dry sown. Trees were also planted: *Lucaena, Eucalyptus, Acacia modesta, A. Arabica, Zisyphus* sp., etc. The result is a dense, thriving, plantation covering, in this instance, some 20 ha.

A watchman has been appointed to try to discourage local people from grazing the area. Local villagers are allowed to cut fodder using a sickle. Unfortunately, some still prefer to cut the wire to provide access for their animals. Indeed, similar problems are experienced at the Missa Experimental Station, where grazing continues even though a watchman is resident on site. During the dry season the plantation's trees stand out as a green island in a desiccated landscape. They offer a rare source of green fodder and lopping becomes a problem. The difficulty is that these plantations cover such small areas and so the pressures tend to become focused. The provision of forest guards is not a totally successful answer. Their presence creates antipathy in the local community, who feel they may be threatened with arrest or be responsible for the arrest of neighbours. This makes the community as a whole much less willing to co-operate with such projects. In addition, overzealous protection can quickly convert a fenced plantation into a fire hazard. Indeed, the Pakistan Forest Institute's Missa afforestation scheme has suffered from unscheduled burns, and it now encourages farmers to cut grasses for fodder to reduce this hazard.

Self-interest also plays an important role in persuading farmers to join afforestation programmes. It is pointed out to landholders that they are not gaining any benefit from this land which is being grazed by others. If the land is planted to forest, then in 10-15 years they will begin to receive some income. Unfortunately, many of the tracts of land won for afforestation by this approach are very small and so are very difficult to manage. Nevertheless, the creation of a nucleus of supporters in the community is a vital first step. Once influential farmers are persuaded of the benefits of the work, they may begin to protect not only their own lands but also those of their neighbours, and they may begin to encourage other farmers to support the land-reclamation programme.

SOCIAL CONSTRAINTS

In India, Hindu social organization gives the village community a strong and unified identity. This social structure has been utilized by some of the more successful non-government organizations who are involved in both the social and physical reclamation of ravinelands (Haigh, 1984). In Pakistan there are few non-government organizations undertaking practical development projects, apart from the Aga Khan Foundation in the far north. There is much less social cohesion in the village community, although large landowners do play a key role, as in the West. Khan (1981) argues that local researchers have paid insufficient attention to the structure of production and interrelationships between the members of the farming community. Once again, it is the larger landowners who hold the key. Khan (1979) notes that there is a positive correlation between the size of a farm, its output per hectare and its level of non-traditional inputs: chemical fertilizer, hired labour and machinery. When subsidies for reclamation lie in the 50 per cent range it is possible to work only with the more prosperous landholders who have the resources to undertake their part of the work.

Another problem for Pakistan arises from the low status awarded to agricultural extension. Pakistan's academics argue that its agricultural extension workers are undertrained and undervalued. Khan (1981) states that current agricultural research tends to be too narrow and technocratic in outlook. Pakistan's agriculture is input limited and economic problems are the reason for many deficiencies in development. Khan (1981) also maintains that it is futile to fight the present land-tenure system. Instead, development workers should ensure that they take the best advantage of whatever opportunities exist in the community. This recommendation is implemented enthusiastically by Soil Conservation Directorate practitioners.

India has created a substantial and self-sustaining tradition of soil-conservation practice and research. This does not exist in Pakistan, which has few watershed management specialists and relies on the United States to train its erosion- and sediment-control professionals. FAO advisors have repeatedly called for increased soil-conservation training (Mitchell, 1987; Heusch, 1988). Presently, students at the University of Agriculture, Faisalabad, receive only token instruction in soil and water conservation (Chaudhry and Shafiq, 1985). There is great potential for establishing postgraduate training in soil and water conservation at Barani College, Rawalpindi. However, it may be many years before Pakistan becomes self-sufficient in soil-conservation expertise.

CONCLUSIONS

It has been suggested that Pakistan needs a national plan for soil conservation to provide for the co-ordination of soil- and water-conservation research, training and management (Mitchell, 1987). The FAO suggestion is that this should involve a Ministerial Soil Conservation Council with representatives from each province to develop policy and a Standing Committee on Soil Conservation (representing each province and organization involved in practical soil-conservation activities) to provide for implementation. Certainly, the establishment of facilities for training soil-conservation practitioners and for the improvement of agricultural extension work need to be high on the agenda for such an organization.

Among the specific problems of reclamation in the Pothwar ravines, soil piping requires immediate specialist attention. The problems of the preservation of the structure and fertility of disturbed soils also need more study. Other difficulties centre on a complex of management issues concerning field expansion and the oversteepening of gully walls; afforestation and selective surface stripping for water harvesting; persuasion and legislation for the promotion of better conservation practice; and of coping with farmers for whom farming is no longer a primary concern. Current economic barriers include a relatively low level of subsidy for reclamation and the expense of fertilizers. Experience shows that fields close to villages and which are provided with nightsoil do not need to be fallowed. Further away, fallowing becomes necessary to preserve soil fertility, but fallow fields are also most prone to soil pipe formation. Despite such difficulties, Pakistan has managed to provide soil conservation treatment across 132 354 ha of eroded lands, and 14 525 ha of gullied land, nationally, in the period 1971–84 (Chaudhry, 1985), and the work of the soil conservationists is widespread in the Pothwar gullied lands.

ACKNOWLEDGEMENTS

We thank the Charles Wallace Trust for a travel award, and colleagues at the Punjab Soil Conservation Directorate, Soil Survey of Pakistan, Pakistan Forest Institute, Pakistan Agricultural Research Council, Pakistan Forest Department, Barani College, Punjab University, and the University of Agriculture, Faisalabad, for their assistance in this project.

REFERENCES

Ahmad, M., Akram, M., Baig, M. S., Javed, M. Y. and Riaz-ul-Amin (1986). *Proceedings of the 12th International Forum on Soil Taxonomy and Agrotechnology Transfer, Pakistan, Second Volume, Field Excursions,* Soil Management Support Services, USA/Soil Survey of Pakistan, Lahore.

Anjum, G. T., Faridi, M. A. F. and Mehmood, T. (1982). Some soil binding algae from Peshawar, Pakistan, *Pakistan Journal of Botany*, **14**(1), 107-9.

Anwar, Ch. M. Shahid Admad, Mohammad Shafiq, Ruhul Amin, Nizami, M. I., Zaheer-ul-Ikram, Ismail Zafar, Iman Ashraf, Abdul Latif, Saleem Akhtar Sial, and Shah Roz Khan (1986). *Management of Gully Eroded Areas in Pothwar*, Pakistan Agricultural Research Council, Islamabad.

Beamish, R. (1969). *Preliminary Work Plan for the Establishment of Gauging Devices on the Missa Comparative Watersheds*, Pakistan Forest Research Institute, Peshawar.

Biswas, A. (1987). Environmental concerns in Pakistan, with special reference to water and forests, *Environmental Conservation*, **14**(4), 319-27.

Chaudhry, M. A. (1985). Present status of soil conservation practices in Pakistan, *Frontier Journal of Agricultural Research* (Peshawar), **11**, 104-14.

Chaudhry, M. A. and M. Shafiq (1985). Soil and water conservation in Pakistan. In *Managing Soil Resources to Meet National Challenges, Proceedings of the First National Congress of Soil Science, Pakistan*. Pakistan Agricultural Research Council, Unpublished photostat in Library, Islamabad.

Faridi, M. A. F., Anjum, G. and T. Mehmood (1980). Some soil binding algae from Lawrencepur, Pakistan, *Journal of Pure and Applied Sciences*, **2**, 1-10.

Haigh, M. J. (1984). Ravine erosion and reclamation in India, *Geoforum*, **15**(4), 543-61.

Haigh, M. J. (1989). Reclamation of gullied lands in northern Pakistan, *Topics in Applied Resource Management* (DITSL, Witzenhausen, FRG.), **1**, 189-201.

Haigh, M. J. and Rydout, G. B. (1987). Erosion pin measurement in a desert gully. In Gardiner, V. (ed.), *International Geomorphology 1986*, Vol. 2, John Wiley, Chichester, pp. 419-36.

Heusch, B. P. (1988). *Soil Conservation in Pakistan*. FAO/UNDP Project PAK 86/012: Field Document 25/Pakistan Forest Institute, Peshawar.

Khan, G. S. and M. Akram (1987). *Why We Fall Short of our Wood Requirements in Punjab*, University of Agriculture, Department of Forestry, Range Management and Wildlife, Faisalabad.

Khan, Kh. (1987). Ravine and gully erosion in the Pothwar plateau, Pakistan. In Shams, F. A. (ed.), *First National Seminar on Mountainous Regions of Pakistan, Abstracts*, Punjab University, Centre for Integrated Mountain Development, Lahore, pp. 81-2.

Khan, M. H. (1979). Farm size and land productivity relationships in Pakistan, *Pakistan Development Review*, **18**(1), 69-77.

Khan, M. H. (1981). Political economy of agricultural research in Pakistan, *Pakistan Development Review*, **20**(2), 191-214.

Mitchell, A. (1986). *Soil and Water Conservation, FAO Technical Assistance to Pakistan Agricultural Research Council—Final Report* UTF/PAK/072/PAK, Food and Agriculture Organization of the United Nations, Rome.

Peev, B. (1988). Changes in the ecological conditions of steeplands after terracing, *International Symposium on Water Erosion, Varna, Bulgaria*, UNESCO IHP/MAB, Paris and Sofia, pp. 355-60.

Ram Babu, Tejwani, K. G., Agarwal, M. C., and L. S. Bhushan (1978). Distribution of erosion index and iso-erodent map of India, *Indian Journal of Soil Conservation*, **6**(1), 1-13.

Subhan, F. (1986). *Evaluation of Hydrologic Performance of Soil Conservation Measures on Comparative Watersheds in the Sub-tropical Scrub Zone*, Pakistan Forest Research Institute (Project PK-Fs-58), Peshawar.

Yair, A. (1983). Hillslope hydrology, water harvesting, and areal distribution of some ancient agricultural systems in the northern Negev desert, *Journal of Arid Environments*, **6**, 283-301.

Wise, S. M., Thornes, J. and A. Gilman (1982). How old are the badlands? A case study from south-east Spain. In Bryan, R. and Yair, A. (eds), *Badland Geomorphology and Piping*, Geobooks, Norwich, pp. 259-77.

CONCLUSION

45 Soil Erosion Studies; Some Assessments

J. BOARDMAN
Countryside Research Unit, Brighton Polytechnic

J. A. DEARING and I. D. L. FOSTER
Geography Department, Coventry Polytechnic

INTRODUCTION

This diverse collection of contributions is the result of convening a broad-based conference on soil erosion on agricultural land, and is an expression of the breadth of perception by researchers as to what constitutes relevant work in this field. Whether one considers this collection as 'state-of-the-art' or not, it exemplifies the nature of soil erosion studies across a large number of agricultural environments which exhibit various physical and socio-economic characteristics. This final chapter reviews some of the major themes to arise from both the contributions presented here and the discussions which followed at the conference. It is hoped that this chapter may open up further discussion on the ways in which soil erosion research may be conducted in the years ahead. Before turning to specific themes, it may first be useful to review both the nature of the problem which the contributers have sought to study and the modes of scientific explanation which they have used.

IDENTIFYING PROBLEMS

Soil erosion research is normally justified in one of two ways; either in terms of explaining geomorphological landforms or of explaining the on- and off-farm aspects of soil movements. Any other justification would elevate the value of methodology and technique above that of explanation and this is considered to be both logically and scientifically unsound. In the context of the present volume it therefore follows that all the contributions should have relevance to a particular requirement of those who deal with soil erosion where it is defined as being detrimental in some way; each of the studies should contain some

---

Soil Erosion on Agricultural Land
Edited by J. Boardman, I. D. L. Foster and J. A. Dearing
©1990 John Wiley & Sons Ltd

reference to a particular problem which they seek to solve either directly or by contribution to a scientific chain of explanation. What the volume clearly shows is that there are two different kinds of perceived problem. To make an analogy with ecological studies (cf. Passmore, 1980), it is possible to distinguish between a 'problem in erosion studies', which is a scientific problem about the factors and processes, and an 'erosion problem', where the concern is with loss of soil which is too high. The phrases 'too high' and 'erosion problem' are often rather loosely defined. However, an 'erosion problem', frequently couched in terms of applied research, is essentially a social problem defined by a society as an undesirable set of processes.

These two different kinds of problems represent the first division in any broad classification of erosion research, and their solution generally serve as ultimate goals in a research project. But how are these problems presented and to what extent are they well defined? Chapters dealing with 'problems in erosion studies' here greatly outnumber those on 'erosion problems', a breakdown that may be cause for some concern, since it implies that within this scientific community there is greater emphasis placed on processes than on the means of conservation and policy. Certainly, there are few generally recognized and accepted goals among this international community, other than the need 'to explain', and individual perceptions and expertise govern the scope and scale of most of the studies. This straightforward classification also belies the fact that the majority of 'problems' are, in fact, either implicit or refer to problems about techniques, rather than specific scientific or environmental concerns. What also comes across strongly in this volume is that it is not always clear how the solution of 'problems in erosion studies' may feed into the solution of 'erosion problems'; contributions which demonstrate this marriage are few in number. If, as a scientific community, we are not articulating clearly the reasons why we are undertaking the work, then why not?

First, it may be that many researchers (especially those who attend a geomorphological research group meeting) are still influenced by geomorphological traditions and do not make the necessary mental jump to the application of theory and data to contemporary 'erosion problems'. Unless field researchers make themselves aware of the requirements of, for example, agriculturalists, erosion modellers or engineering hydrologists, it is unrealistic to expect that the data employed and the environment studied are optimal for reaching full explanations of the erosion process; full, in the sense of all aspects of erosion, from process to policy.

Second, erosion problems frequently exist in regions where the experimental base is poor. It is common that where there exists a large database on soil erosion the rate of soil erosion is low or only locally perceived as being a problem. Those countries and regions which have both recognizable 'erosion problems' and a rich scientific infrastructure, such as the United States and the Low Countries, continue to advance theory and solutions at a fast rate.

Third, and as a consequence of this last point, the argument has been made that to succeed in understanding and explaining soil erosion where it is not a major problem is to provide the strongest of tests for the methodology and techniques. However, in practice, attempts to apply highly tuned methodologies and sophisticated techniques to actual areas where problems exist may fail because, for instance, of the difficulties in compiling sufficient data for input into empirical models, or in linking together vastly different scales of process operation. Arising from this point is the fact that research into soil loss tolerance levels is still underdeveloped. It is not easy to define those agricultural fields or regional environments which, on the basis of generally accepted criteria, do in fact have erosional problems. Yet, except for the purposes of scientific control, there can be no scientific value in studying erosion, in the present context, where it is not defined as a problem.

Finally, it is unfortunately true that too few studies are based in theory, be it the mathematical equation, a published idea or an intuitive feeling. The majority of the field studies in this volume have followed an essentially inductive approach, where aims are descriptive or designed to test not an idea but a technique. As discussed more fully by others (e.g. Church, 1984; Burt and Walling, 1984), it has to be acknowledged that these approaches do not optimize our overall chance of explaining the soil erosion process. Several of these points, in fact, relate to the need for stronger methodological frameworks and in particular the strengths and weaknesses of modelling.

MODELLING

There is no other topic which arouses so much discussion as that of the use, abuse and non-use of models, and researchers are frequently divided in their enthusiasm for them. It is generally accepted, however, that empirical or physical models are the only means for generalizing and predicting future environments. Without them there will be, at best, a large of number of case studies providing local knowledge and partial understanding at both spatial and temporal scales. There is, in modelling, one of the strongest means for bridging the gap between 'problems in erosion studies' and 'erosion problems', a point which raises issues of both scientific approach and public policy. Should academics be interested in developing increasingly sophisticated models of environmental behaviour or providing rapid solutions to severe erosion problems? There is ample evidence in this volume that most erosion problems, and solutions for that matter, are site-specific and that, consequently, the conceptually simpler route of field testing a number of erosion control measures (Chapters 33, 35 and 39) may be more meaningful, quicker and cheaper in a local situation than a new USLE or EPIC style model, the outcome of which can only be evaluated following many years of field trials.

This statement, however, raises two further questions; first, relating to the immediacy of a solution to the erosion problem and, second, relating to the way in which research scientists tackle the problem. Many of the chapters presented in this volume suggest that their erosion problem is an immediate one requiring a rapid response, particularly in the examples from the developing world. There, experience and simple field testing of erosion controls frequently provide a basis for adopting measures which appear to succeed in reducing soil erosion to an 'acceptable level'. Indeed, one is tempted to re-emphasize the point made by Dunne (1984) that in all circumstances there is no substitute for observing and knowing the catchment or environment which one is attempting to research or model. Such a knowledge base, although subjective, may be used to construct more objective conceptual models.

The second point is of more fundamental academic concern. Many of the contributions presented here are evaluating new techniques and models (e.g. ^{137}Cs analysis and the EPIC model), identifying new associated problems (e.g. nutrient and organic matter losses) or applying relatively well-established techniques in different ways (e.g. through laboratory and plot studies). These techniques and models are invariably costly (in terms of field and laboratory equipment or data requirements) and, one might argue, very rarely match environments where the erosion problem is most severe. In the examples found here, researchers working in regions where problems are most severe concentrate upon rapid rather than conceptually elegant solutions.

From an experimental point of view, we can develop both statistical and dynamic models most easily from studies involving contemporary process monitoring in plots, fields or catchments. Here, experiments may be designed to test theories of erosion dynamics or establish a basic principle which might be incorporated into a soil erosion model. The case studies reported here reflect a remarkable diversity of catchment and field scale studies, incorporating a bewildering array of 'so-called' independent variables from which soil losses might be predicted. However, many of the empirically derived models are inadequate for widespread application in that they are both developed and verified in the same physical environment. Once tested, these models are used by many field scientists and advisors in environments in which, first, the model has never been tested and, second, was never intended to be applied: reasons commonly used to discredit models like the USLE.

Historically derived data from deposits, which are now receiving an increasing amount of attention, are probably sub-optimal for model building for three major reasons: first, because the modeller is constrained by data availability, including indeterminate problems like the accuracy and precision of dating; second, because of the constraints imposed upon independent tests of any models developed from such a database and, third, it is unlikely that any model developed from such a database will elucidate the complete nature of controls or dynamics of the processes and transfers involved. Nevertheless, it would seem that the points raised by Thornes (1989) are worth repetition here in that data derived from such historical enquiries might be utilized to test dynamic models developed from contemporary process monitoring.

A further difficulty involves the degree to which modellers and field scientists collaborate over solving erosion problems. This was raised by Professor Kirkby (University of Leeds) in the conference discussion and is a recurrent problem yet to be effectively tackled. For many 'non-modellers' there are still the problems of constructing models directly, or entering into an uneasy collaboration with modellers whose scientific approach and values may appear difficult to follow. Field scientists may need to become more imaginative in the types of questions posed at the beginning of a project; questions which may be tackled within a modeller's framework. However, the research thrust in many countries is one of continually developing new kinds of model rather than exhaustively testing and improving the application of existing models. To this end, more emphasis could be placed on modellers and experimentalists identifying together the right model, or range of models, to suit the particular scale of operation. Many existing models are neither relevant nor calibrated for all situations; rather a surprising fact given the large amount of field data which has never seen a model. Modellers may have to become more pragmatic about what constitutes a successful predictive model, placing greater value on practical application rather than on the quality of the underlying theoretical and mathematical bases. It may be, therefore, that a first approximation of the solution is derived from conceptually less elegant models (such as expert systems), whose application might provide valuable information for optimization of a more sophisticated and accurate dynamic model. Collaboration between modellers and experimentalists is highly desirable, but it may mean that each party relinquishes some of its independence for the sake of scientific explanation.

EROSIONAL THEMES

The papers presented at the conference and in the proceedings volume cover a range of approaches relating to both soil erosion and sediment yield in agricultural areas, though it should be stated that most emphasis was placed on erosion by water rather than by wind. The apparent diversity of interest surrounding water erosion can be given some degree of coherence by considering the main themes which incorporate the questions which we should be attempting to answer.

On-farm Problems

First it is convenient to review on-farm problems, such as the redistribution of soil, the loss of structural stability, organic matter and nutrient elements through both particulate and solutional erosion, and the tendency for intense runoff at certain times of the year to promote rill and gully generation. These studies often define soil erosion in terms of (quasi) economic loss; of cultivable soil depth, loss of fertility; or of the cost of replacing certain essential nutrient elements artificially. However, contributions presented in this volume and elsewhere frequently adopt the case study approach, limiting the analysis to a small number of fields or plots and, usually, drawing conclusions or developing models on the basis of a detailed monitoring or reconnaissance survey. The applicability of such studies is rather narrow, often providing data and a modelling framework which are unique to the area. The types range from empirically-derived models, such as the USLE and EPIC (Chapters 27 and 30), through process-based models (Chapter 28), and eventually to a set of more recently employed models based upon expert systems analysis (Chapter 26). This last strategy involves the collection of general soil erosion data alongside those variables which might have some bearing on soil loss on a wider regional basis. Studies of this type suffer limitations in terms of their explanatory value.

One of the strengths of the chapters dealing with the theme of on-farm effects is the impact of modern agriculture, identifying the 1960s and 1970s in the UK and elsewhere as the period when agricultural intensification and industrialization of the countryside led to the abandonment of more traditional farming methods (Chapters 7, 8 and 11). Changes include an increase in the size of field systems, use of heavier machinery and inorganic fertilizers, a change in crop type exposing soil to erosion at critical periods in the year and increasing cattle and sheep populations in both upland and lowland agricultural systems. The studies demonstrate that increases in soil erosion are real and are not a function of better monitoring and measurement.

Off-farm Problems

Alternative terms of reference have involved the study of what may be loosely defined as off-farm impacts. These studies are divided between those which have attempted to identify the detrimental off-farm effects (Chapter 24) and those which have utilized the off-farm sediment record to learn more about on-farm effects. The latter studies are well represented in the volume, referring to sediment yield records derived from stream monitoring, or valley, lake and reservoir sediments. It is difficult to demonstrate the exact link between land use and sediment yield since the yield of sediment from a drainage basin is a function, at least in part, of the sediment delivery ratio (Roehl, 1962; Trimble, 1981). At best, regional relationships between sediment yield and a process variable such as rainfall may be subdivided by lithology or land use to improve the model.

Figure 45.1 attempts to illustrate in a rather simplistic way the difficulty in interpreting hillslope erosion from sediment yield data alone and taking account of the role of valley storage. Figure 45.1(a) is the simple case where the constant sediment yield from the basin is a direct function of soil erosion on hillslopes.

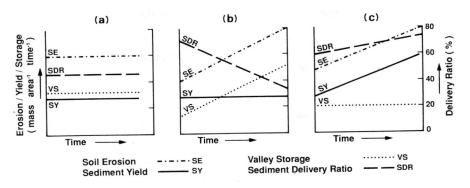

Figure 45.1 Hypothetical relationships between soil erosion (SE), sediment yield (SY), valley storage (VS) and the sediment delivery ratio (SDR). All scales are arithmetic and changes are assumed to be linear through time. (a) With uniform SE, SY and VS, the SDR remains constant. Any change in either SE, SY or VS will involve a change in the other variables, so maintaining a constant SDR. SY can only be used as a surrogate measure of SE if this condition holds true. (b) SY remains constant despite an increase in SE. Both SE and VS change, and the SDR declines over time. SY is therefore not a useful surrogate measure of SE. (c) SY and SE increase at the same rate. Since VS remains constant, the SDR increases to accommodate the increasing yields. Again, SY is not a good surrogate measure for SE because it represents a variable proportion of eroded soil

Valley-side and bottom-storage occurs at a constant rate proportional to that of hillslope erosion and, therefore, the sediment delivery ratio remains constant over time. This assumption is implicit in most studies of land use–sediment yield relationships but few attempts are made to test the assumption. Figure 45.1(b) shows the case where sediment yields remain constant, but where valley storage increases at a rate proportional to hillslope erosion. The sediment delivery now falls through time, and negates the assumption derived from Figure 45.1(a). Where the remobilization of stored valley sediment provides a major source for downstream sediment yield, an increasing sediment yield record would mask a declining rate of hillslope erosion (Figure 45.1(c)); both cases 1(b) and 1(c) could in theory have the same temporal changes in sediment delivery, a declining trend, but for very different reasons.

Such problems are acknowledged in this volume, but the contributors take different approaches to overcome them. First, direct studies of valley sedimentation processes (Chapter 13) suggest that floodplain storage is not only important in terms of the quantity of sediment held in middle reaches but that the rate of storage may be surprisingly high, with complex controls ranging from bridge and road construction to major changes in local agriculture. Second, there are examples of linking complex conceptual frameworks of sediment movements with field data to obtain 'total' sediment budgets (Chapter 11). These are certainly time-consuming projects requiring detailed monitoring of a large number of processes. Such studies generally are site-specific and have not, as yet, led to the development of mathematical and statistical models of the complexity of some of the conceptual models associated with them. Indeed, as mentioned above, it may be argued that the wide diversity of erosion problems presented in this volume would seem to preclude both generalization and modelling other than on a site-specific basis and at a predetermined scale. Third, a number of chapters consider the links between sediment yield and soil loss from the hillslope in situations where valley storage is minimal and data on sediment delivery can be used to interpret soil loss from sediment yield records directly.

In this context, the term 'buffer zone' is used widely to define a strip of land, usually grassed or wooded, which essentially restricts the movement of soil to a channel, either totally or partially. Undoubtedly these zones are relevant in terms of both on- and off-farm impacts (Chapters 6, 7, 10–12, 16 and 23). In each of these studies the buffer zone is shown to limit sediment and nutrient transfer across its boundary. But how do the transfers occur? For each site, they represent a set of identifiable contributions which, in the short term at least, may be spatially fixed. However, the sum of these site-specific transfers, such as breaching by tractor wheelings, rill formation or the frolicking of a cow, may present insurmountable problems in developing dynamic models of the soil and sediment transfer process. In the long term, we have no knowledge of these specific transfers and treat soil erosion as a 'lumped' or aggregate process.

SEDIMENT SOURCES

Finally, a number of chapters use 'fingerprinting' techniques to aid the interpretation of sediment source within the sediment yield record, including ^{137}Cs (Chapters 4, 9 and 10), magnetic parameters (Chapters 12, 14 and 16) and geochemistry (Chapters 16, 20 and 23). Chapter 10 reviews the approaches and techniques, and also provides a comparative study of different techniques at one site. Overall, these techniques hold promise but as yet do not provide a totally quantified breakdown of sediment sources. Common problems seem to be the large variability in sources, as defined by analyses of their properties, and their restrictive use to distinguishing between topsoil and subsoil sources. Although this distinction is very useful in sediment yield–soil erosion studies, the approach is effectively limited to small catchments where soils may be conveniently assumed to be homogeneous in their contribution to the sediment yield. In order to expand 'fingerprinting' techniques to larger catchments it will be necessary to define sediment sources with far greater resolution. At present it would be possible to use source data to test simple predictive models of the contributions of topsoil and channel-derived material to the sediment yield (cf. Chapter 12).

Scale Problems

Much of the diversity prompting the discussions above relates almost solely to the scale of approach. Here, it is convenient to deal with spatial and timescales separately, although many of the issues are interrelated.

Certain common independent variables may be identified, irrespective of the scale of approach. These include gradient (either valley-side or down-valley), vegetation cover and energy input to the system. The last variable may be expressed in many different ways, ranging from estimates of rainfall erosivity on a slope to measures of annual effective rainfall or runoff in a drainage basin. However, some variables are scale dependent. Along a valley-side soil erosion may vary with position in the landscape (for example, whether an erosion plot is set up in a hillslope hollow or on a topographic divide). Controls here are through both variations in variables such as vegetation density and hydrological controls generating surface runoff in different parts of the basin. Similarly, the opportunity for sediment storage to increase, and sediment delivery to decrease, will be a positive function of increasing size of experimental area from plot to drainage basin.

One way of illustrating the scale problem is to consider the answer to a relatively common and simple question: 'How much soil is being lost?' The answer depends to a large extent on the position in the landscape and the scale at which we choose to measure the phenomenon. Application of ^{137}Cs analysis (Chapters 4 and 9) may tell us something about the redistribution of soil over individual fields or even entire hillslopes, although many practical problems have yet to be overcome in equating ^{137}Cs movement quantitatively with soil redistribution. It is difficult to equate such a redistribution with conventional figures of soil erosion presented in units of mass per unit area per unit time, since, strictly speaking, the soil erosion is normally defined as losses measured at the slope foot or channel margin. These difficulties also occur in relation to field plots of different area and, in particular, length. Imposition of artificial boundaries may make such calculations, at best, difficult to compare with other data and, at worst, totally irrelevant to the soil-erosion problem. A further difficulty arises in defining the effective area of certain phenomena which are not bounded by plot sides, field boundary or on easily identifiable topographic divide. This is particularly relevant to the definition of contributing areas of rills and gullies, where calculations in mass per unit area per unit time are meaningless. Despite the obvious difficulties highlighted, the yields estimated by these varying methods cannot be considered comparable, despite the apparent unification implied by common units of measurement. There is undoubtedly a need for scientists researching the soil erosion problem to more clearly define the most fundamental question of measurement units and, therefore, process rates. The scope, purpose and objectives of particular studies, as highlighted earlier, should include a statement and consideration of such problems from both a theoretical and a practical standpoint.

The question of whether or not rates of soil loss have increased or decreased over time again depends on the scale of reference. For example, some studies have tried to examine soil loss at different times during the post-glacial period, through sediment yields derived from lake or reservoir sedimentation (Chapters 11, 12, 14, 18 and 32). Interpretation involves comparison of potential climatic, vegetation and land-use controls, and in some cases using 'finger-printing' techniques to identify shifts in sediment source over time. Other studies have looked at individual events recorded as valley or slope sediments at different times in the post-glacial period (Chapters 15–17). Sediment loss or frequency analysis may be applicable to sedimentation rates and patterns which may be identified on an individual storm event basis (Chapters 11 and 32). Furthermore, rates of soil erosion averaged through long periods of lake and reservoir sedimentation may overcome some of the difficulties in dealing with extreme events in short-term hydrological records (Foster *et al.*, 1985 and Chapter 12). Trends in soil erosion are difficult to determine on the basis of experimental data, since few studies have collected information over a sufficiently long period of time in order to put the contemporary problem in an historical context. Many researchers have blamed our inability to objectively assess the contemporary problem on the lack of a co-ordinated national or international monitoring system (Chapters 7, 14 and 40). Studies of lake and reservoir sedimentation probably offer the best opportunities to obtain surrogate historical data. However, most data collected to date relate to drainage basins contributing to lakes through river system filters rather than directly through an output from hillslope process. Some studies presented in this volume may partly overcome this limitation where recent catchment management strategies have removed channelled inflow to the lake (Chapter 12). However, lake sediment-based studies are suboptimal from three points of view. First, the reconstructed record will usually be an average, over 5–10 years at best, and, at worst, over decades to hundreds of years. Second, models of erosional process may be difficult to build on the basis of these data due to a lack of independent and continuous process and land-use information (annual rainfall records, cattle populations, etc.). Third, sediment yields, as discussed above, are controlled by both hillslope and channel processes, and therefore a sediment budgeting approach based upon sediment source 'fingerprinting' may offer the only opportunity to provide meaningful data that relate soil erosion rates to sediment yield. Where lake basins do not exist, assessment of erosion by studying colluvial and alluvial sediments may provide a valuable alternative. However, in order to more fully evaluate the spatial and temporal variations in sediment transfer, an integrated lake–catchment–hill slope approach may provide the only objective means of examining the soil erosion/sediment budget system at all scales (Chapter 15 and 16).

EROSION RISK

Many of the contributions are, to a greater or lesser extent, concerned with the assessment of erosion risk. This may involve mapping procedures or the application of a model such as the USLE. In each case, data collection and processing is necessary in order to make the assessment. This may be either a major task extending over several years or the work of a rainy morning in the laboratory, where topographic and soil maps are consulted. Chapters 25–27, 29 and 30 use approaches that combine these extremes, for example, supplementing published map data with those from field and laboratory experiments designed to fill in the gaps. The value of such an approach is clear but the problems, particularly those of moving between scales, are discussed in Chapter 27.

The assessment of risk frequently involves a compromise between the demands of scientific rigour and those of cost. Aerial photographs have been used to reduce the cost of assessment of large areas. Repeated survey, or monitoring of erosion risk, has proved to be a difficult concept to sell to decision makers. The justification of such exercises is that they are able to highlight problem areas and allow action to be taken before damage to soils or property occurs. It is, for example, easy to assess off-farm damage to property and costs to local councils on the South Downs in southern England since 1982 as reaching £1 million. There are simple methods of risk assessment that could be used (Boardman, 1988). The problems are threefold. First, the lack of political will to take soil erosion seriously even where off-farm costs are shown to be substantial—and this is not simply a British affliction. Second, the costs of erosion are not primarily borne by the farmer (cf. Chapter 24) but by individual householders, water boards, local councils and insurance companies. Third, there is a failure to take a long-term view of the soil, to regard it as a basic finite resource. Too often in Western societies this position is regarded as the rather crankish prerogative of the organic farming movement. However, unless erosion is monitored using standard scientific procedures it is not possible to assess the scale and extent of the problem.

The use of predictive models, particularly the USLE, has traditionally been a short cut, allowing quick and cheap assessment of the risk of erosion; the risk being equated with the predicted rate of soil loss under given conditions. As discussed above, it is now clear that the USLE has serious drawbacks when used at the field scale. For example, there have been few attempts to validate the model. The results obtained in many situations have an unknown relationship to reality. Validation is, admittedly, difficult because the USLE claims to predict an average rate of erosion over several years, but this does not excuse its widespread and uncritical application. Furthermore, the range of conditions under which the USLE was developed is rather limited; recognition of this has led to the practice of adapting the R factor to local conditions with the assumption that other components are satisfactory. Concentrated flow, which in many field situations is responsible for most soil losses, is not dealt with in the USLE; Chapter 29, for example, proposes a simple method of dealing with the prediction of gully development.

Many contributions in the assessment and modelling sections of the volume reflect the search for models which may replace the USLE (Chapters 25–31). Some, such as EPIC and GAMES, include USLE methodology. One disadvantage of the new generation of models is the amount of data required and the small areas to which they may be applied. A process-based model, as described in Chapter 28, which can be applied to large areas of the agricultural landscape still seems light years away!

POLICIES

Assessment of erosion risk, given the political will, is largely a scientific problem. However, assessment interacts with economic and social considerations, and therefore with the area of policy and implementation. Questions, such as 'Who is responsible for soil conservation?' are raised in different countries, and have been answered in different ways. The American answer has been to set up a Soil Conservation Service; the Canadians, coming to a recognition of the problem later, have preferred to work through existing organizations; the British, in a decade of 'rolling back the frontiers of the state', have dismembered the only organization, the Soil Survey, which had responsibility for monitoring erosion. Signs for the future are that in Europe a supranational organization, the European Community, will have considerable influence on national environmental policies, pollution having been recognized as a transfrontier problem (cf. Chapter 40).

As we move into a period of increased public awareness and environmental concern, where such issues are beginning to feature in political debate, another series of questions that were discussed at the conference emerge. In areas of policy initiation and development, what is the role of scientists? Should they be merely providers of data, moderating the wilder flights of fancy of the (over?) committed environmentalist and the (under?) committed politician, arbitrating between irrational commitment and a rational cost-effective approach? Or is a more involved role obligatory, or at least optional? Should scientists be publicists and propagandists, and remain objective? Perhaps in an ideal world many of these roles would be left to those who are better qualified to fulfil them than the average scientist. The success in making soil degradation a public issue and attracting support and funds at government level has, in Canada, been substantially the work of Senator Herbert Sparrow, but in most societies scientists have had to (or will have to) fulfil that role. In Europe, the setting up of the European Society for Soil Conservation is an attempt to convince both the public and the policy makers of the need for soil conservation (de Ploey, 1989).

Perhaps it is ironic that, as Europe gingerly limbers up to debate conservation policies, the message from the United States is that many of these will not work unless backed by sufficiently attractive financial inducements and some form of coercion (Chapter 43). The 'carrot and stick' might be suitable symbols for our new society, and not simply because they represent the products and tools of agricultural enterprise! American farmers, we are told, understand quite well the 'erosion problem', but choose for financial reasons to opt out of those conservation programmes which are not to their advantage.

In the Third World the cost implications of conservation policies are paramount. Low-cost systems are essential if conservation is to be successful. In this case the best solution may be to look towards traditional methods of soil and water control which under the pressures of modern times have been neglected (Chapter 33). Can they be adapted, not just to traditional peasant agriculture and today's population pressures but also to cash crop and plantation agriculture?

Looking to the future, current speculation with regard to the 'greenhouse effect' has focused on food supply and changes in energy provision, sea level and forests. There has been little recognition of the implications of climatic change for erosion. Shifting foodbelts imply movement of associated problems. In some areas new crops will be grown on soils which are more sensitive to the risk of erosion. The climatic models have emphasized temperature change, but rainfall as a major control on runoff is critical to erosion processes: we need to know far more about changes in rainfall patterns in a warmer world. New crops in some areas also imply new risks. For example, it will only require a slight increase in temperature for maize to be an attractive option in southern England, in which case erosion rates are likely to increase.

Such speculations at least suggest that scientists working on current problems of erosion and conservation can contribute to the debate about the future. Contributions such as Chapters 12 and 15 show that erosion rates have in the past varied with population pressure and changes in farming practice. Climatic change, on a scale and at a rate unknown during the present post-glacial period, constitutes a threat to society but it also offers a new challenge to science.

REFERENCES

Boardman, J. (1988). Public policy and soil erosion in Britain. In Hooke, J. M. (ed.), *Geomorphology in Environmental Planning*, John Wiley, Chichester, pp. 33–50.

Burt, T. P. and Walling, D. E. (1984) Catchment experiments in fluvial geomorphology: a review of objectives and methodology. In Burt, T. P. and Walling, D. E. (eds), *Catchment Experiments in Fluvial Geomorphology*, Geobooks, Norwich.

Church, M. (1984). On experimental method in geomorphology. In Burt, T. P. and Walling, D. E. (eds), *Catchment Experiments in Fluvial Geomorphology*, Geobooks, Norwich.

De Ploey, J. (1989). Why the society; from soil degradation to soil conservation, *European Society for Soil Conservation, Newsletter*, **1**, 3.

Dunne, T. (1984). The prediction of erosion in forests. In O'Loughlin, C. L. and Pearce, A. J. (eds), *Symposium on the Effects of Forest Land Use on Erosion and Slope*

Stability. 7–11 May 1984. Environment & Policy Institute, University Hawaii, Honolulu, pp. 3–11.

Foster, I. D. L., Dearing, J. A., Simpson, A., Carter, A. D., and Appleby, P. G. (1985). Lake catchment based studies of erosion and denudation, *Earth Surfaces Processes and Landforms*, **10**, 45–68.

Passmore, J. (1980). *Man's Responsibility for Nature*, Duckworth, London.

Roehl, J. W. (1962). Sediment source areas, delivery ratios and influencing morphological factors, *IAHS*, **59**, 202–13.

Thornes, J. B. (1987). Models for palaeohydrology in practice. In Gregory, K. J., Lewin, J. and Thornes, J. B. (eds), *Palaeohydrology in Practice*, John Wiley, Chichester, pp. 17–36.

Trimble, S. W. (1981). Change in sediment storage in Coon Creek Basin, Driftless Area, Wisconsin, 1853–1975, *Science*, **214**, 181–3.

Place Names Index

Bold page numbers refer to entries in headings, other page numbers refer to the first entry in a chapter.

Africa   495, **496**, 505, 537
  East   576
  Sahel   505
  Southern   575
  West   505
Asia   645
Australia   119, 205
  Barwon River, NSW   133
  Brisbane   119
  Broken Hill, NSW   205
  Burrunjuck Reservoir, NSW   218
  Canberra   218
  Goodradigbee River, NSW   218
  Hunter Valley, NSW   119
  Lake Burley Griffin, ACT   205
  Merriwa, NSW   121
  Mundi Mundi Plain, NSW   205
  Murrumbidgee River, NSW   218
  New South Wales   119
  Sydney   119
  Umberumberka reservoir, NSW   205
  Yass River, NSW   218

Belgium   513
  Brussels   519, 598
  Middle   388
  Leuven   519
  Loam Belt   **513**
Britain   33, 87, 130, 298, 301, 401, 669
  see also England, Scotland, Ireland, Wales, United Kingdom
Brazil   133
  Amazon River   133
  Sao Francisco   133
Botswana   466
  Hard Veld   471
Burkina Faso   495, **505**
Burma   302

Cameroon   133
  Sanaga River   133
Canada   565, 670
  Guelph   416
  Ontario   417
China   133
  Huangfu River   133
Denmark   313, 595
  Jutland   315
  Lyngbygaard River   315
  Gelbaek River   315
  Gudena (river)   315
Egypt   133
  River Nile   133
England   231, 302, 593
  Albourne, Sussex   91
  Balsdean, Sussex   91, 93
  Bedfordshire   50, 52, 132, 248
  Bevendean, Sussex   93
  Bickley Brook, Cheshire   255
  Bodmin Moor   244
  Bowland Fells   235
  Breaky Bottom, Sussex   91, 93, 101
  Brecon Beacons   235
  Brighton, Sussex   249, **369**
  Brook End Farm, Beds.   39
  Buddington Bottom, Sussex   91
  Cambridgeshire   50, 132, 242
  Castleton Caves, Derby.   307
  Central   235
  Cheshire   255
  Claverley   253
  Coombe Haven, Sussex   273
  Cornwall   232
  Cotswold Hills   235
  Coventry, W. Midlands   157
  Craven Lowlands, Yorks.   236
  Cropston Reservoir, Leics.   298

673

674   Index

England (*continued*)
  Dartmoor 244
  Devon **193**, 232
  Dorset 132
  Downlands 232
  East Anglia 235
  Eastern 235
  Essex 232
  Exeter, Devon 138
  Fenland 245
  Gloucestershire 241
  Hangleton, Sussex 370
  Herefordshire 132, 236
  Highdown, Sussex 93
  Hilton, Shrop. 255
  Hove, Sussex **369**
  Howgill Fells 235
  Jackmoor Brook, Devon 133, 135
  Kent 21, 242
  Lake District 235
  Mangle Hole, Mendips 304
  Mansfield, Notts. 232
  Merevale Lake, Warks. 154, 298
  Midlands **153**, 232
  Mile Oak, Sussex 370, **375**
  Newton Mere, Shrop. 258
  Norfolk 232
  North Yorkshire Moors 233
  Nottinghamshire 232
  Otter Hole, Chepstow 304
  Ouse Washes 245
  Peak District 244, 307
  Pennines 235
  River Clyst, Devon 133
  River Creedy, Devon 133
  River Culm, Devon 133
  River Dart, Devon 133, 136
  River Gara, Devon 70, 194
  River Humber 233
  River Severn, Glouc. 133
  River Start, Devon 70, 194
  River Thames 235
  River Merrifield, Devon 70
  River Worfe, Shrop. 257
  Rottingdean, Sussex 21, 100, 101, **371**
  Rufford Forest Farm, Notts. 39
  Seeswood Pool, Warks. 154
  Sherwood Forest, Notts. 232
  Shropshire 132, 235, 255
  Slapton Ley, Devon **194**
  Slapton Wood, Devon 194
  Somerset 132
  South Downs 87, 369, 401, 569, 594, 669
  South East 15, 88, 283
  Southern 235, **273**, 669
  South West **69**, 241
  Staffordshire 132
  Stokeley Barton, Devon 70
  Strines Reservoir, Yorks. 298
  Surrey 241
  Sussex 232, 273, 369, 403
  Torcross, Devon 195
  Vale of York 238
  Warwickshire 154, 235
  Weald 241
  West Midlands 88
  Worcestershire 50, 132
  Worthing, Sussex 91
  Wye College Farm 16
  Yorkshire Dales 235
  Yorkshire Wolds 241
Europe 120, 137, 401, 575, **591**, 651, 670
  Alpine 593
  Loess Belt 3
  Northern and Central Lowlands 591, **592**
  Northern 217, 387, 591, **592**
  North West 4, 383
  Uplands 591
  Western 217

Federal Republic of Germany 592, 596
  Bayern 594
France 592
  Cévennes 593
  Ile-de-France 387
  Laonnais 387
  Northern 525
  Paris Basin **383**
  Pays de Caux 525
  Pyrenees 593

Golden Triangle (S.E. Asia) 302
Greece 15, 592

Holland
  South Limbourg 3, 8, 10, 11
Hungary **605, 607, 608**
  Komarom County 612, **614**
  Pillismarót 612
  River Danube 614
  Tata Great Lake 617
  Transdanubia 614

Iceland 593
  Gunnarsholt 595
India 646
Indonesia
  Citarun River 133
Iraq 15
Ireland 302
  Burren 302
Israel 495, 648
  Avdat 497
  Beer Sheva 481
  Negev desert **481, 506**
Italy **621**
  Apennines 597
  Cesena 597
  Emilia-Romagna region **621**

Kenya 508, 576

Laos 302
Low Countries 660

Malawi **575**
  Bvumbwe 577
  Mindawo 577
  Mphezo 577
Mediterranean lands 302, **592**

New Zealand
  Waitomo 306
North America 214
  see also United States of America and Canada
Northern Hemisphere 196

Pakistan **645**
  Himalayan Mts. 647
  Indus River 646
  Jhelum River 646
  Pothwar **645, 647**
  Preshawar Valley 653
  Salt Range 646
  Tarbela Dam 645
Papua New Guinea 305
Portugal 592

Rhodesia 501, 537
  see also Zimbabwe
Romania 55
  Carpathian Mts. 55
  Moldavian Mts. 55

Scotland
  Ayrshire 286
  Ballo Reservoir 286
  Carron Valley Reservoir 286
  Central lowlands 285
  Cullaloe Reservoir 286
  Drumain Reservoir 286
  Earlesburn No. 1 Reservoir 286
  Fife 286
  Gargunnock, Stirling 286
  Glenfarg Reservoir 286
  Glenquey Reservoir 286
  Girvan, Ayr. 286
  Harperleas Reservoir 286
  Holl Reservoir 286
  Hopes Reservoir 286
  Kelly Reservoir 286
  Kilsyth Hills, Stirling 286
  Lambielethan Reservoir 286
  Lomond Hills 286
  Lothian 286
  Midland Valley **285**
  N. Esk Reservoir 286
  N. Third Reservoir 286
  Perthshire 286
  Pinmacher Reservoir 286
  Solway Firth 238
  Stirlingshire 286
  Strathclyde 294
South Pacific **339**
  Apia 343
  Togitogiga 340
  W. Samoa 340
  Upolu 340
Spain 15, 592
  Alcantarilla 333
  Murcia 331
  South East 331
South Africa 537, 575
Sweden
  Harlösa, Skåne 108
  Lake Bussjösjön, Skåne 174
  Skåne 174
  Southern 107, 173

Tanzania 495, **498**, 576
  Arusha 496
  Dodoma 496
  Mwanza 496
  Uluguru Mts. 496
  Usambara Mts. 496

Thailand   25, 302
Third World   **495**, 537, 576, 670
Tunisia   495, **508**
  Matmata   497, **508**
  South   **508**

Uganda   576
United Kingdom (UK)   33, **87**, **129**, **130**, 214, 402, **559**, 569, 664
  see also Britain, England, Wales and Scotland
United States of America (USA)   101, 147, 236, 351, 401, **447**, 461, 465, 560, 575, 598, **627**, **639**, 660, 670
  Chesapeake Bay   213
  Chulitna River, Alaska   133
  Coldwater Cave, Iowa   302
  Colorado River, Arizona   133
  Coon Creek   195
  Corn Belt   636
  Door County, Wisconsin   303
  Kentucky   305
  Mississippi   449
  Ohio   637
  Potomac River   214
  Riesel, Texas   403
  Rhode River, Maryland   213
  Mid-West   352, 636, 650
  Oklahoma   352
  South East   **447**
  South West   352
  Texas   352, 403

Wales   **231**
  Cambrian Mts.   236
  Gwent   132
  Llangorse Lake, Brecon.   236
  Mynydd Perseli   244
  Ogof Agen Allwedd   304
  Ogof Craig-ar-Ffynnon   304

Yemen Arab Republic   508
Yugoslavia   302
  Belgrade   531
  Planina Caves   305
  Postojna Caves   305
  Serbia   531

Zimbabwe   495, **500**, 537, **540**, 575
  Bulawayo   500
  Bumburwi   542
  Chabwira   **500**
  Chigwada   542
  Chiwanzamarara   550
  Chizengeni   542
  Harare   542
  Kanjanda   550
  Lowveld   500
  Middleveld   500

Subject Index

Bold page numbers refer to entries in headings, other page numbers refer to the first entry in a chapter.

acidification 203
aerial photographs 34, 93, 176, 196, 297, 482, 500, 538, 569, 668
afforestation 286, 297, 554
aggregate 9, 51, 108, 141, 384, 420
  stability 4, 69, 97, 249, 391, 416, 564
AGNPS model 423
agricultural
  engineering 426
  land 173, **569**
    see also cultivated; farm
  intensification of 154, 607
  landscapes **389, 392, 394**
  methods 256
  productivity of 303, 575
  societies 173
agroecosystem 537
algorithm 110
allochthonous inputs 296
allogenic water 302
allometric analysis 63
alluvial fan 205
alluvium 203, **239**, 237, **240, 244, 246,** 286, 507, 668
almond crops 508
aluminium 261
Anglo-Saxon 235
ANSWERS model 13, 423, 439
antecedent events 220
antecedent moisture 80, 438
apatite 356
apricot crops 508
arable land, *see* cultivated land
archaeological data 89, 190, 231, **283**
arid zones 207
aquifers 301, 500
ARM model 441
atmospheric inputs 119, 160, 216
autochthonous inputs 160, 296

autogenic water 302
autumn-sown crops 5, 232, 563
  see also winter cereals
available water storage 446, **470**

banana plant **339, 343, 345**
barley crops 323, 525
basalt 121, 209
baseline values 137
bathymetry 290
bean crops 157, 579
bench terraces 232, **531**
  see also terraces
biogenic silica 60
biomass **470**
Black Death 233, 592
boreholes 273
break crops 89
breccia 138
Bronze Age 235, **273**, 283, 302
bulldozers 340, 516
bunds 378, 652
buffer strips 164, 188, 363, 666
buried crops 383
burning 236

caesium-137 and -134 33, **119,** 135, 160, 180, 196, 204, 269, 662
  see also radioisotopes
  budgets 141
calcium 261, 347, 533
calcium carbonate 180, 303, 390
Cambrian 174
carbon 16, 143, 351, 464
carbon-14 184, 194, 204, 232, 282
Carboniferous 286
capping 69, 97, 249, 563
cash economy 537

677

catchments 69, 93, 121, **285**, 389, **498**, 517, 576
  characteristics of **285**
  rehabilitation of 153
  small **498**
catena 545
cattle 161, 236, 501, 541
cave systems **306**
  see also karst; limestone
cereal crops 138, 175, 370, 540, 563
  see also winter cereals
channel(s) 132, 233, 385, 555, 647, 666
  bank erosion 82, 166, 179, 244
  bed 164
  deepening 187
  erosion 115, 134, 165, 166, 179, 209
  form 188
  reaction 188
  recovery 188
  scour 184
  sedimentation in 154
  systems 164
  undercutting 168
  widening 221
chalk 233
charcoal 204
Chernobyl nuclear accident 52, 119, 137, 217
chronology 160, 173
clay 44, 138, 166, 182, 211, 390, 484, 646
  see also soils
  minerals 35
    see also kaolinite
  primary 147
  particles 36
  transport 352
climate 55, 158, 193, 397, 437, 498
climatic change 173, 235, 282, 671
  impact 153
  models 671
  triggers 593
  variations 581
clods 7, 249
cluster analysis 213
coarse fraction 134, 290, 434
coastal sediments 201, 245
colluvium 203, 237, **238**, **241**, **243**, **246**, 255, **261**, 668
Compound Topographic Index 448
concentrated flow 98, **387**, 448, 514, 669
conceptual framework 201, 431

conglomerate 138, 646
conservation (of soil) 29, 358, 397, 429, 447, 465, 495, 575, **591**, **599**, **605**, **606**, **607**, **608**, **614**, 633
  see also policy
  coercive approaches to **427**, 633
  effectiveness of **639**
  implications of **639**
  measures 93, 397, 555, 583, **599**, 605, 621
  need for **591**
  planning 440, **612**
  political implications of 598
  practice of **605**
  profitability of **633**
  programmes 115, **628**, **630**, **639**
  strategies 214
  tillage 360
Conservation Compliance (US) **637**
Conservation Reserve Programs (US) **635**
Conservation Title (US) **634**
  limitations **638**
  Sodbuster provisions **634**
  Swampbuster provisions **635**
contour cultivation 482, **561**, 565
cooperative farms 608
copper 261
coring 121, 138, 159, 196
Corn Belt 636
corn crops 10, 332, 462, 533
  see also wheat; maize
correlation analysis 110, 122, 333
CREAMS model 403, **420**, 441, 448
Cretaceous 88, 174
cropland, *see* cultivated land
crop damage 371, 515
  see also on-farm effects; buried crops
crops, *see* name of crop (e.g. wheat); winter cereals; autumn-sown; spring-sown; buried crops
crop rotation 89, 107, 120, 157, 236, 392, 560, **564**, 569, 610
crop yields 364, **531**, 533, 570, 594
cross-profile measurement 57, 306, 447, 513
CRS model (lead-210 dating) 160
crusting 9, 97, 337
cucumber crops 579
cultivation (ed) land 69, **107**, 119, 130, 154, 161, 231, 256, 340, 369, 397, **447**, **605**

changes in   101
continuous   534, 560
impact of   36, 307
intensive   175, 314
methods   571, 575
old   16, 552

dams   498
   see also reservoirs
   earth   372
   subsurface   **503**
   sand storage   **503**
   silting-up of   500
data
   acquisition   **273**
   empirical   52
   presentation   **273**
decision-making   202, 641, 668
deforestation   56, 188, 231, 235, 302, 593
deltaic deposits   162, 295
depression storage   525
derelict land   616
desertification   497, 592
deserts   497
desiccation   290, 484
developing countries   575
Devonian   70, 194, 241, 285
discharge   144, 161, 180, 316, 418, 486
   see also runoff
discriminant analysis   214
disease   503
dispersion ratio   16
ditches   113, 176, 374, 555, 565, 571
   see also drainage
diversion terraces   **565**
dolines (sinkholes)   302
downwashing   303
double split divisor   **25**
drainage   174, **176**, 378, 608
   see also drains; ditches
   density   188, 287, 297
   pattern   113, 538
   pipes   113
   tiles   161
   under-   159, 175
drainage basin   58, 129, 154, 193, 237, **243**, **246**
   see also catchments; watersheds
drains   374
   see also ditches

drilling   90, 404
   direct   563, 572
drought   224, 324, 337, 489, 495, 561
drylands   **495**, 538
   African   **496**
   management   **495**
dry valleys   232, 370, 384
dunes   614
Dust Bowl   627

earthworms   9, 36
echo-soundings   290
ecosystems   149, 473
engineering   495, 546, 660
enclosure   232
energy conditions   187, 306, 362, 466, 515, 667
enrichment ratios   142, 336, 469
environmental management   201
Environmentally Sensitive Areas (UK)   402
Eocene   507
EPIC model   403, **461**, 465, 661
EPROM model   466
erodibility   13, 22, 88, **107**, 187, 244, 390, 420, 468, 609
erosion (water) *see also* conservation; wind erosion; rill erosion; gully erosion
   affected by agricultural landscapes   **389**
   affected by soil properties   **391**
   see also soils
   acceptable level of, *see* tolerance level
   causes of   **96**, 255, 592
   climatic factors of   179
   consequences of   **569**
   control of   297, 339, 536, **559**, **569**, **570**, **571**
      benefits   **559**
      evaluation   **560**
      techniques   **560, 561**
   costs of   369, 402, 461, **559**, 570, 594, **631**, 669
      see also economic losses
   evaluation of   461
   farmers   369
   household   369, **372**, 375
      local authority   369, **374**, **378**
   off-farm   378, 382, 387, 401
      see also off-farm effects
   on-farm   **378**
      see also on-farm effects

erosion (water) (*continued*)
  damage 317
  distribution of **94**
  economic losses from 363, 461, 663
    pressures 94
  effects of 464, **569, 594**
  estimation of **193**
  factors 173, 534, 593, 598
  frequency of **248**
  hazards 51, 605, 624
  impact of **100, 248,** 250, **306,** 401, 429, **461**
  in the past 214, **231,** 233
  inventories 429, 440
  land affected by **621, 624**
  long term 33, 182, 231, 461
    implications 101
  maps 570
  morphological evidence **232**
  needs for research **596**
  productivity index 462
  rates 33, 37, **45,** 48, **193**
  risk 113, 248, 383, 396, 570, 593
    assessment 570, 572
  patterns **33, 37,** 255
  pins 134, 166, 555
  prediction **80**
    see also models
  research in EC 598
  selectivity of 142
  spatial variations in 93
  subsoil 221
    see also subsoil
  susceptibility to 178
  surveys 501, 668
  themes in **663**
  tolerance level 115, 534, 547, 591, 633, 661
  topsoil **246,** 255
    see also topsoil
erosional systems **596**
erosivity 442, 453, 468
  see also rainfall
estuarine sedimentation 213, 273
European Community (EC) 88, 402, 598
eutrophic 296
eutrophication 154, 203, 313, 351, **617**
evapotranspiration 159, 282, 500
experimental farms 600
experiments **15,** 202, 477, **531,** 576, 650
  see also plot experiments

Expert Systems 105, **405**
  evaluation **407**

fallow land **249, 331,** 332
farming, *see* agriculture
farm planning 459
farm type **560**
ferrimagnetic minerals 166, 182, 257
fertilizers 101, 256, 296, 464, 515, 533, 585, 592
fields
  abandoned 336, 651
  boundaries 107, 232, 372, 667
  enlarged 107, **648**
  layout 571
  size 178, 607
field experiments
  see also experiments; plot experiments
  data collection **449**
  sampling 39
  survey/study **261,** 297, **518, 575**
figs 508
fine fractions 124, 133, 290, 434
fingerprinting techniques 135, 190, 666
fire 653
firewood, *see* fuelwood
flooding 90, 134, 195, 369, 383, 489, 594
floodplain 232, 498, 665
  deposits 187, 273
flow velocity 134, 161, 417
flume studies 416
fluvio-glacial deposits 174, 286, 315
fluvial transport 168
fodder 176, 506, **652**
food surpluses 402
forest cover 55, 187
forest clearance 282, **340**
fragipan 514
Froude number 417
fuelwood 496, **652**
furrows 514

GAMES model **421,** 670
GAMESP model 422
garden cultivation 540
geochemistry **256,** 470, 666
geomorphology 153, 203, 301, 470, 576, 593
geomorphic effectiveness 153
geomorphic processes 193
geosynthetics 529

Gerlach troughs   134, 179, 332
GINO-F package   473
GINOSURF package   43
glacial deposits   182
glacial erosion   302
gleying   284
global changes   202
gneiss   500
goethite   182
government   101, 595, 621, 627
granite   209, 500, 545
grass   16, **564**
  see also rotations
grass banks   94
grassland   **69**, 394
grass waterways   164, 437, 525, 563, 571
grazing   82, 124, 286, 473, 496, 540, 653
  heavy   69
  pressure   236
greenhouse effect   671
groundnut crops   579
groundwater   175, 500
gully erosion 33, 221, 298
gullies   **34, 55**, 91, 267, 383, 652, 669
  associated with banks   **516, 527**
  associated with soil layers   **518**
  channel gradient of   519
  control of   **513, 525**
  cross-section of   519
  density of   481
  definition of   **513**
  discontinuous   58
  ephemeral   **437, 447, 449, 513, 514, 525**
  erodibility of   519
  erosion of   **55**, 108, **548**
  expansion of   62
  evolution of   66
  frequency of   58
  initiation of   541
  prevention of   **525**, 513
  risk assessment of   525
  surveys of   549
  typology of   513

haematite   82, 209
headcutting   434
heathland   232, 599
heavy metals   257
hedgerows   44, 256, 266, 388, 571
hillslope   58, 81, 195, 394, 609
historical inertia   631

historical records   154, 190
  of erosion   129
  of site   124
Holocene   216, 250
horticulture   305
hydraulic conductivity   391, 419
hydraulic gradient   306
hydrograph   305, 320
hydrological cycle   **438**
hydrological data   294, 668
hydrothermal coefficient   55
hypsographic curves   293
humus   **469**
  see also soil organic matter
human impact   153, 187, 498

impoundments, see reservoirs
industrial deposition   214
Industrial Revolution   270
infiltration   3, 16, **73**, 79, 97, 159, 249, 336, **419**, 432, 572, 652
  capacity   7, 70, 113, 391
insurance claims   373
Intelligent Knowledge-based Systems   405
interrill erosion   **33**, 385, **432**
interrill flow   416, **417**
interviews   196
iron   143, 261
Iron Age   235, **273**
irrigation   21, 308, 496, 540

Jurassic   241

kaolinite   545
karst   301
K erodibility factor (USLE)   13, **420**, 434, 614

lag deposits   211
lake-catchment framework   154
lakes   116, 255, 313, 461, 570
  see also dams; reservoirs
  levels   198
  marginal erosion of   166
  siltation   **617**
  eutrophication   **617**
lake sediments   154, 160, **193**, **201**, 202, 232, 531, 667
  chronology of   195
  influx events   181
  laminated   204, 481

lake sediments (*continued*)
  multiple-core approaches   195
  polluted   217
  potential of   **202**
  recent   180
  sedimentation rates of   180
  stratigraphy   180
land degradation   29, 129, 465, 495, **621, 624**
landforms   92, 153, 659
L and S slope factors (USLE)   614
land use   **69**, **73**, 107, 124, **176**, **233**, **285**, 287, 307, 313, 389, **561**, **575**, **577**
  change   130, 173, 231, 380, 475
  maps   176
  planning   301
  policy   **501**
  records   176
Langmuir adsorption plot   356
lead   261
lead-210   160, 180, 204, **218**
legal responsibilities   380
legislation   429, 540, 641
ley   176, 236
limestone caves   **301**
linking field to river   **129**, **153**
litter layer   466
local authorities   101
long term records   306
  see also lake sediments; monitoring
lowlands   **69**, 130, 231, 296, 422, 591
low flow   320
lynchets   94, 243, 516

macrofossils   276
macropores   303
maghemite   182
magnesium   261, 347
magnetic measurements   204, **209**, **255**, 666
  see also mineral magnetism
  authigenic phases   216
  correlations   160
  diagenic phases   216
  dissolution   216
  grain-size   208
  mineralogy   **257**
  mixing models   **270**
  remanence   182
  saturation isothermal remanence (SIRM)   166, 257

S-ratio   166
magnetic susceptibility   166, 257
  frequency dependent   166, 182, 258
magnetite   182
magnitude and frequency   87, 97, **153**, 163, 173, 284, 409
maize   501, 540, 579, 594, 671
management programmes   116
manganese   261
manual labour   349
marginal land   567, 629
mass movement   527
meadow   176
mechanical processes   301
Medieval Age   233
meltwater   180
microscale studies   **417**
mineral magnetism, *see* magnetic measurements
mining   160
Ministry of Agriculture, Fisheries and Food, UK (MAFF)   34, 87, 569
model(s)
  see also modelling; ARM; CREAMS; EPIC; GAMES; THEPROM; USLE
  assessment of   **661**
  building   **415**
  comparison of   **423**
  conceptual   153, **203**, 424
  cross-scale fertilization of   **424**
  digital terrain   440
  dynamic   662
  empirically based   173, **429**
  hydrological   303
  mixing   135, 213, 270
  physically based   397
  plot scale   **420**
  predictive   173, 402
  process based   190, 270, 403, **429**, **430**, 441, 669
  semi-logarithmic   66
  watershed   154, **421**
  validation of   403, 415
modelling   52, 146, 261, 465
  see also models; ARM; CREAMS; EPIC; GAMES; THEPROM; USLE
  available water   **470**
  biomass production   **470**
  humus   **469**

multiscale approaches to 416
mineral soil **470**
overland flow 419
phosphorus 352
slope erosion 270
monitoring 129, 193, 329, 403, 549
monoculture 592
molluscs 233
moorlands 286
mulching 348, 498
multiple regression 347, **585**

Napoleonic Wars 236
National Soil Map, UK 87
natural vegetation 554
Neogene 55, 531
Neolithic 89, 250, 256
newspapers 196
nickel 261
nitrogen 143, 351, 464, 533, 579, 594
nomograph **455**
no tillage 525
non-point inputs 69, 107, 149, 351, 422, 461
non-uniform landscapes **439**
Norse times 235
nutrient 339
   budget 136
   flux 329
   limiting 351
   load 69
   losses 537, 579, 662
   sediment-associated 136
   transfer 70
   transport 113, 154

off-farm effects (of erosion) 101, 154, 559, 570, 594, **617**, 629, **631**, 659
oil-seed rape crops 157, 176, 560
on-farm effects (of erosion) 101, 129, 203, 401, 461, 559, 594, **631**, 659, 664
onion crops 508
organic carbon, *see* carbon
organic matter 69, 94, 180, 205, 274, 315, 332, 564
   losses 335, 515, 662
overgrazing 218, 498, 593
overland flow 3, 33, 70, 175, 189, 303, 324, 416, 449

paired catchments 153
palaeoclimate 221, 301
palaeoecology 203
palaeohydrology 193
palaeomagnetism 204
palaeosols 89
parent material 135, 182, 212
partial area contribution 85
partial duration sequence 489
particle size 16, 52, 124, 130, 146, 261
   effective 147
   ultimate 147
particulate matter **324**
pasture land 82, 122, 136, 157, 307, 432
PATN package 213
peanut crops 359
peasant agriculture **537**
peat erosion 236
pediment slopes 498
penetrometer 520
periglacial 190
permeability 159
Permian 138, 232, 302
physical geography 426
   role of **612**
phosphorus 107, 261, **313**, 347, 351, 579, 594
   adsorption 352
   bioavailable 351
   budgets 313
   dissolved 313
   inorganic 315, 351
   model 422
   organic 315, 464
   particulate 313, **320**, 351, **357**
   sediment associated 313, 325
   soluble **356**
   total 143
   transport of **324, 356**
photographs 580
   reference 10
pineapple crops 579
planimetric analysis 290
Pleistocene 209, 286, 647
plot studies 15, 25, 26, 33, 70, **73**, 129, 201, 255, 332, 416, 431, 490, 507, 525, 545, 566, 662
   see also experiments
ploughed layer (A_p) 4, 36, 47, 119, 136, 220, 256, 390
ploughing, *see* cultivation

plough pan   7, 514, 572
ploughs   343
poaching (by livestock)   168
point sources   305
policy (agricultural and conservation)   102, 203, 401, 537, 575, **591**, 596, **605, 607, 627, 628**, 661, **670**
  see also conservation
  adoption of   **627**
  coercion   **627**
  definition of   **594**
  formulation of   **594**
political pressure   596
pollen records   180, 204, **218**, 233, 273, 276
pollution   236, 592, 670
pollutants   70, 170, 423
ponding   420
pool and riffle sequences   306
population   **233**, 575, 621
post-glacial period   186, 668
potash   347
potassium   347, 533, 579
potato crops   560
Precambrian   498
precipitation, see rainfall
prediction   **360, 415, 447, 486**
probability distributions   **481**
productivity index   **466**
profits   629

quartz   261
Quaternary   55
  see also Neogene
questionnaire surveys   371
quickflow   82

rabbits   218
radioisotopes   117, 136, 204, **217**
  see also caesium-137; lead-210; carbon-14; radium-226
radium-226   217
raindrop impact   3, 418, 432, 484, 563
rainfall   17, 55, 90, 122, 332, 579
  distribution   449
  duration   80
  energy   17, 417
  erosivity, see erosivity
  events   93, 204, 481, 527, 581
  index   98, 404

intensity   10, 73, 107, 362, 417, 432, **487**, 531
magnitude   80
momentum   418
records   158
simulation   **417**
simulators   416
rainsplash erosion   33, 555
reclamation   66
reconstruction (of environments)   201
redeposition   44, 134
redistribution (of soil)   **33**, 141, 238
reference site   121
regression analysis   111, 120, 164, 320, 333, 356, 486
relaxation time   153
relief   108, 608
  see also slopes
research priorities   225
reservoirs   293, 333, 364, 431, **438, 501**, 538, 594
  see also lakes; dams
  infilling of   218
  sediments   154, 202, 285, 481, **484**, 667
  small   **501**
  surveys   130, 285
  trap efficiency of   294
reworking   153
Reynold's number   417
rice crops   540
ridge and furrow   39, 235, 337, 432
rill erosion   13, 34, 50, 91, **108**, 138, 179, 267, 372, 425, 431, **434**, 447, 484, 516, 554, 560
  see also erosion; interrill erosion; rill-interrill erosion
rill–interrill erosion   **385, 394**
  see also rill erosion; interrill erosion
riparian zone, see buffer zone
rivers, see channels
Roman Age   233
root crops   236
rotation, see crop rotation
runoff   3, 17, 20, 70, **80**, 81, **85, 99**, 107, 158, 238, 282, 304, 315, 356, 375, 447, 481, 486, 533, 541, 580
  control   571
  desert   **506**
  generation   97, 190, **391**
  index   396
  peak   114

Index

plots, *see* plot studies
troughs, *see* Gerlach trap
rule-based approaches   411

salinization   509, 592
sampling   41, 121, 129, 138, 307, 315, 344
sand fraction   484
sandstone   138, 232, 646
satellite images   500, 538
savanna   474, 538
scale   153, 173, 224, 396, 416, 431, 660, 667
scientific explanation   146, 659
sediment
    see also valley sedimentation
  accumulation   485, **487**, **489**
  availability   164, 325
  budgets   125, 154, 668
  concentrations   164
  contributing area   34, 178, 396
  delivery   **80**, **81**, **129**, 153, 308, 433
  delivery ratio   129, 166, 664
  deposition   124
  dynamics   189
  effective size   **146**
  entrainment   98, 185
  exhaustion   164, 329
  fans   179
  flux   154
  load   180, 308, 419, **423**, 431
  loss   115, 579
  movement   **136**
  pathways   188, 394
  production   **69**
  quality   **180**, 256
  sequences   231
  sinks   137, 164, **195**, 198
  sorting   124
  source   **134**, **164**, **180**, 182, 201, **209**, 245, 384, 666
  source tracing   136
  storage   **136**, 153, 165, 394, 664
  supply   161
  transport   113, 124, 149, **153**, 195, 385
  traps   165, 307
  yield   129, 153, **161**, 180, 193, **195**, 224, **285**, 415, **481**, 547, 663
    event   **481**
    long term records   **159**, 181, 184
    recent records   **173**
sedimentation   198, 488, 496, 594

  see also lakes; reservoirs; valleys
semi-arid   496
sensitivity   153
Set Aside schemes (UK)   402, 564
settling tanks   160
sewage   308
shaft flow   304
shear strength   514
shear stress   385, 417, 434, 514
sheep   236, 379, 595
sheet wash/erosion   33, 98, 122, 165, 179, 209, 263, 305, 348, 385, **470**, 554
shelterbelts   599, 612
shifting cultivation   302, 339, 501, 537
silage   566, 594
silt fraction   166, 211
simulation   **70**, 352, **470**
  see also models; rainfall
slaking   4, 249, 390
SLEMSA model   430, 468, 547
slope   **89**, 108, 121, 134, **241**, 263, 404, 468, 667
  angle/gradient   97, 332, **345**, 387, 471, 531
  form   56, 112, 346
  importance of   115
  length   178, 471
  position on   **345**
slopewash, *see* sheet wash
slumping   303
snowmelt   107, 179
sodium   549
soils (classified types)
  see also soils (general types)
  Soil Series (UK)
    Bridgnorth   263
    Hook   16
    Newport   263
    Panholes   16
    Upton   16
  7th Approximation (US)
    Psammentic Paleustalfs   353
    Typic Ustochrepts   353
    Udertic Paleustolls   353
    Udic Pellusterts   353
  others
    Brown Lithosols   483
    Stony Serozems   483
soils (general types of)
  see also soils (classified types)
  acid brown earth   70

soils (general types of) (*continued*)
  black earths 122
  brown earths 138, 175
  brown forest 616
  brown sandy 263
  calcareous brown earths 175
  chernozems 614
  dambo 498, **537, 542**
  duplex 4
  ferruginious-ferrallitic 577
  immature 186
  loessial 3, 88, 238, 483, 507, 513, 592, 605, 645
  podzolic 232
  rendzinas 404
  saline 483
  sodic 390, 483, 542
  terra rossa 302
  volcanic 347
soil (properties/states)
  see also conservation; erosion; topsoil; subsoil
  amelioration **606**
  argillic horizon 235
  biomass 467
  bulk density 7, 16, 70, **73**, 180
  buried 235
  calcareous **15**, 174
  chalky **15**, 404, 592
  chemistry **340**
  clayey, *see* clays
  compaction 4, 69, 340, 390, 525, 572
  cores 39
  cracking 541
  degradation of 4, 308, **537**
  deposition of 45, 138
  depth of **241**, 245, 380
  detachment of 51, 190, 384, **390**, 433
  dispersive 549
  dispersion of 8, 20, 390
  evolution of 4
  fauna 466
  fertility 175, 308, 337, **345**, 561, **652**
  flora 466
  formation 33, 470
  free draining 217
  frozen 179
  layered 419
  loamy 175, 424, 462, 513, 591
  moisture content 6, 19, 584
  mottles 540
  organic 217
  organic matter 16, 75, 540, **550**
  pipes 484, **649**
  porosity 4, 391
  response to wetting 3
  roughness 6, 97, 391, 527
  sandy 43, 98, 236, 255, 501, 591
  saturated 180
  silty 97, 138
  sinks **195**
  strength 256
  stripping **243**
  structure 256, 461
  texture 391
Soil Bank (USA) 629
Soil Conservation Service (USA) 670
soil loss 33, **93**, 107, **119**, 302, 333, 369, 415, 533, 537
  expression of 93
  ratio 561
  tolerance, *see* erosion, tolerance level
soil management 34
soil maps 237
soil productivity 33, 461, 464, 465, 560, 594, 627, 631
  effects on by erosion **461**
Soil Survey of England and Wales (SSEW) 34, 87, 237, 569
solutional erosion 301
sorghum 359
source partial areas 80, 134
speleothem 306
splash erosion 35, 245
spring-sown crops 5, **563**, 564
statistical analysis 58, 109, 213, 405, 586
stochastic analyses 481, **489**
storm events 17, 82, 218, 244, 270, 315
stratigraphic evidence **232**, 273
streams
  stage 161
  frozen 165
  gauging 72, 651
  source areas 554
stream channels, *see* channels
strip cropping 566, 599
subcutaneous flow 304
subsidence 303
subsistence farming 339, **496**, 575
subsoil 122, 182, 256, 345
subsurface flow 13, 75
sugar beet 176, 592

sulphur 261
surface lowering 238
surface runoff, *see* runoff
surface soil, *see* topsoil
suspended sediment 27, 70, 130, 173, 199, 245, 305, 354, 416
swales 448
sweet potato 579
system evolution 63
systems 212, 466

temporal resolution 201, **205**
terraces 332, 438, 507
 see also bench terraces
terracettes 482
Tertiary 88, 121, 331, 592, 646
thalweg 57, 514
thorium 218
thresholds 77, 100, 173, 195, 387, 407, 454, 487
throughflow 303
THEPROM model **465**, **466**
till 108, 174
tillage 3, 13, 94, 107, **343**, 434
 see also cultivation, autumn-sown and spring-sown crops
timescales 193, 209, 218
tomato 579
topography 109, 448
 index 448
topsoil 107, 119, 135, 182, 341, 353, 540, 645
 stripping 237, **243**, **246**, 270
tracers 119, 217
tramlines 572, 592
trampling 165, 541
transmission losses 486
translocation 137
transport capacity 145
tree-crops 498
Triassic 232, **255**
tropical forest 339
turbidity meter 161

undercutting 434
uranium series 218
urban areas 380, 383
upland areas 130, 231, 298, 422, 525, 591, 600

USLE model (Universal Soil Loss Equation) 13, 135, 390, 402, 421, 429, **442**, 447, 481, 548, 560, 661
 limitations of 430

vadose flow 304
valley bottom 114, 153
 deposition in 89, 154, **193**, 237, **273**, 664
 erosion in **395**
valley sides 108, 282, 372
vegetable crops 540
vegetation cover 77, 96, 138, 164, 362, 396, 514, 571
 effects on erosion 331, **333**, 468, 498
vegetation history **276**, 282
vineyards 381

water erosion, *see* erosion; solutional erosion; rills; gullies
watershed 109, 129, 421, 554, 653
 unit source 431, **438**
water level recorders 73, 579
water
 conservation 29, 465
 harvesting 495
 management **569**, 599
 quality 85, 116, 363, 464
 retention 11
 storage 290, 375, **500**
 supply 540
watering points 85
wave erosion 295
weathering 168, 257, 469
weeds 18, 249, 341, 374
WEPP model 441, 448
wetlands 114, 176, 537, 635
wetted perimeter 434
wheat 122, 235, 358, 403, 533
wheelings 90, 165
wind erosion 44, 87, 249, 599, 605, **610**, 627
wind deposition 211, 244
winter cereals 88, 157, 372, 403, 560, 592
 see also autumn-sown crops
woodland 154, 194, 231, 394, 538

X-ray diffraction 484

zinc 261
zircon 261